Advanced Surfaces for Stem Cell Research

Scrivener Publishing
100 Cummings Center, Suite 541J
Beverly, MA 01915-6106

Advanced Materials Series

The Advanced Materials Series provides recent advancements of the fascinating field of advanced materials science and technology, particularly in the area of structure, synthesis and processing, characterization, advanced-state properties, and applications. The volumes will cover theoretical and experimental approaches of molecular device materials, biomimetic materials, hybrid-type composite materials, functionalized polymers, supramolecular systems, information- and energy-transfer materials, biobased and biodegradable or environmental friendly materials. Each volume will be devoted to one broad subject and the multidisciplinary aspects will be drawn out in full.

Series Editor: Ashutosh Tiwari
Biosensors and Bioelectronics Centre
Linköping University
SE-581 83 Linköping
Sweden
E-mail: ashutosh.tiwari@liu.se

Managing Editors: Sachin Mishra and Sophie Thompson

Publishers at Scrivener
Martin Scrivener (martin@scrivenerpublishing.com)
Phillip Carmical (pcarmical@scrivenerpublishing.com)

Advanced Surfaces for Stem Cell Research

Edited by

Ashutosh Tiwari, Bora Garipcan and Lokman Uzun

WILEY

Copyright © 2017 by Scrivener Publishing LLC. All rights reserved.

Co-published by John Wiley & Sons, Inc. Hoboken, New Jersey, and Scrivener Publishing LLC, Beverly, Massachusetts.
Published simultaneously in Canada.

No part of this publication may be reproduced, stored in a retrieval system, or transmitted in any form or by any means, electronic, mechanical, photocopying, recording, scanning, or otherwise, except as permitted under Section 107 or 108 of the 1976 United States Copyright Act, without either the prior written permission of the Publisher, or authorization through payment of the appropriate per-copy fee to the Copyright Clearance Center, Inc., 222 Rosewood Drive, Danvers, MA 01923, (978) 750-8400, fax (978) 750-4470, or on the web at www.copyright.com. Requests to the Publisher for permission should be addressed to the Permissions Department, John Wiley & Sons, Inc., 111 River Street, Hoboken, NJ 07030, (201) 748-6011, fax (201) 748-6008, or online at http://www.wiley.com/go/permission.

Limit of Liability/Disclaimer of Warranty: While the publisher and author have used their best efforts in preparing this book, they make no representations or warranties with respect to the accuracy or completeness of the contents of this book and specifically disclaim any implied warranties of merchantability or fitness for a particular purpose. No warranty may be created or extended by sales representatives or written sales materials. The advice and strategies contained herein may not be suitable for your situation. You should consult with a professional where appropriate. Neither the publisher nor author shall be liable for any loss of profit or any other commercial damages, including but not limited to special, incidental, consequential, or other damages.

For general information on our other products and services or for technical support, please contact our Customer Care Department within the United States at (800) 762-2974, outside the United States at (317) 572-3993 or fax (317) 572-4002.

Wiley also publishes its books in a variety of electronic formats. Some content that appears in print may not be available in electronic formats. For more information about Wiley products, visit our web site at www.wiley.com.

For more information about Scrivener products please visit www.scrivenerpublishing.com.

Cover design by Russell Richardson

Library of Congress Cataloging-in-Publication Data:

ISBN 978-1-119-24250-5

Printed in the United States of America

10 9 8 7 6 5 4 3 2 1

Contents

Preface		xv
1	**Extracellular Matrix Proteins for Stem Cell Fate**	**1**
	Betül Çelebi-Saltik	
	1.1 Human Stem Cells, Sources, and Niches	2
	1.2 Role of Extrinsic and Intrinsic Factors	5
	1.2.1 Shape	5
	1.2.2 Topography Regulates Cell Fate	6
	1.2.3 Stiffness and Stress	6
	1.2.4 Integrins	7
	1.2.5 Signaling via Integrins	9
	1.3 Extracellular Matrix of the Mesenchyme: Human Bone Marrow	11
	1.4 Biomimetic Peptides as Extracellular Matrix Proteins	13
	References	15
2	**The Superficial Mechanical and Physical Properties of Matrix Microenvironment as Stem Cell Fate Regulator**	**23**
	Mohsen Shahrousvand, Gity Mir Mohamad Sadeghi and Ali Salimi	
	2.1 Introduction	24
	2.2 Fabrication of the Microenvironments with Different Properties in Surfaces	25
	2.3 Effects of Surface Topography on Stem Cell Behaviors	28
	2.4 Role of Substrate Stiffness and Elasticity of Matrix on Cell Culture	31
	2.5 Stem Cell Fate Induced by Matrix Stiffness and Its Mechanism	31
	2.6 Competition/Compliance between Matrix Stiffness and Other Signals and Their Effect on Stem Cells Fate	33

2.7　Effects of Matrix Stiffness on Stem Cells in
         Two Dimensions versus Three Dimensions　33
   2.8　Effects of External Mechanical Cues on Stem
         Cell Fate from Surface Interactions Perspective　35
   2.9　Conclusions　36
   Acknowledgments　36
   References　37

3　**Effects of Mechanotransduction on Stem Cell Behavior**　45
   Bahar Bilgen and Sedat Odabas
   3.1　Introduction　45
   3.2　The Concept of Mechanotransduction　47
   3.3　The Mechanical Cues of Cell Differentiation and Tissue
         Formation on the Basis of Mechanotransduction　48
   3.4　Mechanotransduction via External Forces　49
         3.4.1　Mechanotransduction via Bioreactors　50
         3.4.2　Mechanotransduction via Particle-based Systems　53
         3.4.3　Mechanotransduction via Other External Forces　55
   3.5　Mechanotransduction via Bioinspired Materials　56
   3.6　Future Remarks and Conclusion　56
   Declaration of Interest　57
   References　57

4　**Modulation of Stem Cells Behavior Through
   Bioactive Surfaces**　67
   *Eduardo D. Gomes, Rita C. Assunção-Silva, Nuno Sousa,
   Nuno A. Silva and António J. Salgado*
   4.1　Lithography　68
   4.2　Micro and Nanopatterning　72
   4.3　Microfluidics　73
   4.4　Electrospinning　73
   4.5　Bottom-up/Top-down Approaches　76
   4.6　Substrates Chemical Modifications　77
         4.6.1　Biomolecules Coatings　78
         4.6.2　Peptide Grafting　79
   4.7　Conclusion　80
   Acknowledgements　81
   References　81

5 Influence of Controlled Micro- and Nanoengineered Environments on Stem Cell Fate   87
Anna Lagunas, David Caballero and Josep Samitier
- 5.1 Introduction to Engineered Environments for the Control of Stem Cell Differentiation   88
  - 5.1.1 Stem Cells Niche *In Vivo*: A Highly Dynamic and Complex Environment   88
  - 5.1.2 Mimicking the Stem Cells Niche *In Vitro*: Engineered Biomaterials   90
- 5.2 Mechanoregulation of Stem Cell Fate   91
  - 5.2.1 From *In Vivo* to *In Vitro*: Influence of the Mechanical Environment on Stem Cell Fate   91
  - 5.2.2 Regulation of Stem Cell Fate by Surface Roughness   92
  - 5.2.3 Control of Stem Cell Differentiation by Micro- and Nanotopographic Surfaces   94
  - 5.2.4 Physical Gradients for Regulating Stem Cell Fate   98
- 5.3 Controlled Surface Immobilization of Biochemical Stimuli for Stem Cell Differentiation   102
  - 5.3.1 Micro- and Nanopatterned Surfaces: Effect of Geometrical Constraint and Ligand Presentation at the Nanoscale   102
  - 5.3.2 Biochemical Gradients for Stem Cell Differentiation   109
- 5.4 Three-dimensional Micro- and Nanoengineered Environments for Stem Cell Differentiation   114
  - 5.4.1 Three-dimensional Mechanoregulation of Stem Cell Fate   115
  - 5.4.2 Three-dimensional Biochemical Patterns for Stem Cell Differentiation   121
- 5.5 Conclusions and Future Perspectives   124
- References   124

6 Recent Advances in Nanostructured Polymeric Surface: Challenges and Frontiers in Stem Cells   143
Ilaria Armentano, Samantha Mattioli, Francesco Morena, Chiara Argentati, Sabata Martino, Luigi Torre and Josè Maria Kenny
- 6.1 Introduction   144
- 6.2 Nanostructured Surface   146
- 6.3 Stem Cell   148

6.4	Stem Cell/Surface Interaction	149
6.5	Microscopic Techniques Used in Estimating Stem Cell/Surface	150
	6.5.1 Fluorescence Microscopy	150
	6.5.2 Electron Microscopy	151
	6.5.3 Atomic Force Microscopy	155
	6.5.3.1 Instrument	156
	6.5.3.2 Cell Nanomechanical Motion	158
	6.5.3.3 Mechanical Properties	158
6.6	Conclusions and Future Perspectives	160
References		160

7 Laser Surface Modification Techniques and Stem Cells Applications — 167
Çağrı Kaan Akkan

7.1	Introduction	168
7.2	Fundamental Laser Optics for Surface Structuring	168
	7.2.1 Definitive Facts for Laser Surface Structuring	169
	7.2.1.1 Absorptivity and Reflectivity of the Laser Beam by the Material Surface	169
	7.2.1.2 Effect of the Incoming Laser Light Polarization	170
	7.2.1.3 Operation Mode of the Laser	171
	7.2.1.4 Beam Quality Factor	172
	7.2.1.5 Laser Pulse Energy/Power	173
	7.2.2 Ablation by Laser Pulses	174
	7.2.2.1 Focusing the Laser Beam	174
	7.2.2.2 Ablation Regime	175
7.3	Methods for Laser Surface Structuring	176
	7.3.1 Physical Surface Modifications by Lasers	176
	7.3.1.1 Direct Structuring	177
	7.3.1.2 Beam Shaping Optics	179
	7.3.1.3 Direct Laser Interference Patterning	182
	7.3.2 Chemical Surface Modification by Lasers	183
	7.3.2.1 Pulsed Laser Deposition	183
	7.3.2.2 Laser Surface Alloying	186
	7.3.2.3 Laser Surface Oxidation and Nitriding	188
7.4	Stem Cells and Laser-modified Surfaces	189
7.5	Conclusions	193
References		194

| 8 | Plasma Polymer Deposition: A Versatile Tool for Stem Cell Research | 199 |

M. N. Macgregor-Ramiasa and K. Vasilev

- 8.1 Introduction — 199
- 8.2 The Principle and Physics of Plasma Methods for Surface Modification — 201
  - 8.2.1 Plasma Sputtering, Etching an Implantation — 202
  - 8.2.2 Plasma Polymer Deposition — 203
- 8.3 Surface Properties Influencing Stem Cell Fate — 204
  - 8.3.1 Plasma Methods for Tailored Surface Chemistry — 205
    - 8.3.1.1 Oxygen-rich Surfaces — 206
    - 8.3.1.2 Nitrogen-rich Surfaces — 210
    - 8.3.1.3 Systematic Studies and Copolymers — 212
  - 8.3.2 Plasma for Surface Topography — 213
  - 8.3.3 Plasma for Surface Stiffness — 216
  - 8.3.4 Plasma for Gradient Substrata — 217
  - 8.3.5 Plasma and 3D Scaffolds — 220
- 8.4 New Trends and Outlook — 221
- 8.5 Conclusions — 221
- References — 222

| 9 | Three-dimensional Printing Approaches for the Treatment of Critical-sized Bone Defects | 233 |

Sara Salehi, Bilal A. Naved and Warren L. Grayson

- 9.1 Background — 234
  - 9.1.1 Treatment Approaches for Critical-sized Bone Defects — 234
  - 9.1.2 History of the Application of 3D Printing to Medicine and Biology — 235
- 9.2 Overview of 3D Printing Technologies — 236
  - 9.2.1 Laser-based Technologies — 237
    - 9.2.1.1 Stereolithography — 237
    - 9.2.1.2 Selective Laser Sintering — 238
    - 9.2.1.3 Selective Laser Melting — 238
    - 9.2.1.4 Electron Beam Melting — 239
    - 9.2.1.5 Two-photon Polymerization — 239
  - 9.2.2 Extrusion-based Technologies — 240
    - 9.2.2.1 Fused Deposition Modeling — 240
    - 9.2.2.2 Material Jetting — 240

       9.2.3   Ink-based Technologies   241
                9.2.3.1   Inkjet 3D Printing   241
                9.2.3.2   Aerosol Jet Printing   241
   9.3 Surgical Guides and Models for Bone Reconstruction   242
        9.3.1   Laser-based Surgical Guides   242
        9.3.2   Extrusion-based Surgical Guides   242
        9.3.3   Ink-based Surgical Guides   244
   9.4 Three-dimensionally Printed Implants
        for Bone Substitution   244
        9.4.1   Laser-based Technologies for Metallic
                Bone Implants   246
        9.4.2   Extrusion-based Technologies for Bone Implants   247
        9.4.3   Ink-based Technologies for Bone Implants   248
   9.5 Scaffolds for Bone Regeneration   248
        9.5.1   Laser-based Printing for Regenerative Scaffolds   249
        9.5.2   Extrusion-based Printing for
                Regenerative Scaffolds   249
        9.5.3   Ink-based Printing for Regenerative Scaffolds   252
        9.5.4   Pre- and Post-processing Techniques   253
                9.5.4.1   Pre-processing   253
                9.5.4.2   Post-processing: Sintering   259
                9.5.4.3   Post-processing: Functionalization   259
   9.6 Bioprinting   260
   9.7 Conclusion   264
   List of Abbreviation   265
   References   266

10 **Application of Bioreactor Concept and Modeling
   Techniques to Bone Regeneration and Augmentation
   Treatments**   **279**
   Oscar A. Deccó and Jésica I. Zuchuat
   10.1 Bone Tissue Regeneration   280
        10.1.1 Proinflammatory Cytokines   281
        10.1.2 Transforming Growth Factor Beta   281
        10.1.3 Angiogenesis in Regeneration   282
   10.2 Actual Therapeutic Strategies and Concepts to
        Obtain an Optimal Bone Quality and Quantity   283
        10.2.1 Guided Bone Regeneration Based on Cells   284
                10.2.1.1 Embryonic Stem Cells   284
                10.2.1.2 Adult Stem Cells   284
                10.2.1.3 Mesenchymal Stem Cells   285

		10.2.2	Guided Bone Regeneration Based on Platelet-Rich Plasma (PRP) and Growth Factors	286

     10.2.2 Guided Bone Regeneration Based on
        Platelet-Rich Plasma (PRP) and
        Growth Factors 286
        10.2.2.1 Bone Morphogenetic Proteins 289
     10.2.3 Guided Bone Regeneration Based on Barrier
        Membranes 290
     10.2.4 Guided Bone Regeneration Based on Scaffolds 292
   10.3 Bioreactors Employed for Tissue Engineering in
     Guided Bone Regeneration 293
     10.3.1 Spinner Flask Bioreactors 294
     10.3.2 Rotating Wall Bioreactors 295
     10.3.3 Perfusion Bioreactors 295
   10.4 Bioreactor Concept in Guided Bone Regeneration
     and Tissue Engineering: *In Vivo* Application 296
     10.4.1 Sand Blasting 298
     10.4.2 Chemical Treatment 299
     10.4.3 Heat Treatment 300
   10.5 New Multidisciplinary Approaches Intended to
     Improve and Accelerate the Treatment of Injured
     and/or Diseased Bone 305
     10.5.1 Application of Bioreactor in Dentistry:
        Therapies for the Treatment of Maxillary Bone
        Defects 306
     10.5.2 Application of Bioreactor in Cases
        of Osteoporosis 309
   10.6 Computational Modeling: An Effective Tool
     to Predict Bone Ingrowth 312
   References 313

11 Stem Cell-based Medicinal Products: Regulatory Perspectives **323**
  Deniz Ozdil and Halil Murat Aydin
   11.1 Introduction 323
   11.2 Defining Stem Cell-based Medicinal Products 325
   11.3 Regional Regulatory Issues for Stem Cell Products 328
   11.4 Regulatory Systems for Stem Cell-based Technologies 329
     11.4.1 The US Regulatory System 330
   11.5 Stem Cell Technologies: The European
     Regulatory System 338
   References 342

12 Substrates and Surfaces for Control of Pluripotent Stem Cell Fate and Function   343
Akshaya Srinivasan, Yi-Chin Toh, Xian Jun Loh and Wei Seong Toh

- 12.1 Introduction   344
- 12.2 Pluripotent Stem Cells   344
- 12.3 Substrates for Maintenance of Self-renewal and Pluripotency of PSCs   346
  - 12.3.1 Cellular Substrates   346
  - 12.3.2 Acellular Substrates   347
    - 12.3.2.1 Biological Matrices   347
    - 12.3.2.2 ECM Components   350
    - 12.3.2.3 Decellularized Matrices   352
    - 12.3.2.4 Cell Adhesion Molecules   353
    - 12.3.2.5 Synthetic Substrates   354
- 12.4 Substrates for Promoting Differentiation of PSCs   357
  - 12.4.1 Cellular Substrates   357
  - 12.4.2 Acellular Substrates   358
    - 12.4.2.1 Biological Matrices   358
    - 12.4.2.2 ECM Components   360
    - 12.4.2.3 Decellularized Matrices   364
    - 12.4.2.4 Cell Adhesion Molecules   365
    - 12.4.2.5 Synthetic Substrates   365
- 12.5 Conclusions   368
- Acknowledgments   369
- References   369

13 Silk as a Natural Biopolymer for Tissue Engineering   381
Ayşe Ak Can and Gamze Bölükbaşi Ateş

- 13.1 Introduction   382
  - 13.1.1 Mechanical Properties   383
  - 13.1.2 Biodegradation   384
  - 13.1.3 Biocompatibility   385
- 13.2 SF as a Biomaterial   385
  - 13.2.1 Fibroin Hydrogels and Sponges   386
  - 13.2.2 Fibroin Films and Membranes   388
  - 13.2.3 Nonwoven and Woven Silk Scaffolds   388
  - 13.2.4 Silk Fibroin as a Bioactive Molecule Delivery   388
- 13.3 Biomedical Applications of Silk-based Biomaterials   389
  - 13.3.1 Bone Tissue Engineering   389
  - 13.3.2 Cartilage Tissue Engineering   391

	13.3.3	Ligament and Tendon Tissue Engineering	393

	13.3.3	Ligament and Tendon Tissue Engineering	393
	13.3.4	Cardiovascular Tissue Engineering	393
	13.3.5	Skin Tissue Engineering	395
	13.3.6	Other Applications of Silk Fibroin	395
13.4	Conclusion and Future Directions		395
References			396

14 Applications of Biopolymer-based, Surface-modified Devices in Transplant Medicine and Tissue Engineering — 401
Ashim Malhotra, Gulnaz Javan and Shivani Soni

14.1	Introduction to Cardiovascular Disease	402
14.2	Need Assessment for Biopolymer-based Devices in Cardiovascular Therapeutics	402
14.3	Emergence of Surface Modification Applications in Cardiovascular Sciences: A Historical Perspective	403
14.4	Nitric Oxide Producing Biosurface Modification	405
14.5	Surface Modification by Extracellular Matrix Protein Adherence	406
14.6	The Role of Surface Modification in the Construction of Cardiac Prostheses	407
14.7	Biopolymer-based Surface Modification of Materials Used in Bone Reconstruction	408
14.8	The Use of Biopolymers in Nanotechnology	411

	14.8.1	Protein Nanoparticles		412
		14.8.1.1	Albumin-based Nanoparticles and Surface Modification	413
		14.8.1.2	Collagen-based Nanoparticles and Surface Modification	414
		14.8.1.3	Gelatin-based Nanoparticle Systems	415
	14.8.2	Polysaccharide-based Nanoparticle Systems		415
		14.8.2.1	The Use of Alginate for Surface Modifications	415
		14.8.2.2	The Use of Chitosan-based Nanoparticles and Chitosan-based Surface Modification	416
		14.8.2.3	The Use of Chitin-based Nanoparticles and Chitin-based Surface Modification	418
		14.8.2.4	The Use of Cellulose-based Nanoparticles and Cellulose-based Surface Modification	419
References				420

15 Stem Cell Behavior on Microenvironment Mimicked Surfaces   **425**
M. Özgen Öztürk Öncel and Bora Garipcan

15.1 Introduction   426
15.2 Stem Cells   427
     15.2.1 Definition and Types   427
         15.2.1.1 Embryonic Stem Cells   428
         15.2.1.2 Adult Stem Cells   428
         15.2.1.3 Reprogramming and Induced Pluripotent Stem Cells   429
     15.2.2 Stem Cell Niche   429
15.3 Stem Cells: Microenvironment Interactions   430
     15.3.1 Extracellular Matrix   431
     15.3.2 Signaling Factors   431
     15.3.3 Physicochemical Composition   432
     15.3.4 Mechanical Properties   432
     15.3.5 Cell–Cell Interactions   433
15.4 Biomaterials as Stem Cell Microenvironments   433
     15.4.1 Surface Chemistry   433
     15.4.2 Surface Hydrophilicity and Hydrophobicity   436
     15.4.3 Substrate Stiffness   437
     15.4.4 Surface Topography   437
15.5 Biomimicked and Bioinspired Approaches   438
     15.5.1 Bone Tissue Regeneration   441
     15.5.2 Cartilage Tissue Regeneration   442
     15.5.3 Cardiac Tissue Regeneration   443
15.6 Conclusion   444
References   444

Index   **453**

Preface

Stem cells have attracted much attention in the fields of regenerative medicine and tissue engineering for their important role in the treatment of several diseases. This is due to their unique properties such as their self-renewal capability and ability to differentiate into specific cell types. New research and therapies in these fields are mainly focused on a better understanding of the natural mechanisms of stem cells and the control and regulation of their behavior under *in-vivo* or *in-vitro* conditions. Since a natural and/or synthetic surface is an important physical structure for most of the cells, the effect of surface properties, such as chemistry, charge, energy, hydropathy, pattern, topography, and stiffness, with or without differentiation media, influences stem cell behavior as well as controls and directs stem cell differentiation. Biomaterials that are developed by altering surface properties are a promising challenge for regenerative medicine and tissue engineering fields, drug investigation/toxicity studies and stem cell-based therapies.

This book, *Advanced Surfaces in Stem Cell Research,* part of the Advanced Materials Series, first outlines the importance of extra cellular matrix (ECM), which is a natural surface for most cells, and is discussed in the first chapter entitled "Extracellular Matrix Proteins for Stem Cell Fate." Chapters 2 through 6 discuss the influence of biological, chemical, mechanical, and physical properties on stem cell behavior and fate. The mechanical and physical properties of matrix microenvironment as stem cell fate regulator are reviewed in Chapter 2, followed by a discussion on the effect of mechanotransduction on stem cell behavior in chapter 3. In chapter 4, stem cell modulation on bioactive surfaces is disputed. Since micro- and nanoscale structure and surfaces have an influence on stem cell behavior and fate, these properties are discussed in chapters 5 and 6, respectively entitled "Influence of Controlled Micro- and Nano-Engineered Surfaces on Stem Cell Fate" and "Recent Advances in Nanostructured Polymeric Surface: Challenges and Frontiers in Stem Cells." Chapters 7 through 10 deliberate 2D and 3D surface fabrication and modification using different techniques on stem cell fate. Laser surface modification techniques and stem cell applications and plasma polymer deposition as a versatile tool for stem cell research are discussed in chapters 7 and 8, respectively. The effect of 3D structures and dynamic cell environment, such as bioreactors, on stem cell fate are presented in detail in chapters 9 and 10, respectively entitled "3D Printing Approaches for the Treatment of Critical-Sized Bone Defect" and "Application of Bioreactor Concept and Modeling Techniques in Bone Regeneration and Augmentation Treatments." Chapter 11 is an important and interesting chapter which will inform readers from a different point of view, with regulatory perspectives on medical products as stem cell-based medicinal products. One of the recent stem cell sources, pluripotent stem cells, are discussed in chapter 12, "Substrates and Surfaces for Control of Pluripotent Stem Cell Fate and Function." Surface engineering applications are discussed in tissue engineering, regenerative medicine and different types of biomaterials in chapters 13 and 14, respectively entitled "Silk as a Natural Biopolymer for Tissue Engineering" and "Application of Biopolymer-Based, Surface Modified Devices in Transplant Medicine and Tissue Engineering." Biomimetic and bioinspired approaches are also indicated for developing microenvironment of several tissues in chapter 15, "Stem Cell Behavior on Microenvironment Mimicked Surfaces."

We would like to thank the authors that have contributed to the chapters of this book, including all scientists who have contributed to this topic. We hope and believe that this book will be very useful to those in the biomaterials, tissue engineering, regenerative medicine, stem cell research and material science communities.

Editors
Ashutosh Tiwari, PhD, DSc
Bora Garipcan, PhD
Lokman Uzun, PhD
September 2016

1
Extracellular Matrix Proteins for Stem Cell Fate

Betül Çelebi-Saltik

Graduate School of Health Sciences, Department of Stem Cell Sciences, Hacettepe University, Ankara, Turkey
Center for Stem Cell Research and Development, Hacettepe University, Ankara, Turkey

Abstract

Stem cell-based regenerative medicine aims to repair and regenerate injured and/or diseased tissues by implanting a combination of cells, biomaterials, and soluble factors. Unfortunately, due to an incomplete understanding and knowledge of the interactions between biomaterials and specific stem cell types, and the inability to control the complex signaling pathways ensured by these interactions, the ability to design functional tissue and organ substitutes has been limited. The greatest challenge remains the ability to control stem cells' fate outside of the cell's natural microenvironment or "niche". Stem cell fate is known to be regulated by signals from the microenvironment, such as extracellular matrix (ECM) including glycosaminoglycans and proteoglycans to which stem cells adhere. They represent an essential player in stem cell microenvironment because they can directly or indirectly modulate the maintenance, proliferation, self-renewal, and differentiation of stem cells. The interactions between stem cells and the ECM play a critical role in living tissue development, repair, and regeneration as well. The design of artificial ECM and/or binding site is important in tissue engineering because artificial ECM and/or binding regulates cellular behaviors. Identification of binding sites and key motifs in ECM proteins that interact with cellular receptors can allow researchers to generate small peptides that can mimic the function of large ECM proteins.

Keywords: Extracellular matrix proteins, stem cells, niche, integrin, signaling

Corresponding author: betul.celebi@hacettepe.edu.tr

1.1 Human Stem Cells, Sources, and Niches

Stem cells have two distinct abilities: self-renewal of themselves and differentiation into tissue/organ-specific cells. Based on their differentiation potential, they can be classified as totipotent, pluripotent, multipotent, or unipotent cells. The totipotent fertilized egg exhibits the stem cell that gives rise to all embryonic and extra-embryonic structures of the developing embryo [1]. Human embryonic stem cells (hESCs) derived from the inner cell mass of the blastocyst have the ability to self-renew over a long period without undergoing senescence [1]. Consequently, cells with higher regeneration capacity and plasticity are needed which lead to use of pluripotent stem cells. Identifying suitable source of stem cells is elemental for regeneration of any tissue. Mature and differentiated multipotent stem cells are easily available but least preferred due to their limited cell division and differentiation capacity. Indeed, the number of stem cells in adults is very low, and it depletes with age. Bone marrow (BM)-derived stem cells first described by Friedenstein *et al.* are still the most frequently investigated cell type [2]. These cells are lineage-restricted, and in contrast to hESCs, multipotent adult stem cells undergo replicative senescence and their lifespan in culture is limited. The existence of multipotent postnatal stem cells has been reported in BM, peripheral blood, umbilical cord, umbilical cord blood (UCB), Wharton's jelly, placenta, neuronal, and adipose tissues [3–7]. Takahashi and Yamanaka developed a new technique by describing the reprogramming of human somatic fibroblasts into primitive pluripotent stem cells by over-expressing OCT4, SOX2, KLF4, and MYC [8]. These human induced pluripotent stem cells are similar to hESCs in the sense that they also express pluripotency genes, have telomerase activity, and are able to differentiate into all cell types of the three embryonic germ layers (endoderm, ectoderm, and mesoderm) (Figure 1.1).

Stem cell niche consists of stem cells, supporting cells, extracellular matrix (ECM), soluble factors, and nervous systems. All these factors have an important role in stem cell niche; however, ECM that holds stem cells in a niche and controls their cellular processes plays a critical role in the control of stem cell fate [9]. Since its first definition originally proposed in 1978 by Schofield for the hematopoietic stem cell (HSC) microenvironment, the concept of the niche has increased in complexity [10]. Niches are highly specialized for each type of stem cell, with a defined anatomical localization, and they are composed by stem cells and by supportive stromal cells (which interact each other through cell surface receptors, gap junctions and soluble factors), together with the ECM in which they are located (Figure 1.2). In addition, secreted or cell surface factors, signaling

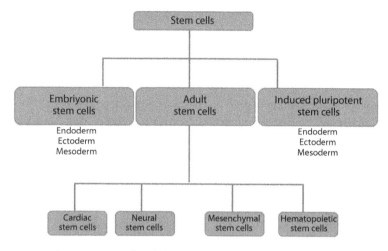

Figure 1.1 Embryonic stem cells, adult stem cells, and induced pluripotent stem cells.

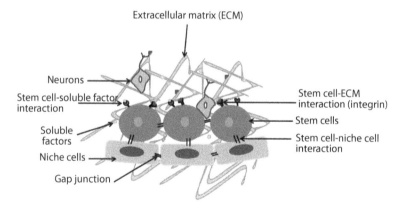

Figure 1.2 Stem cell niche. Adapted from Jhala *et al.* [9].

cascades and gradients, as well as physical factors such as shear stress, oxygen tension, and temperature, promote to control stem cell behavior [11].

Stem cell niche cells mesenchymal and HSCs play an important role in many regeneration processes in human body. HSCs are blood forming cells which were recognized more than 50 years ago [12]. They are produced during embryogenesis in a complex developmental process in several anatomical sites (the yolk sac, the aorta-gonad-mesonephros region, the placenta, and the fetal liver), after which HSCs colonize the BM at birth [13]. These cells are divided into two major progenitor lineages; common myeloid (CMP) and common lymphoid progenitors (CLP). CMPs give rise to megakaryocyte (MK)/erythroid and granulocyte/macrophage progenitors developing into platelet producing MKs, erythrocytes, mast cells, neutrophils, eosinophils, monocytes, and macrophages. CLPs will mature into B-lymphocytes, T-lymphocytes, and natural killer cells. They are positive to cell surface markers such as CD34, CD45, and CD117 [14]. HSCs have the ability to leave their tissue of origin, enter circulation, identify, and relocate to an available niche elsewhere during early development [15]. In the adult, they can leave the BM and return back to it through homing mechanisms [16].

Since the identification of mesenchymal stem cells (MSCs) as colony-forming unit fibroblasts (CFU-Fs) by Friedenstein et al. in 1970 and the first detailed description of the tri-lineage capacity by Pittenger et al., our understanding of these cells has taken great strides forward [17]. These plastic adherent multipotent cells are able to differentiate into bone, cartilage, and fat cells and can be isolated from many adult tissue types. They have been obtained from a wide variety of fetal and adult tissues: adipose tissue, placenta, umbilical cord, Wharton's Jelly, synovial membrane, and dental pulp [18]. Specifically, a recent study reported plastic-adherent MSC-like colonies derived from the non-mesodermal tissues such as brain, spleen, liver, kidney, lung, BM, muscle, thymus, and pancreas [19]. They are positive to cell surface markers such as CD29, CD44, CD105, CD90, and CD73 [20]. MSCs secrete cytokines, chemokines, growth factors, and ECM either spontaneously or after induction by other cytokines [21]. The mesenchymal ECM consists of various proteins such as collagens types I, III, IV, V, and VI, laminin, fibronectin, proteoglycans, such as syndecan, Pln, decorin, and the glycosaminoglycan (GAG) hyaluronan (Figure 1.3) [22]. Proteoglycans are macromolecules consisting of a core protein attached to several polysaccharides (GAGs). They are able to bind cells to the matrix and bind growth factors to GAGs, thus preventing extracellular protease degradation of growth factors [23].

The ECM is a highly dynamic structure that is constantly being remodeled, either enzymatically or non-enzymatically, and its molecular components are subjected to many post-translational modifications [24]. ECM proteins vary not only in composition but also in physical parameters including elasticity and topography. Such physical properties are known to affect the micro- to nanotopography of integrin receptors (alpha and beta subunits) and to influence a range of cellular processes through changes in cell shape and the actin cytoskeleton [25–27]. Cell adhesion to the ECM is mediated by ECM receptors, such as integrins, discoidin domain receptors, and syndecans [24]. ECM properties including stiffness, fiber orientation, and ligand presentation dimensionality provoke specific cellular behaviors [24]. It is clear that ECM-based control of the

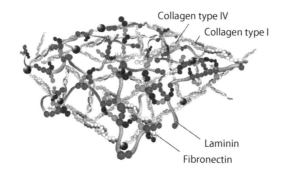

Figure 1.3 The mesenchymal ECM proteins network.

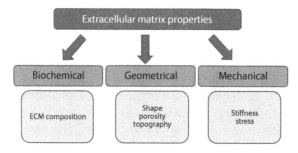

Figure 1.4 ECM-based control of the cell.

cell may also occur through multiple physical mechanisms, such as ECM geometry at the micro- and nanoscale (shape, porosity, and topography), ECM elasticity or mechanical signals (stiffness and stress) transmitted from the ECM to the stem cells that turn into biochemical response (Figure 1.4). An understanding of the interaction of these mediators with signaling pathways may provide new insights into the regulation of self-renewal and differentiation of stem cells [28].

1.2 Role of Extrinsic and Intrinsic Factors

1.2.1 Shape

The interactions between many extrinsic and intrinsic factors that govern cell shape are varied and may involve relatively long-term interactions with the cellular niche, as well as more acute changes due to physical factors such as mechanical or osmotic stress. In 1911, Carrel and Burrows showed that cells were responsive to shape cues [29] and over the last decade, the effects of surface shape have been well documented. Chen *et al.* performed a study to clarify how cells can differentiate on different surface shape via molecular signaling pathways. Based on their research, human MSCs were induced to take on either round or spread out shapes by fibronectin-patterned surfaces. The cells on the round islands underwent adipogenesis but those on spread out islands underwent osteogenesis. This phenomenon was reported due to the activation of RhoA which is induced by surface shape [30]. In other research, MSCs that cultured on small 30 nm nanotubes showed no differentiation whereas MSCs on 70–100 nm nanotubes showed cytoskeletal stress and osteoblastic lineage differentiation. This change was also reported due to the activation of RhoA and its effector ROCK kinase [31].

1.2.2 Topography Regulates Cell Fate

Cells are encountering different topographies sized clues *in vivo* from macro- (bone and ligament) to micro- (other cells) and nanotopography (proteins and ligands), all of which influence cell behavior and cell fate. It is known clear that topographical clues alone can produce the same effect as chemical induction including growth factors, chemokines and cytokines [32]. Topographical (physical) cues in the sub-micrometer range such as increasing the roughness or displaying specific geometrical shapes elicit specific cell responses. Pattering techniques such as soft lithography, photolithography, electrospinning, layer-by-layer microfluidic patterning, three-dimensional (3D) printing, ion milling, and reactive ion etching create scaffolds with precise controlled geometry, texture, porosity, and rigidity [32]. Micro-topographies that include micropits, microgrooves, and micropillars induce the cell body by physical confinement or alignment [33]. Focal adhesions (FAs), the sites of cell attachment to the underlying surfaces, play a pivotal role in all cell actions in response to nano- and microtopography. These dynamic adhesions are subject to complex regulation involving integrin binding to ECM components and the reinforcement of the adhesion plaque by recruitment of additional proteins [33].

1.2.3 Stiffness and Stress

The mechanical properties of the ECM not only allow such tissues to cope with stresses but also regulate various cellular functions such as spreading, migration, proliferation, stem cell differentiation, and apoptosis [28, 34]. Gels based on natural ECM, such as collagen type I, Matrigel, and fibrin, were the first materials used to suggest an impact of stiffness on cell fate [34]. Increasing the cross-linking of these matrixes that modulates matrix stiffness impacts integrin signaling and acto-myosin-mediated cellular tension [35]. For example, when a low concentration of cross-linker was used to attach collagen to stiff hydrogels, epidermal stem cells were stimulated to terminally differentiate, and MSCs differentiated into adipocytes rather than osteoblasts [36]. In the case of MSCs, osteogenic differentiation is favored by stiff substrates, whereas adipogenic differentiation is promoted by soft substrates [25]. Increases in microenvironment stiffness led to greater ECM and adhesion protein expression in cardiac stem cells [37]. Substrate stiffness also directs skeletal muscle stem cells and neuronal stem cells to either self-renew or differentiate. It has been also reported that cell spreading, self-renewal, and differentiation were inhibited on soft substrates (10 Pa), whereas with moduli of 100 Pa or greater, cells exhibited peak levels of a neuronal marker, beta-tubulin III, on substrates that had the approximate stiffness of brain tissue. Softer substrates (100–500 Pa) promoted neuronal differentiation (neural stem cells cultured on fibronectin peptide containing hydrogels in serum-free neuronal differentiation medium exhibit maximal differentiation at 500 Pa), whereas stiffer substrates (1,000–10,000 Pa) led to glial differentiation [38].

1.2.4 Integrins

Stem cells interact with ECM proteins via a family of surface receptors called integrins. Their types and quantities on each stem cell are specific to the cell and tissue type. Integrins facilitate the interactions between stem cells and ECM proteins and transduce chemical and physical signals from the matrix to the cells. They are involved in many cellular functions, including quiescence, cell cycle progression, cell adhesion, migration, and survival. Integrins are heterodimeric transmembrane receptors. An integrin molecule is mainly composed of two glycoprotein subunits, alpha (α) and beta (β). In vertebrates, 18 α subunits and 8 β subunits have been found, and they can form 24 different heterodimeric structures [39]. α Subunit has four Ca^{2+}-binding domain on its extracellular part of polypeptide chain, and β subunit bears a number of cysteine-rich domains on its extracellular part of polypeptide chain. The extracellular domain of integrins binds to ECM ligands, such as laminin, fibronectin, collagen, and vitronectin, while the intracellular domain connects with cytoskeletal proteins, such as α-actinin and talin, as well as regulatory proteins, such as calreticulin and cytohesin. Integrins interact with ligands through weak interactions, but the ligand-binding affinity may be modulated by intracellular signals. After the ligand binds between the two subunits, the induced conformational change physically pushes the two subunits apart and initiates downstream signaling [e.g. activation of cytoplasmic TPK (FAK) and serine/threonine kinases, activation of small GTPases, induction of calcium transport, or changes of phospholipid synthesis]. This transmembrane linker makes integrins important in cell–cell and cell–ECM adhesion, signal transduction, and growth factor receptor responses [40]. Basement membranes are typically rich in laminins and non-fibrillar collagen type IV, whereas soft connective tissue is dominated by the presence of fibrillar collagens types I and III [41]. Cells respond to the mechanical and biochemical changes in ECM through the cross-talk between integrins and the actin cytoskeleton [42].

Integrins have been recognized as a key regulator in embryogenesis. Endoderm is one of the three primary germ layers. The expression of integrin subunits $\alpha 3$, $\alpha 6$, and $\beta 4$ decreased in definitive endoderm (DE) compared to undifferentiated cells, but αV and $\beta 5$ are highly expressed in undifferentiated hESCs, and this expression is increased after DE formation. When ESCs are directed to differentiate toward endoderm lineages, laminin substrates are useful during the differentiation toward DE. This response is mediated by integrins αV and $\beta 5$. Besides, when cells are differentiated toward pancreatic or hepatic lineages, integrin $\beta 1$ becomes important [40]. Mesoderm derived from ESC has potential to regenerate multiple tissues and organs. Integrins such as $\alpha 5\beta 1$ and $\alpha 6\beta 1$ may be useful in mesoderm induction. The activation of $\alpha 5\beta 1$ modulated bone morphogenetic protein (BMP)-4 expression and BMP-4 induction during differentiation activates BMP, Wnt, fibroblast growth factor (FGF), and transforming growth factor (TGF)-β/nodal/activin signaling [43]. The generation of neuroectoderm lineages from ESC becomes the main focus of present ectoderm differentiation study. Deficiency of the $\alpha 3$ integrin subunit has been observed in mice with defective neuron migration. On the other hand, the interaction of laminin and fibronectin with $\beta 1$ integrins promotes the maintenance and migration of neural precursor cells [44]. It has been reported that $\alpha 1\beta 1$, $\alpha 6\beta 1$, and $\alpha 5\beta 1$ are crucial in the differentiation of specific neural crest lineages [40]. Skin substitutes derived from ESC differentiation may serve as a continual source for the treatment of wound healing. $\beta 1$ Integrin is responsible for melanocyte proliferation, $\alpha 2\beta 1$-collagen IV, $\alpha 3\beta 1$-laminin, and $\alpha 6\beta 4$-laminin are responsible for keratinocytes differentiation [45], $\alpha 6\beta 1$-laminin and $\alpha 2$, $\alpha 5$, $\alpha v\beta 3$-collagen type IV are responsible for melanocytes differentiation (Table 1.1) [46].

Table 1.1 Adhesive sequences in matrix proteins and their integrin receptor.

ECM proteins	Sequence	Integrins
Collagens	GFOGER	$\alpha1\beta1, \alpha2\beta1, \alpha10\beta1, \alpha11\beta1$
Fibronectin	RGD	$\alpha5\beta1, \alpha V\beta3, \alpha8\beta1, \alpha V\beta1, \alpha V\beta6, \alpha IIb\beta3$
	LDV	$\alpha4\beta1, \alpha4\beta7$
	REDV	$\alpha4\beta1$
Laminins	E1 fragment	$\alpha1\beta1, \alpha2\beta1, \alpha10\beta1$
	E8 fragment	$\alpha3\beta1, \alpha6\beta1, \alpha7\beta1, \alpha6\beta4$
Vitronectin	RGD	$\alpha V\beta3, \alpha IIb\beta3, \alpha V\beta5, \alpha V\beta1, \alpha V\beta8$
Fibrinogen	RGD	$\alpha V\beta3$
	KQAGV	$\alpha IIb\beta3$
Von Willebrand factor	RGD	$\alpha V\beta3, \alpha IIb\beta3$

1.2.5 Signaling via Integrins

Expression of integrins is regulated by ECM to cell signaling. It has been reported that TGF-β1 induced elevated expression of $\alpha2\beta1$ integrin in fibroblasts. Therefore, cytokines regulate the expression of integrins [47]. FGFs and vascular endothelial growth factors (VEGFs) have been shown to bind to heap and can be cleaved off from the GAG components of heparan sulfate proteoglycans (HSPG) by the enzyme heparanase and released as soluble ligand [48]. Fibronectin and tenascin-C bind to VEGF, which potentiates the VEGF-mediated signaling through its receptor VEGFR2 (Table 1.2) [49].

Several tyrosine kinases and phosphatases are necessary for the regulation of mechano-sensory activity of FAs; such as FA kinase (FAK), Src, receptor type tyrosine protein phosphatase α (RPTP α) and SH2 domain containing protein tyrosine phosphatase 2 (SHP2) [50]. FAK localizes at FAs on integrin clustering to regulate cell adhesion, migration and mechanotransduction [51]. FAK is a non-receptor tyrosine kinase which is activated upon integrin binding to autophosphorylate Y397—this induces subsequent binding of Src by the SH2 domain, leading to stable and increased activation of Src–FAK complex [42]. Src is a non-receptor tyrosine kinase associated with the cytoplasmic tail of β3 integrins via its SH3 domains. Overexpression and phosphorylation of FAK correlate with the increase in cell motility and invasion. Adhesion and spreading of cells on a variety of ECM proteins, including collagen type IV, lead to an increase

Table 1.2 Activation of several signaling pathways by integrins with growth factor receptors.

Growth factor receptors	Ref.
Insulin receptor	[56]
Type 1 insulin-like growth factor (IGF1) receptor	[57]
VEGF receptor	[58]
TGF-β receptor	[59]
Hepatocyte growth factor (HGF) receptor	[60]
Platelet-derived growth factor-β (PDGF-β) receptor	[61]
Epidermal growth factor (EGF) receptor	[62]

in tyrosine phosphorylation and activation of FAK. Furthermore, suppression of adhesion-induced tyrosine phosphorylation of FAK may interrupt cancer cell–ECM interactions and affect the invasive and metastatic potential of cancer cells [52]. Phosphatidylinositol-4,5-bisphosphate 3-kinase (PI3K) has also been shown to bind FAK in a cell adhesion-dependent manner at the major autophosphorylation site Y397 leading to activation of AKT, a ubiquitously expressed serine/threonine kinase that regulates integrin-mediated cell survival [53]. The AKT pathway is essential for proliferation because a dominant negative mutant of PI3K prevents cyclin D1 expression. However, proliferation also requires ERK signaling. Both AKT and ERK phosphorylations are required to induce cell growth when stimulated with mitogens [54]. Members of the GTPase family, Rac1, RhoA and Cdc42, are associated with adhesion-dependent cell cycle regulation. Rac and Rho are activated by integrin and they control cell cycle. Besides, Rac1 has a central role in linking cell adhesion to G1/S transition by activating mitogenic pathways via a range of effectors. In contrast, RhoA controls cell cycle in part via the organization of actin stress fibres and cell shape [54]. To promote quiescence, stem cells express high levels of cyclin-dependent kinase inhibitors (CDKIs) such as p21, p27, p57, and p15. Following anti-mitogenic signals, p21 and p27 bind to cyclin–CDK complexes to inhibit their catalytic activity and induce cell cycle arrest [55].

The alignment and orientation of ECM molecules contributes to mechanosensing. Progenitor stem cells enhanced cell cycle on aligned ECM proteins, and this effect is mediated by β1 integrin and ERK activation [63]. It has been reported by Roncoroni *et al.* that BM MSC express the α1, α2, α3, α6, α7, α9, α11, and β1 subunits of integrins, and TGF-β1 regulates the expression of α2, α6, and β1 integrins, thus helps forward the attachment of MSC to ECM proteins [64]. Therefore, cytokines regulate the expression of integrins too. ECM remodeling takes place by means of proteases like matrix metalloproteases, serine and cysteine proteases. These interactions stimulate or inhibit various signaling pathways in the stem cell niche.

1.3 Extracellular Matrix of the Mesenchyme: Human Bone Marrow

In adult, hematopoiesis is restricted to the extravascular compartment where HSCs are in contact or close proximity with a heterogeneous population of stromal cells in the niche. Cellular interactions between HSCs and stromal cells involve various cell surface molecules, including integrins, selectins, sialomucins, and the immunoglobulin gene superfamily, that are subsequently translated into cell signaling regulating the localization and function of the cells within the niche [65].

In BM, integrin receptors interact with ligands that include ECM proteins, immunoglobulin superfamily members, and vascular cell adhesion protein 1 (VCAM-1) and, therefore, they play an important role in cell adhesion and signaling. Several integrins have been identified in the BM microenvironment and more specifically on HSCs [66, 67]. Functional integrins depend on the cytokine medium in which they reside, making adhesive events regulatable by cytokines. Among the different integrin subtypes, the $\beta1$ integrins have shown to play an important role in HSC migration and homing to the BM [67]. Although $\beta1$ integrin is generally known to promote proliferation, it can also be inhibitory, constituting a dominant negative signal over the stimulatory effects [54]. In the HSCs, the $\alpha4$, $\alpha5$, $\alpha6$, $\alpha7$, and $\alpha9$ integrin subunits are expressed [68]. A study on the spatial location of ECM proteins including fibronectin, collagen types; I, III, and IV, and laminin in murine femoral BM by immunofluorescence has revealed distinct locations for each protein, supporting the notion that they have an important role in the homing and lodgment of transplanted cells (Table 1.3, [69]). Collagen type IV, laminin, nidogen/entactin, and perlecan are the major components of this network [70]. Minor components bind to the major components in a tissue-specific manner via their chains, and they contribute to BM's heterogeneity [70]. Many reports have documented the importance of $\alpha4\beta1$ and $\alpha5\beta1$, in modulating adhesive interactions between HPCs and ECM components that comprise the stem cell niche [71, 72]. A study conducted by Van Der Loo *et al.* demonstrated that $\alpha5\beta1$ is expressed on mouse and human long-term repopulating hematopoietic cells and binds to fibronectin in the ECM and that

Table 1.3 ECM proteins and their localization within the BM.

	Bone compact and/or trabeculae	Endosteal surface/ marrow	Periosteal surface	Central marrow	Marrow vessel: sinuses and arteries
Fibronectin	+++	+++	+++	++	−
Collagen type I	+++	+++	−	−	−
Collagen type II	−	−	++	−	−
Collagen type IV	+	++	++	−	++
Laminin	+	−	+++	+	+++

+++, bright expression; ++, moderate expression; +, faint expression; −, absent. Adapted from [69].

disruption of this binding can lead to decreased engraftment in the BM [73]. Also, expression of the laminin receptor has been found on erythroid progenitors and its inhibitions blocks BM homing of Burst forming unit-erythroid (BFU-E) [74]. MK maturation and platelet generation are consequent to MK migration from the osteoblastic to the vascular niche, where MKs extend proplatelets, and newly generated platelets are released into the blood [75].

It has been demonstrated that interactions of MKs with fibrinogen or von Willebrand factor in vascular space, are able to sustain MK maturation and proplatelets, whereas collagen type I suppresses these events and prevents premature platelet release in the osteoblastic niche [76]. The negative regulation of proplatelets by collagen type I is mediated by the interaction with the integrin $\alpha 2\beta 1$. Hence, recent studies have demonstrated that fibronectin may represent a new regulator of MK maturation and platelet release [77]. Collagen fibers stimulate platelet activation, leading to inside out regulation of the integrin GP IIb–IIIa, secretion from dense and alpha granules, generation of thromboxanes, and expression of procoagulant activity, all of which support the hemostatic process. ECM proteins such as laminin, collagen type IV self-assemble in the BM into a honeycomb-like polymer. The role of collagen in supporting platelet adhesion to the endothelium is mediated through indirect and direct interactions [78]. Another ECM protein, osteopontin, is a phosphorylated glycoprotein that can be produced by a variety of BM cells, especially bone surface. Osteopontin has the ability to bind multiple integrins as well as CD44. It has some roles affecting many physiological and pathological processes including chemotaxis, adhesion, apoptosis, inflammatory responses, and tumor metastasis. Several integrins with potential binding to osteopontin are expressed by HSCs, making them molecular candidates in the cross-talk between osteoblasts and HSCs at the level of BM stem cell niches [79]. We previously investigated whether coating of culture surfaces with ECM proteins normally present in the marrow microenvironment could benefit the *ex vivo* expansion of UCB CD34+ hematopoietic progenitor cells (HPCs). Toward this, collagen types I, IV, laminin, and fibronectin were tested individually or as component of two ECM-mix complexes. Individually, ECM proteins had both common and unique properties on the growth and differentiation of UCB CD34+ cells; some ECM proteins favored the differentiation of some lineages over that of others (e.g. fibronectin for erythroids), some the expansion of HPCs (e.g. laminin and MK progenitor), while others had less effects. Next, two ECM-mix complexes were tested; the first one contained all four ECM proteins (4ECMp), while the second 'basement membrane-like structure' was without collagen type I (3ECMp). Removal of collagen type I led to strong reductions in cell growth and HPCs expansion. Interestingly, the 4ECMp-mix complex reproducibly increased CD34+ and CD41+ (MK) cell expansions and induced greater myeloid progenitor expansion than 3ECMp. In conclusion, these results suggest that optimization of BM ECM protein complexes could provide a better environment for the *ex vivo* expansion of hematopoietic progenitors than individual ECM protein [80].

1.4 Biomimetic Peptides as Extracellular Matrix Proteins

Physicochemical and biological functions of the cell membrane create new opportunities for developing bioactive peptides for biomedical applications. Membrane proteins are asymmetrically distributed in the lipid bilayer of all biological membranes. Transmembrane proteins have some specific orientations; one side of the protein might be shaped such that it can act as a receptor for a signaling molecule, while the other side changes shape in response to the binding signal [81]. The communication of cells with surfaces is mediated by pre-adsorbed protein layers. Understanding the interactions of the ECM on stem cell fate will provide the framework for designing and engineering ECMs to control cell behavior that can be used to direct stem cell expansion and differentiation.

The bioactive peptides can perform the desired function of ECM proteins and have additional advantages such as they are smaller than parent proteins, more stable, precisely synthesized, and have tunable properties such as selectivity and solubility [82]. Peptide sequences with different properties such as cell adhesion, growth factor binding, and enzymatically degradable sequences are most commonly used [9]. The commonly used arginine–glycine–aspartic acid (RGD) peptides for cell adhesion have shown to affect differentiation of various stem cells but inhibit adhesion when in solution by blocking integrin binding sites. RGD peptides do not support the growth of hESCs as these cells utilize non-RGD-binding integrins for attachment to adherent substrates. On the other hand, laminin, fibrinogen, vitronectin, tenascin, thrombospondin, entactin, bone sialoprotein, osteopontin, and certain collagens contain RGD sites [83]. It has been reported that RGD plays an important role in initiating chondrogenic differentiation [84]. RGD peptides can be used to provide cell attachment to non-adhesive surfaces and can be easily combined with scaffolds or biomaterials [85, 86]. Naghdi et al. demonstrated that polyethylene glycol hydrogel conjugated with RGD improved neurite outgrowth [87]. Another amino acid sequence, YIGSR, resides on the β1-chain of the laminin. The YIGSR peptide can be used to mimic the properties of soluble laminin, such as inhibiting neurite growth, while the IKVAV peptide antagonizes the activity of soluble laminin and activates cell spreading, neurite outgrowth, and angiogenesis [88]. The peptide sequence GFOGER in collagen type I specifically binds the $\alpha 2\beta 1$ receptor and promotes osteoblastic differentiation [89]. It has been previously shown that GFOGER significantly enhanced the migration, proliferation, and osteogenic differentiation of hMSCs on nanofiber meshes [90]. Connelly et al. explained that RGD peptides but not GFOGER increased collagen type I expression of hMSCs. Moreover, the GFOGER and RGD peptides enhanced osteocalcin at similar levels [91]. It has been previously reported that MSCs cultured in DGEA-collagen type I mimetic peptide exhibited increased levels of osteocalcin production and mineral deposition. These data suggest that the presentation of the DGEA ligand is a feasible approach for selectively inducing an osteogenic phenotype in encapsulated MSCs [92]. Hyaluronan, also known as hyaluronic acid (HA), is a large GAG that typically surrounds non-polarized and migratory cells. The various activities of HA are a consequence of its binding to hyaluronan binding proteins. It contains the 9–11-residue long B-X_7-B motif, in which B represents a basic amino acid (arginine or lysine) and X consists of a non acidic amino acid residue [93]. The B-X7-B motif can be used to induce proliferation, adhesion, or other signaling events supported by hyaluronan. In general, the overall effect of Perlecan/HSPG 2 domain I contains a 22 kDa protein core, comprised of three SDG motifs that function as attachment sites for GAGs. This site serves as a high-density binding repository for VEGF, FGF-2, and TGF-β that regulates many biological functions such as angiogenesis, cell growth, development, and tissue regeneration [94].

References

1. Jung K.W. Perspectives on human stem cell research. *J. Cell Physiol.* 220, 535–7, 2009.
2. Hass R., Kasper C., Bohm S., Jacobs R. Different populations and sources of human mesenchymal stem cells (MSC): A comparison of adult and neonatal tissue-derived MSC. *Cell Commun. Signal* 9, 12, 2011.
3. Fraser J.K., Wulur I., Alfonso Z., Hedrick M.H. Fat tissue: an underappreciated source of stem cells for biotechnology. *Trends Biotechnol.* 24, 150–4, 2006.
4. Pittenger M.F., Mackay A.M., Beck S.C., Jaiswal R.K., Douglas R., Mosca J.D., et al. Multilineage potential of adult human mesenchymal stem cells. *Science* 284, 143–7, 1999.
5. Kassis I., Zangi L., Rivkin R., Levdansky L., Samuel S., Marx G., et al. Isolation of mesenchymal stem cells from G-CSF-mobilized human peripheral blood using fibrin microbeads. *Bone Marrow Transpl.* 37, 967–76, 2006.
6. Anzalone R., Lo Iacono M., Corrao S., Magno F., Loria T., Cappello F., et al. New emerging potentials for human Wharton's jelly mesenchymal stem cells: immunological features and hepatocyte-like differentiative capacity. *Stem Cells Dev.* 19, 423–38, 2010.
7. Barlow S., Brooke G., Chatterjee K., Price G., Pelekanos R., Rossetti T., et al. Comparison of human placenta- and bone marrow-derived multipotent mesenchymal stem cells. *Stem Cells Dev.* 17, 1095–107, 2008.
8. Takahashi K., Yamanaka S., Induction of pluripotent stem cells from mouse embryonic and adult fibroblast cultures by defined factors. *Cell* 126, 663–76, 2006.
9. Jhala D., Vasita R., A review on extracellular matrix mimicking strategies for an artificial stem cell niche. *Polym. Rev.* 55, 561–95, 2015.
10. Schofield R. The relationship between the spleen colony-forming cell and the haemopoietic stem cell. *Blood Cells* 4, 7–25, 1978.
11. Gattazzo F., Urciuolo A., Bonaldo P. Extracellular matrix: a dynamic microenvironment for stem cell niche. *Biochim. Biophys. Acta* 1840, 2506–19, 2014.
12. Becker A.J., Mc C.E., Till J.E. Cytological demonstration of the clonal nature of spleen colonies derived from transplanted mouse marrow cells. *Nature* 197, 452–4, 1963.
13. Mikkola H.K., Orkin S.H. The journey of developing hematopoietic stem cells. *Development* 133, 3733–44, 2006.

14. Dao M.A., Arevalo J., Nolta J.A. Reversibility of CD34 expression on human hematopoietic stem cells that retain the capacity for secondary reconstitution. *Blood* 101, 112–8, 2003.
15. Quesenberry P.J., Colvin G., Abedi M. Perspective: fundamental and clinical concepts on stem cell homing and engraftment: a journey to niches and beyond. *Exp. Hematol.* 33, 9–19, 2005.
16. Bhattacharya D., Rossi D.J., Bryder D., Weissman I.L. Purified hematopoietic stem cell engraftment of rare niches corrects severe lymphoid deficiencies without host conditioning. *J. Exp. Med.* 203, 73–85, 2006.
17. Kolf C.M., Cho E., Tuan R.S. Mesenchymal stromal cells. Biology of adult mesenchymal stem cells: regulation of niche, self-renewal and differentiation. *Arthritis Res. Ther.* 9, 204, 2007.
18. Amable P.R., Teixeira M.V., Carias R.B., Granjeiro J.M., Borojevic R. Protein synthesis and secretion in human mesenchymal cells derived from bone marrow, adipose tissue and Wharton's jelly. *Stem Cell Res. Ther.* 5, 53, 2014.
19. da Silva Meirelles L., Chagastelles P.C., Nardi N.B. Mesenchymal stem cells reside in virtually all post-natal organs and tissues. *J. Cell Sci.* 119, 2204–13, 2006.
20. Law S, Chaudhuri S. Mesenchymal stem cell and regenerative medicine: regeneration versus immunomodulatory challenges. *Am. J. Stem Cells* 2, 22–38, 2013.
21. Bernardo M.E, Fibbe W.E. Mesenchymal stromal cells: sensors and switchers of inflammation. *Cell Stem Cell* 13, 392–402, 2013.
22. Gordon M.Y. Extracellular matrix of the marrow microenvironment. *Br J. Haematol.* 70, 1–4, 1988.
23. van Winterswijk PJ, Nout E. Tissue engineering and wound healing: an overview of the past, present, and future. *Wounds* 19, 277–84, 2007.
24. Frantz C., Stewart K.M., Weaver V.M. The extracellular matrix at a glance. *J. Cell Sci.* 123, 4195–200, 2010.
25. Engler A.J, Sen S., Sweeney H.L., Discher D.E. Matrix elasticity directs stem cell lineage specification. *Cell* 126, 677–89, 2006.
26. Discher D.E., Janmey P., Wang Y.L. Tissue cells feel and respond to the stiffness of their substrate. *Science* 310, 1139–43, 2005.
27. Pelham R.J., Jr., Wang Y. Cell locomotion and focal adhesions are regulated by substrate flexibility. *Proc. Natl. Acad. Sci. U S A* 94, 13661–5, 1997.
28. Guilak F., Cohen D.M., Estes B.T., Gimble J.M., Liedtke W., Chen C.S. Control of stem cell fate by physical interactions with the extracellular matrix. *Cell Stem Cell* 5, 17–26, 2009.
29. Carrel A., Burrows M.T. Cultivation *in vitro* of malignant tumors. *J. Exp. Med.* 13, 571–5, 1911.
30. Chen G., Ito Y., Imanishi Y. Regulation of growth and adhesion of cultured cells by insulin conjugated with thermoresponsive polymers. *Biotechnol. Bioeng.* 53, 339–44, 1997.
31. Oh S., Brammer K.S., Li Y.S., Teng D., Engler A.J., Chien S., *et al.* Stem cell fate dictated solely by altered nanotube dimension. *Proc. Natl. Acad. Sci. U S A* 106, 2130–5, 2009.

32. Griffin M.F., Butler P.E., Seifalian A.M., Kalaskar D.M. Control of stem cell fate by engineering their micro and nanoenvironment. *World J. Stem Cells* 7, 37–50, 2015.
33. McNamara L.E., McMurray R.J., Biggs M.J., Kantawong F., Oreffo R.O., Dalby M.J. Nanotopographical control of stem cell differentiation. *J. Tissue Eng.* 2010, 120623, 2010.
34. Trappmann B., Chen C.S. How cells sense extracellular matrix stiffness: a material's perspective. *Curr. Opin. Biotechnol.* 24, 948–53, 2013.
35. Levental K.R., Yu H., Kass L., Lakins J.N., Egeblad M., Erler J.T., *et al.* Matrix cross-linking forces tumor progression by enhancing integrin signaling. *Cell* 139, 891–906, 2009.
36. Trappmann B., Gautrot J.E., Connelly J.T., Strange D.G., Li Y., Oyen M.L., *et al.* Extracellular-matrix tethering regulates stem-cell fate. *Nat. Mater.* 11, 642–9, 2012.
37. Qiu Y., Bayomy A.F., Gomez M.V., Bauer M., Du P., Yang Y., *et al.* A role for matrix stiffness in the regulation of cardiac side population cell function. *Am. J. Physiol. Heart. Circ. Physiol.* 308, H990–7, 2015.
38. Saha K., Keung A.J., Irwin E.F., Li Y., Little L., Schaffer D.V., *et al.* Substrate modulus directs neural stem cell behavior. *Biophys. J.* 95, 4426–38, 2008.
39. Barczyk M., Carracedo S., Gullberg D. Integrins. *Cell. Tissue Res.* 339, 269–80, 2010.
40. Wang H., Luo X., Leighton J. Extracellular matrix and integrins in embryonic stem cell differentiation. *Biochem. Insights* 8, 15–21, 2015.
41. Watt F.M., Huck W.T. Role of the extracellular matrix in regulating stem cell fate. *Nat. Rev. Mol. Cell Biol.* 14, 467–73, 2013.
42. Kim S.H., Turnbull J., Guimond S. Extracellular matrix and cell signalling: the dynamic cooperation of integrin, proteoglycan and growth factor receptor. *J. Endocrinol.* 209, 139–51, 2011.
43. Zhang P., Li J., Tan Z., Wang C., Liu T., Chen L., *et al.* Short-term BMP-4 treatment initiates mesoderm induction in human embryonic stem cells. *Blood* 111, 1933–41, 2008.
44. Anton E.S., Kreidberg J.A., Rakic P. Distinct functions of alpha3 and alpha(v) integrin receptors in neuronal migration and laminar organization of the cerebral cortex. *Neuron* 22, 277–89, 1999.
45. Eckes B., Krieg T., Wickstrom SA. Role of integrin signalling through integrin-linked kinase in skin physiology and pathology. *Exp. Dermatol.* 23, 453–6, 2014.
46. Hara M., Yaar M., Tang A., Eller M.S., Reenstra W., Gilchrest B.A. Role of integrins in melanocyte attachment and dendricity. *J. Cell Sci.* 107 (Pt 10), 2739–48, 1994.
47. Arora P.D., Narani N., McCulloch C.A. The compliance of collagen gels regulates transforming growth factor-beta induction of alpha-smooth muscle actin in fibroblasts. *Am. J. Pathol.* 154, 871–82, 1999.
48. Patel V.N., Knox S.M., Likar K.M., Lathrop C.A., Hossain R., Eftekhari S., *et al.* Heparanase cleavage of perlecan heparan sulfate modulates FGF10 activity

during *ex vivo* submandibular gland branching morphogenesis. *Development* 134, 4177–86, 2007.
49. Wijelath E.S., Rahman S., Namekata M., Murray J., Nishimura T., Mostafavi-Pour Z., et al. Heparin-II domain of fibronectin is a vascular endothelial growth factor-binding domain: enhancement of VEGF biological activity by a singular growth factor/matrix protein synergism. *Circ. Res.* 99, 853–60, 2006.
50. Tilghman R.W., Parsons J.T. Focal adhesion kinase as a regulator of cell tension in the progression of cancer. *Semin. Cancer Biol.* 18, 45–52, 2008.
51. Seong J., Ouyang M., Kim T., Sun J., Wen P.C., Lu S., et al. Detection of focal adhesion kinase activation at membrane microdomains by fluorescence resonance energy transfer. *Nat. Commun.* 2, 406, 2011.
52. Sawai H., Okada Y., Funahashi H., Matsuo Y., Takahashi H., Takeyama H., et al. Activation of focal adhesion kinase enhances the adhesion and invasion of pancreatic cancer cells via extracellular signal-regulated kinase-1/2 signaling pathway activation. *Mol. Cancer.* 4, 37, 2005.
53. Guan J.L. Role of focal adhesion kinase in integrin signaling. *Int. J. Biochem. Cell Biol.* 29, 1085–96, 1997.
54. Moreno-Layseca P., Streuli C.H. Signalling pathways linking integrins with cell cycle progression. *Matrix Biol.* 34, 144–53, 2014.
55. Coqueret O. New roles for p21 and p27 cell-cycle inhibitors: a function for each cell compartment? *Trends Cell Biol.* 13, 65–70, 2003.
56. Schneller M., Vuori K., Ruoslahti E. Alphavbeta 3 integrin associates with activated insulin and PDGFbeta receptors and potentiates the biological activity of PDGF. *EMBO. J.* 16, 5600–7, 1997.
57. Zhang X., Yee D. Tyrosine kinase signalling in breast cancer: insulin-like growth factors and their receptors in breast cancer. *Breast Cancer Res.* 2, 170–5, 2000.
58. Ruoslahti E. Specialization of tumour vasculature. *Nat. Rev. Cancer* 2, 83–90, 2002.
59. Scaffidi A.K., Petrovic N., Moodley Y.P., Fogel-Petrovic M., Kroeger K.M., Seeber R.M., et al. alpha(v)beta(3) Integrin interacts with the transforming growth factor beta (TGF beta) type II receptor to potentiate the proliferative effects of TGF beta 1 in living human lung fibroblasts. *J. Biol. Chem.* 279, 37726–33, 2004.
60. Sridhar S.C., Miranti C.K. Tetraspanin KAI1/CD82 suppresses invasion by inhibiting integrin-dependent cross-talk with c-Met receptor and Src kinases. *Oncogene* 25, 2367–78, 2006.
61. DeMali K.A., Balciunaite E., Kazlauskas A. Integrins enhance platelet-derived growth factor (PDGF)-dependent responses by altering the signal relay enzymes that are recruited to the PDGF beta receptor. *J. Biol. Chem.* 274, 19551–8, 1999.
62. Bill H.M., Knudsen B., Moores S.L., Muthuswamy S.K., Rao V.R., Brugge J.S., et al. Epidermal growth factor receptor-dependent regulation of integrin-mediated signaling and cell cycle entry in epithelial cells. *Mol. Cell Biol.* 24, 8586–99, 2004.

63. Wang Y., Yao M., Zhou J., Zheng W., Zhou C., Dong D., et al. The promotion of neural progenitor cells proliferation by aligned and randomly oriented collagen nanofibers through beta1 integrin/MAPK signaling pathway. *Biomaterials* 32, 6737–44, 2011.
64. Warstat K., Hoberg M., Rudert M., Tsui S., Pap T., Angres B., et al. Transforming growth factor beta1 and laminin-111 cooperate in the induction of interleukin-16 expression in synovial fibroblasts from patients with rheumatoid arthritis. *Ann. Rheum. Dis.* 69, 270–5, 2010.
65. Lam BS, Adams GB. Hematopoietic stem cell lodgment in the adult bone marrow stem cell niche. *Int. J. Lab. Hematol.* 32, 551–8, 2010.
66. Klein G. The extracellular matrix of the hematopoietic microenvironment. *Experientia* 51, 914–26, 1995.
67. Potocnik A.J., Brakebusch C., Fassler R. Fetal and adult hematopoietic stem cells require beta1 integrin function for colonizing fetal liver, spleen, and bone marrow. *Immunity* 12, 653–63, 2000.
68. ter Huurne M., Figdor C.G., Torensma R. Hematopoietic stem cells are coordinated by the molecular cues of the endosteal niche. *Stem. Cells Dev.* 19, 1131–41, 2010.
69. Nilsson S.K., Debatis M.E., Dooner M.S., Madri J.A., Quesenberry P.J., Becker P.S. Immunofluorescence characterization of key extracellular matrix proteins in murine bone marrow in situ. *J. Histochem. Cytochem.* 46, 371–7, 1998.
70. LeBleu V.S., Macdonald B., Kalluri R. Structure and function of basement membranes. *Exp. Biol. Med. (Maywood)* 232, 1121–9, 2007.
71. Levesque J.P., Haylock D.N., Simmons P.J. Cytokine regulation of proliferation and cell adhesion are correlated events in human CD34+ hemopoietic progenitors. *Blood* 88, 1168–76, 1996.
72. Carstanjen D., Gross A., Kosova N., Fichtner I., Salama A. The alpha-4beta1 and alpha5beta1 integrins mediate engraftment of granulocyte-colony-stimulating factor-mobilized human hematopoietic progenitor cells. *Transfusion* 45, 1192–200, 2005.
73. van der Loo J.C., Xiao X., McMillin D., Hashino K., Kato I., Williams D.A. VLA-5 is expressed by mouse and human long-term repopulating hematopoietic cells and mediates adhesion to extracellular matrix protein fibronectin. *J. Clin. Invest.* 102, 1051–61, 1998.
74. Bonig H., Chang K.H., Nakamoto B., Papayannopoulou T. The p67 laminin receptor identifies human erythroid progenitor and precursor cells and is functionally important for their bone marrow lodgment. *Blood* 108, 1230–3, 2006.
75. Junt T., Schulze H., Chen Z., Massberg S., Goerge T., Krueger A., et al. Dynamic visualization of thrombopoiesis within bone marrow. *Science* 317, 1767–70, 2007.
76. Sabri S., Jandrot-Perrus M., Bertoglio J., Farndale R.W., Mas V.M., Debili N., et al. Differential regulation of actin stress fiber assembly and proplatelet

formation by alpha2beta1 integrin and GPVI in human megakaryocytes. *Blood* 104, 3117–25, 2004.
77. Cuvelier D., Thery M., Chu Y.S., Dufour S., Thiery J.P., Bornens M., *et al.* The universal dynamics of cell spreading. *Curr. Biol.* 17, 694–9, 2007.
78. Watson S.P. Collagen receptor signaling in platelets and megakaryocytes. *Thromb. Haemost.* 82, 365–76, 1999.
79. Aguila H.L., Rowe D.W. Skeletal development, bone remodeling, and hematopoiesis. *Immunol. Rev.* 208, 7–18, 2005.
80. Celebi B., Mantovani D., Pineault N. Effects of extracellular matrix proteins on the growth of haematopoietic progenitor cells. *Biomed. Mater.* 6, 055011, 2011.
81. Gong Y.K., Winnik F.M. Strategies in biomimetic surface engineering of nanoparticles for biomedical applications. *Nanoscale* 4, 360–8, 2012.
82. Collier J.H., Segura T. Evolving the use of peptides as components of biomaterials. *Biomaterials* 32, 4198–204, 2011.
83. Pradhan S., Farach-Carson M.C. Mining the extracellular matrix for tissue engineering applications. *Regen. Med.* 5, 961–70, 2010.
84. DeLise A.M., Fischer L., Tuan R.S. Cellular interactions and signaling in cartilage development. *Osteoarthritis Cartilage* 8, 309–34, 2000.
85. Shin H., Zygourakis K., Farach-Carson M.C., Yaszemski M.J., Mikos A.G. Attachment, proliferation, and migration of marrow stromal osteoblasts cultured on biomimetic hydrogels modified with an osteopontin-derived peptide. *Biomaterials* 25, 895–906, 2004.
86. Wang X., Ye K., Li Z., Yan C., Ding J. Adhesion, proliferation, and differentiation of mesenchymal stem cells on RGD nanopatterns of varied nanospacings. *Organogenesis* 9, 280–6, 2013.
87. Naghdi P., Tiraihi T., Ganji F., Darabi S., Taheri T., Kazemi H. Survival, proliferation and differentiation enhancement of neural stem cells cultured in three-dimensional polyethylene glycol-RGD hydrogel with tenascin. *J. Tissue. Eng. Regen. Med.* 10, 199–208, 2016.
88. Ratcliffe E.M., D'Autreaux F., Gershon M.D. Laminin terminates the Netrin/DCC mediated attraction of vagal sensory axons. *Dev. Neurobiol.* 68, 960–71, 2008.
89. Wojtowicz A.M., Shekaran A., Oest M.E., Dupont K.M., Templeman K.L., Hutmacher D.W., *et al.* Coating of biomaterial scaffolds with the collagen-mimetic peptide GFOGER for bone defect repair. *Biomaterials* 31, 2574–82, 2010.
90. Kolambkar Y.M., Bajin M., Wojtowicz A., Hutmacher D.W., Garcia A.J., Guldberg R.E. Nanofiber orientation and surface functionalization modulate human mesenchymal stem cell behavior *in vitro*. *Tissue. Eng. Part. A* 20, 398–409, 2014.
91. Connelly J.T., Petrie T.A., Garcia A.J., Levenston M.E. Fibronectin- and collagen-mimetic ligands regulate bone marrow stromal cell chondrogenesis

in three-dimensional hydrogels. *Eur. Cell. Mater.* 22, 168–76, discussion 76–7, 2011.
92. Mehta M., Madl C.M., Lee S., Duda G.N., Mooney D.J. The collagen I mimetic peptide DGEA enhances an osteogenic phenotype in mesenchymal stem cells when presented from cell-encapsulating hydrogels. *J. Biomed. Mater. Res. A.* 103, 3516–25, 2015.
93. Yang B., Yang B.L., Savani R.C., Turley E.A. Identification of a common hyaluronan binding motif in the hyaluronan binding proteins RHAMM, CD44 and link protein. *EMBO. J.* 13, 286–96, 1994.
94. Kirn-Safran C.B., Gomes R.R., Brown A.J., Carson D.D. Heparan sulfate proteoglycans: coordinators of multiple signaling pathways during chondrogenesis. *Birth Defects Res. C Embryo Today* 72, 69–88, 2004.

2

The Superficial Mechanical and Physical Properties of Matrix Microenvironment as Stem Cell Fate Regulator

Mohsen Shahrousvand[1]*, Gity Mir Mohamad Sadeghi[1]* and Ali Salimi[2]

[1]Department of Polymer Engineering and Color Technology, Amirkabir University of Technology, Tehran, Iran
[2]Nanobiotechnology Research Center, Baqiyatallah University of Medical Science, Tehran, Iran

Abstract

All of the cells' interactions are done through their surfaces. So, whatever associated with cell surface can affect its behavior. Stem cells show more significant sensitivity to surface interactions than other cells in the body because of their stemness. Thus, a series of extrinsic and intrinsic cell mechanisms lead to regulated unique cell behavior, including the following: the capacity for proliferation without loss of potency and the ability to differentiate into specialized cell type(s). Therefore, understanding those surface mechanisms underlying self-renewal and differentiation of stem cells including the study of cell communication pathways and endogenous stem cell niche is in the center of attention. Cells can sense the surface properties of substrate that are located on their surfaces by their receptors. Recent progresses are in this field to understanding the importance and influence of the surface properties on the fate. To develop better techniques for controlling cell fate, cell- and matrix-based assays for evaluating the effectiveness of engineered surfaces are critical. This chapter is an introduction to the effects of surface properties of the polymer matrices that can be used in tissue engineering and the impact of these properties on the cell fates. This can open new prospects in tissue engineering based on engineered surface structures of scaffolds as microenvironments.

Keywords: Self-renewal and differentiation, superficial mechanical and physical properties, matrix microenvironment, stem cell fate

---

*Corresponding authors: mohsen.shahrousvand@gmail.com; gsadeghi@aut.ac.ir

2.1 Introduction

The cellular niche is a microenvironment or sheltering environment including from intrinsic and extrinsic mechanisms to control self-renewal and differentiation of stem cells (embryonic stem cells, tissue stem cells, and induced pluripotent stem cells) [1]. The cellular niche maintains the stem cells from differentiation stimuli, apoptotic stimuli, and other stimuli that would challenge stem cell reserves. On the other hand, niche prevents uncontrolled growth of cells that leads to cancer [2]. Thus, maintaining a balance of stem cell quiescence and activity is a hallmark of a functional niche. The relationship between stem cells in a niche with differentiated cells is established out by surface receptors. These receptors are establishing cellular communication. Therefore, niche as an interactive structural unit determines cell fate [3].

Niche structure is composed of a fixed part [such as extracellular matrix (ECM)] and a non-fixed part (such as soluble factors and body fluids). All of the responsibilities of a niche for stem cells are done by surface cell signaling of both parts. The effects of soluble parameters in the microenvironment on differentiated cells and stem cells behavior have long been established, but a new attractive challenge is the direct impact of physical attributes of microenvironments on cell behavior which include: wettability and hydrophilicity, surface topography, roughness, surface stiffness, surface energy, etc. [4].

The cells release secreting molecules on exposure to various matrices and other cells in the ECM that helps more compatibility and remodeling with the microenvironments [5]. Therefore, artificial ECMs must be modified in confronting with any types of stem cells in order to minimize remodeling [6–8]. On the other side, it has been confirmed that the substrate surface properties alter cell behavior such as cell adhesion, cell proliferation, and cell differentiation [9–11]. Cellular fate is controlled by the following parameters [12]:

- Interaction with matrix proteins (surface of matrix),
- Soluble factors and secreted factors,
- Other cell types in stem cell niches or microenvironment.

All of existing parameters are converted into biological encodes by surface communications and are transferred to the cell nucleus. Therefore, proper interactions between the microenvironment and its components with stem cells are very important.

2.2 Fabrication of the Microenvironments with Different Properties in Surfaces

Preparation of suitable substrates and preset features for specific applications is one of the most influential parameters in tissue engineering. Thus, understanding of geometry and tissue properties that should be repaired or replaced is the first step in the preparation of tissue engineering scaffolds. Each body tissue of the body has its own unique characteristics, the recognition of which can help replacement and repair. Including materials used in the construction of scaffolds can be noted to polymers (natural and/or synthetic), metals, ceramics, and combinations of these materials. From these materials, use of polymers is more fortunate because of their controllable features [13–18]. There are various methods to prepare controlled materials with suitable properties that affect the physical, mechanical, biochemical, and biological properties of scaffolds as microenvironments. Effective and engineered procedures in preparation of superficial properties in polymeric microenvironments are as follow: physical and chemical quiddity of polymers, polymer blending, the use of polymer composites (nano and/or micro), cross-linked polymer structures, control of the size and morphology of crystals in the polymer structures, changes in the molecular weight of the polymers, control of the phase separation in microstructures and nanostructures, polymer functionalization (on the backbone and/or branches), mechanical and thermal treatments during manufacturing, etc. [19–27].

Physical, mechanical, and biochemical surface properties can be changed with a polymer blending. Alloying polymer(s) with poor mechanical properties and polymer(s) with better mechanical behavior strengthens its mechanical properties. For example, poly(lactic-co-glycolic acid) (PLGA) as synthetic biodegradable polymer and two natural proteins gelatin (denatured collagen) and α-elastin were blended to improve elastin elasticity [28]. Microscopic evaluation confirmed that myoblasts reached confluence on the scaffold surfaces while simultaneously growing into the scaffolds. This polymeric blend can be useful for engineering soft tissues, such as heart, lung, and blood vessels. To improve the water absorption of poly(ε-caprolactone) (PCL) scaffolds, this polymer were blended with polyglycolide (PGA), and poly(ethylene oxide) (PEO) [29]. The scaffolds displayed high mechanical properties, water uptake, and wettability, in addition to a remarkably fast degradation rate.

Blending hydrophobic polymers with hydrophilic polymers increase the water absorption and cell adhesion to the scaffolds. Porous PLGA/PVA scaffolds were fabricated by blending PLGA with polyvinyl alcohol (PVA) to improve the hydrophilicity and cell compatibility of the scaffolds for tissue-engineering applications. The PLGA/PVA blend scaffolds with PVA compositions more than 5% were easily wetted in cell culture medium without any prewetting treatments, which is highly desirable for tissue-engineering applications [19].

Polymer blends can be miscible or immiscible with phase separation in micro- or nanoscale that can be controlled by the application. Changes in the structure of the used polymers in preparation of scaffolds can help cell adhesion, induced growth and cell differentiation, structural strength, degradation rate of the scaffold, etc. For example, the blending of collagen with chitosan gives the possibility of producing new bespoke materials for potential biomedical applications that collagen/chitosan blends are miscible at the molecular level and exhibit interactions between the components [30].

The use of polymer composites can create special properties in the polymer matrix and provide navigating behavior in cell proliferation and cell differentiation. With the help of polymer composites can be prepare scaffolds with controlled properties such as degradability, hydrophilicity, water permeability, surface and bulk modulus, and other chemical and physical properties [31–33]. Particle–matrix compatibility is very important in the preparation of nontoxic polymer composites. For instance, to better mimic the mineral component and the microstructure of natural bone, nano-hydroxyapatite (NHAP)/polymer composite scaffolds with high porosity and well-controlled pore architectures were prepared using thermally induced phase separation (TIPS) techniques [34]. The introduction of hydroxyapatite greatly increased the mechanical properties and improved the protein adsorption capacity.

Cross-linking is the other way to increase the mechanical properties of the scaffolds that makes also scaffold degradation to longer time period. So, the rate of degradation and mechanical properties of the cross-linked scaffold can be controlled by density of cross-linking in the polymer scaffold [35, 36]. For example, chitosan substrates are cross-linked with a functional diepoxide (1,4-butanediol diglycidyl ether) to alter its mechanical property, and the viability and proliferation of the canine articular chondrocytes seeded on the cross-linked surface are further assayed. However, networked structures reduce water penetration and consequently the rate of degradation will be low. But a new idea in the development of biodegradable cross-linked scaffolds is that the cross-linker linkages consists of biodegradable groups, and then in addition to improving the mechanical properties of scaffolds, scaffold degradation will not be a problem [36, 37]. Developed biodegradable polyurethane films based on hexamethylene diisocyanate (HDI) and glycerol as the hard segment (HS), and poly(caprolactone) triol (PCL triol) and low-molecular-weight poly(ethylene glycol) (PEG) as the soft segment (SS) without the use of a catalyst synthesized. As a consequence, a homogeneous structure without distinct hard and SSs was formed in the samples which allows water to penetrate into the structure and increases hydrolytic degradation [37].

If the structure of a hydrophilic polymer to be cross-linked, hydrogel is obtained which is used in tissue engineering widely and maintains its geometry and appropriate water absorption without collapse *in vitro* and *in vivo* environments [38, 39].

Crystalline polymers used in the preparation of scaffolds for tissue engineering are considerable interest for two reasons: on the one hand, due to its crystalline structure reduce water penetration to the crystalline structure which hydrolytic degradation will be later. On the other hand, crystalline structures in tissue-engineering scaffolds affect on bulk and surface modulus, cell adhesion [40], and cell differentiation [41, 42]. Many current scaffold manufacturing techniques induce random porosity in bulk materials, requiring high porosities (>95%) to guarantee complete interconnectivity, but the high porosity sacrifices mechanical properties. Additionally, the stochastic arrangement of pores causes scaffold-to-scaffold variation. So, a biodegradable PLGA scaffold with an inverted colloidal crystal (ICC) structure that provides a highly ordered arrangement of identical spherical cavities [43].

Size and morphology of crystallinity can be controlled with change in crystallization kinetics [44–46]; therefore, properties in bulk and surface of microenvironments regulate the stem cell fates for any tissue.

Copolymerization is as synthetic trick for preparation of microenvironments. Since the polymerization procedures are very divers, any monomer can synthesized by special method(s) for preparation of scaffolds with the predetermined properties [47]. For example, with copolymerization of gelatin/chondoitin-6-sulfate/hyaluronan tri-copolymer mimicked natural cartilage matrix for use as a scaffold for cartilage tissue engineering which this ECM has potential for use as a cartilage tissue engineering scaffold [48].

Functionalization is another strategy to change matrix properties that can be done both polymer structure (on the homopolymers, copolymers, or polymer blends) and micro- and/or nanoparticles that dispersed in polymer matrix. Functionalization is for specific purposes, including polymer compatibility in blending, compatibility of particles dispersed in the polymer matrix, change the surface energy, creation biological structures on the surface and bulk polymer matrix, improved cell adhesion and cell differentiation, pharmaceutical properties, etc. [26, 49–53].

The cases listed above are considered as raw material for manufacturing of tissue-engineering scaffolds. These materials in combination with various methods in scaffold fabrication can express a new idea in regenerative medicine.

2.3 Effects of Surface Topography on Stem Cell Behaviors

Polymer surface topography affects both on the scaffold properties (such as hydrophilicity and degradation) and on cell behaviors (such as cell attachment, cell migration, cell proliferation, and cell differentiation). In some applications require that the cells attached to the surface of the polymer. In the cellular niche, if integrin-based adhesions are not in microenvironment surface, cell stickiness and other vital behavior are being compromised. Thus, surface topography changes cell adhesion and this alter cell shape. Cell shape is significant potential in regulation of cell growth and its physiology. For example, changes in cell shape lead to myocardial development [54, 55]. Topography, external forces, geometry, and physical interactions in scaffolds alter the cell shape; mutually, changes in cell shape will change the cell fate. Whatever cell stickiness increases, variation in internal and external forces have more effect on the cell shape [56]. Studies showed whatever micropatterning be small island, the cell genotype is to orbicular. Also, large island in micropatterning makes cell flattened shape [57]. Size and array of surface topography effect on cell adhesion/attachment, cell migration, cell proliferation, and cell differentiation. The depth of topographies changes cytoskeletal tension and the transmission and transduction of other molecular and biomechanical signals by matrix type and surface interactions. If the balance of surface interactions does not change, just the size of the topography will be discussed [58, 59]. This finding suggests that although cell size is about of several nanometers but is able to sense the surface nanostructure. For this reason, human MSCs grown on nanoscale grooves showed alignment of their cytoskeleton and nuclei of MSCs along the grooves and even nanotopographies have been shown stronger effects than biochemical effects (Figure 2.1) [60].

Cells stretched multidirectionally to follow underlying 283 nm fibers but when grown on larger fibers, extended along a single fiber axis. With

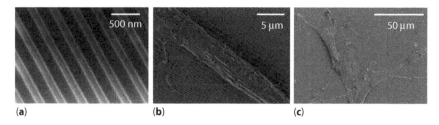

Figure 2.1 Changes in morphology and proliferation of human mesenchymal stem cells (hMSCs) cultured on nanogratings. Scanning electron micrographs (SEM) of (a) PDMS nanopatterned by replica molding; hMSCs cultured on (b) nanopatterned PDMS and (c) unpatterned PDMS. Adapted from Yim EK *et al.* [60].

decreasing fiber diameter, a higher degree of proliferation and cell-spreading and lower degree of cell aggregation were observed (Figure 2.2) [56, 61]. So, fiber diameter in electrospun scaffolds and pore size and pore distribution in porous scaffolds is important.

In some applications, cell should not adhere to the substrate such as cardiovascular applications. For example, influence of surface topography of PLGA films on the absorption fibrinogens cells and platelets were evaluated which their surface topographies were fabricated by solvent-mediated polymer casting [62]. Blood is the first fluid contacted with foreign substances (scaffolds or implants) in the body and blood coagulation created due to rapid absorption of plasma proteins on the surfaces leading to thrombus formation. Therefore, achieving blood compatible surfaces with minimal protein adsorption and platelet adhesion is considered. Including strategies for minimization of protein adsorption and platelet adhesion, the following can be noted [63–67]:

- Inorganic coating (Carbone based),
- Heparin immobilization,
- Covalently surface modification,
- Chemical composition modifications of polymer surfaces,
- Patterned surfaces and immobilization of biological molecules with controllable positioning and size, etc.

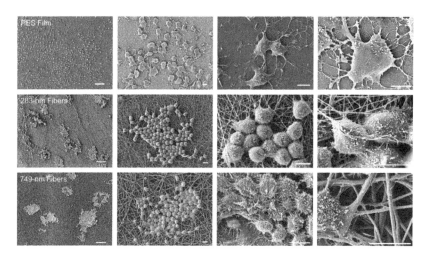

Figure 2.2 SEM images of rat neural stem/progenitor cells (NSCs) cultured in serum-free medium and 20 ng/mL FGF-2 for 5 days. Scale bars for low-magnification images (first column) are 100 mm; all other scale bars are 10 mm. Adapted from Christopherson et al. [61].

For example, created sharp topography duo to existence of Multiwall Carbon Nanotubes (MWCNT) in PLGA–MWCNT reduced platelet adhesion [68]. According to microenvironment conditions, surface topographies can be fixed or vary, for instance with degradation or mechanical stimuli scaffold surface will be delaminated, such as intramuscular movements and body fluid/blood flow shear stress. Size, aspect ratio, and density of topography in surface of scaffolds are important parameters that prepared by following methods [69–72]:

- Polymer demixing,
- Colloidal-derived topographies,
- Dual-scale diameter topographies,
- Polymers including hot embossing,
- Nanoimprint lithography.

Platelet adhesion will be decreased on substrates with increasing aspect ratio. It appears that platelets are able to interact only with the tips of the pillars when interspacing becomes sufficiently small. Pillars with high aspect ratio are likely to get deformed and bent in a liquid media even under slight agitation during incubation (Figure 2.3) [62]. Width and height (i.e. aspect ratio) have seen to considerably impact on the platelet stickiness as follows:

- Increasing the height of the structures results in lower platelet attachment when the width and interspacing are decreased,
- Increasing the width increases the likelihood of platelet adherence, due to the increase in contact area at the tips of the pillars and presumably via the conformation of the adsorbed,

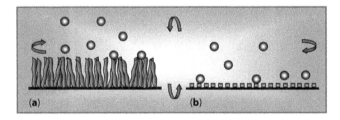

Figure 2.3 Illustration on the effect of aspect ratio on platelet adhesion. (a) The bending movements of the high-aspect-ratio flexible pillars prevent the platelets from making a successful contact on the topographic surfaces compared to (b) the low-aspect-ratio pillars where they remain stiff and upright in liquid media even under agitation, thereby permitting platelet attachment at a greater ease. Dimensions are not drawn to scale. Adapted from Koh *et al.* [62].

- Increased interspacing to dimension >200 nm results in the entrapment of platelets and its subsequent activation due to increased surface area of contact and entrapment,
- Increased aspect ratio in submicron structures reduces the number of adherent platelets due to the inability of the platelets to form stable contacts with the reduced surface area for attachment.

2.4 Role of Substrate Stiffness and Elasticity of Matrix on Cell Culture

When cells are placed in microenvironment at the first times will be planned by soluble factors, but later establishes specified lineage by the matrix elasticity [4].

The stiffness of the surface to which cells adhere can have a profound effect on cell structure and protein expression, but these mechanical effects vary with different cell types and depend on the nature of the adhesion receptors by which the cells bind their substrate. Fibroblasts and endothelial cells develop a spread morphology and actin stress fibers only when grown on surfaces with an elastic modulus greater than 2,000 Pa, with a greater effect seen when bound to fibronectin compared to collagen. In contrast, neutrophils appear to be insensitive to stiffness changes over a very wide range. The stiffness dependence of fibroblasts and endothelial cells is no longer evident when cells become confluent, or in the case of fibroblasts even when two cells make contact suggesting either that mechanosensing uses the cells' internal stiffness as a criterion or else that signaling from cadherins in cell–cell contacts overrides signals from the cell–matrix adhesion complexes [73].

The speed of cell proliferation on stiff substrates is more than soft substrates. The substrate elasticity influences both myoblast proliferation and differentiation, whereas Bone morphogenetic proteins (BMPs) protein coating of these substrates only has an additive effect on the differentiation/maturation of these cells [74].

2.5 Stem Cell Fate Induced by Matrix Stiffness and Its Mechanism

The moduli of various body tissues are different together, such as brain, muscle, and bone modulus is 1, 8–17, and 100 kPa, respectively (Figure 2.4). Cells sense matrix stiffness by mechano-transducer and prepare their

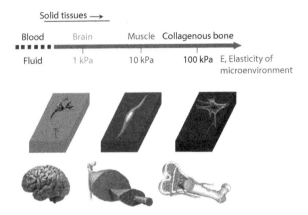

Figure 2.4 Solid tissues exhibit a range of stiffness, as measured by the elastic modulus, E.

morphology and specific lineage with transduce this information to cell nuclear. Cell shapes in different stiffness surface (brain, muscle, and bone-like) are branched shape, spindle shape, and polygonal shape, respectively (Figure 2.4).

Stem cells have an important interaction with surrounding heterogeneous cells in their niche by cell–cell signaling. Since the mechanical properties of the different tissues of the body are different, the responses of stem cells to surface modulus of microenvironment will be different. Initially, stem cells attracted to surface of ECM by integrin and receive microenvironment cues, and these cues convert to sensible signals for cell nucleus [75]. The ability to better engineer artificial ECMs that can control cell behavior, through physical as well as molecular interactions may further extend our capabilities in engineering tissue substitutes from adult or embryonic stem cells [76]. One of the regulator cues is elasticity of surrounding cell and microenvironment surface. For example, Engler *et al.* hypothesized that matrix elasticity is particularly instructive in lineage specification, such as bone-like stiffness hydrogels led to osteogenic differentiation. Similarly, the effective stiffness of the underlying substrate has been shown to regulate the differentiation of neural stem cells [4]. Saha *et al.* have shown in contact with hard substrates GFAP-expressing astrocytes differentiation is cured on the soft substrates β-III-tubulin expressing neurons differentiation is supported [59].

In general, it seems in design of tissue-engineering scaffold besides the solution and secreted factors it is necessary to design substrates with the regulator modules for engineered surfaces that simulate a cell niche in best situation.

2.6 Competition/Compliance between Matrix Stiffness and Other Signals and Their Effect on Stem Cells Fate

The effects of the surface module of substrates can be used as a complementary and synergistic effect alongside other factors in the niche impact on stem cell fate. It is expressed that the matrix compliances can lead to the development of lineages, but it is not enough for terminal differentiation in comparison with other parameters [4]. Therefore, congruence of scaffolds properties with growth factors is a synergistic effect in expression of cell differentiation marker. This leads to parallel changes in the matrix properties and the use of other factors according to the characteristics desired tissue and even enter a new idea in mind that prepared the scaffolds that are self-differentiating by soluble factors which been entered in the backbone of matrix structures. The role of other parameters alongside the module of substrates when becomes apparent that we know some body tissues have near or identical module (E). So, in such a situation, other parameters that affect the fate of cells play an important role in determining correct destiny. An interaction between the cell and ECM is established by biochemical factors that controls proliferation and differentiation on biopolymer substrates [77–81]. A property that has been shown to regulate SCs activity is chemical functionalisation of the substrate surface, for example, by anchoring monomers representing the ECM binding sites, including RGD and IKVAV [82–85], or other functional groups, for example, $-CH_3$, $-NH_2$, $-SH$, $-OH$, and $-COOH$ [85]. Along with strategies to control cell fate by chemical agents, biophysical cues also affect the cell fate *in vitro* or *in vivo*. For instance, it has been demonstrated that the matrix elasticity can influence the lineage commitment of MSCs into neurons, osteoblasts, and myoblasts [4].

2.7 Effects of Matrix Stiffness on Stem Cells in Two Dimensions versus Three Dimensions

Providing two-dimensional substrates is important because of two reasons. On the one hand knowledge of the parameters that affect the fate of cells in 2D microenvironments is underlying for recognition of 3D factors. On the other hand, there are some tissues such as the skin that grow in two-dimensional platform. Use 3D and 2D scaffolds cell are effect on stem cell fate, even if the other parameters are fixed, leading to changes in cell shape and cell behavior are mutually different. For example, chondrocytes cultured in a 2D scaffolds differentiated to the fibroblasts or cultured in 3D that maintain their normal phenotype [86]. Stem cell culture on flat substrates coated with materials such as collagen or laminin as feeder cell layers or hydrogel [87]. Although to this day most of the materials used for cell culture were polystyrene rigid substrates which may be coated with a thin layer of gelatin. Biomaterials approaches have been explored to define the identity, concentration and patterns of soluble or tethered ECM molecules singly and in combination. Biomechanical gradients in ECM not only lead to cell developing but also in homeostasis and regeneration during lead to cell migration towards more concentration. For 2D mediums the best and the most accurate method of preparation of gradient biomolecular in a tissue is used by microfluidic technology but in 3D systems these gradients are created through the molecules attached and soluble molecules [88, 89]. The difference in the density of molecules (either soluble or sticking) leads differences in the cell signaling and cell adhesion. Separation of the role of these molecules on the cell growth, cell adhesion and cell differentiation is difficult. In the cellular niche, secreted growth factors and cytokines are tethered mostly. For example, covalent attachment of fibroblast growth factor 2 (FGF2) to a synthetic polymer stabilized the growth factor and increased its potency 100-fold relative to FGF2 in solution [90]. This shows whatever cell adherent molecules in the matrix structure and cell contact with these molecules be further, the cell behavior can be further strengthened. Cell shape is determined by the limits of the surrounding ECM [55, 91]. Even shown physical control of cell shape alone is a strong factor in the regulation of cell signaling and cellular fate [92]. So 3D scaffolds compared to 2D scaffolds are more complex in regenerative medicine. Two-dimensional scaffolds provide evaluation and control of cell behavior for identifying the influence factors and Three-dimensional scaffolds are more similar to natural tissue in tissue engineering. Cell viability on the scaffolding is the first and most important step in the construction of three-dimensional scaffold. Non-toxic materials used in the manufacture of artificial microenvironments with 3D communicates is an important feature of the 3D scaffolds. So there are two constraints to build three-dimensional scaffolds:

- biochemical constraints
- biophysical and biomechanical constraints

Physical limitations inhibit the cell proliferation, cell migration, cell morphology, cell shape and cell differentiation; so by making connected porous scaffolds destroyed these restrictions largely [93].

2.8 Effects of External Mechanical Cues on Stem Cell Fate from Surface Interactions Perspective

When a living organ is formed, it is exposure to mechanical stimuli. Blood flow to the stimuli of external stimuli, such as external physical forces. From blood flow stimuli to external stimulus such as muscle forces and external forces impact on stem cell fate. Mechanical stimuli can effect on many of cellular interactions and play an important role in a variety of disease states such as atherosclerosis, osteoarthritis, and osteoporosis [94]. However, researchers have found it for more than half a century. For example, in a series of experiments in the 1930s and 1940s, Glucksmann demonstrated that cultured chick rudiments under static compression following displacement of the periosteum and perichondrium resulted in cartilaginous tissue formation, whereas tensile stresses promoted bone formation [95]. The main problem in the simulation of mechanical effects *in vitro* is complexed flows such as blood flow, and the other hand mechanical interactions may lead to changes in surface structures such as alter the structure of ECM proteins and cytokines.

There is a bilateral balance between cells and microenvironment. As far as the cell change microenvironment conditions to suitable position for growth and differentiation and if cells not be able to change desired conditions in ECM, itself will change.

If cell adhesion with surface of scaffold be a suitable interaction, the cells under the imposed forces will be elongated and develop their shapes. For example, strains of 1% or 15% failed to either induce commitment to the myogenic lineage or caused a decrease in smooth muscle cell markers, pointing to the importance of the magnitude of strain during differentiation. The strain-induced myogenic phenotype was dependent on the protein to which the cells were attached [96, 97]. The strains can be pulsated, unidirectional, bidirectional, or multidirectional (anisotropic/isotropic cytoskeletal tension phenomena) (Figure 2.5). Amount of imposed forces, type of stem cell, type of biomaterial for ECM, soluble factors, etc. regulate the response of stem cell to imposed forces. Mechanical loading on stem cell response appears to depend on the type of stem cell as well as the state of (pre-)differentiation. For example, dynamic mechanical compression can significantly increase the chondrocytic expression (e.g. Sox-9, type II collagen, and aggrecan) of bone-marrow-derived MSCs encapsulated in a hydrogel, irrespective of the presence of chondrogenic growth factors [98]. Under the same conditions, embryonic stem cell-derived embryoid bodies exhibit significant

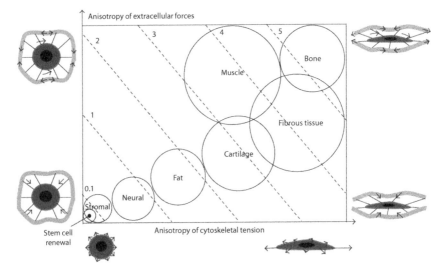

Figure 2.5 Effects of mechanical cues and oxygen tension on the fate of mesenchymal stem cells derived from the endosteal compartment. Adapted from Nava et al. [99].

downregulation of cartilage-specific genes in response to mechanical compression.

2.9 Conclusions

Understanding the effects of physical and mechanical superficial properties of tissue-engineering substrates that are in contact with the cells is important in two respects. On the one hand, the design of substrates with synergistic effect in mechanical and physical properties with the biological and biochemical cues can control stem cell fate with more quality and accurate. On the other hand, with a more detailed understanding of biological microenvironments and mimic their structures can prevent since many of the diseases. So, alterations in superficial cues (such as physical and mechanical properties on the surfaces of substrates) could not only lead to aberrant growth or differentiation of stem cells but also provide a potential therapeutic target for regenerative medicine.

Acknowledgments

The authors thank Mrs. Maryam Jafari and Dr. Hamid Shahrousvand for their useful comments and special thanks to Mr. Mahmoud Shahrousvand for designing schematic pictures.

References

1. Moore K.A., Lemischka I.R. Stem cells and their niches. *Science* 311, 1880–5, 2006.
2. Yan K., Wu Q., Yan D.H., Lee C.H., Rahim N., Tritschler I., et al. Glioma cancer stem cells secrete Gremlin1 to promote their maintenance within the tumor hierarchy. *Genes & Development* 28, 1085–100, 2014.
3. Ordonez P., Di Girolamo N. Limbal epithelial stem cells: role of the niche microenvironment. *Stem Cells* 30, 100–7, 2012.
4. Engler A.J., Sen S., Sweeney H.L., Discher D.E. Matrix elasticity directs stem cell lineage specification. *Cell* 126, 677–89, 2006.
5. Bobrie A., Colombo M., Raposo G., Théry C. Exosome secretion: molecular mechanisms and roles in immune responses. *Traffic* 12, 1659–68, 2011.
6. Xu J., Li S., Hu F., Zhu C., Zhang Y., Zhao W. Artificial biomimicking matrix modifications of nanofibrous scaffolds by hE-Cadherin-Fc fusion protein to promote human mesenchymal stem cells adhesion and proliferation. *Journal of Nanoscience and Nanotechnology* 14, 4007–13, 2014.
7. Artner L.M., Merkel L., Bohlke N., Beceren-Braun F., Weise C., Dernedde J., et al. Site-selective modification of proteins for the synthesis of structurally defined multivalent scaffolds. *Chemical Communications* 48, 522–4, 2012.
8. Nakaoka R., Hirano Y., Mooney D.J., Tsuchiya T., Matsuoka A. Study on the potential of RGD-and PHSRN-modified alginates as artificial extracellular matrices for engineering bone. *Journal of Artificial Organs* 16, 284–93, 2013.
9. Fiedler J., Özdemir B., Bartholomä J., Plettl A., Brenner R.E., Ziemann P. The effect of substrate surface nanotopography on the behavior of multipotnent mesenchymal stromal cells and osteoblasts. *Biomaterials* 34, 8851–9, 2013.
10. Fischer R.S., Myers K.A., Gardel M.L., Waterman C.M. Stiffness-controlled three-dimensional extracellular matrices for high-resolution imaging of cell behavior. *Nature Protocols* 7, 2056–66, 2012.
11. Trappmann B., Gautrot J.E., Connelly J.T., Strange D.G., Li Y., Oyen M.L., et al. Extracellular-matrix tethering regulates stem-cell fate. *Nature Materials* 11, 642–9, 2012.
12. Eshghi S., Schaffer D.V. Engineering microenvironments to control stem cell fate and function. 2008.
13. Lu, Tingli, Yuhui Li, and Tao Chen. "Techniques for fabrication and construction of three-dimensional scaffolds for tissue engineering." *International journal of nanomedicine* 8, 337, 2013.
14. Bose S., Tarafder S. Calcium phosphate ceramic systems in growth factor and drug delivery for bone tissue engineering: a review. *Acta Biomaterialia* 8, 1401–21, 2012.
15. Mallick S., Tripathi S., Srivastava P. Advancement in scaffolds for bone tissue engineering: a review. *IOSR Journal of Pharmacy and Biological Sciences* 10, 37–54, 2015.

16. Fallahiarezoudar E., Ahmadipourroudposht M., Idris A., Yusof NM. A review of application of synthetic scaffold in tissue engineering heart valves. *Materials Science and Engineering: C* 48, 556–65, 2015.
17. Sarker B., Hum J., Nazhat S.N., Boccaccini A.R. Combining collagen and bioactive glasses for bone tissue engineering: a review. *Advanced Healthcare Materials* 4, 176–94, 2015.
18. Hutmacher D.W. Scaffolds in tissue engineering bone and cartilage. *Biomaterials* 21, 2529–43, 2000.
19. Oh S.H., Kang S.G., Kim E.S., Cho S.H., Lee J.H. Fabrication and characterization of hydrophilic poly (lactic-co-glycolic acid)/poly (vinyl alcohol) blend cell scaffolds by melt-molding particulate-leaching method. *Biomaterials* 24, 4011–21, 2003.
20. Rezwan K., Chen Q., Blaker J., Boccaccini A.R. Biodegradable and bioactive porous polymer/inorganic composite scaffolds for bone tissue engineering. *Biomaterials* 27, 3413–31, 2006.
21. Cheung HY., Lau KT., Lu T.P., Hui D. A critical review on polymer-based bio-engineered materials for scaffold development. *Composites Part B: Engineering* 38, 291–300, 2007.
22. Ramakrishna S., Mayer J., Wintermantel E., Leong K.W. Biomedical applications of polymer-composite materials: a review. *Composites Science and Technology* 61, 1189–224, 2001.
23. Trenor S.R., Shultz A.R., Love B.J., Long T.E. Coumarins in polymers: from light harvesting to photo-cross-linkable tissue scaffolds. *Chemical Reviews* 104, 3059–78, 2004.
24. Smith I., Liu X., Smith L., Ma P. Nanostructured polymer scaffolds for tissue engineering and regenerative medicine. *Wiley Interdisciplinary Reviews: Nanomedicine and Nanobiotechnology* 1, 226–36, 2009.
25. Liu X., Holzwarth J.M., Ma P.X. Functionalized synthetic biodegradable polymer scaffolds for tissue engineering. *Macromolecular Bioscience* 12, 911–9, 2012.
26. South C.R., Burd C., Weck M. Modular and dynamic functionalization of polymeric scaffolds. *Accounts of Chemical Research* 40, 63–74, 2007.
27. Meredith J.C., J Amis E. LCST phase separation in biodegradable polymer blends: poly (D, L-lactide) and poly (e-caprolactone). *Macromolecular Chemistry and Physics* 201, 733–9, 2000.
28. Li M., Mondrinos M.J., Chen X., Gandhi M.R., Ko F.K., Lelkes P.I. Co-electrospun poly (lactide-co-glycolide), gelatin, and elastin blends for tissue engineering scaffolds. *Journal of Biomedical Materials Research Part A* 79, 963–73, 2006.
29. Allaf, Rula M., Iris V. Rivero, and Ilia N. Ivanov. "Fabrication of co-continuous poly (ε-caprolactone)/polyglycolide blend scaffolds for tissue engineering." *Journal of Applied Polymer Science* 132(35), 2015.
30. Sionkowska A., Wisniewski M., Skopinska J., Kennedy C., Wess T. Molecular interactions in collagen and chitosan blends. *Biomaterials* 25, 795–801, 2004.

31. Yunos D.M., Bretcanu O., Boccaccini A.R. Polymer-bioceramic composites for tissue engineering scaffolds. *Journal of Materials Science* 43, 4433–42, 2008.
32. Habraken W., Wolke J., Jansen J. Ceramic composites as matrices and scaffolds for drug delivery in tissue engineering. *Advanced Drug Delivery Reviews* 59, 234–48, 2007.
33. Venkatesan J., Bhatnagar I., Manivasagan P., Kang K-H., Kim S-K. Alginate composites for bone tissue engineering: a review. *International Journal of Biological Macromolecules* 72, 269–81, 2015.
34. Wei G., Ma P.X. Structure and properties of nano-hydroxyapatite/polymer composite scaffolds for bone tissue engineering. *Biomaterials* 25, 4749–57, 2004.
35. Rowland C.R., Lennon D.P., Caplan A.I., Guilak F. The effects of cross-linking of scaffolds engineered from cartilage ECM on the chondrogenic differentiation of MSCs. *Biomaterials* 34, 5802–12, 2013.
36. Subramanian A., Lin H.Y. Cross-linked chitosan: its physical properties and the effects of matrix stiffness on chondrocyte cell morphology and proliferation. *Journal of Biomedical Materials Research Part A* 75, 742–53, 2005.
37. Barrioni B.R., de Carvalho S.M., Oréfice R.L., de Oliveira A.A.R., de Magalhães Pereira M. Synthesis and characterization of biodegradable polyurethane films based on HDI with hydrolyzable cross-linked bonds and a homogeneous structure for biomedical applications. *Materials Science and Engineering: C* 52, 22–30, 2015.
38. Drury J.L., Mooney D.J. Hydrogels for tissue engineering: scaffold design variables and applications. *Biomaterials* 24, 4337–51, 2003.
39. Peppas N.A., Hilt J.Z., Khademhosseini A., Langer R. Hydrogels in biology and medicine: from molecular principles to bionanotechnology. *Advanced Materials-Deerfield Beach Then Weinheim* 18, 1345, 2006.
40. Fujii Y., Okuda D., Fujimoto Z., Horii K., Morita T., Mizuno H. Crystal structure of trimestatin, a disintegrin containing a cell adhesion recognition motif RGD. *Journal of Molecular Biology* 332, 1115–22, 2003.
41. Kuo Y.C., Chung C.Y. TATVHL peptide-grafted alginate/poly (γ-glutamic acid) scaffolds with inverted colloidal crystal topology for neuronal differentiation of iPS cells. *Biomaterials* 33, 8955–66, 2012.
42. Kotov N.A., Liu Y., Wang S., Cumming C., Eghtedari M., Vargas G., et al. Inverted colloidal crystals as three-dimensional cell scaffolds. *Langmuir* 20, 7887–92, 2004.
43. Cuddihy M.J., Kotov N.A. Poly (lactic-co-glycolic acid) bone scaffolds with inverted colloidal crystal geometry. *Tissue Engineering Part A* 14, 1639–49, 2008.
44. Niu L., Kvarnström C., Fröberg K., Ivaska A. Electrochemically controlled surface morphology and crystallinity in poly (3, 4-ethylenedioxythiophene) films. *Synthetic Metals* 122, 425–9, 2001.
45. Reiter G., Castelein G., Sommer J.U., Röttele A., Thurn-Albrecht T. Direct visualization of random crystallization and melting in arrays of nanometer-size polymer crystals. *Physical Review Letters* 87, 226101, 2001.

46. Reiter G. Some unique features of polymer crystallisation. *Chemical Society Reviews* 43, 2055–65, 2014.
47. Odian G. Chain copolymerization. *Principles of Polymerization, Fourth Edition* 464–543, 2004.
48. Chang C.H., Liu H.C., Lin C.C., Chou C.H., Lin F.H. Gelatin–chondroitin-hyaluronan tri-copolymer scaffold for cartilage tissue engineering. *Biomaterials* 24, 4853–8, 2003.
49. Guo B., Lei B., Li P., Ma P.X. Functionalized scaffolds to enhance tissue regeneration. *Regenerative Biomaterials* 2, 47–57, 2015.
50. Bisson I., Kosinski M., Ruault S., Gupta B., Hilborn J., Wurm F., et al. Acrylic acid grafting and collagen immobilization on poly (ethylene terephthalate) surfaces for adherence and growth of human bladder smooth muscle cells. *Biomaterials* 23, 3149–58, 2002.
51. Miyashita H., Shimmura S., Kobayashi H., Taguchi T., Asano-Kato N., Uchino Y., et al. Collagen-immobilized poly (vinyl alcohol) as an artificial cornea scaffold that supports a stratified corneal epithelium. *Journal of Biomedical Materials Research Part B: Applied Biomaterials* 76, 56–63, 2006.
52. Ragothaman M., Palanisamy T., Kalirajan C. Collagen–poly (dialdehyde) guar gum based porous 3D scaffolds immobilized with growth factor for tissue engineering applications. *Carbohydrate Polymers* 114, 399–406, 2014.
53. Yoo H.S., Kim T.G., Park T.G. Surface-functionalized electrospun nanofibers for tissue engineering and drug delivery. *Advanced Drug Delivery Reviews* 61, 1033–42, 2009.
54. Manasek F.J., Burnside M.B., Waterman R.E. Myocardial cell shape change as a mechanism of embryonic heart looping. *Developmental Biology* 29, 349–71, 1972.
55. Folkman J., Moscona A. Role of cell shape in growth control. 1978.
56. Guilak F., Cohen D.M., Estes B.T., Gimble J.M., Liedtke W., Chen C.S. Control of stem cell fate by physical interactions with the extracellular matrix. *Cell Stem Cell* 5, 17–26, 2009.
57. McBeath R., Pirone D.M., Nelson C.M., Bhadriraju K., Chen C.S. Cell shape, cytoskeletal tension, and RhoA regulate stem cell lineage commitment. *Developmental Cell* 6, 483–95, 2004.
58. Discher D.E., Janmey P., Wang Y.-L. Tissue cells feel and respond to the stiffness of their substrate. *Science* 310, 1139–43, 2005.
59. Saha K., Keung A.J., Irwin E.F., Li Y., Little L., Schaffer D.V., et al. Substrate modulus directs neural stem cell behavior. *Biophysical Journal* 95, 4426–38, 2008.
60. Yim E.K., Pang S.W., Leong K.W. Synthetic nanostructures inducing differentiation of human mesenchymal stem cells into neuronal lineage. *Experimental Cell Research* 313, 1820–9, 2007.
61. Christopherson G.T., Song H., Mao H-Q. The influence of fiber diameter of electrospun substrates on neural stem cell differentiation and proliferation. *Biomaterials* 30, 556–64, 2009.

62. Koh L.B., Rodriguez I., Venkatraman S.S. The effect of topography of polymer surfaces on platelet adhesion. *Biomaterials* 31, 1533–45, 2010.
63. Cui F., Li D. A review of investigations on biocompatibility of diamond-like carbon and carbon nitride films. *Surface and Coatings Technology* 131, 481–7, 2000.
64. Liu L., Ito Y., Imanishi Y. Synthesis and antithrombogenicity of heparinized polyurethanes with intervening spacer chains of various kinds. *Biomaterials* 12, 390–6, 1991.
65. Chen H., Zhang Z., Chen Y., Brook M.A., Sheardown H. Protein repellant silicone surfaces by covalent immobilization of poly (ethylene oxide). *Biomaterials* 26, 2391–9, 2005.
66. Abbasi F., Mirzadeh H., Katbab A.A. Modification of polysiloxane polymers for biomedical applications: a review. *Polymer International* 50, 1279–87, 2001.
67. Lim J.Y., Donahue H.J. Cell sensing and response to micro- and nanostructured surfaces produced by chemical and topographic patterning. *Tissue Engineering* 13, 1879–91, 2007.
68. Koh L.B., Rodriguez I., Venkatraman S.S. A novel nanostructured poly (lactic-co-glycolic-acid)–multi-walled carbon nanotube composite for blood-contacting applications: Thrombogenicity studies. *Acta Biomaterialia* 5, 3411–22, 2009.
69. Sutherland D.S., Broberg M., Nygren H., Kasemo B. Influence of nanoscale surface topography and chemistry on the functional behaviour of an adsorbed model macromolecule. *Macromolecular Bioscience* 1, 270–3, 2001.
70. Ye X., Shao Y.l., Zhou M., Li J., Cai L. Research on micro-structure and hemocompatibility of the artificial heart valve surface. *Applied Surface Science* 255, 6686–90, 2009.
71. Koerner T., Brown L., Xie R., Oleschuk R.D. Epoxy resins as stamps for hot embossing of microstructures and microfluidic channels. *Sensors and Actuators* B: Chemical 107, 632–9, 2005.
72. Pozzato A., Dal Zilio S., Fois G., Vendramin D, Mistura G, Belotti M, et al. Superhydrophobic surfaces fabricated by nanoimprint lithography. *Microelectronic Engineering* 83, 884–8, 2006.
73. Yeung T., Georges P.C., Flanagan L.A., Marg B., Ortiz M., Funaki M., et al. Effects of substrate stiffness on cell morphology, cytoskeletal structure, and adhesion. *Cell Motility and the Cytoskeleton* 60, 24–34, 2005.
74. Boonen K.J., Rosaria-Chak K.Y., Baaijens F.P., van der Schaft D.W., Post M.J. Essential environmental cues from the satellite cell niche: optimizing proliferation and differentiation. *American Journal of Physiology-Cell Physiology* 296, C1338–C45, 2009.
75. Daley W.P., Peters S.B., Larsen M. Extracellular matrix dynamics in development and regenerative medicine. *Journal of Cell Science* 121, 255–64, 2008.
76. Metallo C.M., Mohr J.C., Detzel C.J., de Pablo J.J., Van Wie B.J., Palecek S.P. Engineering the stem cell microenvironment. *Biotechnology Progress* 23, 18–23, 2007.

77. Awad H.A., Wickham M.Q., Leddy H.A., Gimble J.M., Guilak F. Chondrogenic differentiation of adipose-derived adult stem cells in agarose, alginate, and gelatin scaffolds. *Biomaterials* 25, 3211–22, 2004.
78. Yokoyama A., Sekiya I., Miyazaki K., Ichinose S., Hata Y., Muneta T. *In vitro* cartilage formation of composites of synovium-derived mesenchymal stem cells with collagen gel. *Cell and Tissue Research* 322, 289–98, 2005.
79. Datta N., Pham Q.P., Sharma U., Sikavitsas V.I., Jansen J.A., Mikos A.G. *In vitro* generated extracellular matrix and fluid shear stress synergistically enhance 3D osteoblastic differentiation. *Proceedings of the National Academy of Sciences of the United States of America* 103, 2488–93, 2006.
80. Nöth U., Rackwitz L., Heymer A., Weber M., Baumann B., Steinert A., et al. Chondrogenic differentiation of human mesenchymal stem cells in collagen type I hydrogels. *Journal of Biomedical Materials Research Part A* 83, 626–35, 2007.
81. Chung C., Burdick J.A. Influence of three-dimensional hyaluronic acid microenvironments on mesenchymal stem cell chondrogenesis. *Tissue Engineering Part A* 15, 243–54, 2008.
82. Pierschbacher M.D., Ruoslahti E. Influence of stereochemistry of the sequence Arg-Gly-Asp-Xaa on binding specificity in cell adhesion. *Journal of Biological Chemistry* 262, 17294–8, 1987.
83. Drumheller P.D., Hubbell J.A. Polymer networks with grafted cell adhesion peptides for highly biospecific cell adhesive substrates. *Analytical Biochemistry* 222, 380–8, 1994.
84. Wan Y., Yang J., Yang J., Bei J., Wang S. Cell adhesion on gaseous plasma modified poly-(L-lactide) surface under shear stress field. *Biomaterials* 24, 3757–64, 2003.
85. Salinas C.N., Anseth K.S. Decorin moieties tethered into PEG networks induce chondrogenesis of human mesenchymal stem cells. *Journal of Biomedical Materials Research Part A* 90, 456–64, 2009.
86. Holtzer H., Abbott J., Lash J., Holtzer S. The loss of phenotypic traits by differentiated cells *in vitro*, I. Dedifferentiation of cartilage cells. *Proceedings of the National Academy of Sciences of the United States of America* 46, 1533, 1960.
87. Lutolf MP, Gilbert PM, Blau HM. Designing materials to direct stem-cell fate. *Nature* 462, 433–41, 2009.
88. Choi N.W., Cabodi M., Held B., Gleghorn J.P., Bonassar L.J., Stroock A.D. Microfluidic scaffolds for tissue engineering. *Nature materials* 6, 908–15, 2007.
89. Chung B.G., Flanagan L.A., Rhee S.W., Schwartz P.H., Lee A.P., Monuki E.S., et al. Human neural stem cell growth and differentiation in a gradient-generating microfluidic device. *Lab on a Chip* 5, 401–6, 2005.
90. Irvine D.J., Hue K-A., Mayes A.M., Griffith L.G. Simulations of cell-surface integrin binding to nanoscale-clustered adhesion ligands. *Biophysical Journal* 82, 120–32, 2002.
91. Chen C.S., Mrksich M., Huang S., Whitesides G.M., Ingber D.E. Geometric control of cell life and death. *Science* 276, 1425–8, 1997.

92. Wozniak M.A., Chen C.S. Mechanotransduction in development: a growing role for contractility. *Nature Reviews Molecular Cell Biology* 10, 34–43, 2009.
93. Silva G.A., Czeisler C., Niece K.L., Beniash E., Harrington D.A., Kessler J.A., et al. Selective differentiation of neural progenitor cells by high-epitope density nanofibers. *Science* 303, 1352–5, 2004.
94. Ingber D. Mechanobiology and diseases of mechanotransduction. *Annals of Medicine* 35, 564–77, 2003.
95. Glucksmann A. The role of mechanical stresses in bone formation *in vitro*. *Journal of Anatomy* 76, 231, 1942.
96. Gong Z., Niklason L.E. Small-diameter human vessel wall engineered from bone marrow-derived mesenchymal stem cells (hMSCs). *The FASEB Journal* 22, 1635–48, 2008.
97. Yang Y., Beqaj S., Kemp P., Ariel I., Schuger L. Stretch-induced alternative splicing of serum response factor promotes bronchial myogenesis and is defective in lung hypoplasia. *Journal of Clinical Investigation* 106, 1321, 2000.
98. Terraciano V., Hwang N., Moroni L., Park H.B., Zhang Z., Mizrahi J., et al. Differential response of adult and embryonic mesenchymal progenitor cells to mechanical compression in hydrogels. *Stem Cells* 25, 2730–8, 2007.
99. Nava M.M., Raimondi M.T., Pietrabissa R. Controlling self-renewal and differentiation of stem cells via mechanical cues. *BioMed Research International* 2012, p. 797410. 2012.

3
Effects of Mechanotransduction on Stem Cell Behavior

Bahar Bilgen[1] and Sedat Odabas[2]*

[1]*Department of Orthopaedics, The Warren Alpert Medical School of Brown University, Providence, RI, USA*
[2]*Department of Chemistry, Faculty of Science, Ankara University, Ankara, Turkey*

Abstract

Cells are exposed to several physical forces in their natural environment. These physical and mechanical cues are processed and converted to biochemical signals as a cellular response. This action–reaction mechanism, called "mechanotransduction", can regulate the basic cellular metabolic activities, such as proliferation and differentiation, while also altering the cell behavior at a molecular level to initiate tissue formation and organogenesis. Generation of mechanical cues *in vitro* has allowed researchers to study the effects of mechanotransduction on cell behavior and/or enhance stem cell differentiation for therapeutic purposes. This chapter comprehensively discusses the theories about the effects of mechanotransduction on stem cell behavior and presents different methods of application of biophysical stimuli on cells *in vitro* such as bioreactors that provide compression, shear, tension, electromagnetic fields via magnetic nanoparticles, and bioinspired biomaterials as substrates for cells.

Keywords: Mechanotransduction, stem cells, biomaterials, bioreactor, magnetic particles tension, compression, shear, mechanical loading

3.1 Introduction

"Cell theory" describes the basic phenomenon of being a living unit with the propositions of "all living organisms are composed of one or more cells", "the cell is the most basic unit of life", and "all cells arise only from

---

*Corresponding author: sedatodabas@gmail.com

Ashutosh Tiwari, Bora Garipcan and Lokman Uzun (eds.) *Advanced Surfaces for Stem Cell Research*, (45–66) © 2017 Scrivener Publishing LLC

preexisting cells" [1, 2]. Although one organism has the same genetic instructions throughout all its cells, cells can be specialized individually and gather together to carry out different or common functions. In accordance with this theory, today we know that multiple cells can organize tissue formations, organoids, organ systems, and eventually a living complex organism [1–3].

In a multicellular organism, tissues comprise two basic units: cells and the material between the cells, which is called the extracellular matrix (ECM) [4–6]. ECM is a mixture of biochemicals secreted by the cells. This chemical composition enables ECM to regulate cell behavior. ECM has several roles for tissue formation such as providing scaffolding for cell adherence, providing essential and crucial chemicals for cell proliferation, growth, differentiation, and cell–cell interactions. ECM is generally composed of water, structural proteins like collagen, elastin, and glycoproteins such as fibrillin, hyaluronic acid, chondroitin sulfate, keratin sulfate, heparan sulfate proteoglycans and other proteoglycans such as aggrecan, biglycan, syndecan, perlecan, etc. ECM has a dynamic composition, which varies based on the tissue types. This chemical composition helps cells for tissue organization, cell proliferation, and differentiation and signal transduction process [4–9].

Earlier research about cells and ECM relations reveal that the topography of the ECM macromolecular structure can also modulate cell behavior via the transduction of the biophysical signals through the ECM [6–10].

Cells exhibit several distinct behavioral patterns such as dynamic interaction with the surrounding cells and microenvironment, proliferation, growth, and differentiation that leads to new tissue formation. All these patterns are strongly associated with the tissue type, tissue-specific ECM composition, topography, and the external and outsourcing effects to the tissue [11, 12].

Physical properties of ECM like stiffness which can be defined as a rigidity of a material or substrate, pore size, pore structure, and the porosity (fractional void volume of a material) and solubility can influence the biological function of the cells [13–18]. In a physiological point of view, cells interact both with each other and the surrounding ECM to proliferate, carry out their functional duties, and differentiate into several other cells. This phenomenon comprises all cell types from the basic fibroblast and myoblast to the more characteristic cells like stem cells.

Stem cells are the unique cells in an organism that have the potential to differentiate and develop into many types of cells originated or not originated from its germ layer. In addition to this property, stem cells have self-renewal capability to generate daughter cells to retain stemness. Stem cells can also be classified according to their potency, the capacity of self-renewal and the colony-forming ability [19–23]. Hematopoietic stem cells (HSCs) are multipotent stem cells and can generate all mature blood cells to maintain hematopoietic hemostasis. On the other hand, mesenchymal stem cells (MSCs) can generate mainly mesenchyme originated tissues such as bone, cartilage, and fat.

Stem cells respond to different biochemical signals according to their own fate. These cells can stay in undifferentiated/self-renewal state or differentiate into specific cell types to carry out the functions in the body. These behavioral changes are controlled by the intracellular signals integrated with extrinsic cues and factors by their surrounding microenvironment. The exact niche of a stem cell can be exploited with different chemical, biological, or mechanical stimuli. Under specific influences, stem cells can alter their state to promote differentiation [24–30].

3.2 The Concept of Mechanotransduction

In a biochemical point of view, mechanotransduction is a process by which cells transform physical forces into intracellular biochemical signals. These signals can trigger defined activities such as repair and regeneration. How does the mechanical stimulus influence the repair and regeneration? Studies about the effects of mechanotransduction on tissues reveal that repair and regeneration process are strongly related to the geometry, shape, topography, texture, and functionality of the materials of which cells are interacted within their microenvironment [31].

Mechanical stimulation of cells can be viewed in both macro and micro-perspectives. Macro-perspective is dealing with the system that mechanical stimuli mainly formed by the stretch or the compression on cell membrane interfaces or cells between neighboring cells. On the other hand, micro-perspective is focused on the molecular signaling pathways in specialized tissues. Both micro- and macro-perspectives are involved in particular theories. In tensegrity theory, the architecture and the tension of the cytoskeleton are the key determinants [32, 33]. Mechanical stimuli are transmitted to the nucleus by actin cytoskeleton. The whole cell can act as a single mechanotransducer and cells have protein-based mediators (e.g. integrins) which also act as a mechanoreceptor of the cell, setting up integration with neighboring cells and the ECM. The mechanical stimuli can be sensed by the integrins and produce a response in stress-sensitive ion channels, protein kinases, or other signaling proteins inside the cells [34]. Here, integrins act as a mechanosensor that physically connect to

48 Advanced Surfaces for Stem Cell Research

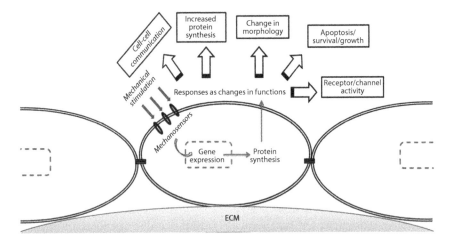

Figure 3.1 Behavioral changes of cells under mechanical stimulation.

the ECM, activating a cascade of signaling pathways [35]. Mechanosome theory is another theory, which proposes mechanosensitive multiprotein complexes called mechanosomes, located in the cell membrane. These mechanosomes are linked to specific membrane proteins that can induce a biochemical response, altering the chemical activity of the cells related to the changes in protein conformations and cytoplasm changes [36].

Eventually, mechanical signals can be transmitted through ECM and intracellular pathways and the cell responds with changes in cell behavior. This chapter focuses on the behavioral changes of cells, in particular stem cells, under specific mechanotransduction signals and the several strategies for generating mechanical stimuli (Figure 3.1).

3.3 The Mechanical Cues of Cell Differentiation and Tissue Formation on the Basis of Mechanotransduction

Physical forces such as gravity, compression, pressure, tension, and shear stress can generate mechanical stimuli which can alter the behavior of stem cells under physiological conditions, simultaneously initiating the chemical activities through ion channels, adhesion molecules, and the cytoskeleton [37]. We do not have a clear understanding of how a stem cell decides to differentiate or stay in an undifferentiated state after a physical stimulus [38]. Physical forces via mechanotransduction could lead to the adaptation of the tissue and eventually the tissue can physically and dynamically remodel itself. This phenomenon is called Wolff's law that explains the adaptation of bone tissue when subjected to loads. Thus, bone tissue formation, bone resorption, retention, and bone loss can be regulated by dynamic mechanical loads. Bone tissue comprises of several types of cells: active osteoblasts, inactive lining cells, osteoclasts, osteocytes, and MSCs. Each of these cells has specific roles in osteogenesis and bone regeneration. When a mechanical load is applied on the bone tissue, it generates a potent osteogenic stimulus throughout the osteocytes, osteoblasts, and osteoclasts which in turn regulates absorption, resorption, and formation cycle of bone formation. However, this external mechanical stimulus should be continuous, cyclic, or periodic in order to be more effective [39–41]. Evidently, the rate of formation and regeneration is strongly related to the age, gender, and type of the defect [42]. However, in all cases of bone formation, MSCs play key roles to regulate osteogenesis.

Studies report that MSC can remodel the ECM around the injury site, triggering the neighboring cells by releasing growth factors [43]. Moreover, mediator proteins like extracellular signal-regulated kinase (ERK) synthesis in MSC are upregulated in osseous regeneration characterized by constant distraction. Other proteins can also be induced through this mechanotransduction signaling. Bone morphogenetic proteins (BMPs) are a group of regulatory proteins that belong to the transforming growth factor-beta (TGF-β) superfamily that have critical roles and functions in early development as well as regeneration and repair of the tissue and organ systems. Several studies suggest that a cross-talk between BMP proteins and mechanical stimulus may exist, also triggering Smad complexes [44, 45]. BMP-2 and BMP-4 have important roles in bone and cartilage development. These protein expressions are increased under proto-oncogene tyrosine-protein kinase Src-dependent mechanotransduction pathway [45]. Although mechanical stimuli can be sensed by all the cells in a transduction flow, stem cells can be defined as at the center of these sequences of events.

3.4 Mechanotransduction via External Forces

Cells sense and respond to certain mechanical stimuli. Drachman and Sokoloff had made the seminal observation of abnormal formation of joints in paralyzed chick embryos [46]. This demonstrated that mechanical stimuli through the movement were essential for the stem cell differentiation pathways. This phenomenon can be extended to the macro role of physical exercise for the musculoskeletal development at the organ level. Carter *et al.* proposed that intermittent mechanical stress affected tissue differentiation in early fracture healing [47]. In keeping with this concept, many research groups have investigated how different modes of physical stimuli during *in vitro* cultivation modulate the biochemical and mechanical properties of three-dimensional cell–biomaterial biocomposites. The concept of functional tissue engineering has evolved into the *in vitro* use of mechanical stimuli to enhance the functional properties of tissues, e.g. the mechanical properties that enable cartilage to withstand the expected *in vivo* stress and strain [48]. Mauck *et al.* suggest that mechanical loading alone can induce chondrogenesis of MSC even in the absence of growth factors that are considered vital [49].

This biochemical sensing and biochemical transduction can be performed via different strategies involving different types of external mechanical stimuli, some of which are discussed below. Application of intermittent mechanical loading such as compression or shear during tissue cultivation has been shown to modulate differentiation and tissue formation. However, the type of mechanical stimuli is dictated by the type of cell and the functional properties of the tissue of interest for the direction of stem cell differentiation.

3.4.1 Mechanotransduction via Bioreactors

Bioreactors have been used to provide the necessary nutrients and gases while exposing growing cells and tissues to biophysical stimuli. The physical stimuli provided by bioreactors can be in the form of compressive or tensile strain, shear stress, and hydrodynamic forces.

Stirred tank batch bioreactors have historically been used for cultivation of mammalian cells in large-scale dynamic suspension cultures in industry. In order to minimize nutrient diffusion limitations, an important function of bioreactors is to increase mixing, which in turn may lead to high shear forces that may be detrimental to cells. For suspension cultures of either single cells, aggregates, or microcarrier cell cultures, research has focused on minimizing shear [50, 51]. Hydrodynamics characterized by turbulent fluid flow have higher shear and laminar flow environments have low shear which may be more favorable to minimize shear stress on cells. Embryonic stem cell-derived HSCs have expressed distinct differentiation patterns when cultured in the highly turbulent spinner flask and the lower shear rotating wall bioreactor [52]. Shear stress also plays a central role in the endothelial and osteogenic differentiation of stem and progenitor cells [53].

Bioreactors are also used in the field of tissue engineering to provide mechanical stimuli. In order to maximize the mechanical strength of engineered cartilage, researchers have utilized bioreactors such as spinner flask [54], rotating wall vessel bioreactor [55], perfusion bioreactors [56], compression [57], wavy walled bioreactor [58], and biaxial loading bioreactor [59, 60]. Perfusion systems have been proposed for vascular tissues such as liver [61]. Each of these bioreactors offers a combination of different biophysical forces, leading to variations in cell responses and thus engineered tissue properties. Differences in flow and mixing conditions around the developing engineered tissue affect tissue structure by inducing changes in cell shape and function, as well as by changing mass transfer rates of nutrients and metabolites [62]. Mechanotransduction by the bioreactor happens via the cells on the edge of these growing tissues, which sense the physical forces and signal the cells in the interior regions of the tissue to modify their fate: growth, differentiation, or death. The *in vitro* cultivation of three-dimensional cell–ECM constructs in different bioreactor environments has been employed to direct stem cell differentiation in many different applications.

Several investigators have demonstrated that hydrodynamic environment induces mechanotransduction in chondrocytes [60, 63]. This phenomenon has led to the investigation of different bioreactor hydrodynamic environments. Spinner flasks provided a turbulent high-mixing hydrodynamic environment, which enhanced cell growth rates and cell density, ECM secretion when compared to static cultures [62]. While this environment enhanced synthesis of glycosaminoglycans (GAGs) in engineered cartilage, the high shear environment also led to the release and thus loss of these molecules into the culture medium. The high shear forces produced by the mixed bioreactor environment also induced cell morphology changes. Chondrocytes typically have round morphology and predominantly secrete type II collagen. However, on the surface of engineered cartilage cultivated in high shear environments, a fibrous capsule was formed, characterized by elongated chondrocytes and type I collagen formation [58, 62].

Compressive loading has been widely demonstrated to be influential on stem cell differentiation. Differentiation toward the cell phenotypes of load bearing tissues such as cartilage and bone have benefited most from dynamic compressive loading. Static compressive loading appears to be detrimental while dynamic compressive loading has been shown to induce chondrogenesis of MSCs. Huang *et al.* used intermittent cyclic compressive loading to induce chondrogenic differentiation in rabbit MSC [64]. Similarly, bovine MSCs respond to the application of dynamic compressive loading during *in vitro* cultivation by upregulating chondrogenic gene expression as well as increasing GAG production over long term (4-week) culture [49]. Induced pluripotent stem cells (IPSCs) were also demonstrated to differentiate toward osteoblastic and chondrocytic lineage in the presence of compressive loading [65].

Compressive bioreactor systems have enabled the investigation of the synergy between biochemical and biophysical signals. Cyclic compressive loading was demonstrated to directly influence the early BMP signaling pathways in osteogenic differentiation [44]. Similarly, compressive loading is suggested to influence the TGF-β pathways in chondrogenic differentiation of bone marrow-derived MSC [66].

Tensile strain is another type of mechanical stimuli that is used for mechanotransduction. Tensile forces typically induce fibroblastic morphology, cell alignment over the direction of the forces, and secretion of ECM along the direction of the forces. Cyclic tensile loading was shown to increase fibroblastic gene expression, type I collagen synthesis and tensile modulus of engineered fibrocartilage tissue using MSC embedded in nanofiber scaffolds [67]. This study shows that tensile loading is a viable method for the cultivation of tissue-engineered meniscus fibrocartilage from MSC [67]. Uniaxial static and cyclic tension has also been used to cultivate cardiomyocytes using a co-culture of human embryonic stem cells and IPS-derived cardiomyocytes in collagen type I scaffolds [68]. Both static and cyclic stress promoted cell alignment and both ESC and IPSC exhibited increased fibril alignment in the presence of static tensile stress.

Pulsed electromagnetic fields (PEMFs) are another type of biophysical stimuli that is shown to regulate the differentiation, mainly studied for its role in osteogenic differentiation of stem cells. Bioreactors with controlled dosimetry of PEMF allow the investigation of cellular and tissue responses *in vitro* [69]. Reports of the osteogenic effect of PEMF in concert with BMP-2 on MSC have initiated many clinical trials of this therapy to induce bone repair and regeneration [70]. Current clinical trials investigate the efficacy of PEMF on cartilage regeneration in osteoarthritic joints.

Bioreactors not only enhance stem cell differentiation but also provide *in vitro* tools to study stem cell differentiation. This has led to a new field of commercial production of specialized bioreactors for research use. Herein we list some examples of mechanical loading bioreactors developed by researchers. Flexcell® provides compression, fluid shear and tension systems for studying mechanotransduction mechanisms in numerous types of cells. Their bioreactor products have been used to study mechanotransduction of cells derived from various tissues such as muscle (cardiac, skeletal, smooth), lung, endothelium, skin, bone, cartilage, ligament, and tendon [71].

Tissue growth technologies (BISS TGT) is another supplier of mechanical loading bioreactors that incorporate Instron mechanical testing and control technology in their *in vitro* tissue growth bioreactors. Their product line includes bioreactors that provide compression, tension, pulsatile pressure and shear stress, and perfusion for tissues that include but not limited to cardiac muscle, ligament, cartilage, bone, and skin [72, 73].

Rotating wall bioreactors have been developed to cultivate many different types of cells in a low shear and low turbulence controlled fluid flow environment, originally thought to mimic the microgravity environment [74]. Synthecon has adapted this technology into many products geared toward small-scale stem cell cultivation [52].

3.4.2 Mechanotransduction via Particle-based Systems

Mechanical stimuli can be generated by applying external physical forces using instruments such as bioreactors or can be generated via specialized particles, such as magnetic particles. So far, magnetic particles found a broad spectrum of use in microfluidics, data storage, separation and purification processes, magnetic imaging, etc. [75–80]. There are several techniques for producing magnetic nanoparticles (MNPs) depending on the needs of the application. For example, in data storage, the particles should be stable enough to not be affected by temperature changes. On the other hand, for cellular applications MNPs should retain their chemical integrity and stability in tissue's physiological environment (e.g. pH 7.4; 37 °C) [79, 81]. In some applications, MNPs can be coated with a polymer shield to form a more biocompatible active complex. In biomedical and therapeutic applications of superparamagnetic magnetite (Fe_3O_4) or maghemite ($\gamma\text{-}Fe_2O_3$), nanoparticles are commonly used [77, 82, 83].

MNPs have been demonstrated to be useful for specific cell-tracking applications [81, 84–86]. Discovery of their possible role in tissue-specific cell tracking has led to the possibility of use as a targeted cancer therapy. These studies demonstrate that by binding these magnetic particles to the cell surfaces and applying an external magnetic field, the cell surface activity can be manipulated to trigger a mechanical stimulus [87–90].

Studies on MNPs as mechanotransduction agents date back to the 1980s. The initial efforts were to investigate the structural response and mechanical properties of cytoskeleton and the cell membrane [91–94]. In these studies, mechanosensitive ion channel kinetics and the alteration in cell membrane were evaluated. It was shown that MNPs were attached to the cell membrane receptors or proteins like integrin and evaluation were performed under cyclic or pulsatile external magnetic forces [91, 94].

Magnetic core can be coated with biocompatible polymers. The polymeric layer itself allows binding of several specific ligands for promoting cell membrane or triggering intracellular signaling pathways [90, 95, 96]. The idea is based on the working principal of the ion channels. Cells have mechanosensitive ion channels, and under mechanical stimulations in the form of stretch, pressure, or twist, these channels play key roles on cell behavior. It is shown by Dobson *et al.* that magnetic particles can actuate ion channels on the cell membrane. This actuation leads a potential change in the membrane, which stimulates intracellular calcium storage and upregulates gene expression, eventually promoting MSC differentiation. The actuation mechanism is based on the possible stretching activation mechanism of the ion channels on the cell membrane [97–100].

Moreover, another remarkable application of MNPs is using magneto-sensitive scaffolds (ferroscaffolds) to affect cell behavior. The technique is based on the combination of MNPs, bare or coated with a functional polymeric shell, with polymeric scaffold to form a biohybrid biomaterial [101]. Studies on ferroscaffolds suggest induction of bone formation and the regeneration by MNPs embedded in polymeric or inorganic scaffolds with or without application of an external magnetic force [102–104]. In theory, superparamagnetic nanoparticles produce a low magnetic field in culture or in tissue microenvironment that stretches the cell membrane and activate specific membrane channels and the receptors [82]. Magnetization of scaffolds in *in vivo* can induce angiogenesis or can set off several cellular reactions, which eventually lead to increase in cell proliferation and intracellular activities [105–107]. The nanoparticles' superparamagnetic ability is activated by exposure to a magnetic force. The small size of the particles (mostly around 10 nm) may present a disadvantage such that, the generated magnetic force may not be sufficient to overcome other forces (such as Brownian forces) to create a proper magnetic field and to form mechanically stable ferroscaffolds. One interesting study proposed that, multidomain MNPs with the size of 50–100nm can generate a stronger magnetic field energy that can overcome Brownian forces when they are integrated in polymeric biomaterials [108]. On the other hand, high-strength magnetic fields, with a magnetic flux over 2 Tesla (T) can cause serious health problems such as cancer and development problems [109].

With the ability of magnetic guidance, ferroscaffolds draw attention in the field of tissue engineering and regenerative medicine. Traditional approaches usually carry some concerns about mimicking tissue's own microenvironment, homing stem cells and progenitor cells to the target site [110]. Studies reveal that low-frequency magnetic flux (1–100 mT) may have therapeutic potential with the ability to trigger MSC proliferation and the differentiation into hepatocytes to form micro-liver construct or differentiate into osteoblast like cells to improve bone regeneration [110, 111]. It is worth to note that, magnetic flux, even having a low field density, may alter cell morphology *in vitro* [106].

3.4.3 Mechanotransduction via Other External Forces

Vibrational force is a type of physical force that can affect cell behavior. Vibrational forces come from the periodic motions (oscillations) of the particles, which have a strong relation with sound and the pressure. Vibration frequency of 30–40 Hz can induce adipose-derived stem cell (AdSC) differentiation in bone or can induce the proliferation of MSC by upregulating the expression of collagen types I and II, vascular endothelial growth factor (VEGF), TGF-β, and other related proteins like BMPs and alkaline phosphatase [112, 113]. Nanovibrational stimulation at around 200 Hz could affect MSC to differentiate into vocal fold fibroblasts [114].

Ultrasound stimulation has also been proposed as a tool to induce chondrogenic differentiation of MSC. Low-intensity ultrasound (200 mW/cm^2, 1 MHz) in a continuous wave for 10 min every 12 h was shown to enhance chondrogenic differentiation of rabbit bone marrow-derived MSC in the presence of TGF-β3 [115]. Another study reports enhanced chondrogenic differentiation of human MSCs with ultrasound stimulation 5 MHz in concert with TGF-β3 [116]. It is suggested that the ultrasound stimulation triggers integrin-mediated mechanotransduction pathways and cell migration activity.

Magnetic resonance (MR) imaging technique is based on using a magnetic fields and radio waves to take diagnostic images [117]. This technique needs an external magnetic field near 2 T. Beyond its diagnostic use, clinical observations revealed that MR can help to treat musculoskeletal disorders [118–122]. Mechanical stimulation using MR can trigger several biochemical pathways in cells and can eventually lead to upregulation of gene expression, stimulation of integrins and other crucial ligands or components. Integrins have critical roles in cellular behavior which they can modulate differentiation, matrix remodeling, and gene expression as a response to a mechanical stimulus [118, 123].

Similarly, magnetic field therapies have been applied for years. Generally, they can be categorized in five groups: permanent magnetic fields, low-frequency magnetic waves, pulsed radiofrequency fields, PEMFs, and transcranial magnetic/electric stimulation [124]. All these techniques create mechanical stimuli on cells, which eventually affect cellular behavior. The effect on the cellular response depends on several parameters such as the dose of the application, time of exposure, and physiological condition of the tissue [124, 125].

3.5 Mechanotransduction via Bioinspired Materials

There are also several other strategies to alter cell behavior using mechanotransduction pathways. Using active nanopatterned topographical surfaces could be one of these strategies. Studies about the effect of topography showed that, active nanotopography-based surfaces can change cell behavior by upregulating gene and protein expressions and affect their adhesion kinetics [126–130]. Moreover, the shear stress that is directly related to the nanotopography can also change cell adhesion behavior [131].

The actual reason for this behavioral change is still unclear but in theory, when cells interact with these nanopatterned surfaces, they have experienced forces in nanoscale similar to the mechanical forces in their biological microenvironment, which cause changes in gene and protein expressions. Thus, cells like MSC can change their state from "quiescent" to "active" [130, 131].

MSC can respond patterned topographical surfaces. Certain micropatterned groove surfaces can induce differentiation to cells that have neural characteristic. This can be associated with the changes in cell alignment and elongation, changes in microtubule organization, upregulation of specific transcription factors through integrins, and G-protein-coupled receptors [132].

3.6 Future Remarks and Conclusion

Stem cells have a great potential for tissue engineering and regenerative medicine. We still do not fully understand their specific niche and the mechanism of how these cells can decide their fate. Mechanical stimulation may be the key action for the modulation of stem cell behavior. It can be achieved by bioreactor systems, particle-based systems, or other mentioned strategies; however, their clinical efficiency is yet to be proved. Until now, researchers mainly focused on the bulk biomaterial properties. Future strategies should focus on the specific interactions between the cells and the biomaterials, and the integration of specific biophysical stimulation with inherent material properties. The ultimate goal should be to reveal the mechanisms of how the material properties alter the behavior of cells in concert with mechanical stimuli. Eventually, controlled modulation of stem cell behavior by mimicking their own environment will help us to achieve clinically relevant repair, remodeling, and regeneration.

Declaration of Interest

The authors have no competing interests or other interests that might be perceived to influence the discussion reported in this chapter.

References

1. Baluska, F., D. Volkmann, and P.W. Barlow, Eukaryotic cells and their cell bodies: cell theory revised. *Ann. Botany*, 94(1), p. 9–32, 2004.
2. Mazzarello, P., A unifying concept: the history of cell theory. *Nat. Cell Biol.*, 1(1), p. E13–5, 1999.
3. Bryant, D.M. and K.E. Mostov, From cells to organs: building polarized tissue. *Nat. Rev. Mol. Cell Biol.*, 9(11), p. 887–901, 2008.
4. Davies, J.A., *Extracellular Matrix*, in eLS. 2001, John Wiley & Sons, Ltd.
5. Frantz, C., K.M. Stewart, and V.M. Weaver, The extracellular matrix at a glance. *J. Cell Sci.*, 123(24), p. 4195–200, 2010.
6. Gattazzo, F., A. Urciuolo, and P. Bonaldo, Extracellular matrix: a dynamic microenvironment for stem cell niche. *Biochim. Biophys. Acta*, 1840(8), p. 2506–19, 2014.
7. Alberts B., J.A., Lewis J., et al., *Cell Junctions, Cell Adhesion, and the Extracellular Matrix 4th ed. Molecular Biology of the Cell.* New York: Garland Science, 2002.
8. Hay, E.D., *Cell Biology of Extracellular Matrix.* 2nd ed. Springer, Newyork US, 1991.
9. Rozario, T. and D.W. DeSimone, The extracellular matrix in development and morphogenesis: a dynamic view. *Dev. Biol.*, 341(1), p. 126–40, 2010.
10. Lu, P., K. Takai, V.M. Weaver, and Z. Werb, Extracellular matrix degradation and remodeling in development and disease. *Cold Spring Harb. Perspect. Biol.*, 3(12), 2011.
11. Lu, P., V.M. Weaver, and Z. Werb, The extracellular matrix: a dynamic niche in cancer progression. *J. Cell Biol.*, 196(4), p. 395–406, 2012.
12. Magno, R., V.A. Grieneisen, and A.F.M. Maree, The biophysical nature of cells: potential cell behaviours revealed by analytical and computational studies of cell surface mechanics. *BMC Biophysics*, 8, p. 1–37, 2015.

13. Clark, P., P. Connolly, A.S.G. Curtis, J.A.T. Dow, and C.D.W. Wilkinson, Topographical control of cell behavior. 1. Simple step cues. *Development*, 99(3), p. 439–48, 1987.
14. Discher, D.E., P. Janmey, and Y.-l. Wang, Tissue cells feel and respond to the stiffness of their substrate. *Science*, 310(5751), p. 1139–43, 2005.
15. Fletcher, D.A. and R.D. Mullins, Cell mechanics and the cytoskeleton. *Nature*, 463(7280), p. 485–92, 2010.
16. Garcia, J.R. and A.J. Garcia, Cellular mechanotransduction: sensing rigidity. *Nat. Mater.*, 13(6), p. 539–40, 2014.
17. Lo, C.M., H.B. Wang, M. Dembo, and Y.L. Wang, Cell movement is guided by the rigidity of the substrate. *Biophys. J.*, 79(1), p. 144–52, 2000.
18. Mammoto, T. and D.E. Ingber, Mechanical control of tissue and organ development. *Development*, 137(9), p. 1407–20, 2010.
19. Alhadlaq, A. and J.J. Mao, Mesenchymal stem cells: isolation and therapeutics. *Stem Cells Dev.*, 13(4), p. 436–48, 2004.
20. Bernardo, M.E., F. Locatelli, and W.E. Fibbe, Mesenchymal stromal cells. *Ann. N. Y. Acad. Sci.*, 1176, p. 101–17, 2009.
21. Dominici, M., K. Le Blanc, I. Mueller, I. Slaper-Cortenbach, F. Marini, D. Krause, R. Deans, A. Keating, D. Prockop, and E. Horwitz, Minimal criteria for defining multipotent mesenchymal stromal cells. The International Society for Cellular Therapy position statement. *Cytotherapy*, 8(4), p. 315–7, 2006.
22. Quesenberry, P., Stem cell plasticity: an overview. *Blood Cells Mol. Dis.*, 32(1), p. 1–4, 2004.
23. Weissman, I.L., Stem cells: units of development, units of regeneration, and units in evolution. *Cell*, 100(1), p. 157–68, 2000.
24. Lv, H., L. Li, M. Sun, Y. Zhang, L. Chen, Y. Rong, and Y. Li, Mechanism of regulation of stem cell differentiation by matrix stiffness. *Stem Cell Res. Ther.*, 6, p. 103, 2015.
25. Melton A.D., Cowen, C., "Stemness": Definitions, Criteria, and Standards. *Essentials of Stem Cell Biology*, p 23–29, 2009.
26. Mitsiadis, T.A., O. Barrandon, A. Rochat, Y. Barrandon, and C. De Bari, Stem cell niches in mammals. *Exp. Cell Res.*, 313(16), p. 3377–85, 2007.
27. Morrison, S.J. and A.C. Spradling, Stem cells and niches: mechanisms that promote stem cell maintenance throughout life. *Cell*, 132(4), p. 598–611, 2008.
28. Ohlstein, B., T. Kai, E. Decotto, and A. Spradling, The stem cell niche: theme and variations. *Curr. Opin. Cell Biol.*, 16(6), p. 693–9, 2004.
29. Paluch, E.K., C.M. Nelson, N. Biais, B. Fabry, J. Moeller, B.L. Pruitt, C. Wollnik, G. Kudryasheva, F. Rehfeldt, and W. Federle, Mechanotransduction: use the force(s). *BMC Biol.*, 13, p. 47, 2015.
30. S. Deshpande, S., Mesenchymal stem cell transplantation: new avenues for stem cell therapies. *J. Transplant. Technol. Res.*, 03(02), 2013.
31. Huang, C. and R. Ogawa, Mechanotransduction in bone repair and regeneration. *FASEB J.*, 24(10), p. 3625–32, 2010.

32. Ingber, D.E., Tensegrity II. How structural networks influence cellular information processing networks. *J. Cell Sci.*, 116(8), p. 1397–408, 2003.
33. Ingber, D.E., Cellular mechanotransduction: putting all the pieces together again. *FASEB J.*, 20(7), p. 811–27, 2006.
34. Ingber, D.E., Tensegrity-based mechanosensing from macro to micro. *Prog. Biophys. Mol. Biol.*, 97(2–3), p. 163–79, 2008.
35. Stupack, D.G., The biology of integrins. *Oncology (Williston Park)*, 21(9 Suppl 3), p. 6–12, 2007.
36. Pavalko, F.M., S.M. Norvell, D.B. Burr, C.H. Turner, R.L. Duncan, and J.P. Bidwell, A model for mechanotransduction in bone cells: The load-bearing mechanosomes. *J. Cellular Biochem.*, 88(1), p. 104–12, 2003.
37. Wang, J.H.C, Thampatty, B.P, Mechanobiology of adult andstem cells, *Inter. Rev. Cell Mol. Biol.* 271, p. 301–346, 2008.
38. Wolf, C.B. and M.R.K. Mofrad, Mechanotransduction and its role in stem cell biology. p. 389–403, 2009.
39. Santos, A., A.D. Bakker, and J. Klein-Nulend, The role of osteocytes in bone mechanotransduction. *Osteoporos Int.*, 20(6), p. 1027–31, 2009.
40. Saxon, L.K., A.G. Robling, I. Alam, and C.H. Turner, Mechanosensitivity of the rat skeleton decreases after a long period of loading, but is improved with time off. *Bone.*, 36(3), p. 454–64, 2005.
41. Zeng, S.W.S.C.C.Y., A model for the excitation of osteocytes by mechanical loading-induced bone fluid shear stresses. *J. Biomech.*, 27(3), p. 339–60, 1994.
42. Owan, C.H.T.Y.T.I., Aging changes mechanical loading thresholds for bone formation in Rats. *J. Bone. Mineral. Res.*, 10(10), p. 1544–49, 1995.
43. Kurpinski, K., J. Chu, C. Hashi, and S. Li, Anisotropic mechanosensing by mesenchymal stem cells. *Proc. Natl. Acad. Sci. U S A*, 103(44), p. 16095–100, 2006.
44. Kopf, J., A. Petersen, G.N. Duda, and P. Knaus, BMP2 and mechanical loading cooperatively regulate immediate early signalling events in the BMP pathway. *BMC Biol.*, 10, p. 1–12, 2012.
45. Papachroni, K.K., D.N. Karatzas, K.A. Papavassiliou, E.K. Basdra, and A.G. Papavassiliou, Mechanotransduction in osteoblast regulation and bone disease. *Trends. Mol. Med.*, 15(5), p. 208–16, 2009.
46. Drachman, D.B. and L. Sokoloff, The role of movement in embryonic joint development. *Devl. Biol.*, 14, p. 401–20, 1966.
47. Carter, D.R., P.R. Blenman, and G.S. Beaupre, Correlations between mechanical stress history and tissue differentiation in initial fracture healing. *J. Orthop. Res.*, 6(5), p. 736–48, 1988.
48. Butler, D.L., S.A. Goldstein, and F. Guilak, Functional tissue engineering: the role of biomechanics. *J. Biomech. Eng.*, 122, p. 570–5, 2000.
49. Mauck, R.L., B.A. Byers, X. Yuan, and R.S. Tuan, Regulation of cartilaginous ECM gene transcription by chondrocytes and MSCs in 3D culture in response to dynamic loading. *Biomech. Model. Mechanobiol.*, 6(1–2), p. 113–25, 2007.

50. Bueno, E.M., B. Bilgen, R.L. Carrier, and G.A. Barabino, Increased rate of chondrocyte aggregation in a wavy-walled bioreactor. *Biotechnol. Bioeng.*, 88(6), p. 767–77, 2004.
51. Chattopadhyay, D., M. Garcia-Briones, R.V. Venkat, and J.J. Chalmers, Hydrodynamic properties in bioreactors, in *Mammalian Cell Biotechnology in Protein Production*, H. Hauser and R. Wagner, Editors. Walter de Gruyter: Berlin. p. 491, 1997.
52. Fridley, K.M., I. Fernandez, M.T. Li, R.B. Kettlewell, and K. Roy, Unique differentiation profile of mouse embryonic stem cells in rotary and stirred tank bioreactors. *Tissue. Eng. Part A*, 16(11), p. 3285–98, 2010.
53. Stolberg, S. and K.E. McCloskey, Can shear stress direct stem cell fate? *Biotechnol. Prog.*, 25(1), p. 10–9, 2009.
54. Freed, L.E., G. Vunjak-Novakovic, and R. Langer, Cultivation of cell-polymer cartilage implants in bioreactors. *J. Cell. Biochem.*, 51, p. 257, 1993.
55. Freed, L.E. and G. Vunjak-Novakovic, Cultivation of cell-polymer tissue constructs in simulated microgravity. *Biotechnol. Bioeng.*, 46(4), p. 306–13, 1995.
56. Carver, S.E. and C.A. Heath, Increasing extracellular matrix production in regenerating cartilage with intermittent physiological pressure. *Biotechnol. Bioeng.*, 62(2), p. 166–74, 1999.
57. Mauck, R.L., M.A. Soltz, C.C. Wang, D.D. Wong, P.H. Chao, W.B. Valhmu, C.T. Hung, and G.A. Ateshian, Functional tissue engineering of articular cartilage through dynamic loading of chondrocyte-seeded agarose gels. *J. Biomech. Eng.*, 122(3), p. 252–60, 2000.
58. Bueno, E.M., B. Bilgen, and G.A. Barabino, Wavy-walled bioreactor supports increased cell proliferation and matrix deposition in engineered cartilage constructs. *Tissue. Eng.*, 11(11–2), p. 1699–709, 2005.
59. Bilgen, B., D. Chu, R. Stefani, and R.K. Aaron, Design of a biaxial mechanical loading bioreactor for tissue engineering. *J. Vis. Exp.*, 2013(74), p. e50387.
60. Grodzinsky, A.J., M.E. Levenston, M. Jin, and E.H. Frank, Cartilage tissue remodeling in response to mechanical forces. *Ann. Rev. Biomed. Eng.*, 2, p. 691–713, 2000.
61. Uygun, B.E., A. Soto-Gutierrez, H. Yagi, M.L. Izamis, M.A. Guzzardi, C. Shulman, J. Milwid, N. Kobayashi, A. Tilles, F. Berthiaume, M. Hertl, Y. Nahmias, M.L. Yarmush, and K. Uygun, Organ reengineering through development of a transplantable recellularized liver graft using decellularized liver matrix. *Nat. Med.*, 16(7), p. 814–20, 2010.
62. Vunjak-Novakovic, G., L.E. Freed, R.J. Biron, and R. Langer, Effects of mixing on the composition and morphology of tissue-engineered cartilage. *AIChE J.*, 42(3), p. 850–60, 1996.
63. Smith, R.L., B.S. Donlon, M.K. Gupta, M. Mohtai, P. Das, D.R. Carter, J. Cooke, G. Gibbons, N. Hutchinson, and D.J. Schurman, Effects of fluid-induced shear on articular chondrocyte morphology and metabolism *in vitro*. *J. Orthop. Res.* 13, 824–31, 1995.

64. Huang, C.Y., K.L. Hagar, L.E. Frost, Y. Sun, and H.S. Cheung, Effects of cyclic compressive loading on chondrogenesis of rabbit bone-marrow derived mesenchymal stem cells. *Stem Cells*, 22(3), p. 313–23, 2004.
65. Yanagisawa, M., N. Suzuki, N. Mitsui, Y. Koyama, K. Otsuka, and N. Shimizu, Effects of compressive force on the differentiation of pluripotent mesenchymal cells. *Life Sci.*, 81(5), p. 405–12, 2007.
66. Mouw, J.K., J.T. Connelly, C.G. Wilson, K.E. Michael, and M.E. Levenston, Dynamic compression regulates the expression and synthesis of chondrocyte-specific matrix molecules in bone marrow stromal cells. *Stem Cells*, 25(3), p. 655–63, 2007.
67. Baker, B.M., R.P. Shah, A.H. Huang, and R.L. Mauck, Dynamic tensile loading improves the functional properties of mesenchymal stem cell-laden nanofiber-based fibrocartilage. *Tissue. Eng. Part A*, 17(9–10), p. 1445–55, 2011.
68. Tulloch, N.L., V. Muskheli, M.V. Razumova, F.S. Korte, M. Regnier, K.D. Hauch, L. Pabon, H. Reinecke, and C.E. Murry, Growth of engineered human myocardium with mechanical loading and vascular coculture. *Circ. Res.*, 109(1), p. 47–59, 2011.
69. Mognaschi, M.E., P. Di Barba, G. Magenes, A. Lenzi, F. Naro, and L. Fassina, Field models and numerical dosimetry inside an extremely-low-frequency electromagnetic bioreactor: the theoretical link between the electromagnetically induced mechanical forces and the biological mechanisms of the cell tensegrity. *Springerplus.*, 3, p. 473, 2014.
70. Schwartz, Z., B.J. Simon, M.A. Duran, G. Barabino, R. Chaudhri, and B.D. Boyan, Pulsed electromagnetic fields enhance BMP-2 dependent osteoblastic differentiation of human mesenchymal stem cells. *J. Orthop. Res.*, 26(9), p. 1250–5, 2008.
71. Banes, A.J., Out of academics: education, entrepreneurship and enterprise. *Ann. Biomed. Eng.*, 41, p. 1926–1938, 2013.
72. Klein, T.J., J. Malda, R.L. Sah, and D.W. Hutmacher, Tissue engineering of articular cartilage with biomimetic zones. *Tissue. Eng. Part B Rev.*, 15(2), p. 143–57, 2009.
73. Porter, B.D., A.S. Lin, A. Peister, D. Hutmacher, and R.E. Guldberg, Noninvasive image analysis of 3D construct mineralization in a perfusion bioreactor. *Biomaterials*, 28(15), p. 2525–33, 2007.
74. Schwarz, R.P., T.J. Goodwin, and D.A. Wolf, Cell culture for three-dimensional modeling in rotating-wall vessels: an application of simulated microgravity. *J. Tissue. Cult. Methods*, 14(2), p. 51–7, 1992.
75. Kavaz, D., S. Odabas, E. Guven, M. Demirbilek, and E.B. Denkbas, Bleomycin loaded magnetic chitosan nanoparticles as multifunctional nanocarriers. *J. Bioactive Compatible Polym.*, 25(3), p. 305–18, 2010.
76. Odabas, S., F. Sayar, G. Guven, G. Yanikkaya-Demirel, and E. Piskin, Separation of mesenchymal stem cells with magnetic nanosorbents carrying CD105 and CD73 antibodies in flow-through and batch systems. *J. Chromatogr. B Analyt. Technol. Biomed. Life Sci.*, 861(1), p. 74–80, 2008.

77. O'Grady, K., Biomedical applications of magnetic nanoparticles. *J Phys D Appl. Phys.*, 36(13), 2003.
78. Reiss, G. and A. Hutten, Magnetic nanoparticles – applications beyond data storage. *Nat. Mater.*, 4(10), p. 725–6, 2005.
79. Safarik, I. and M. Safarikova, Use of magnetic techniques for the isolation of cells. *J. Chromatogr B Anal. Technol. Biomed. Life Sci.*, 722(1–2), p. 33–53, 1999.
80. Singamaneni, S., V.N. Bliznyuk, C. Binek, and E.Y. Tsymbal, Magnetic nanoparticles: recent advances in synthesis, self-assembly and applications. *J. Mater Chem.*, 21(42), p. 16819–45, 2011.
81. Akbarzadeh, A., M. Samiei, and S. Davaran, Magnetic nanoparticles: preparation, physical properties, and applications in biomedicine. *Nanoscale Res. Lett.*, 7(1), p. 144, 2012.
82. Dobson, J., Remote control of cellular behaviour with magnetic nanoparticles. *Nat. Nanotechnol*, 3(3), p. 139–43, 2008.
83. Pankhurst, Q.A., J. Connolly, S.K. Jones, and J. Dobson, Applications of magnetic nanoparticles in biomedicine. *J. Phys. D Appl. Phys.*, 36(13), p. R167–R181, 2003.
84. Cromer Berman, S.M., P. Walczak, and J.W. Bulte, Tracking stem cells using magnetic nanoparticles. *Wiley. Interdiscip Rev. Nanomed Nanobiotechnol*, 3(4), p. 343–55, 2011.
85. Hachani, R., M. Lowdell, M. Birchall, and N.T.K. Thanh, Tracking stem cells in tissue-engineered organs using magnetic nanoparticles. *Nanoscale*, 5(23), p. 11362–73, 2013.
86. Lalande, C., S. Miraux, S.M. Derkaoui, S. Mornet, R. Bareille, J.C. Fricain, J.M. Franconi, C. Le Visage, D. Letourneur, J. Amedee, and A.K. Bouzier-Sore, Magnetic resonance imaging tracking of human adipose derived stromal cells within three-dimensional scaffolds for bone tissue engineering. *Eur. Cells Mater.*, 21, p. 341–54, 2011.
87. Banobre-Lopez, M., A. Teijeiro, and J. Rivas, Magnetic nanoparticle-based hyperthermia for cancer treatment. *Rep. Pract. Oncol. Radiother*, 18(6), p. 397–400, 2013.
88. Giustini, A.J., A.A. Petryk, S.M. Cassim, J.A. Tate, I. Baker, and P.J. Hoopes, Magnetic nanoparticle hyperthermia in cancer treatment. *Nano Life*, 1, p. 17–32, 2010.
89. Mahmoudi, K. and C.G. Hadjipanayis, The application of magnetic nanoparticles for the treatment of brain tumors. *Front Chem*, 2, p. 109, 2014.
90. Umut, E., Surface Modification of Nanoparticles Used in Biomedical Applications, Modern Surface Engineering Treatments, Dr. M. Aliofkhazraei Ed.), InTech, Croatia, 2013.
91. Meyer, C.J., F.J. Alenghat, P. Rim, J.H. Fong, B. Fabry, and D.E. Ingber, Mechanical control of cyclic AMP signalling and gene transcription through integrins. *Nat. Cell Biol.*, 2(9), p. 666–8, 2000.
92. Valberg, P.A. and J.P. Butler, Magnetic particle motions within living cells – physical theory and techniques. *Biophys. J.*, 52(4), p. 537–50, 1987.

93. Valberg, P.A. and H.A. Feldman, Magnetic particle motions within living cells – measurement of cytoplasmic viscosity and motile activity. *Biophys. J.*, 52(4), p. 551–61, 1987.
94. Wang, N., J.P. Butler, and D.E. Ingber, Mechanotransduction across the cell-surface and through the cytoskeleton. *Science*, 260(5111), p. 1124–7, 1993.
95. Berry, C.C. and A.S.G. Curtis, Functionalisation of magnetic nanoparticles for applications in biomedicine. *J. Phys. D Appl. Phys.*, 36(13), p. R198–R206, 2003.
96. Sniadecki, N.J., A tiny touch: activation of cell signaling pathways with magnetic nanoparticles. *Endocrinology*, 151(2), p. 451–7, 2010.
97. Cartmell, S.H., J. Dobson, S.B. Verschueren, and A.J. El Haj, Development of magnetic particle techniques for long-term culture of bone cells with intermittent mechanical activation. *IEEE Trans. Nanobiosci.*, 1(2), p. 92–7, 2002.
98. Dobson, J., Z. Stewart, and B. Martinac, Preliminary evidence for weak magnetic field effects on mechanosensitive ion channel subconducting states in Escherichia coli. *Electromagn. Biol. Med.*, 21(3), p. 89–98, 2002.
99. Henstock, J. and A. El Haj, Controlled mechanotransduction in therapeutic MSCs: can remotely controlled magnetic nanoparticles regenerate bones? *Regen. Med.*, 10(4), p. 377–80, 2015.
100. Tseng, P., J.W. Judy, and D. Di Carlo, Magnetic nanoparticle-mediated massively parallel mechanical modulation of single-cell behavior. *Nat. Methods.*, 9(11), p. 1113–9, 2012.
101. Tampieri, A., M. Iafisco, M. Sandri, S. Panseri, C. Cunha, S. Sprio, E. Savini, M. Uhlarz, and T. Herrmannsdorfer, Magnetic bioinspired hybrid nanostructured collagen-hydroxyapatite scaffolds supporting cell proliferation and tuning regenerative process. *ACS Appl. Mater. Interfaces.*, 6(18), p. 15697–707, 2014.
102. Hu, S.H., T.Y. Liu, C.H. Tsai, and S.Y. Chen, Preparation and characterization of magnetic ferroscaffolds for tissue engineering. *J. Magn. Magn. Mater.*, 310(2), p. 2871–3, 2007.
103. Panseri, S., C. Cunha, T. D'Alessandro, M. Sandri, A. Russo, G. Giavaresi, M. Marcacci, C.T. Hung, and A. Tampieri, Magnetic hydroxyapatite bone substitutes to enhance tissue regeneration: evaluation *in vitro* using osteoblast-like cells and *in vivo* in a bone defect. *PLoS One*, 7(6), p. e38710, 2012.
104. Xu, H.Y. and N. Gu, Magnetic responsive scaffolds and magnetic fields in bone repair and regeneration. *Front. Mater. Sci.*, 8(1), p. 20–31, 2014.
105. Bock, N., A. Riminucci, C. Dionigi, A. Russo, A. Tampieri, E. Landi, V.A. Goranov, M. Marcacci, and V. Dediu, A novel route in bone tissue engineering: magnetic biomimetic scaffolds. *Acta. Biomater.*, 6(3), p. 786–96, 2010.
106. Kim, J-J., R-K. Singh, S-J. Seo, T-H. Kim, J-H. Kim, E-J. Lee, and H-W. Kim, Magnetic scaffolds of polycaprolactone with functionalized magnetite nanoparticles: physicochemical, mechanical, and biological properties effective for bone regeneration. *RSC. Adv.*, 4(33), p. 17325, 2014.

107. Tampieri, A., E. Landi, F. Valentini, M. Sandri, T. D'Alessandro, V. Dediu, and M. Marcacci, A conceptually new type of bio-hybrid scaffold for bone regeneration. *Nanotechnology*, 22(1), p. 015104, 2011.
108. Lopez-Lopez, M.T., G. Scionti, A.C. Oliveira, J.D. Duran, A. Campos, M. Alaminos, and I.A. Rodriguez, Generation and characterization of novel magnetic field-responsive biomaterials. *PLoS One*, 10(7), p. e0133878, 2015.
109. Adair, R.K., Static and low-frequency magnetic field effects: health risks and therapies. *Rep. Progr. Phys.*, 63(3), p. 415–54, 2000.
110. Bishi, D.K., S. Guhathakurta, J.R. Venugopal, K.M. Cherian, and S. Ramakrishna, Low frequency magnetic force augments hepatic differentiation of mesenchymal stem cells on a biomagnetic nanofibrous scaffold. *J. Mater. Sci. Mater. Med.*, 25(11), p. 2579–89, 2014.
111. Nakamae, T., N. Adachi, T. Kobayashi, Y. Nagata, T. Nakasa, N. Tanaka, and M. Ochi, The effect of an external magnetic force on cell adhesion and proliferation of magnetically labeled mesenchymal stem cells. *Sports Med Arthrosc. Rehabil Ther Technol.*, 2(1), p. 5, 2010.
112. Kim, I.S., Y.M. Song, B. Lee, and S.J. Hwang, Human mesenchymal stromal cells are mechanosensitive to vibration stimuli. *J. Dent. Res.*, 91(12), p. 1135–40, 2012.
113. Pre, D., G. Ceccarelli, G. Gastaldi, A. Asti, E. Saino, L. Visai, F. Benazzo, M.G.C. De Angelis, and G. Magenes, The differentiation of human adipose-derived stem cells (hASCs) into osteoblasts is promoted by low amplitude, high frequency vibration treatment. *Bone*, 49(2), p. 295–303, 2011.
114. Gaston, J., B.Q. Rios, R. Bartlett, C. Berchtold, and S.L. Thibeault, The response of vocal fold fibroblasts and mesenchymal stromal cells to vibration. *PloS One*, 7(2), p. 1–9, 2012.
115. Lee, H.J., B.H. Choi, B.H. Min, Y.S. Son, and S.R. Park, Low-intensity ultrasound stimulation enhances chondrogenic differentiation in alginate culture of mesenchymal stem cells. *Artif. Organs.*, 30(9), p. 707–15, 2006.
116. Guha Thakurta, S., G. Budhiraja, and A. Subramanian, Growth factor and ultrasound-assisted bioreactor synergism for human mesenchymal stem cell chondrogenesis. *J. Tissue. Eng.*, 6, p. 1–13, 2015.
117. Edelman, R.R. and S. Warach, Medical progress. 1. Magnetic-resonance-imaging. *N. Eng. J. Med.*, 328(10), p. 708–16, 1993.
118. Bibiane Steinecker-Frohnwieser, G.W., Georg Kress, Werner Kullich, Lukas Weigl, Hans Georg Kress, The influence of nuclear magnetic resonance therapy (NMRT) and interleukin IL1-β stimulation on cal 78 chondrosarcoma cells and C28I2 chondrocytes. *J. Orthoped. Rheumatol.*, 1(3), p1-9, 2014.
119. Faria, S.C., K. Ganesan, I. Mwangi, M. Shiehmorteza, B. Viamonte, S. Mazhar, M. Peterson, Y. Kono, C. Santillan, G. Casola, and C.B. Sirlin, MR Imaging of liver fibrosis: current state of the art. *Radiographics*, 29(6), p. 1615–36, 2009.
120. Kullich, W., Ausserwinkler, M, Functional Improvement in Finger Joints Arthrosis by Magnetic Resonance, *Orthopadische Praxis*, 6, p287–290, 2008.

121. W. Kullich. H. Schwann, K.M.M.A., Additional outcome improvement in the rehabilitation of chronic low back of chronic low back pain after nuclear resonance therapy. *POVODNA PRACA*, 1, p. 7–12, 2006.
122. Yan, J., L. Dong, B. Zhang, and N. Qi, Effects of extremely low-frequency magnetic field on growth and differentiation of human mesenchymal stem cells. *Electromagn Biol Med*, 29(4), p. 165–76, 2010.
123. Humphries, J.D., A. Byron, and M.J. Humphries, Integrin ligands at a glance. *J Cell Sci*, 119(19), p. 3901–3, 2006.
124. Markov, M.S., Magnetic field therapy: a review. *Electromagn Biol Med*, 26(1), p. 1–23, 2007.
125. Colbert, A.P., H. Wahbeh, N. Harling, E. Connelly, H.C. Schiffke, C. Forsten, W.L. Gregory, M.S. Markov, J.J. Souder, P. Elmer, and V. King, Static magnetic field therapy: a critical review of treatment parameters. *Evid Based Complement Alternat Med*, 6(2), p. 133–9, 2009.
126. Christopher J Bettinger, R.L., Jeffrey T Borenstein, Engineering substrate micro- and nanotopography to control cell function. *Angew Chem Int Ed Engl*, 48, p. 5406–15, 2009.
127. Curtis, A.S.G., B. Casey, J.O. Gallagher, D. Pasqui, M.A. Wood, and C.D.W. Wilkinson, Substratum nanotopography and the adhesion of biological cells. Are symmetry or regularity of nanotopography important? *Biophys Chem*, 94(3), p. 275–83, 2001.
128. Curtis, A.S.G., S. Reid, I. Martin, R. Vaidyanathan, C.A. Smith, H. Nikukar, and M.J. Dalby, Cell interactions at the nanoscale: piezoelectric stimulation. *IEEE Trans Nanobiosci*, 12(3), p. 247–54, 2013.
129. Lin, C.C., C.C. Co, and C.C. Ho, Micropatterning proteins and cells on polylactic acid and poly(lactide-co-glycolide). *Biomaterials*, 26(17), p. 3655–62, 2005.
130. Pemberton, G.D., P. Childs, S. Reid, H. Nikukar, P.M. Tsimbouri, N. Gadegaard, A.S.G. Curtis, and M.J. Dalby, Nanoscale stimulation of osteoblastogenesis from mesenchymal stem cells: nanotopography and nanokicking. *Nanomedicine*, 10(4), p. 547–60, 2015.
131. McMurray, R.J., M.J. Dalby, and P.M. Tsimbouri, Using biomaterials to study stem cell mechanotransduction, growth and differentiation. *J Tissue Eng Regen Med*, 9(5), p. 528–39, 2015.
132. D'Angelo, F., I. Armentano, S. Mattioli, L. Crispoltoni, R. Tiribuzi, G.G. Cerulli, C.A. Palmerini, J.M. Kenny, S. Martino, and A. Orlacchio, Micropatterned Hydrogenated Amorphous Carbon Guides Mesenchymal Stem Cells Towards Neuronal Differentiation. *Eur Cells Mater*, 20, p. 231–44, 2010.

4
Modulation of Stem Cells Behavior Through Bioactive Surfaces

Eduardo D. Gomes[1,2,#], Rita C. Assunção-Silva[1,2,#], Nuno Sousa[1,2], Nuno A. Silva[1,2] and António J. Salgado[1,2]*

[1]*Life and Health Sciences Research Institute (ICVS), School of Health Sciences, University of Minho, Campus de Gualtar, Braga, Portugal*
[2]*ICVS/3B's – PT Government Associate Laboratory, Braga/Guimarães, Portugal*

Abstract

Stem cells have been investigated for the treatment of diseases and the regeneration of tissues where complex surgical treatments or tissue transplantation is far from success. The current research on stem cell-based therapies relies on controlling their potency and self-renewal capacity, to provide new insights into their application in clinics. Stem cells' properties such as migration, proliferation, viability, and differentiation are highly dependent on their surrounding environment. In this sense, the use of surfaces with precise architecture and textures has been demonstrated to modulate their behavior. The fabrication of such materials is accomplished using advanced techniques such as lithography, micro/nanopatterning, electrospinning, microfluidics, and bottom-up/top-down approaches. For poor or non-adhesive surfaces, fabrication technologies may profit with biochemical functionalization to present specific physical cues to cells. An example is the use of calcium phosphates coating [1, 2] or adhesive motifs grafting. Overall, the goal of these approaches is to provide platforms that mimic the extracellular matrix environment for stem cells integration. Playing with the available materials and techniques, embryonic and adult stem cells have cartilage, cardiac, connective and nervous tissue regeneration. Some of these were already tested in animal models with very promising results that may be explored for the development of next-generation surfaces for regenerative medicine. In summary, the focus of this chapter is on the current fabrication strategies used for the development of bio-inspired patterned substrates and their potential in guiding stem cell behavior in a tissue-engineering context.

Corresponding author: asalgado@ecsaude.uminho.pt
#These authors contributed equally to this work.

Ashutosh Tiwari, Bora Garipcan and Lokman Uzun (Eds.) Advanced Surfaces for Stem Cell Research, (67–86) © 2017 Scrivener Publishing LLC

Keywords: Stem cells, bioactive surfaces, lithography, micro/nanopatterning, microfluidics, electrospinning, bottom-up/top-down approaches, substrate chemical modifications

The application of biomaterials in tissue engineering has been extensively explored for the study of cellular and molecular biology. The strategy mostly used in these applications is seeding cells on different substrates and investigate materials/cell interactions at cellular and subcellular levels. Within the past 15 years, stem cells have been one of the most used cell populations in tissue-engineering applications [3–7]. Stem cells are very appealing for this purpose mainly due to their self-renewal capacity and differentiation potential [8–10]. Besides this, they have also been described to provide a large repertoire of signaling molecules, including anti-inflammatory cytokines and growth factors, which may modulate the inhibitory environments while increasing the trophic support to the resident cells in damaged tissues [11–16]. So far, stem cells from different origins have been tested for tissue regeneration, such as embryonic, induced pluripotent, neural, and mesenchymal stem cells (MSCs) [17].

As it has been stated that cells behave according to the materials' physical and biomolecular features [18], a set of novel fabrication methods are currently in use to enhance the bioactivity of such materials and control the biological processes within it [19]. Lithography, micro/nanopatterning, microfluidics, electrospinning, bottom-up/top-down approaches, and surface chemical modification with biomolecules coatings or peptide grafting are some of these methods (summarized in Table 4.1), which will be herein described in detail. They offer the possibility to develop hierarchical structures exhibiting precise features at the nano-, micro-, and macroscale levels, essential to control the pattern of different biomolecular and physical signals, thus allowing us to gain fundamental biological insights into cellular functions.

4.1 Lithography

Lithography is a precise and versatile tool to develop well-controlled surfaces and exhibits several variants, such as photolithography, soft lithography [20], e-beam lithography [21], focused ion beam lithography [22], and others. Among these, soft lithography has been used by some groups to fabricate patterned surfaces that require scalable and highly standardized systems for stem cell aggregation [20] and growth [23]. In line with this,

Table 4.1 Advanced technologies for the fabrication of bioactive surfaces for stem cell research.

Techniques	Tested approach	In vitro	In vivo	TE application	Refs
Lithography	CA + human CTPs	x	-	Bone TE	[23]
	DOPA-coated PLGA nanopatterned substrates + human NSCs	x	-	Neural TE	[26]
	PMMA substrates with channels + rat MSCs	x	-	n.d.	[19]
Micro/ nanopatterning	Anodization of titanium substrates to create pillar-like nanostructures + human MSCs	x	-	Bone TE	[35]
	Nanofibrillar collagen scaffolds + iPSCs-ECs	x	x	Vascular TE	[37]
	PDMS substrates with arrays of parallel channels + human MSCs	x	-	Tendon TE	[38]
Microfluidics	Microfluidic chambers with collagen cylinders + ECs + MSCs	x	-	n.d.	[40]
	Microchannels + neuronal-like cells + biomolecules	x	-	Neuronal TE	[41, 42]
Electrospinning	PCL nanofibers + rat MSCs	x	-	Bone TE	[55]
	PLLA/Col nanofibers + Col/HA sponge + rabbit BM-MSCs	x	-	Cartilage TE	[58]
	PPC microfibers + rat BM-MSCs	x	-	Vascular TE	[60]
	PLLA fibers + mouse NSCs	x	-	Neural TE	[61]
Bottom-up/ top-down	Hydrolytically degradable hydrogels + MSCs	-	x	Bone TE (Periosteum allograft)	[68]
	Self-assembly nanofibers of peptide amphiphile molecules + laminin-IKVAV + NPCs	x	-	Neural TE	[69]

(Continued)

Table 4.1 Cont.

Techniques	Tested approach	*In vitro*	*In vivo*	TE application	Refs
Biomolecules coating	C-PBS scaffolds with Ca-P + mouse bone marrow-derived mesenchymal progenitor cells	x	-	Cartilage	[75]
	Fibronectin-coated substrates + BMSCs	x	-	Bone	[71]
	Titanium-coated hemisphere-like topographic nanostructures + human MSCs	x	-	Bone	[72]
Peptide grafting	GG-GRGDS hydrogels + neural stem/progenitor cells + BM-MSCs	x	-	Spinal cord injury	[74]
	GG-GRGDS hydrogels + ASCs + GDNF-NPs	x	-	Peripheral nerve injury	[78]
	Elastin-like recombinamer (ELR) membranes + RGG/REDV/HAP/HAP-RGDS motifs + channels/holes/posts + primary rat MSC	x	-	Bone	[82]

x Studies performed for the described strategy.
- Studies not performed.
n.d., Non defined.

patterned agarose–DMEM microwell surfaces were developed by Gruh et al. [20]. In this work, murine induced pluripotent stem cells (iPSCs) and human embryonic stem cells (ESCs) were used and demonstrated to aggregate to the agarose microwell plates. Both cell types rapidly formed homogeneous embryoid bodies and aggregates, respectively. Besides this, it is of note that ESCs were also able to differentiate into cardiomyocytes [20]. In other study, Kim et al. [23] fabricated cellulose acetate (CA) scaffolds to analyze human marrow-derived connective tissue progenitor cells (CTPs) growth. These scaffolds were developed comprising microtextures or smooth surfaces. The authors pointed out the importance of having microtextures in the surface of the developed scaffolds, since cell growth, migration, and extracellular matrix (ECM) production were increased in this environment, when compared to their growth on smooth surfaces. This system was also important for the study of bone tissue-engineering applications, as osteocalcin mRNA expression was increased on microtextured surfaces [23]. Other authors have also studied differences in cell morphology, directed cell migration, and organization. Using the multiphoton ablation lithography method, nanoscale crater-containing surfaces were developed [24]. The authors played around with nanocraters of different aspect ratio and pitches, which were shown to affect cell's focal adhesion size and distribution along the surfaces. Similar outcomes were shown in another study, where 3,4-dihydroxy-L-phenylanine (DOPA)-coated poly(lactic-co-glycolic acid) (PLGA) nanopatterned substrates constructed using solving-assisted capillary force lithography [25], were used to guide human neural stem cell (NSC) differentiation and neurite extension [26]. Apart from the presence of biophysical cues, nerve growth factor was added to the culture, further enhancing human NSC differentiation. These results indicate that synergistic effects of biophysical and biochemical cues can be advantageous for neural tissue-engineering applications [26].

This type of devices can also be fabricated using a combination of different techniques. For example, Mata et al. [19] used a variety of lithographic techniques, namely photolithography, reactive ion etching, focused ion beam lithography, electron beam lithography, and soft lithography to develop polymethylmethacrylate (PMMA) substrates comprising channels with different sizes and placed in different directions, for cell attachment studies. It was described that rat MSCs interacted differently with the tested substrates, which influenced their alignment and morphology [19]. Furthermore, lithographic techniques were also shown to successfully produce substrates by combination with other techniques such as sol–gel chemistry [27], colloidal self-assembly and glancing angle deposition [28], magnetic tweezer techniques [29], surface micro-wrinkling method [25], and micropatterning [30-32]. All of these studies have the common feature of providing precise surface engineered substrates for cell culture, giving insights into cellular bioassays, such as the study of cell adhesion, proliferation, differentiation, and even molecular signaling pathways.

4.2 Micro and Nanopatterning

Micro- and nanopatterning techniques are strongly based on different lithography methods. However, there are other techniques designed to specifically create micro- and nanotextures in order to modulate cellular behavior. Traditional nanopatterning techniques such as lithography have been used on polymer-based materials [33, 34]. However, its application to materials such as titanium is limited, and this kind of source material is highly relevant in orthopedics and dentistry fields [35]. Taking this into account, Sjöström et al. [36] fabricated pillar-like titania nanostructures and evaluated their interaction with human MSCs. The nanopatterning was obtained using anodization through a porous alumina mask. Titanium scaffolds with pillar structures of 15 nm induced higher levels of MSCs adhesion and spreading, in comparison to scaffolds with polished surfaces. At the same time, there was an increase in the expression of osteopontin and osteocalcin, both osteoblast markers [36]. In a work performed by the same authors [31, 35], anodization of titanium surfaces with block copolymer templates was done, avoiding therefore the presence of residual amounts of aluminum. Titanium microbeads patterned with this process proved to be good 3D surfaces for human MSCs adhesion and spreading. Osteocalcin was also detected in these cell-scaffold conjugates, after 21 days in culture [36]. Another frequent approach is the use of nanopatterning strategies based on nanofibrillar structures [37]. Aligned nanofibrillar collagen scaffolds that mimic collagen bundles in blood vessels were developed, and its effects were tested on endothelial cells (ECs) alignment, function, and in vivo survival [37]. ECs derived from human iPSCs cultured on aligned scaffolds were capable of surviving over 28 days when implanted into ischemic tissues. In contrast, ECs–iPSCs cultured in nonpatterned scaffolds were not detected after 4 days in vivo. Furthermore, aligned scaffolds with 30 nm diameter guided ECs cellular organization as well as modulated their inflammatory response [37]. Finally, a study performed by Iannone et al. [38] showed the importance of nanopatterning in tissue genesis. By cultivating human MSCs on polydimethylsiloxane (PDMS) substrates with simple arrays of parallel channels, it was shown that it was possible to control the initial positioning and growth of focal adhesion points, which ultimately dictated tissue formation, in this case tenogenesis [38].

4.3 Microfluidics

Polymeric nanodevices can also be developed using fluidic-based approaches [22], namely microfluidics technique. A microfluidic device commonly contains a chamber where it is possible to apply standard culture protocols using a minimum amount of resources such as cell culture medium [39]. The basis of these applications is the flow of fluids or particles in the chamber in order to create gradients within these microdevices [22] by the delivery of chemical and mechanical cues into the system. Furthermore, there is a modular design that enables combinations of different substrates, cell cultures, and biomolecules [39]. Cylindrical structures are many times introduced on these chambers in order to perform cell migration and chemoattraction studies. In a study performed by Sefton et al. [40], ECs-coated collagen cylinders were created into a microfluidic chamber, upon which MSCs were randomly embedded within the EC channels. The purpose of this study was to observe if the presence of ECs in the channels would influence the migration and differentiation of the MSCs through the formation of functional vasculature by the ECs in the constructs. A constant flow through the channels increased EC–MSCs interactions, leading to an increase in MSC differentiation smooth muscle cell- or pericyte-like cells, apart from stimulating MSCs to create strong proteoglycan-based ECM, indicative of MSC migration throughout the channels [40]. Axonal biology is also extensively studied using microfluidic chambers. Such studies are performed either by using microchannels, which serve to analyze the ability of neuronal-like cells to migrate and re-arrange their axonal pathways between the two compartments of the chambers connected by the channel [41], or by using chemoattractive components (e.g. biomolecules) in one or both compartments of the microfluidic chambers to influence axonal guidance [42].

These features make microfluidics a powerful tool to control a cell culture environment, offering the possibility to perform a set of cellular studies.

4.4 Electrospinning

Electrospinning techniques are based on a process involving electrostatic forces with the objective of forming synthetic fibers from a given polymer solution or melt. It comprises the use of high-voltage sources to inject charge of a certain polarity into a polymer solution, which is then accelerated toward a collector of opposite polarity. During its path to the collector, the solvent is evaporated, which allows obtaining solid polymer fibers at the nanometer scale [43].

Electrospinning has been known for more than 100 years, but the first major breakthrough on this technique was obtained by the work of Formhals in the 1930s [44–46]. In the late 1960s, Taylor studied in more detail the fiber jet that is formed in the electrospinning process [47], and after that, others started to investigate the properties of the fibers produced and their relation to the polymer source as well as different process parameters [48–50]. Among the different processing parameters that can influence the quality of the obtained fibers, one should consider the applied voltage, the polymer flow rate, and capillary–collector distance [43]. Solution parameters are also important since polymer concentration, solvent volatility, and solution conductivity are crucial points for the final product. For more details about different factors affecting the electrospinning process, please consult Sill *et al.* [43].

During the 1980s, the first works exploring electrospinning's potential to tissue-engineering applications emerged [51, 52], but it was only in the late 1990s, this technique started to be looked as a valid alternative in this field. At the same time, stem cell research gained increased attention, and taking into account the potentialities of these cells, its application to tissue-engineering strategies became obvious. The combination of cells and electrospun materials for tissue-engineering strategies started more than a decade ago [53, 54]. One of the first works involving stem cells and electrospinning processes was published by Yoshimoto *et al.* [55]. Nanofiber meshes were produced from a poly(ε-caprolactone) (PCL) solution and combined with MSCs derived from the rat bone marrow. MSCs were seeded on PCL scaffolds (with an average fiber diameter of 400 nm) and cultured with osteogenic supplements. The main results showed that MSCs were capable of penetrating the polymer construct and produce a significant amount of ECM. After 4 weeks of culture, cells were still viable and mineralization and type I collagen were detected. By analyzing the results obtained, the high surface area and porosity of the fibers produced were described as key events for the results obtained [55]. In a recent study, similar PCL nanofibrous scaffolds were developed by an innovative approach, by thermally inducing nanofiber self-agglomeration followed by freeze drying [56]. This resulted in a scaffold with hierarchically structured and interconnected pores with physical properties very similar to the natural ECM. The scaffold allowed high cellular viability of mouse bone marrow MSCs and promoted a potent BMP2-induced chondrogenic differentiation. *In vivo*, these electrospun scaffolds revealed to be highly favorable for functional bone regeneration, by acting in the endochondral ossification process [56]. Another recent work on bone regeneration combined the highly promising iPSCs with a composite of hydroxyapatite/

collagen/chitosan (HAp/Col/CTS) nanofibers (190–230 nm diameter) [57]. Murine iPSCs were first differentiated into iPSCs–MSCs, and then its interaction with Hap/Col/CTS scaffolds was characterized. The newly developed composite improved cellular attachment and proliferation of iPSCs–MSCs, while increasing the expression of several osteogenic genes such as *Ocn*, *Alp*, or *Runx2*. In an *in vivo* model of mouse cranial defects, the cell–scaffold construct under study was capable of effectively promoting bone regeneration and increasing bone mineral density [57].

In a similar application, MSCs from rabbit bone marrow were cultured in nanofiber scaffolds for cartilage regeneration [58]. The skeleton of the scaffold was composed by oriented poly(L-lactic acid)-co-poly(ε-caprolactone) [P(LLA-CL)] nanofibers that were blended with collagen type I by a dynamic, liquid electrospinning process. Then, after mixing with a solution of collagen type I and hyaluronic acid (HA), the scaffold was subjected to a freeze-drying process. This complex structure improved MSCs adhesion, proliferation, and orientation. Moreover, the scaffold induced chondrogenic differentiation, as demonstrated by a large increase in the glycosaminoglycan content and expression of collagen type II [58].

In another application context, human bone marrow MSCs were seeded on scaffolds made of PCL electrospun nanofibers, vitamin B12, aloe vera, and silk fibroin, further embedded with gold nanoparticles [59]. The objective was to differentiate MSCs into cardiac lineage for the regeneration of the infarcted myocardium. This complex scaffold construct was found to present mechanical strength matching that of the native myocardium, while enhancing MSCs proliferation and differentiation into cardiogenesis. Differentiated MSCs were capable of producing contractile proteins, achieving the typical cardiac phenotype. The chemical composition, as well as the stiffness of the scaffold, was found to be crucial for the effects observed [59].

MSCs from the rat bone marrow have also been tested with poly(propylene carbonate) (PPC) fibers for vascular tissue engineering [60]. The fibers with an average diameter of 5 μm were collected in a way such they formed a tubular structure with a diameter of 2 mm. Then, MSCs were further modified to express the vasculoprotective gene endothelial nitric oxide synthase (eNOS). In summary, MSCs integrated well within the microfibers of the scaffold and in addition, modified MSCs presented high levels of nitric oxide (NO) production, even comparable to NO levels present in native blood vessels [60].

Other particular stem cells that have been conjugated with electrospun materials are NSCs [61]. Fibers made from poly(L-lactic acid) (PLLA) were used as scaffolds for testing NSCs morphology, differentiation, and neurite outgrowth [61]. In this study, the influence of fiber alignment and diameter was evaluated. In summary, it was shown that the direction of NSCs elongation and neurite outgrowth were parallel to the direction of the PLLA fibers, but only when they were aligned. Fiber diameter had no influence in cellular orientation. Curiously, NSC differentiation was higher in nanofibers in comparison to microfibers, while fiber alignment did not affect this event [61]. These results support the use of aligned nanofibers for neural tissue engineering. Still on the topic of neural tissue applications, mouse ESCs were incorporated into electrospun PCL nanofiber scaffolds in order to evaluate neural differentiation [62]. Embryoid bodies were cultured on random or aligned fiber scaffolds, and their effects on neural differentiation were compared. Both scaffolds presented neural markers expression at the protein level. However, aligned fibers induced longer neurite extensions, enhancing neuron outgrowth along the fibers. Additionally, aligned fiber scaffolds induced higher expression of mature specific neural markers such as β-III tubulin and MAP-2. This highlights the importance of fiber alignment in achieving neural maturity during the differentiation process [62].

Besides the aforementioned examples, there are several other works based on electrospinning techniques in tissue engineering of skin [63], musculoskeletal [64], or dental tissues [65]. However, the most common trait among all of these studies is the similarity of the electrospun scaffolds to the natural fibrils found in the ECM. The nano- or microscale of these fibers, together with specific parameters such as fiber alignment or porosity allows a close interaction between cells and scaffold. Furthermore, its reproducibility and tailoring capacity make electrospinning one of the most important techniques for creating a suitable environment for stem cell growth and exploration of their different potentialities.

4.5 Bottom-up/Top-down Approaches

One important factor to consider upon the fabrication of a substrate or matrix for stem cell seeding approaches is the need of mimicking the cell's natural extracellular environment. In nature, tissues are three-dimensional matrices, in which cells search for a specific spatial organization to migrate, proliferate, and differentiate according to their needs [17].

One approach currently used to create 3D-patterned bioactive matrices is the integration of top-down and bottom-up techniques. The top-down approaches focus merely on designing a scaffold in which cells can be subsequently seeded. However, 3D structures with a controlled and functional architecture are not obtained with this technique [66]. On the other hand, bottom-up approaches address this limitation, under the basis of modular tissue engineering that assembles individual modular microscale units to create a macroscale structure [67], besides supporting the construct vascularization and functional organization [66]. By doing so, the cells seeded on the fabricated scaffolds can profit with the synergistic potential of both strategies, that will guide their behavior. In this methodology, self-assembly gels are normally used to incorporate stem cells. One study comprising the development of a tissue-engineered periosteum through a bottom-up approach, reported the use of hydrolytically degradable hydrogels for MSCs transplantation to periosteum allograft surfaces [68]. In comparison to allografts with no MSCs, it was possible to observe that cells were responsible for graft vascularization, endochondral bone formation, and increased biomechanical strength [68]. In another work based on self-assembly strategies, neural progenitor cells (NPCs) were encapsulated within a 3D network of nanofibers formed by peptide amphiphile molecules [69]. In this case, the nanofibers were designed to present laminin–IKVAV epitopes to cells, which are associated with neural differentiation. The main results showed that the self-assembly scaffolds induced a rapid and selective differentiation of NPCs into neurons, which was favored over astrocyte differentiation [69].

As we get more knowledge about stem cells behavior and their interactions with the environment, bottom-up and top-down approaches will emerge as the strategies to follow since these systems can better mimic the complexity existing in living tissues and organisms.

4.6 Substrates Chemical Modifications

As previously mentioned, the fact that the cellular behavior is highly dependent on the surrounding microenvironment is of common knowledge [18]. Thus, the presence of biological cues in patterned surfaces/substrates with poor or nonadhesive characteristics are crucial for cell responsiveness and interaction with the material in which they are seeded, as well as with the surrounding cells. In this sense, several attempts have been made, so patterning approaches can profit with biochemical functionalization [70]. This is normally accomplished by chemical modification of the substrates or scaffolds [18, 70]. Some investigated strategies consist on using the coating of calcium phosphates [1, 2] or other compounds [71, 72] on the substrate's surfaces or the grafting of adhesive motifs within it [73, 74].

These techniques are used to improve the performance and bioactivity of the developed materials, while influencing cell adhesion, proliferation, and differentiation.

4.6.1 Biomolecules Coatings

The use of calcium phosphates (Ca-P) for the coating of biomaterials surface has been widely used in bone tissue-engineering applications. Salgado *et al.* [2] characterized the effects of Ca-P treatment on corn starch/ethylene-vinyl alcohol (SEVA-C)-based scaffolds. Human osteoblast-like cells (SaOS-2) seeded on the pre-mineralized starch-based scaffolds were shown to remain viable in culture and to rapidly contribute to the mineralization of the coated scaffolds by the production of osteopontin and collagen type I [2]. Later on, the performance of these SEVA-C-based scaffolds containing a layer of Ca-P was compared with corn starch and ethylene–vinyl alcohol (SEVA-C), and corn starch, and cellulose acetate (SCA)-based ones *in vivo* [1]. After scaffolds implantation in rat's distal femurs, bone formation was observed around all the tested scaffolds, accompanied by a rapidly connective tissue formation [1].

The use of Ca-P was also used in the context of spinal cord injury regeneration [75]. For that, a hybrid system consisting of starch/poly-ε-caprolactone (SPCL) semi-rigid tubular porous structure filled by a gellan gum (GG) hydrogel concentric core was developed [73, 76]. First, the hybrid scaffolds were found to be noncytotoxic, as supported the *in vitro* culture of oligodendrocyte-like cells. Moreover, preliminary *in vivo* studies on hemisection rat SCI model showed that scaffolds integrated well in the injured spinal cord [76]. In this construct, the GG was expected to promote axonal regeneration in the interior, while the purpose of its external surface was to promote osteogenic activity. For that, SPCL surface was pre-mineralized and the formation of a crystalline carbonated apatite layer was observed, at which bone MSCs adhered and facilitated bone ingrowth [75].

Other compounds such as fibronectin or titanium are also reported as good candidates for the coating of substrates. Regarding fibronectin-coated substrates developed for bone tissue engineering, bone MSCs have been reported to respond specifically to intermittently exposure to shear stress when cultured in a radial-flow chamber. In such dynamic conditions, osteocalcin production was stimulated by the flow, suggesting osteoblastic maturation, though bone marrow-derived MSCs proliferation was not observed [71]. In a very different study, titanium-coated hemisphere-like topographic nanostructures were developed with different sizes by using colloidal lithography in combination with coating technologies. The author's hypothesis was based on the use of different topographic features at the nanoscale level to modulate human MSCs. What they observed was that both proliferation and osteogenic differentiation of hMSCs varied in a size-dependent fashion [72].

4.6.2 Peptide Grafting

The attachment of adhesive motifs to the surface of substrates or to the backbone of scaffolds is also a common technique in the design of biomaterials. The procedure mostly used is the attachment of an Arginylglycylaspartic acid (RGD) tripeptide to biomaterials, in order to promote cell adhesion.

A good example of the use of RGD peptides is the work recently performed by Silva et al. [73, 74]. In these studies, the author used GG as a template for the repair and regeneration of spinal cord injuries. As an inert hydrogel [77], GG by itself did not promote cell adhesion [73]. Therefore, Diels-Alder click chemistry was used to graft fibronectin-derived peptides (GRGDS) onto GG structure in order to enhance its bioactivity. The influence of GG modification with GRGDS motifs (GG-GRGDS) was tested in terms of cell survival and proliferation. The interactions of cell/GG-GRGDS were clearly observed when neural stem/progenitor cells [73] and, further on, bone marrow MSCs (BM-MSCs) [74] presented increased cell survival, metabolic activity, and proliferation within the modified hydrogels. Moreover, BM-MSCs secretome was also shown to be significantly affected by hydrogel modification [74]. GG-GRGDS hydrogels containing adipose tissue-derived stem cells (ASCs) were also used for peripheral nerve regeneration by Assunção-Silva et al. [78]. The aim of this work was to assess the potential of ASCs in promoting axonal growth. To potentiate this, glial cell-line-derived factor (GDNF) covalently linked to iron oxide nanoparticles [79] was added to the cell culture environment to evaluate whether ASCs and GDNF would provide higher levels of axonal outgrowth. No cumulative effect was observed by using these two elements when comparing to the use of the growth factor alone. Nevertheless, the authors pointed out to the well-described non-inflammatory [14], non-apoptotic, and angiogenic [80] character of ASCs, not excluding its use on this type of application, in which the cells will exert a more protective role to the injured tissues, creating a favorable environment for regeneration [78].

The study of the effect of linking peptidic sequences on biomaterials is not only directed to their performance as cell-seeding matrices, but also on the conformational changes that may occur at the hydrogel structural level. In fact, a recent study has shown that upon RGD peptide conjugation to polysaccharides such as alginate, chitosan, and HA in aqueous solution, structural differences on the conformational state of their individual chains occur, which influences the fraction of peptide that bounds to the hydrogel scaffolds, thus determining their behavior [81].

RGD is not the only peptide currently used in biomaterial-based tissue-engineering approaches. Bioactive elastin-like recombinamer (ELR) membranes containing not only RGD but also REDV motifs for MSCs and ECs adhesion, respectively, and DDDEEKFLRRIGRFG peptide (HAP) for mineralization and HAP–RGDS for both cell adhesion and mineralization were synthesized [82]. However, in addition to biomolecular signals, physical cues were also used in this study, by incorporating topographical micropatterns such as channels, holes, and posts in the biomembranes. Subsequently, the effect of these signals was evaluated in primary rat MSCs behavior, which have shown to adapt their morphology according to surface topography, while their proliferation and differentiation were dependent on the biomolecules present in the ERL membranes. As these membranes were developed for bone tissue-engineering purposes, MSCs osteogenic differentiation was screened in the different conditions, which was significantly enhanced on HAP membranes [82].

4.7 Conclusion

Stem cell behavior highly depends not only on chemical signals but also on physical cues. The understanding of this paradigm is a crucial point for the development of strategies concerning stem cell-based approaches. The mechanical properties of the scaffold, its size (micro/nanoscale), or even the alignment of their structure are important aspects to take in consideration when combining stem cells with a material for tissue-engineering applications. Depending on the surface characteristics, stem cells might change their adherence, spreading, or differentiation patterns. The different techniques described in this chapter allow us to recognize essential mechanisms underlying cellular functions and how to use the different above-referred tissue-like constructs for pharmacological and therapeutic applications. The technology behind the basis of these techniques enables researchers to design surfaces with specific textures at the micro- and nanoscale levels. This unprecedented capacity approximates the scaffolds produced to the micro and nanostructures of the natural ECM. Further on, the next steps will be to understand which physical and chemical cues are more suitable for each type of application intended for stem cell-based therapies. The pursuit of the ideal environment to extract all the potential from stem cells will be the future of tissue-engineering field along with stem cell research.

Acknowledgements

Financial support from Prémios Santa Casa Neurociências - Prize Melo e Castro for Spinal Cord Injury Research; Portuguese Foundation for Science and Technology [Doctoral fellowship (SFRH/BD/103075/2014) to E. D. Gomes; IF Development Grant to A. J. Salgado; Post-Doctoral fellowship (SFRH/BPD/97701/2013) to N. A. Silva]; This article has been developed under the scope of the projects NORTE-01-0145-FEDER-000013, supported by the Northern Portugal Regional Operational Programme (NORTE 2020), under the Portugal 2020 Partnership Agreement, through the European Regional Development Fund (FEDER); This work has been funded by FEDER funds, through the Competitiveness Factors Operational Programme (COMPETE), and by National funds, through the Foundation for Science and Technology (FCT), under the scope of the project POCI-01-0145-FEDER-007038.

References

1. Salgado, A.J., *et al.*, In vivo response to starch-based scaffolds designed for bone tissue engineering applications. *J. Biomed. Mater. Res. A*, 80a(4), pp. 983–9, 2007.
2. Salgado, A.J., *et al.*, Biological response to pre-mineralized starch based scaffolds for bone tissue engineering. *J. Mater. Sci. Mater. Med.*, 16(3), pp. 267–75, 2005.
3. Friedenstein, A.J., R.K. Chailakhjan, and K.S. Lalykina, The development of fibroblast colonies in monolayer cultures of guinea-pig bone marrow and spleen cells. *Cell Tissue Kinet.*, 3(4), pp. 393–403, 1970.
4. Ramon-Cueto, A. and M. Nieto-Sampedro, Regeneration into the spinal cord of transected dorsal root axons is promoted by ensheathing glia transplants. *Exp. Neurol.*, 127(2), pp. 232–44, 1994.
5. Liu, S., *et al.*, Embryonic stem cells differentiate into oligodendrocytes and myelinate in culture and after spinal cord transplantation. *PNAS*, 97(11), pp. 6126–31, 2000.
6. Iwanami, A., *et al.*, Transplantation of human neural stem cells for spinal cord injury in primates. *J. Neurosc. Res.*, 80(2), pp. 182–90, 2005.

7. Zhao, T., et al., Immunogenicity of induced pluripotent stem cells. *Nature*, 474(7350), pp. 212–5, 2011.
8. Hoffman, J.A. and B.J. Merrill, New and renewed perspectives on embryonic stem cell pluripotency. *Front. Biosci.*, 12, pp. 3321–32, 2007.
9. Weiss, S., et al., Multipotent CNS stem cells are present in the adult mammalian spinal cord and ventricular neuroaxis. *J. Neurosci.*, 16(23), pp. 7599–609, 1996.
10. Kobayashi, Y., et al., Pre-evaluated safe human iPSC-derived neural stem cells promote functional recovery after spinal cord injury in common marmoset without tumorigenicity. *PLoS One*, 7(12), p. e52787, 2012.
11. Lu, P., et al., Neural stem cells constitutively secrete neurotrophic factors and promote extensive host axonal growth after spinal cord injury. *Exp. Neurol.*, 181(2), pp. 115–29, 2003.
12. Salgado, A.J., et al., Adipose tissue derived stem cells secretome: soluble factors and their roles in regenerative medicine. *Curr. Stem. Cell. Res. Ther.*, 5(2), pp. 103–10, 2010.
13. Meyerrose, T., et al., Mesenchymal stem cells for the sustained *in vivo* delivery of bioactive factors. *Adv. Drug Deliv. Rev.*, 62(12), pp. 1167–74, 2010.
14. Teixeira, F.G., et al., Mesenchymal stem cells secretome: a new paradigm for central nervous system regeneration?. *Cell. Mol. Life Sci.*, 70(20), pp. 3871–82, 2013.
15. Boruch, A.V., et al., Neurotrophic and migratory properties of an olfactory ensheathing cell line. *Glia*, 33(3), pp. 225–9, 2001.
16. Woodhall, E., A.K. West, and M.I. Chuah, Cultured olfactory ensheathing cells express nerve growth factor, brain-derived neurotrophic factor, glia cell line-derived neurotrophic factor and their receptors. *Brain Res. Mol. Brain Res.*, 88(1–2), pp. 203–13, 2001.
17. Assuncao-Silva, R.C., et al., Hydrogels and Cell Based Therapies in Spinal Cord Injury Regeneration. *Stem Cells Int.*, 2015, p. 948040, 2015.
18. Nakanishi, J., et al., Recent advances in cell micropatterning techniques for bioanalytical and biomedical sciences. *Analytical Sciences*, 24(1), pp. 67–72, 2008.
19. Lopez-Bosque, M.J., et al., Fabrication of hierarchical micro-nanotopographies for cell attachment studies. *Nanotechnology*, 24(25), p. 255305, 2013.
20. Dahlmann, J., et al., The use of agarose microwells for scalable embryoid body formation and cardiac differentiation of human and murine pluripotent stem cells. *Biomaterials*, 34(10), pp. 2463–71, 2013.
21. McMurray, R.J., et al., Nanoscale surfaces for the long-term maintenance of mesenchymal stem cell phenotype and multipotency. *Nat. Mater.*, 10(8), pp. 637–44, 2011.
22. Mills, C.A., et al., Nanoembossed polymer substrates for biomedical surface interaction studies. *J. Nanosci. Nanotechnol.*, 7(12), pp. 4588–94, 2007.
23. Kim, E.J., et al., Modulating human connective tissue progenitor cell behavior on cellulose acetate scaffolds by surface microtextures. *J. Biomed. Mater. Res. A.*, 90(4), pp. 1198–205, 2009.
24. Jeon, H., et al., Directing cell migration and organization via nanocrater-patterned cell-repellent interfaces. *Nat. Mater.*, 14(9), pp. 918–23, 2015.

25. Kim, J., et al., Multiscale patterned transplantable stem cell patches for bone tissue regeneration. *Biomaterials*, 35(33), pp. 9058–67, 2014.
26. Yang, K., et al., Biodegradable nanotopography combined with neurotrophic signals enhances contact guidance and neuronal differentiation of human neural stem cells. *Macromol. Biosci.*, 15(10), pp. 1348–56, 2015.
27. Pelaez-Vargas, A., et al., Isotropic micropatterned silica coatings on zirconia induce guided cell growth for dental implants. *Dent. Mater.*, 27(6), pp. 581–9, 2011.
28. Wang, P.Y., et al., Modulation of human mesenchymal stem cell behavior on ordered tantalum nanotopographies fabricated using colloidal lithography and glancing angle deposition. *ACS Appl. Mater. Interfaces*, 7(8), pp. 4979–89, 2015.
29. Roberts, S.P., et al., Magnetically actuated single-walled carbon nanotubes. *Nano. Lett.*, 15(8), pp. 5143–8, 2015.
30. Mata, A., et al., Micropatterning of bioactive self-assembling gels. *Soft Matter.*, 5(6), pp. 1228–36, 2009.
31. Sjostrom, T., et al., 2D and 3D nanopatterning of titanium for enhancing osteoinduction of stem cells at implant surfaces. *Adv. Health Mater.*, 2(9), pp. 1285–93, 2013.
32. Carvalho, A., et al., Micropatterned silica thin films with nanohydroxyapatite micro-aggregates for guided tissue regeneration. *Dent. Mater.*, 28(12), pp. 1250–60, 2012.
33. Dalby, M.J., et al., The control of human mesenchymal cell differentiation using nanoscale symmetry and disorder. *Nat. Mater.*, 6(12), pp. 997–1003, 2007.
34. Lim, J.Y., et al., The regulation of integrin-mediated osteoblast focal adhesion and focal adhesion kinase expression by nanoscale topography. *Biomaterials*, 28(10), pp. 1787–97, 2007.
35. Sjostrom, T., et al., Fabrication of pillar-like titania nanostructures on titanium and their interactions with human skeletal stem cells. *Acta Biomater.*, 5(5), pp. 1433–41, 2009.
36. Sjostrom, T., et al., Novel anodization technique using a block copolymer template for nanopatterning of titanium implant surfaces. *ACS Appl. Mater. Interfaces*, 4(11), pp. 6354–61, 2012.
37. Huang, N.F., et al., The modulation of endothelial cell morphology, function, and survival using anisotropic nanofibrillar collagen scaffolds. *Biomaterials*, 34(16), pp. 4038–47, 2013.
38. Iannone, M., et al., Nanoengineered surfaces for focal adhesion guidance trigger mesenchymal stem cell self-organization and tenogenesis. *Nano. Lett.*, 15(3), pp. 1517–25, 2015.
39. Reichen, M., et al., Microfabricated modular scale-down device for regenerative medicine process development. *PLoS One*, 7(12), p. e52246, 2012.
40. Khan, O.F., M.D. Chamberlain, and M.V. Sefton, Toward an *in vitro* vasculature: differentiation of mesenchymal stromal cells within an endothelial cell-seeded modular construct in a microfluidic flow chamber. *Tissue Eng. Part A*, 18(7–8), pp. 744–56, 2012.

41. Shin, H.S., et al., Compartmental culture of embryonic stem cell-derived neurons in microfluidic devices for use in axonal biology. *Biotechnol. Lett.*, 32(8), pp. 1063–70, 2010.
42. Shi, P., et al., Combined microfluidics/protein patterning platform for pharmacological interrogation of axon pathfinding. *Lab on a Chip*, 10(8), pp. 1005–10, 2010.
43. Sill, T.J. and H.A. von Recum, Electrospinning: applications in drug delivery and tissue engineering. *Biomaterials*, 29(13), pp. 1989–2006, 2008.
44. Formhals, A., inventor. Process and apparatus for preparing artificial threads. US Patent No. 1,975,504; 1934.
45. Formhals, A., inventor. Method and apparatus for spinning. US Patent No. 2,169,962; 1939.
46. Formhals, A., inventor. Artificial thread and method of producing same. US patent No. 2,187,306; 1940.
47. Taylor, G., Electrically driven jets. *Proc. R. Soc. London Ser. A*, 313(1515), pp. 453–75, 1969.
48. Baumgart.Pk, Electrostatic spinning of acrylic microfibers. *J. Colloid Interface Sci.*, 36(1), pp. 71–79, 1971.
49. Larrondo, L. and R.S.J. Manley, Electrostatic fiber spinning from polymer melts. 1. Experimental-observations on fiber formation and properties. *J. Polym. Sci. Polym. Phy. Ed.*, 19(6), pp. 909–920, 1981.
50. Larrondo, L. and R.S.J. Manley, Electrostatic fiber spinning from polymer melts. 2. Examination of the flow field in an electrically driven jet. *J. Polym. Sci. Polym. Phy. Ed.*, 19(6), pp. 921–932, 1981.
51. Annis, D., et al., Elastomeric vascular prosthesis. *Trans. Am. Soc. Artif. Intern. Organs*, 24, pp. 209–214, 1978.
52. Fisher, A.C., et al., Long term *in-vivo* performance of an electrostatically-spun small bore arterial prosthesis: the contribution of mechanical compliance and anti-platelet therapy. *Life Support Syst.*, 3 Suppl 1, pp. 462–5, 1985.
53. Matthews, J.A., et al., Electrospinning of collagen nanofibers. *Biomacromolecules*, 3(2), pp. 232–8, 2002.
54. Li, W.J., et al., Electrospun nanofibrous structure: a novel scaffold for tissue engineering. *J. Biomed. Mater. Res.*, 60(4), pp. 613–21, 2002.
55. Yoshimoto, H., et al., A biodegradable nanofiber scaffold by electrospinning and its potential for bone tissue engineering. *Biomaterials*, 24(12), pp. 2077–82, 2003.
56. Xu, T., et al., Electrospun polycaprolactone 3D nanofibrous scaffold with interconnected and hierarchically structured pores for bone tissue engineering. *Adv. Health Mater.*, 4(15), pp. 2238–46, 2015.
57. Xie, J., et al., Osteogenic differentiation and bone regeneration of the iPSC-MSCs supported by a biomimetic nanofibrous scaffold. *Acta Biomater.*, 29, p. 365–79, 2016.
58. Zheng, X., et al., Enhancement of chondrogenic differentiation of rabbit mesenchymal stem cells by oriented nanofiber yarn-collagen type I/hyaluronate hybrid. *Mater. Sci. Eng. C. Mater. Biol. Appl.*, 58, pp. 1071–6, 2016.

59. Sridhar, S., et al., Cardiogenic differentiation of mesenchymal stem cells with gold nanoparticle loaded functionalized nanofibers. *Colloids Surf. B Biointerfaces*, 134, pp. 346–54, 2015.
60. Zhang, J., et al., Engineering of vascular grafts with genetically modified bone marrow mesenchymal stem cells on poly (propylene carbonate) graft. *Artif. Organs*, 30(12), pp. 898–905, 2006.
61. Yang, F., et al., Electrospinning of nano/micro scale poly(L-lactic acid) aligned fibers and their potential in neural tissue engineering. *Biomaterials*, 26(15), pp. 2603–10, 2005.
62. Abbasi, N., et al., Influence of oriented nanofibrous PCL scaffolds on quantitative gene expression during neural differentiation of mouse embryonic stem cells. *J. Biomed. Mater. Res. A*, 2015.
63. Steffens, D., et al., Development of a biomaterial associated with mesenchymal stem cells and keratinocytes for use as a skin substitute. *Regen. Med.*, 2015.
64. Orr, S.B., et al., Aligned multilayered electrospun scaffolds for rotator cuff tendon tissue engineering. *Acta Biomater.*, 24, pp. 117–26, 2015.
65. Munchow, E.A., et al., Development and characterization of novel ZnO-loaded electrospun membranes for periodontal regeneration. *Dent. Mater.*, 31(9), pp. 1038–51, 2015.
66. Guillotin, B. and F. Guillemot, Cell patterning technologies for organotypic tissue fabrication. *Trends Biotechnol.*, 29(4), pp. 183–90, 2011.
67. Nichol, J.W. and A. Khademhosseini, Modular tissue engineering: engineering biological tissues from the bottom up. *Soft Matter.*, 5(7), pp. 1312–9, 2009.
68. Hoffman, M.D., et al., The effect of mesenchymal stem cells delivered via hydrogel-based tissue engineered periosteum on bone allograft healing. *Biomaterials*, 34(35), pp. 8887–98, 2013.
69. Silva, G.A., et al., Selective differentiation of neural progenitor cells by high-epitope density nanofibers. *Science*, 303(5662), pp. 1352–5, 2004.
70. Falconnet, D., et al., Surface engineering approaches to micropattern surfaces for cell-based assays. *Biomaterials*, 27(16), pp. 3044–63, 2006.
71. Kreke, M.R. and A.S. Goldstein, Hydrodynamic shear stimulates osteocalcin expression but not proliferation of bone marrow stromal cells. *Tissue Eng.*, 10(5–6), pp. 780–8, 2004.
72. de Peppo, G.M., et al., Osteogenic response of human mesenchymal stem cells to well-defined nanoscale topography *in vitro*. *Int. J. Nanomed.*, 9, pp. 2499–515, 2014.
73. Silva, N.A., et al., The effects of peptide modified gellan gum and olfactory ensheathing glia cells on neural stem/progenitor cell fate. *Biomaterials*, 33(27), pp. 6345–54, 2012.
74. Silva, N.A., et al., Modulation of bone marrow mesenchymal stem cell secretome by ECM-like hydrogels. *Biochimie*, 95(12), pp. 2314–9, 2013.
75. Oliveira, A.L., et al., Peripheral mineralization of a 3D biodegradable tubular construct as a way to enhance guidance stabilization in spinal cord injury regeneration. *J. Mater. Sci. Mater. Med.*, 23(11), pp. 2821–30, 2012.

76. Silva, N.A., et al., Development and characterization of a novel hybrid tissue engineering-based scaffold for spinal cord injury repair. Tissue Eng. Part A, 16(1), pp. 45–54, 2010.
77. Jansson, P.E., B. Lindberg, and P.A. Sandford, Structural studies of gellan gum, an extracellular polysaccharide elaborated by Pseudomonas elodea. Carbohydr. Res., 124(1), pp. 135–9, 1983.
78. Assuncao-Silva, R.C., et al., Induction of neurite outgrowth in 3D hydrogel-based environments. Biomed. Mater., 10(5), p. 051001, 2015.
79. Ziv-Polat, O., et al., The role of neurotrophic factors conjugated to iron oxide nanoparticles in peripheral nerve regeneration: in vitro studies. Biomed. Res. Int., 2014, p. 267808, 2014.
80. Rehman, J., et al., Secretion of angiogenic and antiapoptotic factors by human adipose stromal cells. Circulation, 109(10), pp. 1292–8, 2004.
81. Bernstein-Levi, O., G. Ochbaum, and R. Bitton, The effect of covalently linked RGD peptide on the conformation of polysaccharides in aqueous solutions. Colloids Surf. B Biointerfaces., 2015.
82. Tejeda-Montes, E., et al., Bioactive membranes for bone regeneration applications: effect of physical and biomolecular signals on mesenchymal stem cell behavior. Acta Biomater., 10(1), pp. 134–41, 2014.

5
Influence of Controlled Micro- and Nanoengineered Environments on Stem Cell Fate

Anna Lagunas[1,2]*, David Caballero[2,1] and Josep Samitier[2,1,3]

[1]Networking Biomedical Research Center (CIBER), Madrid, Spain
[2]Nanobioengineering Group, Institute for Bioengineering of Catalonia (IBEC), Barcelona, Spain
[3]Department of Engineering: Electronics, University of Barcelona (UB), Barcelona, Spain

Abstract

The development of micro- and nanotechnologies has allowed the fabrication of novel engineered environments for the study of cell–surface interaction at the micro- and nanoscales. Exerting control through surface design at this scale range, which comprises the size of the cells, sub-cellular structures, and extracellular matrix components, has a great influence in multiple aspects of cell behavior, such as proliferation, migration, and differentiation. Studies on cell adhesion behavior have shown the enormous sensitivity of cells to environmental features such as the nature of surface adhesive molecules, their precise spatial distribution at the micro- and nanoscales, and the physical properties of the surface (stiffness, topography, and dimensionality). Environmental features have also been shown to impact on stem cells differentiation. *Smart* surfaces with controlled stiffness, and/or micro- and nanoscale topographies, and chemistries mimicking the architecture of native cellular niche have been developed to direct stem cells fate into specific lineages. This regulation of stem cell fate offers unprecedented possibilities in regenerative medicine and stem cell-based therapies to restore the normal function of organs replacing malfunctioning cells and tissues by healthy surface-induced specialized cells. In this chapter, we discuss about the main strategies typically used to engineer surfaces at the micro- and nanoscale levels to control stem cell proliferation and differentiation. We pay special attention to the next generation of developed

---

*Corresponding author: alagunas@ibecbarcelona.eu

Ashutosh Tiwari, Bora Garipcan and Lokman Uzun (eds.) *Advanced Surfaces for Stem Cell Research*, (87–142) © 2017 Scrivener Publishing LLC

surfaces expected to become a solid alternative to conventional biochemical methods and discuss about the challenges to be addressed in the near future. Finally, we hope that this chapter will provide the readers with interesting new insights and help them to reveal key mechanisms involved in stem cell differentiation.

Keywords: Stem cells, differentiation, mechanoregulation, roughness, micro/nanotopography, micro/nanopatterning, gradients, 3D culture

5.1 Introduction to Engineered Environments for the Control of Stem Cell Differentiation

5.1.1 Stem Cells Niche *In Vivo*: A Highly Dynamic and Complex Environment

Stem cells are essential during development and in adulthood in most multicellular organisms as they are responsible for the generation of tissue-specific cell types. Homeostatic and regenerative capacities of mature tissues under normal physiological conditions can be attributed partially to the presence of resident stem cell populations within the tissue [1–6]. Adult stem cells are found in specific locations of a tissue that provide a unique microenvironment, or "niches", which are ultimately responsible for the maintenance of stem cell populations, their controlled proliferation, and the differentiation of their progeny into multiple cellular lineages.

The stem cell and the niche hypothesis was first developed in the hematopoietic system in mammals by Shofield in 1978 [7, 8], with the first example of stem cell niche described for *Caenorhabditis elegans* [9]. In this small nematode, distal tip cell (DTC) membrane presenting Notch ligands to adjacent germline stem cells (GSCs) defines the niche. The DTC and Notch signaling establish and maintain the pattern of maturation of GSCs [9, 10]. While the study of stem cells and their respective niches for the invertebrates could be addressed at single cell resolution, the size of mammalian tissues and the rare occurrence of stem cells within them, limited the *in vivo* identification of stem cells in mammals. The skin, with a high demand for tissue regeneration, is the most abundant source of stem cells in mammals. The skin contains hair follicle stem cells and melanocyte stem cells that reside in the outermost layer of the bulge, just below the sebaceous glands of the follicles [11, 12]. A pool of stem cells is known to reside in the crypt base of the epithelium of mammalian small intestine [13, 14], and the tissue-specific satellite cell niche is found as a satellite stem cell niche close to the muscle fiber, under the basal lamina that separates individual muscle fibers from the interstitial space [15, 16]. Although the central nervous system does not regenerate in any significant extent in mammals, a population of progenitor cells was found in the regions near the lateral ventricles of the forebrain (the subventricular zone) and a part of the dentate gyrus of the hypocampus (the subgranular zone) that sustain neurogenesis and neuronal regeneration [17, 18]. The study of these stem cell populations and the environmental cues characteristic of their respective niches that allow retaining stem cell identity, guide cell fate, and maintain stem cell population, revealed some of the general aspects of the stem cell niche.

Stem cell niche can be composed by populations of support cells denoted as cellular niches. Cellular niches can be either postmitotic or retain proliferative capacity. Cells of the stem cell niche (support cells) interact with the stem cells to maintain them undifferentiated or promote their differentiation. Therefore, an efficient spatiotemporal dialog takes place between stem cells and support cells in order to fulfill lifelong demands for differentiated cells in the tissue. The niche can also be composed only by a basement membrane. These acellular niches are characterized by the absence of a dedicated set of support cells and by the presence of a specialized extracellular matrix (ECM) that is in charge of the maintenance and proliferation of stem cells. Finally, more complex niches composed of both support cells and an interacting ECM can also be found [5]. The diverse and dynamic composition of the ECM, which is under continuous remodeling, provides controlled biochemical and physical (*e.g.* structural, mechanical) properties to the different niches. Since primary function of the niche is to anchor stem cells, adhesion proteins such as integrins play an important role in the microenvironment-stem cell interaction. Studies in *Drosophila* ovary showed that the maintenance and proliferation of somatic follicle stem cells are indeed controlled through integrin-mediated interaction with the basement membrane [19]. Integrins can directly activate downstream signaling pathways regulating self-renewal and proliferation [20, 21]. In particular, $\beta 1$ integrins have been shown to be actively involved in preserving the pool of different types of stem cells [22, 23]. Proteins and proteoglycans of the ECM also act as distributors of growth factors, regulating their local availability, establishing gradients, or maintaining them inactive until they become activated upon enzymatic ECM remodeling [24–26].

The ECM organization and composition are not only sensed as a mere biochemical effect on stem cells but also as a physical input on the stem cell behavior. Cells in the niche sense and respond to the morphology and mechanical properties of their environment, which arise from the compression exerted by neighboring cells and the stiffness of the ECM [27].

The study of cellular niche characteristics and their application on *in vitro* engineered niches will not only improve the knowledge of stem cell biology, but also contribute to the development of reliable cell-based therapies.

5.1.2 Mimicking the Stem Cells Niche *In Vitro*: Engineered Biomaterials

Stem cells with a vast proliferative capacity and differentiation potential are in the focus of research to be used in clinical applications for regenerative medicine. With this aim, significant challenges in identifying stem cell populations, controlling their fate, and characterizing their phenotype need to be addressed. The interaction of stem cells with their complex surrounding environment can be mimicked *in vitro* using physiologically relevant environments. Biocompatible materials – biomaterials – can be engineered with specific biochemical and physical characteristics to provide stem cells with the stimuli needed to guide their fate or self-renewal.

The mechanical properties of biomaterials have been shown to directly influence stem cell fate acting alone or in combination with other relevant surface parameters, such as surface chemistry, to produce specific cell responses [28–32] (see Sections 5.2 and 5.3). Engineered biomaterials commonly include the modification of their surfaces with proteins from the ECM to increase their biocompatibility. To support the self-renewal and differentiation of stem cells, feeder cell layers and co-cultures have been used as a source of ECM proteins and growth factors [33–35]. Some cell-free culture systems, such as Matrigel™ coating, which consist of a complex mixture of various ECM proteins, proteoglycans, and growth factors, demonstrated to support the long-term self-renewal of stem cell populations *in vitro* [36]. Nevertheless, concerns over xenobiotic contamination have prompted the development of growth substrates containing purified proteins or synthetic peptide mimics. Multiple studies have been performed using coatings of typical ECM proteins such as collagen, fibronectin, or laminin showing influence on stem cell differentiation decisions [37]. Protein-based biomaterials and synthetic peptides containing the cell adhesive sequence arginylglycylaspartic acid (RGD) were used as surface coatings to promote stem cells differentiation [38, 39]. Since the composition of the stem cell niche varies with tissue type, protein coatings should be adapted to the different requirements of the stem cells to be cultured, incorporating relevant signaling molecules such as growth factors [40]. Different technologies have been developed to incorporate drug delivery function into biomaterials allowing growth factors to be released in a controlled manner [41, 42]. Also, growth factor immobilization on the biomaterials surface has been reported to trigger specific differentiation pathways through their interaction with cell membrane receptors [43, 44]. More recently, decellularized extracellular matrices containing all the biochemical and structural cues of the original cell microenvironments have been used to guide stem cell fate *in vitro* [45].

Experimental evidence of ECM organization at the micro- and nanoscales – the size of cells and sub-cellular structures – has steered the production of synthetic micro- and nanoengineered surfaces directed toward identifying the geometric cues that control stem cell fate [46]. The development of micro- and nanotechnologies has allowed the fabrication of engineered environments which have proved a valuable tool in guiding their expansion and lineage-specific differentiation. In the following sections, we will introduce the latest advances in the development of micro- and nanoengineered environments to influence stem cell fate, and we will discuss on how such a regulation can offer new possibilities in regenerative medicine and stem cell-based therapies.

5.2 Mechanoregulation of Stem Cell Fate

5.2.1 From *In Vivo* to *In Vitro*: Influence of the Mechanical Environment on Stem Cell Fate

In vivo, during morphogenesis and tissue repair, both physical (mechanical and morphological) and biochemical (cytokines, cell–cell, and cell–ECM interactions) cues guide the differentiation and/or self-renewal of stem cells. While the biomolecular factors have been extensively studied [47], less is known about the physical mechanisms that drive stem cell differentiation and/or self-renewal, especially at the micro- and nanometric scales. This is in part due to the high complexity of the cellular microenvironment [48]. Indeed, stem cells, when removed from the *in vivo* niche and cultured *in vitro*, tend to differentiate spontaneously in an uncontrolled way. In order to understand the influence of the physical environment on stem cell fate in a more controllable way, a diverse variety of engineered *in vitro* assays have been proposed. The mechanical properties and architecture of the stem cell microenvironment have been shown to be an important mediator for regulating cell fate, and it can be as important as biomolecular cues (see Section 5.3) [49–51]. This mechanoregulation process which includes the modulation of the mechanical forces that stem cells exert on their environment activates downstream transmembrane integrin signaling pathways that influence the cell fate [51].

Micro- and nanofabrication technologies have been applied to provide a set of engineered 2D surfaces and 3D scaffolds to influence stem cells with mechanostimulatory signals that mimic the architecture and mechanical properties of the ECM of the tissue. In this way, the physical (and biochemical) mechanisms involved in guiding differentiation can be elucidated. In a seminal work, the role of surface rigidity in stem cell lineage specification was unraveled. Mesenchymal stem cells (MSCs) cultured on polyacrylamide substrates with elastic modulus mimicking the stiffness of collagenous bone, muscle, and brain tissues became osteogenic, myogenic, and neurogenic, respectively [30]. This pioneering work confirmed that the stiffness of the cell environment induced differentiation by activating specific signaling pathways. Similarly, other cell types such as neural stem cells (NSCs) undergo either neuronal or glial differentiation when cultured on hydrogel surfaces of different stiffness [52]. On soft surfaces (100–500 Pa) NSCs differentiated mostly in neurons, whereas harder surfaces (1000–10000 Pa) promoted glial cultures. However, recent evidences using human MSCs (hMSCs) showed that cell lineage depended on the previous culture physical environment where cells were cultured, suggesting that stem cells possess mechanical memory that influences stem cells' fate [53]. In this regard, muscle stem cells cultured on standard Petri dishes were also found to lose their regenerative potential [54].

Together, this draws a picture of stem cell fate resulting from a multi-step bio-mechano-molecular process where stem cells differentiate into specific lineages and highlights the need of using appropriate engineered micro- and nanoenvironments to regulate the growth, proliferation, and differentiation of stem cells [55]. This is critically important for clinical applications. In the following sections, we will review the most recent work and techniques applied for unraveling the role of physical cues on stem cell fate. We will mainly focus on the impact of surface roughness, surfaces with micro- and nanostructured features and surfaces presenting gradients of specific physical properties.

5.2.2 Regulation of Stem Cell Fate by Surface Roughness

The ECM of many tissues displays features of pores, fibers, and ridges in the micro- and nanometer range with significant effects on cell behavior *in vivo* [56, 57]. At this level, the surface features are of similar size of the individual molecular receptors (*e.g.* integrins), which target receptor-driven pathways managing the response of cells [58]. *In vitro*, micro- and nanoscale properties of surfaces also have a huge impact on cell phenotype and cellular functions, such as adhesion, polarization, orientation, proliferation, motility, and gene expression [59–67]. In particular, surface roughness has a great influence on cell behavior. It mimics the nanotopography encountered by cells *in vivo* and controls cell adhesion, morphology, and motility [60]. This mechanosensitive phenomenon has also a critical impact on stem cell growth and differentiation. Cell fate is mediated by the alteration of cell adhesion through the interaction of the nanotopographical features of the surface with the integrin receptors, inducing changes in the inner biochemical machinery and morphology of the cell [58].

In vivo studies have suggested that the control of surface roughness is one of the most important parameters governing stem cell differentiation [68]. *In vitro*, the impact of surface roughness on the regulation of stem cell fate has been extensively investigated, most notably during osteogenesis (bone formation) [69–72]. Micro- and nanoscale roughness features can be fabricated by different means, including mechanical polishing, etching, blasting, and anodizing methods. Interaction of stem cells with sub-micron and nanosized roughness materials can also be investigated by means of 'disordered patterns'. This type of surface was demonstrated to influence on cell fate. An example around this topic was provided using hMSCs deposited on nanoscale disordered substrates in the absence of differentiation supplements. Rapid osteogenesis was observed with a demonstrated efficiency equal to chemical stimulation [73]. It was demonstrated that if MSCs were spread, long focal adhesions were observed and osteogenesis was favored. On the contrary, if adhesions were smaller, adipogenesis was observed [74]. Finally, upregulation of specific endoderm marker gene expression and cell adhesion molecules were observed on human embryonic stem cells (hESCs) cultured on a porous polyethylene membrane substrate [75]. The interaction of hESCs with the nanoroughness of the surface resulted in the augmentation of the Wnt signaling pathway, which impacted on the hESCs fate. The sensitivity to the roughness cue can nevertheless vary depending on the cell types [76].

However, there is a lot of controversy about the actual effects of features' size on stem cell behavior with some contradictory results being reported [58]. In order to shed light into this discrepancy, gradients of polycaprolactone (PCL) with roughness features varying from the sub-micron to the micrometer range were used to direct human bone marrow mesenchymal stem cells (hBMSCs) differentiation into osteogenic lineage [69] (see also Section 5.2.4). The variation in roughness modulated cell spreading and shape, and consequently, osteogenic commitment and expression, and most notably, in the total absence of standard soluble differentiation supplements [70]. Osteogenic differentiation was assessed in terms of cytoskeleton organization, alkaline phosphatase activity, collagen type I formation, and mineralization of the matrix [69].

Surface roughness can be also used for long-term maintenance of undifferentiated embryonic stem cells *in vitro* avoiding the uncontrolled differentiation of isolated stem cells [77–79], and the use of feeder cell layer or stemness compounds such as leukemia inhibitor factor (LIF) or LIF/2i media [80]. In this regard, hierarchical micro- and nanoroughness surfaces were shown to promote long-term self-renewal of mouse embryonic stem cells (mESCs) [81]. Stemness was monitored by the levels of expression of pluripotency markers, such as Oct4, Nanog, or alkaline phosphatase. On the contrary, culturing mESCs on smooth or nanoroughness surfaces, induced fast differentiation. Similar works showed that nanoroughness surfaces promoted embryonic stem cell pluripotency to a higher level compared to surfaces with microroughness or flat substrates, in feeder-free conditions [78]. Again, the influence of micro- and nanoscale surface topography of the maintenance of pluripotency state is controversial, and contradictory results are often encountered [82, 83], most probably because the effect is cell dependent or is masked by the surface fictionalization with ECM proteins, which can compete or synergize with the mechanical cues. Solid evidences postulate that cell adhesions size repress key metabolic and biochemical pathways below a critical threshold required for active differentiation [84]. Translated to *in vivo* scenario, the stem cell niche containing micro- and nanometric features possess distinct regions of cell quiescence and self-renewing populations.

5.2.3 Control of Stem Cell Differentiation by Micro- and Nanotopographic Surfaces

The topography of the ECM has a direct impact on cell behavior [85]. Indeed, cells sense surface patterns ranging from 10 nm to 100 μm [86]. As mentioned above, this mechanosensitive interaction initiates a cascade of signaling pathways within the cell leading to inner cytoskeleton rearrangements. Similarly, stem cells fate is also influenced by the physical features of the ECM, as well as cell–cell interactions, which guide cell differentiation or self-renewal. The stem cell niche is composed of a diverse variety of ECM topologies, enabling cells to better adhere to the substrate determining their spatial orientation. This physical niche can be mimicked *in vitro* using micro- and nanostructured biomaterials which recapitulate the topographical landscape of the native tissue.

State-of-the-art micro- and nanotechnologies have also enabled the development of a diverse variety of micro- and nanostructured surfaces to 'stimulate' stem cell differentiation into desired phenotypes [58, 87–93]. Typically, physical structuring of surfaces can be undertaken using serial fabrication methods, such as electrospun nanofibers (see Section 5.4.1), electron beam or ion beam lithography [94–96], or parallel methods such as hot embossing, imprint lithography, conventional UV photolithography, or UV nanoimprinting lithography. All of these techniques offer miniaturization, precision, and reproducibility of the fabrication process [59, 97, 98]. The selection of the fabrication methodology truly relies on the final application and selected material. Typical micro- and nanostructured materials include hydrogels, which better mimic the high water content of the ECM [99]. Hydrogels can be of natural origin (*e.g.* Matrigel™, collagen, fibrin, gelatin, elastin, hyaluronic acid (HA), and derivatives of natural materials such as chitosan or alginate, among others) or synthetic hydrogels (*e.g.* polyacrylamide, poly(vinyl alcohol), poly(ethylene glycol)diacrylate). The formers are bioactive which make them ideal candidates for cell bioengineering studies, but they are uncontrollable in terms of composition and, in general, mechanical properties. On the contrary, the mechanical properties of the latter are controllable and high control on their composition and morphology is easily achievable, but they are inert and coating of the fibers with a natural ECM protein is needed. Other typically used materials are synthetic polymers (*e.g.* polydimethylsiloxane – PDMS, polymethylmethacrylate – PMMA, polystyrene – PS). They can be easily engineered using standard micro- and nanofabrication techniques.

Numerous works have been reported during the past decade on the effect of surface topography in stem cell differentiation. Multiple combinations of materials and physical parameters have been used to assess the role of topographical features in cell fate (see Figure 5.1 and Table 5.1) [86, 87, 90, 100, 101]. A large plethora of topographically patterned arrays with different geometries and sizes have been proposed aiming at controlling the mechanical stimuli on stem cells. This includes posts [100], pits [102], ridges [88], grooves [103], or gratings [104, 105]. As an example, MSCs cultured on polyimide micro- and nanosized grooves/ridges induced osteogenesis and adipogenesis differentiation. When MSCs were deposited on small surface pattern with 5 µm grooves (2 µm ridges), cells showed an elongated phenotype and differentiated mainly toward an osteogenic lineage. On the contrary, when MSCs were deposited on larger motifs (15 µm ridges), cells elongated less and adipogenesis was favored. This might be a consequence of the elongated morphology of cells. In addition to the cell spreading and orientation-induced changes, these structures also affected the morphology and size of focal adhesions, impacting on integrin-mediated signaling pathways that drive the MSCs fate toward osteogenic or adipogenic lineage. Surprisingly, nanosized grooves did not affect lineage-specific differentiation, but only enhanced the differentiation

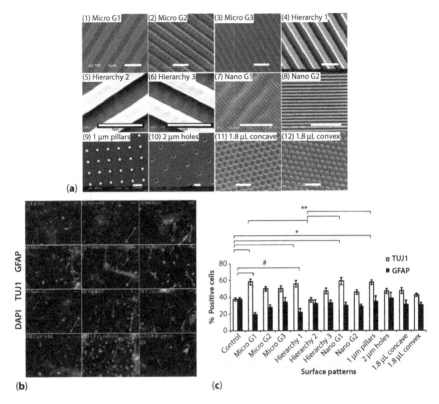

Figure 5.1 Effect of topographical cues on stem cell fate. (a) Scanning electron microscopy images of microsized feature arrays used for directing primary mNPCs differentiation. Scale bars: 5 μm. (b) Immunofluorescence staining of TUJ1 (β-tubulin III)-positive neuronal populations and glial fibrillary acidic protein (GFAP)-positive cells (astrocytes). Cells were cultured on PDMS replicas of the structures shown in (a). (c) Quantification of TUJ1- and GFAP-positive cells for each surface topography. Anisotropic structures significantly promoted differentiation of mNPCs while isotropic topographies enhance glial differentiation on the same conditions. Reproduced with permission of John Wiley & Sons Ltd. [90].

process initiated by the differentiation media [86]. Topography feature size and dimensions have also been shown to affect the proliferation and differentiation on adult NSCs and human induced pluripotent stem cells (hiPSCs) [90, 105, 106]. In this case, a direct correlation between the feature sizes and the signaling pathways involved was revealed. This is probably related with the concentration and distribution of adhesion molecules (*e.g.* integrin clusters) which significantly affect cell behavior [107].

The topographic feature arrays include both isotropic and anisotropic structures, aspect ratios, and stiffness [108] (see Figure 5.1 and Table 5.1). Interestingly, anisotropic topographies have been shown to enhance

Table 5.1 Stem cells response on micro- and nanostructured 2D surfaces.

Structure	Feature size	Material	Cell type	Differentiation	Ref.
Roughness	0.5–4.7 µm	PCL	hMSCs	Osteogenic	[70]
Porous	In the order of 10–100 of nm	Si	rMSCs	Osteogenic Adipogenic	[113]
Rounded Square	x–y: 50–2 µm z: 1 µm	PMMA	rMSCs	Osteogenic	[100]
Square	5–300 µm	CnT	hNSCs	Neuron	[114]
Grooves	400 nm–5 µm Aspect ratio = 1	PLGA	hNSCs	Neurons	[103]
Ridges	350 nm	PUA	hESCs	Neurons	[88]
Pillars	x–y: 2–4 µm z: 4 µm	PS	hMSC	Osteogenic	[102]
Anisotropic dots, circles, lines	2–10 µm	Si	NSCs	Neurons	[106]
Grating	Width: 350 µm/ 2 µm/5 µm Height: 300 nm	PDMS	hiPSCs	Neurons	[105]
Grid lines	100 nm	GO	hADMSCs	Ectoderm Osteoblasts	[104]
Distinct	x–y: 0.15–2 µm different aspect ratios	PDMS	mNPCs	Neurons Glial	[90]

hADMSCs: human adipose-derived mesenchymal stem cells; hESCs: human embryonic stem cells; hiPSCs: human induced pluripotent stem cells; hMSCs: human mesenchymal stem cells; hNSCs: human neuronal stem cells; mNPCs: mouse neural precursor cells; NSCs: neuron stem cells; rMSCs: rat mesenchymal stem cells. CnT: carbon nanotubes; GO: graphene oxide; PCL: poly(e-caprolactone); PDMS: polydymethylsiloxane; PLGA: poly(lactic-co-glycolic acid); PMMA: poly(methyl methacrylate); PS: polysterene; PUA: polyurethane acrylate; Si: silicon.

neuronal differentiation, while isotropic topographies enhance glial differentiation of primary murine neural progenitor cells (mNPCs) [90]. This provides a complex landscape where asymmetry may play an important role on the determination of cell lineage. Indeed, asymmetry has been shown to impact on multiple biological phenomena, such as cell migration [61–63, 109]. Altogether, the unlimited combination of feature designs and sizes has led to the development of a mathematical algorithm to generate a chip with thousands of topographies with random surface features [110]. Human mesenchymal stromal cells grown on these high-throughput chips made of poly(lactic acid), combined with high-content-screening studies, revealed those topographies which induced osteogenesis or self-renewal. This algorithm could facilitate the selection of the optimal configuration which induces the highest rate of specific lineage, and therefore, could have important implications for regenerative medicine and tissue-engineering applications.

Micro- and nanostructured surfaces can also be used to study the role of mechanical cues in cell self-renewal or reprogramming [111]. In fact, the mechanical regulation of stem cells self-renewal *in vivo* is not yet well understood. To shed light on the importance of mechanical factors, researchers have used multiple micro- and nanofabricated assays to study how the mechanical environment influences these processes. Strikingly, microsized grooves were observed to increase significantly the efficiency of reprogramming of mouse fibroblasts [112]. The mechanical origin of this reprogramming was found to be correlated to cytoskeletal tension induced by the microgrooves.

Altogether, differentiation of stem cells by means of arrays of topographic micro- and nanoscale features has provided important insights into the mechanical mechanism of the determination of stem cell fate. However, these engineered substrates do not recreate completely the complexity of the native tissue environment. To solve this, novel engineered approaches which better mimic the morphology of the native ECM have been proposed.

5.2.4 Physical Gradients for Regulating Stem Cell Fate

Gradients of signaling molecules are inherently found *in vivo* during all stages of morphological and pathological events [115–117], and they have been extensively used *in vitro* to study the biological phenomena in which they are involved [109, 118–120] (see Section 5.3.2). Physical gradients are also found *in vivo* [121]. They regulate various important cell behaviors, such as proliferation, migration, wound healing, cancer expansion, and differentiation. As an example, during osteogenesis the pore size of the bone increases from inside to outside which impacts on the differentiation lineage [122]. Further, soft-to-hard interfacial tissues (*e.g.* ligament-to-bone, cartilage-to-bone, or tendon-to-bone interface) are particular regions within the body where physical gradients (stiffness, porosity, etc) are also present [123]. Therefore, micro- and nanoengineered biomaterials with graded physical properties are of critical importance to recreate *in vitro* these *in vivo* scenarios [124].

Surfaces with physical gradients are referred as those materials which possess anisotropic physical properties, which can be in terms of spatial distribution of features, mechanical properties (stiffness), porosity, and or stress/strain. These materials have numerous and obvious advantages compared to their standard monophasic counterparts, such as to enable the screening of cellular responses in a high-throughput manner. Multiple strategies for the development of biomaterials with surface (and 3D scaffolds) containing physical gradients have been extensively described in the literature [125–127], and nicely documented in Ref. [123]. Briefly, on polymeric materials, gradients of pore size can be created by, *e.g.* a centrifugation method [128], a combination of melt pressing and porogen leaching [129], a spinning technique [130], or by a cryogenic method [131]. These gradients in porosity can be then applied for studies in cell invasion, motility, proliferation, and differentiation [132]. Gradients of other physicochemical characteristics of a surface such as spatial distribution of topographical features, roughness, wettability, or stiffness can also be achieved by means of different techniques. As an example, gradients in roughness can be easily obtained by UV-ozone [133] and plasma treatment [134]. Recently, a stiffness gradient on polyvinyl alcohol (PVA) hydrogel was also fabricated by a freezing–thawing method to investigate stem cell differentiation behavior [135]. Diffusion and controlled dipping into cross-linking solution, photopolymerization using a gradually darkening or sliding mask, or a gradient-maker device, are other commonly used methods for creating stiffness gradients, among many others [136–140]. Finally, gradients in thickness and nanostructures can be easily produced using thin diblock co-polymer films casting [141].

Applied for stem cells research, these assays permit the rapid assessment of the optimal physical environment which promotes the higher differentiation or self-renewal yield. Table 5.2 summarizes some examples of physical gradients applied for regulating stem cell fate. In this regard, interfacial tissues were engineered *in vitro* using a nanofibrous scaffold with gradients in amorphous calcium phosphate nanoparticles distribution [142]. Osteogenesis of MC3T3-E1 murine pre-osteoblasts was enhanced in the gradient regions that contained higher nanoparticles density, yielding a graded osteoblast response. Gradients in surface stiffness were also created to study the effect on stem cell differentiation. Valve interstitial cells (VICs) were cultured on a dynamic photodegradable PEG hydrogel exhibiting elasticity gradients in the range of few kPa [143]. Myofibroblast

Table 5.2 Examples of stem cell response on physical gradients.

Gradient type	Technique	Material	Cell type	Differentiation	Ref.
Stiffness	Freezing–thawing	PVA	hBMSCs	−: Neurogenesis +: Osteogenesis	[135]
Stiffness	Gradually dark photomask	PA	C2C12 hMSC	+: Myogenesis	[145]
Roughness	Sandblasting Chemical polishing	PCL	hMSCs	−: Osteogenesis	[70]
Porosity	Anodization	Si	rMSCs	−: Osteogenesis −/+: Adipogenesis	[113]
Porosity	Spinning Fibril bonding	PCL	ASCs	+: Chondrogenesis	[146]
Topography	Nanoimprint Lithography	PS	hESCs	−: Self-renewal	[79]
Topography	Electrospinning	$Ca_3(PO_4)_2$	MC3T3-E1	+: Osteogenesis	[90]

−/+: Smallest/largest features size and/or physical properties of the gradient, respectively.
ASCs: adipose stem cells; MC3T3-E1: murine calvarial pre-osteoblastic cell line; hBMSCs: human bone marrow stem cells; hESCs: human embryonic stem cells; rMSCs: rat mesenchymal stem cells. PA: polyacrylamide; PCL: polycaprolactone; PE: polyethylene; PS: polysterene; PVA: polyvinyl alcohol; Si: silicon.

differentiation was observed mostly in the direction of higher elasticity modulus. Similarly, a PVA cylindrical hydrogel with a stiffness gradient ranging from 1 to 24 kPa was used with hBMSCs to promote effective neurogenesis and osteogenesis of the cells [135].

Surface porous dimensions have been shown to be an important regulator to control stem cells fate (see Sections 5.2.2 and 5.2.3). For this reason, the effect of gradients of surface porosity on stem cell fate has also been investigated during the last years. Rat mesenchymal stem cells (rMSCs) cultured on stable porous silicon (Si) surfaces with pore sizes ranging from tens to hundreds of nanometers in diameter, and changes in the ridge nanoroughness from a few to tens nanometers, induced differentiation into osteogenic and adipogenic lineages [113] (see Tables 5.1 and 5.2). It was observed that osteogenesis was dependent on both topography and local cell density, whereas adipogenesis was only dependent on topography. In a similar work, hMSCs cultured on porous alumina, with pore sizes ranging from 50 nm to 3 μm, differentiated into osteoblasts in the gradient region of 120–230 nm [144]. Large pores resulted in a low hMSCs attachment density and spreading area which resulted in low differentiation yield.

The effect of surface topography can also be applied for the maintenance and self-renewal of undifferentiated stem cells. In this context, hESCs cultured on a substrate featuring arrays of increasing nanopillar diameter, in the absence of feeder cells, demonstrated a tendency to express higher levels of undifferentiated markers toward regions with smaller nanopillar diameter range [79]. Taken together, the above-described results encourage the combination of surface micro- and nanotopography with physical gradients as a promising strategy in order to regulate the differentiation of stem cells into specific lineages or promote self-renewal and reprogramming.

Overall, physical gradients can recreate with higher fidelity specific regions on the native tissue, in terms of morphology and mechanical properties. Therefore, these engineered biomaterials allow the study of a wide range of biological phenomena *in vitro*. In particular, we have shown that physical gradients are particularly interesting to study the mechanoregulation of stem cell fate. Novel micro- and nanotechnologies are continuously emerging aiming at recreating the *in vivo* scenario. Finally, combination of physical and chemical gradients will contribute to the development of a new generation of 'smart' materials which will facilitate the study of cellular phenomena. We anticipate that physicochemical gradients will contribute to reveal key mechanisms in stem cell differentiation and in the development of new strategies for tissue engineering and regenerative medicine applications.

5.3 Controlled Surface Immobilization of Biochemical Stimuli for Stem Cell Differentiation

5.3.1 Micro- and Nanopatterned Surfaces: Effect of Geometrical Constraint and Ligand Presentation at the Nanoscale

As seen in Section 5.1, the multifaceted extracellular milieu presents biochemical stimuli that influence stem cell differentiation [147, 148]. The need for controlling these microenvironmental cues *in vitro* fostered the engineering of two-dimensional (2D) surface biochemical patterns. Since many signaling molecules can function under restricted diffusion conditions, surface confinement does not compromise the biological relevance of surface-bound assays. Therefore, chemical patterns were envisioned as a high-throughput screening alternative to conventional cell culture. Combinatorial chemical microarrays have been developed to rapidly interrogate the interactions between the chemistry of biomaterials and cells [149]. As summarized in Table 5.3, cell microarrays allow the screening of the effect of signaling molecules printed alone or in combination, and significantly reduce the amount of reagents needed and the inter-experimental variability of conventional microwell plate tests [150]. Moreover, 2D micropatterning offers the possibility of testing the geometrical guidance effects in collective cell behavior. In this context, hESCs colonies were deposited onto defined adhesive circular islands of Matrigel™ with controlled diameter (200, 400, and 800 μm) and pitch (the distance between colonies; fixed at 500 or 1000 μm) to quantitatively interrogate cell-specific localized signaling activation at early time points. Results demonstrated the spatial control of the activation of Smad1 and consequently hESCs fate, with larger colonies promoting the maintenance of the undifferentiated phenotype by suppressing Smad1 activation [151]. Similarly, hiPSCs were cultured onto micropatterned circular fibronectin islands of 80, 140, 225, and 500 μm of diameter. Cells differentiated toward a bicellular population of endothelial cells (ECs) and pericytes depending on feature sizes. Smaller islands promoted EC differentiation efficiency, yielding a derived population composed of 70% ECs, with a greater sprouting propensity. Differentiation on the largest feature size exhibited a smaller EC yield, similar to that on nonpatterned substrates (Figure 5.2) [152].

During embryonic patterning, cells are allocated to the three germ layers in a spatially ordered sequence. It has been shown *in vitro* that human pluripotent stem cells (hPSCs) differentiated in suspended cell aggregates,

Table 5.3 Examples of stem cells response on micropatterned 2D surfaces.

Feature shape	Feature size	Technique	Patterned compound	Cell type	Stem cell response	Ref.
Circular	200–800 μm diam.	μCP	Matrigel™	hESCs	Maintenance of undifferentiated state	[151]
Circular	80–500 μm diam.	Deep UV photopatterning	Fibronectin	hiPSCs	ECs/perycites differentiation	[152]
Circular	100–400 μm diam.	Maskless photolithography	Fibronectin	mESCs aggregates	Cardiomyocyte differentiation	[154]
Circular	200–1200 μm diam.	μCP	Matrigel™	hESCs	Mesoderm/endoderm fates	[155]
Circular	250–1000 μm diam.	Deep UV photopatterning	PDL/Matrigel™	hESCs	Self-organized patterning	[156]
Circular	400 μm diam.	Inkjet printing	Laminin+(Wnt/Notch ligands)	hNPCs	Neural/glial differentiation	[157]
Square	1024–10,000 μm²	μCP	Fibronectin	hMSCs	Adipogenic/osteogenic differentiation	[159]
Square, rectangle, star and flower	1000–5000 μm²	μCP	Fibronectin	hMSCs	Adipogenic/osteogenic differentiation	[74]
Circular, square, rectangle, triangular and star	900 μm²	Photolithography and transfer lithography	RGD	rMSCs	Adipogenic/osteogenic differentiation	[160]
Circular	40–80 μm diam.	Photolithography	PAAc/polystyrene	hMSCs	Adipogenic differentiation	[161]

μCP: microcontact printing; PDL: poly(d-lysine); RGD: arginylglycylaspartic acid; hESCs: human embryonic stem cells; hiPSCs: human induced pluripotent stem cells; hMSCs: human mesenchymal stem cells; hNPCs: human neural precursor cells; mESCs: mouse embryonic stem cells; rMSCs: rat mesenchymal stem cells.

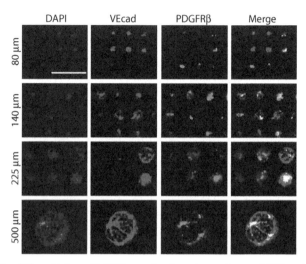

Figure 5.2 Co-differentiation of hiPSCs on circular micropatterns of different diameter sizes. Differentiated hiPSC cells on micropatterns were assessed for endothelial and pericyte differentiation, indicated by VEcad (red) and PDGFR β (green), respectively (nuclei in blue; scale bar: 500 μm). Reproduced with permission of John Wiley & Sons Ltd. [152].

or embryoid bodies (EBs). EBs have been used as a reliable model for embryonic development, since they largely exhibit heterogeneous patterns of differentiated cell types enabling differentiation and morphogenesis and yielding microtissues that are similar to native tissue structures. EBs exhibit size-dependent differentiation decisions: EBs diameters of 450 μm have been shown to promote cardiac lineage, whereas smaller diameters (~150 μm diameter) favored endothelial lineages [153]. Similarly, when constrained using adhesive patches, EBs differentiate into specific lineages. As an example, mouse ESCs aggregates were cultured onto micropatterned surfaces containing fibronectin circular microdomains instead of using the traditional floating EBs format. The greatest potential of mESCs to differentiate toward cardiomyocytes was exhibited on patches of 200 μm in diameter, compared to 100, 300, and 400 μm diameters [154].

Two-dimensional micropatterning offers the possibility to recapitulate the embryonic spatial ordering *in vitro*, paving the way toward more systematic studies of embryogenesis and cell-to-cell communication during patterning. The cellular organization typical of pluripotent cell differentiation along the mesoderm and endoderm lineages was evidenced using micropatterning. Human ESCs cultured onto Matrigel™ micropatterns in the presence of bone morphogenetic protein 2 (BMP-2) and activin A differentiated toward mesoderm on larger islands (1200 μm in diameter), whereas smaller islands (200 μm in diameter) favored endoderm fates [155]. In fact, the geometric confinement was sufficient to trigger self-organized patterning in hESCs. Cells confined on cell-adhesive circular patches of 1000, 500, and 250 μm in diameter, and stimulated with bone morphogenetic protein 4 (BMP-4), organized into an ordered array of germ layers along the radial axis of the colony. This order resulted from self-organized signaling that confined the response to the BMP-4 to the colony border while inducing a broader gradient of Activin-Nodal signaling to pattern mesendodermal fates. Since control of fates is established from the border of the colony, small colonies are equivalent to the edges of large colonies and central fates are lost [156].

Two-dimensional micropatterns have also been used to explore cell regulation using arrays of signaling microenvironments. As an example, primary neural precursor cells were cultured on an array composed of mixtures of ECM proteins, morphogens, and other signaling proteins presented individually and in combinations. Quantitative high-throughput analysis of multiple phenotypic outcomes revealed that certain combinations of molecular signals influenced the balance between differentiating neural and glial cells. In particular, Wnt and Notch signaling pathways co-stimulation were found to favor undifferentiated-like proliferative state in cells, whereas BMP-4 induced an "indeterminate" differentiation phenotype with simultaneous expression of glial and neuronal markers [157].

Besides the advantages inherent to the microarray format (*e.g.* multivariable testing in a single experiment with low reagent consumption), biochemical patterning also allows for single cell resolution experiments in spatially defined geometries with several micrometers of resolution. The most widely used technique for the fabrication of few micrometer-sized patches is microcontact printing (μCP). This technique uses a PDMS stamp soaked with the cell adhesive compound of interest, which is patterned in self-assembled monolayers (SAMs) onto the substrate. Lateral dimensions from ten to hundreds of micrometers can be obtained by μCP, which corresponds to the range comprising cell sizes [158]. In a seminal work, μCP was applied to analyze the effect of spreading area on stem cell differentiation at the single cell level. hMSCs seeded onto fibronectin micropatterns were exposed to competing soluble differentiation signals toward adipogenesis and osteogenesis. Cells adhered, spread, and differentiated toward the osteogenic lineage on 10,000 μm^2 square patches, adipogenic lineage on 1024 μm^2 square patches, and a mixture of both lineages on intermediate patches of 2025 μm^2 [159]. To test the effect of cell geometry in stem cell differentiation, μCP was used to produce micrometer-sized fibronectin patterns (1000, 2500, and 5000 μm^2) of different geometries (square, rectangle, star, and flower). Seeded individual hMSCs adhered and spread adopting the shape of the underlying patches. When the patterning area was kept constant but with varying geometries, cells exposed to competing adipogenic/osteogenic signals differentiated preferentially toward the osteogenic fate in geometric features that caused an increase of actomyosin contractility [74]. A similar study carried out using RGD micropatterns on antifouling poly(ethylene glycol) (PEG) hydrogels showed that shape anisotropy also favored osteoblastic differentiation against adipogenesis. When global isotropic circular, square, triangular, and star micropatterns were used, optimal adipogenic and osteogenic differentiations were enhanced in circular and star cells, respectively, highlighting the effect of local anisotropy (cell roundness and perimeter) in stem cell fate [160].

Two-dimensional micropatterned substrates have also been used to evaluate some other relevant effects in stem cell differentiation. Circular micropatterns of 40, 60, and 80 μm in diameter of negatively charged poly(acrylic acid) (PAAc) and neutral polystyrene obtained with UV photolithography were used to investigate the electrostatic effect of chemical groups on the response of individual hMSCs. Results showed that the adipogenic differentiation at the single-cell level was enhanced on the PAAc micropatterns, and decreased as the diameter of micropattern increased [161].

Experimental evidence of ECM organization at the nanoscale (67 nm banding periodicity of collagen fibers [162] and the nanoscale order of epitope presentation found on fibronectin fibers [163, 164]), steered the production of synthetic nanopatterned surfaces directed toward identifying the geometric cues that initiate and guide cell adhesion. Micellar lithography technique was used to generate mesoscopic quasi-hexagonal 2D arrays of nanometer-sized gold particles on flat substrates. The interparticle distance with this technique could be varied between 30 and 140 nm by adjusting the lengths of the blocks in the initial block copolymer template [165]. Further conjugation of the gold nanopatterns with RGD, and the corresponding passivation of unpatterned regions, allowed for the interrogation of cell adhesion mechanisms at the single molecule level. Since the size of a single RGD receptor, integrin, is approximately 12 nm [166], each RGD-coated gold nanoparticle of approximately 10 nm can interact with a single integrin molecule. Experiments using micellar lithography-based RGD nanopatterns showed that RGD nanospacing plays a crucial role in cell adhesion, and revealed a threshold value of around 70 nm above which cell adhesion process is significantly delayed [167–170].

Since nanospacing influences cell adhesion, it is predictive that nanopatterns of different nanospacings might influence the lineage commitments of stem cells (see Table 5.4). In this regard, micellar lithography was

Table 5.4 Examples of stem cells response to nanopatterned RGD on 2D surfaces.

Feature size	Feature spacing	Technique	Cell type	Stem cell response	Refs.
~ 10 nm	68 nm	Micellar lithography	hMSCs	Maintenance of undifferentiated state	[172]
~ 10 nm	20, 32, 58, and 90 nm	Micellar lithography	HSCs	Lipid raft clustering	[174]
~ 10 nm	37, 53, 77, 87, and 124 nm	Micellar lithography/ transfer lithography	rMSCs	Adipogenic/ osteogenic differentiation	[175, 176]
~ 10 nm	63 and 161 nm	Micellar lithography/ transfer lithography	rMSCs	Chondrogenic differentiation	[181]
~ 50 nm	Not reported	Adsorption	hMSCs	Osteogenic differentiation	[182]
~ 10 nm	46 and 96 nm, within micropans of 35 and 65 μm side lengths	Micellar lithography, photolithography, and transfer lithography	hMSCs	Adipogenic/ osteogenic differentiation	[183]

hMSCs: human mesenchymal stem cells; HSCs: hematopoietic stem cells; rMSCs: rat mesenchymal stem cells.

used to create RGD nanopatterns on β-type Ti-40Nb alloys, a material typically used in clinical applications as implant [171]. The interaction of MSCs with the material improved with the presence of defined RGD patterns (RGD nanopatterns of 68 nm interparticle distance). Nanopatterns of RGD regulate cell adhesion locally and reduce population heterogeneity by maintaining the typical phenotype of noncommitted stem cells [172]. The significance of integrin ligand nanopatterning on lipid raft clustering in hematopoietic stem cells (HSCs) was also investigated on nanopatterned surfaces of cyclic RGD (cRGD) produced by micellar lithography with particle spacings of 20, 32, 58, and 90 nm. Lipid rafts are known to play an important role in the differentiation process of HSCs [173]. Cultured HSCs on cRGD nanopatterned surfaces proved to be sensitive to the lateral distance between the presented ligands with regard to adhesion and lipid raft clustering. Spacing of 32 nm appeared to be the critical, maximum tolerated distance between ligands for cell adhesion to nanopatterned cRGD-functionalized surfaces and signal transduction in response to this ligand [174].

Micellar lithography combined with transfer lithography has also been used to evaluate MSCs behavior toward adipogenic and osteogenic fates in response to RGD nanospacing on bioinert PEG hydrogels. Using this methodology, nanopatterns of RGD on a persistently nonfouling biocompatible environment were prepared with interparticle spacings ranging from 37 to 124 nm. Nanospacing of RGD was found to affect the lineage commitment of MSCs. Although previous reports indicated that cell spreading favors osteogenesis [159, 161], in this case both osteogenic and adipogenic inductions were enhanced on patterns of large nanospacings. After discarding other effects influencing cell adhesion as cell density, shape, and cell–cell contact, authors conclude that RGD nanospacing may act as an inherent regulator of differentiation of stem cells beyond cell spreading [175, 176].

Considered a model for stem cell differentiation process [177, 178], chondrogenesis of MSCs was studied on RGD nanopatterns created in PEG hydrogels. The effect of two RGD nanospacings of 63 and 161 nm (below and over the 70 nm threshold value for efficient cell adhesion) was examined [167–170]. Expressions of collagen II proteins and chondrocyte-specific genes (SOX9, aggrecan and collagen II) were detected. Results showed that large nanospacing led to a reduced spreading area and a higher chondrogenic induction. This is in agreement with previous studies indicating that a relatively weak cytoskeleton is beneficial for the chondrogenic differentiation of stem cells [179, 180]. Further tests by the addition of the mitogen-activated protein kinase inhibitor SB203580 confirmed the positive regulation of the p38 phospho-relay cascade on the chondrogenic induction in this model system [181].

Recently, some reports brought the attention into the effect of substrate coupling strength with the immobilized ligands. The interplay between substrates stiffness and cell adhesive coatings has an impact in the mechanical feedback received by the exposed cells, affecting stem cell fate (see Section 5.2) [31, 32]. The coupling strength of RGD-coated gold nanoparticles adsorbed on glass has been modulated in order to investigate how it influenced hMSCs behavior. Results showed that RGD coupling strength with the substrate influenced adhesion, spreading, and differentiation in hMSCs, which exhibited enhanced differentiation ability toward osteogenesis in high-coupling-strength surfaces. This observation pointed out the relevance of considering ligands' linkage to the surface in the experimental design of stem cell differentiation studies [182].

Even when the coupling strength with the substrate is controlled, nanopatterning differentiation experiments can be subjected to the influence of some other relevant parameters. As described above, single cell experiments in micropatterns revealed the cell spreading area (geometrical constrain) as a paramount variable influencing stem cell differentiation [74, 159–161]. To investigate the effect of cell spreading area on hMSCs differentiation on nanopatterns, the osteogenic/adipogenic differentiation in micro/nanopatterns of RGD with nanospacings of 46 and 96 nm, and micropans of 35 and 65 µm side lengths (a total of four substrate types) was studied. Cells were subjected to osteogenic or adipogenic induction in the different substrates, respectively, and results showed that larger RGD nanospacing promoted both osteogenesis and adipogenesis [175, 176]. Moreover, larger RGD nanospacing, independently from cell spreading size, leads to higher osteogenesis induction. This later observation highlights the significant effect of adhesive ligand presentation at the nanoscale on stem cell differentiation (Figure 5.3) [183].

5.3.2 Biochemical Gradients for Stem Cell Differentiation

Biochemical gradient surfaces are described as surfaces with a gradually varying composition along their length. During embryo development, cell patterning is governed by underlying morphogen gradients. Morphogen gradients provide cells with positional information: each cell can read its position in the gradient and respond accordingly toward a particular ligand concentration [184]. Therefore, control of ligand dosage is critical in the evaluation of ligand effects on cells. Microarrays have been used to create microenvironments with spatially defined gradients of immobilized ligands to direct stem cell differentiation into multiple subpopulations of different lineages [185]. As an example, inkjet printing technique was used to create concentration gradients of ciliary neurotrophic factor (CNTF) on poly-acrylamide-based hydrogels. Concentrations of the printed factors were selected based on the known cellular effects (40 ng/µL for CNTF). Primary fetal NSCs cultured on the gradient exhibited a linear increase in the number of cells expressing the astrocytic marker glial fibrillary acidic protein (GFAP) with the increasing CNTF concentration [186].

Discrete microenvironments of different concentrations of BMP-2 on fibrin substrates have also been produced by inkjet printing. Micropatterned arrays of BMP-2 (10 µg/mL), with 2, 8, 14, and 20 overprints to modulate the surface concentrations, were used for the study of primary muscle-derived stem cells (MDSCs) differentiation. Results showed that MDSCs

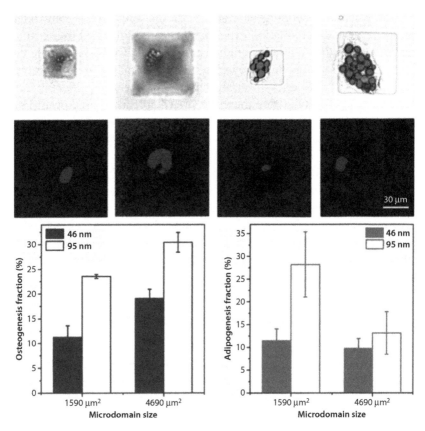

Figure 5.3 Cell differentiation on micro/nanopatterns. (Upper) Typical differentiated hMSCs on the microdomains of RGD nanoarrays of small and large microdomain areas. The left image shows osteogenically differentiated cells stained positively by Fast Blue RR. The right image shows adipogenically differentiated cells stained positively by Oil Red O. (Mid) Fluorescence images of DAPI-stained cells corresponding to the upper row, indicating the single cells included in the statistics. (Lower) Osteogenic and adipogenic fractions with different micropan sizes and RGD nanospacings. Reprinted with permission from Wang, 5., Li, S., Yan, C., Liu, P., Ding, J. Fabrication of RGD Micro/Nanopattern and corresponding study of stem cell differentiation. *Nano Lett.*, 15, 1457, 2015. Copyright 2015 American Chemical Society. [183].

can be patterned toward the osteogenic and myogenic lineages simultaneously on the same printed array. Cells cultured under myogenic conditions differentiated toward the osteogenic linage on BMP-2 patterns, while cells cultured off the patterned areas differentiate toward the myogenic lineage. For 250 ng/10 μL and 500 ng/10μL BMP-2 concentrations, there was a significant upregulation of the osteogenic marker alkaline phosphatase (Alp) gene with no significant differences between the two concentrations.

Nevertheless, only with the highest deposited BMP-2 concentration, MDSCs showed expression of the early osteogenic marker osterix (Osx gene). Results demonstrated that the fate of MDSCs could be modulated by means of BMP-2 arrays and induce dose-dependent osteogenic lineage progression within spatially restricted areas (those containing the BMP-2 pattern) [187]. An analogous microarray platform was developed to test the effects of different concentrations of surface-bound fibroblast growth factor-2 (FGF-2) used to direct tendon specification of multipotent mouse C3H10T1/2 cells in a dose-dependent manner. Square patterns (1 mm × 1 mm) of increasing concentrations of FGF-2 were inkjet-printed onto fibrin substrates. Doses of FGF-2 from 122.4 pg/mm^2 resulted in the upregulation of the tendon marker Scleraxis (Scx) relative to the negative control, therefore demonstrating C3H10T1/2 cells dose dependence to FGF-2 [188].

The microarray system has also been applied to the study of stem cell collective migration under the influence of the underlying patterns of low-to-high, high-to-low, and uniform concentrations of the heparin-binding epidermal growth factor-like growth factor (HB-EGF), which is known for its role in directing proliferation and migration of MSCs [189]. Spots size of HB-EGF remained constant around 50–60 μm in diameter. Mouse MSCs were seeded at the pattern origin to simultaneously initiate cell diffusion. Surprisingly, all printed patterns directed net collective cell guidance with comparable responses. It was suggested that cell diffusion maintained by HB-EGF acted as the main force driving cell migration, and not the HB-EGF gradients [190].

Despite the high versatility of the microarray format to influence stem cell fate under different ligand concentrations, even if a large number of ligand concentrations can be included, these are inherently discrete. Since *in vivo* cells respond to small changes in tiny amounts of signaling molecules, a more accurate screening could be provided by continuous biochemical gradients. Therefore, continuous gradients of surface-bound molecular ligands provide an unmatched setup for the high-throughput screening of stem cell responses to matrix-bound ligands (Table 5.5). One of the most common techniques to create chemical gradients is plasma polymerization [191]. Plasma polymers provide smooth coatings that can be deposited onto any surface without changing its topography, and therefore, their effects on cell response can be attributed solely to the changes produced in the surface chemistry. As an example, the capacity of mouse ESCs self-renewal was examined on a gradient of carboxylic acid concentrations (1.9–10.4%) created from plasma polymerized acrylic acid (AA). Cell area of ESCs cultured on the gradient increased with surface acid density

Table 5.5 Examples of stem cells response to 2D biochemical gradients.

Gradient type	Technique	Immobilized compound (concentration range)	Cell type	Stem cell response	Ref.
Discrete	Inkjet printing	CNTF (not determined).	rNSCs	Astrocyte differentiation	[186]
Discrete	Inkjet printing	BMP-2 (125 ng/10 µL–500 ng/10 µL)	MDSCs	Osteogenic/myogenic differentiation	[187]
Discrete	Inkjet printing	FGF-2 (40.8–244.8 pg/mm^2)	C3H10T1/2	Tendon differentiation	[188]
Discrete	Inkjet printing	HB-EGF	mMSCs	Collective cell migration	[190]
Continuous	Plasma polymerization	OD/AA (1.9–10.4% acid groups)	mESCs	Self-renewal capacity	[192]
Continuous	Plasma polymerization	DG/AA (1–10% acid groups)	mESCs	Retention of stem cell markers	[194]
Continuous	Plasma polymerization	OD/AAm (0–12% atomic nitrogen)	mEBs derived cells	Spontaneous differentiation	[195]
Continuous	Plasma polymerization	OD/AA (2.4–11.3% acid groups) and DG/AA (1–10% acid groups)	rMSCs	Adipogenic/osteogenic differentiation	[198]
Continuous	Plasma polymerization	NGF (39–57.5 ng/cm^2)	mEB-derived cells	Neural differentiation	[199]
Continuous	Microfluidics	RGD	rMSCs	Adhesion and spreading	[201]

AA: acrylic acid; AAm: allylamine; BMP-2: bone morphogenetic protein 2; CNTF: ciliary neurotrophic factor; DG: diethylene glycol dimethyl ether; EGF: heparin-binding epidermal growth factor-like growth factor; FGF-2: fibroblast growth factor-2; HB-NGF: nerve growth factor; OD: octadiene; RGD: arginylglycylaspartic acid. MDSCs: primary muscle-derived stem cells; mEBs: mouse embryonic bodies; mESCs: mouse embryonic stem cells; mMSCs: mouse mesenchymal stem cells; rMSCs: rat mesenchymal stem cells; rNSCs: rat neural stem cells.

showing strong alkaline phosphatase enzyme (ALP) staining (a marker for undifferentiated cells) for cell areas of less than 120 μm² [192]. Cell attachment, colony size, and retention of stem cell markers in mouse ESCs were evaluated in plasma polymer gradients of AA created using two exposure times to evaluate the influence of plasma polymer thickness. Cell adhesion, colony formation rate, and colony size were lower when the number of acid groups decreased. Polymer gradients from shorter exposure times showed higher levels of protein adsorption compared to longer ones. Such differences in protein adsorption were reflected in cell adhesion experiments and in downstream cell responses [193]. Colony size and morphology, which vary differently along gradient distance of gradients produced at different exposure times, correlated with ES cell marker retention: both Oct4 and alkaline phosphatase stem cell markers staining decreased in intensity as colony size increased and colony architecture was lost. Marked differentiation of ES cells occurred from Days 3 to 5 in culture for both gradient types, and stem cell markers and morphology were retained only on regions of reduced AA concentration [194].

Recent studies tested the influence of surface chemistry gradients on the differentiation of stem cells. Gradients of plasma polymerized allylamine (AAm) were produced for the study of spontaneous differentiation of cells derived from differentiating mouse EBs into endoderm, mesoderm, and ectoderm. Results showed that high AAm concentrations triggered differentiation toward both mesoderm and ectoderm [195]. Plasma polymer gradients of AA created on two different backgrounds of plasma polymerized octadiene (OD) and diethylene glycol dimethyl ether (DG), respectively, were used for the screening of rat bone marrow MSCs differentiation. Cell density was found to be higher in the AA-rich regions of plasma polymers with the DG background due to the reinforced cell migration provided by the DG nonfouling environment. Cell differentiation toward the osteogenic commitment was found to be favored by the increased cell density, in well agreement with previous reports [196–198].

Plasma polymer gradients of surface chemistry can be used for the selective immobilization of bioactive compounds. High hydroxyl to high aldehyde density group gradients were used for the immobilization of increasing densities of nerve growth factor (NGF) through reductive amination with the aldehyde groups. The critical growth factor density required to support neural lineage generation was assessed from mouse EB differentiation on the gradient surface. Mouse EB cells experienced significant enhancements in attachment, proliferation, and neural differentiation up to an immobilized NGF density of 52.9 ng/cm². A further increase in NGF density did not convey a greater proliferation or differentiation stimulus [199].

The use of continuous biochemical gradients brought to light some threshold concentrations of signaling molecules triggering significant changes in cell response that could barely be detected in discrete microarray analysis [119, 200]. A threshold range of RGD concentrations of 0.107–0.143 mM was revealed for MSCs adhesion on microfluidic-generated RGD gradient in hydrogels. Both cell adhesion and spreading were found to increase with the increasing RGD concentration up to the threshold value [201].

Altogether, continuous biochemical gradients covering a biologically relevant range of concentrations have proved as efficient screening platforms for the optimization of culture substrates. With continuously varying surface density of matrix-bound biomolecules, they allow determining the optimal concentration for a specific cell response. Nonlinear responses to continuous linear gradients, revealing threshold concentrations, could be associated to the distribution of signaling molecules at the nanoscale [119, 200]. Therefore, it can be envisioned that the production of controlled density gradients of nanopatterned signaling molecules could effectively contribute to the understanding of ligand effects on stem cell behavior at the nanometer scale.

5.4 Three-dimensional Micro- and Nanoengineered Environments for Stem Cell Differentiation

Although much easier to maintain and culture, 2D systems resulted reduced models for complex tissues in the body. Stem cell culture in traditional 2D systems fails to provide an accurate representation of physiological stem cell environment of the native tissue [202]. In general, cells *in vitro* are grown on flat, solid, and impermeable 2D surfaces systems which suffer from inherent heterogeneity, limited scalability, reproducibility, as well as limited expansion toward the z-direction. Cells cultured under these conditions show extreme phenotypes and artifacts which trigger nonphysiological features, such as enhanced stress fibers [203, 204]. For stem cells, they are in general impeded to mature and differentiate as they do in their native state. Furthermore, the impermeability of the substrate prevents cells from taking or secreting signaling molecules and nutrients from both basal and apical surfaces causing cells to deviate from their natural environment [75]. Altogether, 2D cell culture systems are becoming a bottleneck for reproducing with high fidelity the complexity of tissue environment. On the contrary, 3D cell culture assays provide multiple advantages when compared to 2D environments [203, 205–209], even with identical biochemical composition [204].

Therefore, much effort has been directed to combine 3D physiological relevance with 2D convenience. Mimicking the *in vivo* 3D microenvironment is a challenging task in engineering *in vitro* cell models. Since stem cell fate is strongly influenced by the interactions with the niche, it is important to have the capacity to control the complexity of these interactions with 3D culture systems both at the mechanical and biochemical levels.

5.4.1 Three-dimensional Mechanoregulation of Stem Cell Fate

The stem cell niche is a complex three-dimensional (2D) physical environment with a wide range of topographical features and architectures [47, 210]. As mentioned above, this complex 3D cell microenvironment can influence cell shape and affect stem cell proliferation and differentiation [211]. The importance of mimicking *in vitro* with higher fidelity the architecture and the mechanical stimuli encountered by stem cells *in vivo* has led scientists to develop novel micro- and nanoengineered environments which recapitulate the morphological (and partially biochemical) 3D structure of the cell niche [212]. The design of such 3D scaffolds is of critical importance since multiple parameters must be considered. This includes, for instance, the selection of the biomaterial, an optimal design of the micro- and nanoarchitecture reproducing the morphology of the native tissue, the capability of nutrient transport within the scaffold, and the cell–cell and cell–matrix interactions. Multiple approaches and materials have been proposed in this regard. Here, we will restrict ourselves in describing the latest and most promising advances on 3D stem cell culture systems: biomimetic nanofibers, hydrogel scaffolds, and 3D bioprinted (hydrogel) scaffolds.

Nanofibers: The 3D nanofibrous structure of the ECM offers a vital network to support cells and guide cell behavior. Natural and synthetic manufactured nanofibers are attracting the interest of several groups in stem cell research due to their potential use as artificial scaffolds mimicking the fibrillary nanostructures of the stem cell niche [213, 214] (see Table 5.6). Nanofibers have been shown to have a great impact on stem cell adhesion, survival, migration, proliferation, and differentiation. Nanofibers range from tens of nanometers to a few micrometers and are commonly fabricated by means of electrospinning method which permits a high control of their physical properties, such as stiffness, morphology, dimension,

Table 5.6 Stem cell response on 3D scaffolds.

Environment	Technique	Material	Cell type	Differentiation	Ref.
Nanofibers	Electrospinning	PCL	hESCs	Endoderm	[213]
		Collagen	NSCs	Neurons/Glial	[215]
		PLLA	hMSCs	Osteogenic	[216]
		PU	mESCDCs	Cardiomyocyte	[243]
Hydrogel	µ/n-Structures molding	Gelatin	Myoblasts	Muscle	[226]
		OMA/PEG	hASCs	Chondrogenesis	[229]
	Photopolymerization	MAC	NSPCs	Neurons/Glial	[227]
	Stereolithography	PDLLA-PEG:HA	hASCs	Chondrogenesis	[228]
	3D Bioprinting	Agarose	mCpOCs	Osteogenic	[239]
		Alginate	hESCs	Hepatocyte	[241]
		PU	mNSCs	Neural	[240]

µ/n: Micro-/nanostructures; hASCs: human adipose-derived stem cells; hESCs: human embryonic stem cells; mCpOCs: mouse calvarial pre-osteoblast cells; mESCDCs: murine ESC-derived cardiomyocytes; mNSCs: murine neural stem cells; NSCs: neural stem cells; NSPCs: neural stem/progenitor cells; MAC: methacrylamide chitosan; OMA/PEG: oxidized, methacrylated alginate and 8-arm poly(ethylene glycol) amine; PCL: polycaprolactone; PDLLA-PEG:HA: Poly-d,l-lactic acid/polyethylene glycol/poly-d,l-lactic acid/hyaluronic acid; PLLA: Poly(l-lactic acid); PU: polyurethane.

alignment, or pore size [213, 215–217], offering a high versatility in material choice from a variety of polymer solutions or melts. Other methods such as phase separation and self-assembly are also typically used both for *ex vivo* stem cell culture and *in vivo* stem cell delivery. Many material and processing parameters must be considered to obtain the desired fiber size and arrangement. Nanofibers can be designed with different architectures (*e.g.* disordered, aligned fibers) to facilitate stem cell expansion of induce differentiation along specific lineages. Electrospun fibers have been widely applied as tissue engineering scaffolds for various types of stem cells. Recently, a 3D scaffold made of electrospun PCL nanofibers was developed showing interconnected and hierarchically structured pores similarly to the natural ECM [218]. Mouse bone marrow mesenchymal stem cells (mBMSCs) deposited on this scaffold enhanced chondrogenic lineage when stimulated with BMP-2. Finally, *in vivo* results suggested that the developed scaffold acts as a favorable synthetic ECM for the physiological process of ossification. Similarly, electrospun nanofibers made of silk fibroin/chitosan/nanohydroxyapatite tailored with BMP-2-induced osteogenesis on hBMSCs [219]. Interestingly, *in vivo* experiments implanting the developed nanofibers confirmed ectopic bone formation and osteogenesis. Other works have reported the use of biochemically functionalized nanofibers, made of either natural or synthetic materials, to serve as artificial stem cell niches that presents topographical and biochemical cues to influence the local regulation of stem cell fate [220–222] (see Table 5.6).

A nanofiber environment without any biochemical cues is less likely to exert significant influence to stem cells cultured in the matrix. In this regard, nanofibers tailored with cell-specific bioactive ligands (*e.g.* ECM proteins, growth factors), treated with plasma to introduce surface charge, or grafting polymerization to generate surface functional groups, have been used to facilitate cell adhesion and synergize with physical cues enhancing proliferation and differentiation. Specific stem cell biochemical signals should be an integral component of artificial nanofibrous matrices together with an optimal configuration of topographical cues. These nanofibers matrices are intended to provide provisional support and facilitate tissue organization, prior to (or following) *in vivo* implantation.

Altogether, nanofibers provide a unique environment to modulate cellular behavior and tissue regeneration at the nanometer scale, recreating same scale where molecular signaling transduction occurs. A recent review discussing in further detail on this topic can be found in Ref. [223].

Hydrogels: Hydrogels are polymeric materials capable of holding large amounts of water (or physiological solutions). They are used by most research groups to study stem cell fate [224] due to their manufacturing simplicity and increased cellular response. They can be easily tailored with functional groups and engineered to control their architecture, hydrophobicity, porosity, degradation kinetics, mimicking the stem cell niche environment. Hydrogels can be classified for their origin (natural, synthetic, semi-synthetic), their durability, composition, structure, charge, and/or their response to stimuli [225]. Techniques such as molding with sacrificial layers, stereolithograhy, photopolymerization, and bioprinting (see below) have been extensively used for shaping the hydrogel morphology with specific architecture [226–229] (see Table 5.6). The architecture and mechanical properties of hydrogels comparable to physiological ones can be easily achieved by modifying the content of cross-linker agent or ratio between components.

Typically, simple encapsulation of cells within the hydrogel is the most common strategy followed for studying stem cell differentiation in 3D hydrogel-based environments. Encapsulation allows assessing the interaction between stem cells in a native-like environment and specific signaling sequences where the ligand density can be precisely controlled. Examples of typical hydrogel scaffolds include collagen I, alginate, polyethylene glycol diacrylate (PEGDA) or HA (or a combination of them) [230]. Recently, hPSCs and hiPSCs were cultured on a thermo-responsive hydrogel (PNIPAAm-PEG) [231]. This environment provided cells with multiple advantages compared to its 2D counterpart, such as 3D morphology, aggregate-free, shear stress-free, and oxygen–nutrient exchange. Cells expanded and differentiated into multiple lineages of the three germ layers, such as neuron progenitors, endoderm progenitors, and cardiomyocites, among others. Selection of the cell fate was determined by a simple replacement of the expansion with the differentiation medium. Also recently, hMSCs were encapsulated on synthetic photocurable polymer hydrogel (pHEMA-co-APMA) grafted with polyamidoamine (PAA) [232]. Cells differentiated toward chondrogenic lineage confirmed by the elevated expression levels of aggrecan and collagen II. Together with differentiation media and hypoxia conditions, the 3D structure of the hydrogel was thought to govern cell fate. Similarly, chondrogenic lineage was also obtained i using human adipose-derived stem cells (hASCs) deposited on oxidized, methacrylated alginate and 8-arm poly(ethylene glycol) amine (OMA/PEG) 3D hydrogel [229] (see Table 5.6). The size of the 3D micropatterned structures was shown to be critical to influence chondrogenic lineage. Finally, neuronal differentiation was achieved by depositing human MSCs in soft Col:HA 3D scaffold (1000 Pa) and confirmed by the upregulation of neuronal mid- and late protein markers. Interestingly, stiffer values (10000 Pa) of the Col:HA elastic modulus showed glial differentiation within 1 week [212].

Note also that similar results were described for NSCs when deposited on 2D hydrogels (see Ref. [52]).

However, the formation of functional, higher-ordered tissues for clinical tissue regeneration applications will likely require greater sophistication in scaffold architecture. The control of mesh size within 3D hydrogel scaffolds will univocally contribute to determine the effects of specific cell morphologies on differentiation and proliferation. Similarly, the combination of biochemical stimuli together with 3D morphology will provide a more reliable reproduction of the stem cell–niche interactions.

3D Bioprinting: The ability to design and fabricate complex, 3D biomedical devices is critical in tissue engineering and regenerative medicine. Bioprinting of 3D hydrogels has emerged as a powerful alternative for the fabrication of 3D structures (typically collagen, HA, alginate, photocured acrylates, or modified copolymers, among others [233]) which faithfully mimic the structure of the native tissue [233, 234]. This methodology allows a more finely organization of cells and supporting structures in a precise 3D configuration gaining critical knowledge of cell–cell communications and cell–environment interactions. Recent reviews describing in detail the performance of 3D bioprinting have been published elsewhere [233, 235–238]. We will restrict ourselves to this specific type of 3D structures and exclude from our review scaffolds made by standard 3D printing layer-by-layer techniques (based on synthetic polymers). Nevertheless, it is worth mentioning that the fabrication of complex scaffolds, such as internal channels, can be easily performed with these techniques.

Combined with stem cells, 3D bioprinted environments can reproduce with the highest fidelity the complexity of the *in vivo* cell niche. Customized shapes with the desired mechanical and biochemical properties can be easily manufactured to stem cells differentiation or self-renewal. Altogether, this could have important implications for the development of complex 3D tissue-like models and tissue-engineering applications. In this regard, bioprinted agarose scaffolds including microchannels mimicking the blood vessels diameter (150–1000 μm) were used to culture mouse calvarial preosteoblast cells (see Table 5.6) [239]. Osteogenic differentiation assessed by alkaline phosphatise assay, indicated a larger differentiation yield of cells within the microchannels compared to cells within the gel block. Similarly, a biodegradable polyurethane (PU) hydrogel was used to embed NSCs (see Figure 5.4 and Table 5.6) [240]. NSCs proliferated and differentiated in PU hydrogels with a varying stiffness of 680–2400 Pa. The hydrogels were injected into the zebrafish embryo neural injury model. Interestingly, the brain function of a zebrafish with impaired nervous system, as well as with traumatic brain injury, was rescued.

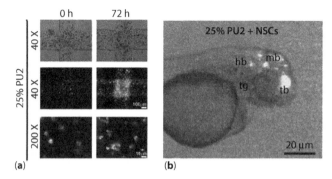

Figure 5.4 Three-dimensional bioprinting technique applied for tissue regeneration applications. (a) Optical (upper) and fluorescence images (mid and lower) of NSCs labeled with PKH26 fluorescent dye, embedded in a 3D bioprinted PU hydrogel (2 layers). Cells were imaged at 0 and 72 h confirming cell proliferation and viability. (b) Zebrafish embryo injected with NSCs-laden PU hydrogel. Implanted embryos showed high rescue rates in locomotion after traumatic injury. (Key: hb, hindbrain; mb, midbrain; fb, forebrain; tg, trigeminal ganglion). Reproduced with permission of Elsevier Ltd. [240].

The combination of induced pluripotent stem cells and bioprinting technologies paves the way toward the production of organs and tissues on demand. On this regard, hPSCs were bioprinted using an alginate hydrogel matrix [241]. Cells differentiated into hepatocyte-like cells (HLCs) which could be applied for the generation of mini-livers in 3D. Prior differentiation, bioprinted hPSCs maintained their pluripotent state before being forced to differentiate into HLCs which confirms that stem cells self-renewal can also be achieved by means of 3D bioprinting technology. In this context, embryonic stem cells with hydrogels were bioprinted into 3D macroporous structure by means of an extrusion-based temperature-sensitive technology. Under controlled conditions, 98% of cells maintained their pluripotent state as examined by the expression of pluripotency markers (Oct4, SSEA1) [242].

Together, these studies demonstrate that, compared with other techniques, 3D bioprinted environments can mimic with higher fidelity the biophysical environment encountered by stem cells. Finally, this new emerging technology needs to face technological limitations if planned to be used of regenerative medicine and tissue engineering applications. Increased resolution, limited geometries, speed, compatibility with biologically relevant materials, and sufficient vascularization, are a challenge that researchers need to address in the near future. Interdisciplinary efforts will be needed in order to address these challenges. We anticipate that 3D bioprinting technologies will have a great impact on the future of biomedicine given the multitude of applications, from tissue and organ development to complex *in vitro* models of human diseases [238].

5.4.2 Three-dimensional Biochemical Patterns for Stem Cell Differentiation

In vivo, 3D biochemical patterns are generated and sustained with precisely controlled spatial and temporal profiles on a variety of lengths and time scales [55]. Controlled immobilization of bioactive molecules for site-specific activation of cell signaling is required for the study of stem cells–ECM interactions occurring *in vivo*. Therefore, the combination of surface functionalization together with 3D micro- and nanopatterning could provide a more reliable reproduction of the stem cell niche interactions. Patterns of tendon-promoting FGF-2 were created by inkjet printing in oriented submicron polystyrene fibers scaffolds with around 655 nm of diameter, coated with serum or fibrin. The deposited concentration of patterned FGF-2 ranged from 40 to 650 pg/mm^2. Multipotent C3H10T1/2 mesenchymal fibroblasts cultured in the fibers showed cell alignment along the fiber length and promoted the tenocyte fate. Cells showed a dose dependence to FGF-2 concentration with FGF-2 doses higher than 245 pg/mm^2 resulting in an increase in the tendon marker Scx expression relative to control regions [244].

More sophisticated 3D stem cells microenvironments can be designed in polymer-based hydrogels [245–247]. Cell studies on proliferation, migration, and differentiation inside these materials have been performed by incorporating physiologically relevant cues [248]. Two-photon irradiation of an agarose gel modified with coumarin-caged thiols was used to pattern simultaneously (due to the orthogonal chemistry of the used peptide binding pairs) the differentiation factors sonic hedgehog (SHH) and CNTF [249]. Coumarin-caged thiols were uncaged upon two-photon irradiation to expose reactive thiol groups. These thiols act as anchoring sites for the immobilization of barnase and streptavidin proteins containing thiol reactive maleimides. Barstar-SHH and biotin-CNTF were soaked simultaneously, and specifically bound to immobilized barnase and streptavidin, respectively. Two-photon patterning afforded 3D control because of the limited excitation volume. A 5 μm resolution in the *x–y* plane could be achieved, while the profile of fluorescence of patterned fluorescent proteins broadened along the *z*-axis with scan number. The amount of irradiation correlates with the amount of protein immobilized. The bioactivity of the immobilized differentiation factors was tested in retinal precursor cells (RCPs). Immobilized proteins showed no cytotoxicity and preserved their bioactivity. A gradient of barstar-SHH (100–500 ng/mL over the first 100 μm depth) was also produced and tested cell migration into the patterned agarose hydrogel with adult NPCs (known to migrate along SHH gradients) [250]. Cell migration was observed predominantly into patterns containing SHH gradients when compared with controls [251].

The biochemical cues present in the stem cell niche are temporally variable. Therefore, a system that permit the temporally controlled manipulation of the exposed biochemical signals, will more closely recapitulate the stem cell microenvironment. Chemical patterning of hydrogel scaffolds have progressively evolved to permit the introduction of localized biochemical signals that can be activated at different time points to dynamically affect cell function [252, 253]. In this regard, hydrogel-based scaffolds that permit the 3D photoreversible patterning of RGD were produced. In an example, the chemical composition of the hydrogel matrix has been tuned to allow the selective immobilization of the fluorescent cell adhesive peptide Ac-C-(PL)-RGDSK(AF$_{488}$)-NH$_2$ through a thiol-ene photoreaction, which is initiated by visible light (λ = 490–650 nm). The degree of thiol-ene photoconjugation was controlled by varying the exposure time of the irradiation or controlling the amount of the photoinitiator. The resulting patterned peptide concentrations obtained are within a biological relevant range (0–1.0 mM) [252, 254]. Also, the photolabile o-nitrobenzyl moiety in Ac-C-(PL)-RGDSK(AF$_{488}$)-NH$_2$ can be cleaved through the photoscission of upon exposure to UV light (λ = 365 nm). Bio-orthogonality among thiol-ene addition and photocleavage reactions allow them to be performed independently by irradiating the system with different light sources. When initiated with multiphoton light, both reactions can be performed in 3D, enabling the reversible spatial presentation of the cell adhesive RGD sequence within the hydrogel matrix. A patterning resolution of 1 μm in the x–y plane, and 3–5 μm in the z plane were obtained (Figure 5.5). A part from single-concentration patterns, nonlinear gradients of RGD were created by exposing the hydrogel to a gradient of UV light. Since the described photopatterning technique (wavelengths, exposure times and intensity of light) is cytocompatible, mouse embryonic fibroblasts (NIH 3T3) were cultured on exposed patterns of RGD. Posterior irradiation with UV light allowed the selectively detachment of cells from the irradiated areas of the pattern. This methodology provides and efficient way of sampling complex cell populations, as those derived from stem cells, which could be selectively collected from their artificial 3D microenvironment and further expanded [255].

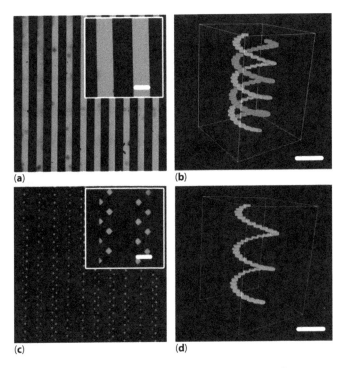

Figure 5.5 Thiol-ene patterning of the fluorescent peptide Ac-C-(PL)-RGDSK(AF488)-NH$_2$ into the hydrogel (a) in 2D throughout the gel after exposure to masked visible light and (b) in 3D after exposure to focused pulsed laser light. False coloring is used for enhanced visualization. (c and d) Subsets of the pre-patterned cues were removed by exposure to UV light to modify the original chemical pattern and yield new 2D and 3D patterns. Scale bars: 200 μm. Reproduced with permission of John Wiley & Sons Ltd. [255].

A very similar strategy have been used in the study of chondrogenic differentiation of MSCs. Hydrogels synthesized from the copolymerization of nondegradable PEG diacrylate (PEGDA) and a photodegradable acrylate containing the photolabile nitrobenzyl ether moiety and the adhesion peptide sequence Arg–Gly–Asp–Ser (RGDS) were used in the dynamic tuning of RGDS presentation to encapsulated MSCs. A portion of the cell–gel constructs containing the photolabile RGDS tether were irradiated to release RGDS from the encapsulated cell microenvironment at day 10 in culture. By Day 21, a four-fold statistical increase in the production of the ECM type II collagen and glycosaminoglycans (GAGs) typical from chondrogenic differentiation occurred, indicating further differentiation of the hMSCs down the chondrogenic pathway in response to the microenvironment modification [256].

5.5 Conclusions and Future Perspectives

In this chapter, we showed how the production of micro- and nanoengineered environments can mechanically and biochemically influence stem cell fate and self-renewal. The interplay between the physical and biochemical stimulation, together with the chosen cell culture media, may influence the determination of stem cell fate. The physical properties of the micro- and nanoengineered domains have a critical effect on lineage determination. We showed solid evidences that this conversion between a physical signals to a biochemical response are carried out by the mechanical interaction between cells and their micro- and nanoenvironments. We also reviewed the effects of controlled surface immobilization of biochemical stimuli in stem cell differentiation, highlighting the geometrical restrictions involved. We presented some works which pointed to integrins as the main determinants of stem cell fate. Therefore, the dimensions, morphology, and/or stiffness of the biomaterials, and their biochemical composition at the micro- and nanoscales must be chosen with great care for particular stem cell responses. The optimal equilibrium between all these cues will result in a high control of cell behavior. The selected engineered material, ranging from micro- and nanostructured surfaces, fibrous environments, 3D hydrogels, or bioprinted scaffolds must be selected depending on the user requirements. Note finally that, 3D engineered biomaterials are becoming a powerful tool for building on-demand tissues in the field of tissue engineering. The future advances of stem cells applications and 3D environments in particular include the integration of different printing mechanisms to engineer multiphasic tissues with optimized scaffolds to increase our understanding of stem cell biology.

References

1. Bianco, P., Riminucci, M., Gronthos, S., Robey, P.G. Bone marrow stromal stem cells: Nature, biology, and potential applications. *Stem Cells*, 19, 180, 2001.
2. Li, L., Clevers, H. Coexistence of quiescent and active adult stem cells in mammals. *Science*, 327, 542, 2010.
3. Cheung, T.H., Rando, T.A. Molecular regulation of stem cell quiescence, *Nature Rev. Mol. Cell Biol.*, 14, 329, 2013.
4. Morrison, S.J., Kimble, J. Asymmetric and symmetric stem-cell divisions in development and cancer. *Nature*, 441, 1068, 2006.
5. Rojas-Ríos, P., González-Reyes, A. Concise review: The plasticity of stem cell niches: a general property behind tissue homeostasis and repair. *Stem Cells*, 32, 852, 2014.

6. Lander, A.D., Kimble, J., Clevers, H., Fuchs, E., Montarras, D., Buckingham, M., Calof, A.L., Trumpp, A., Oskarsson, T. What does the concept of the stem cell niche really mean today? *BMC Biol.*, 10, 19, 2012.
7. Schofield, R. The relationship between the spleen colony-forming cell and the haemopoietic stem cell. *Blood Cells*, 4, 7, 1978.
8. Park, D., Sykes, D.B., Scadden, D.T. The hematopoietic stem cell niche. *Front. Biosci.*, 17, 30, 2012.
9. Kimble JE, White JG: On the control of germ cell development in *Caenorhabditis elegans*. *Dev Biol* 1981, 81:208–219.
10. Austin, J., Kimble, J. glp-1 is required in the germ line for regulation of the decision between mitosis and meiosis in *C. elegans*. *Cell*, 51, 589, 1987.
11. Oshima, H., Rochat, A., Kedzia, C., Kobayashi, K., Barrandon, Y. Morphogenesis and renewal of hair follicles from adult multipotent stem cells. *Cell*, 104, 233, 2001.
12. Tanimura, S., Tadokoro, Y., Inomata, K., Binh, N.T., Nishie, W., Yamazaki, S., Nakauchi, H., Tanaka, Y., McMillan, J.R., Sawamura, D., Yancey, K., Shimizu, H., Nishimura, E.K. Hair follicle stem cells provide a functional niche for melanocyte stem cells. *Cell Stem Cell*, 8, 177, 2011.
13. Marshman, E., Booth, C., Potten, C.S. The intestinal epithelial stem cell. *Bioessays*, 24, 91, 2002.
14. Barker, N., van Es, J.H., Kuipers, J., Kujala, P., van den Born, M., Cozijnsen, M., Haegebarth, A., Korving, J., Begthel, H., Peters, P.J., Clevers, H. Identification of stem cells in small intestine and colon by marker gene Lgr5. *Nature*, 449, 1003, 2007.
15. Mauro, A. Satellite cell of skeletal muscle fibers. *J. Biophys. Biochem. Cytol.*, 9, 493, 1961.
16. Pallafacchina, G., François, S., Regnault, B., Czarny, B., Dive, V., Cumano, A., Montarras, D., Buckingham, M. An adult tissue-specific stem cell in its niche: a gene profiling analysis of *in vivo* quiescent and activated muscle satellite cells. *Stem Cell Res.*, 4, 77, 2010.
17. Doetsch, F. A niche for adult neural stem cells. *Curr. Opin. Genet. Dev.*, 13, 543, 2003.
18. Ihrie, R.A., Alvarez-Buylla, A. Lake-front property: a unique germinal niche by the lateral ventricles of the adult brain. *Neuron*, 70, 674, 2011.
19. O'Reilly, A.M., Lee, H.H., Simon, M.A. Integrins control the positioning and proliferation of follicle stem cells in the Drosophila ovary. *J. Cell Biol.*, 182, 801, 2008.
20. Simmons, P.J., Levesque, J.P., Zannettino, A.C. Adhesion molecules in haemopoiesis. *Baillieres Clin. Haematol.*, 10, 485, 1997.
21. Buitenhuis, M. The role of PI3K/protein kinase B (PKB/c-akt) in migration and homing of hematopoietic stem and progenitor cells. *Curr. Opin. Hematol.* 18, 226, 2011.
22. Suh, H.N., Han, H.J. Collagen I regulates the self-renewal of mouse embryonic stem cells through $\alpha 2\beta 1$ integrin- and DDR1-dependent Bmi-1. *J. Cell. Physiol.*, 226, 3422, 2011.

23. Chen, S., Lewallen, M., Xie, T. Adhesion in the stem cell niche: biological roles and regulation. *Development*, 140, 255, 2013.
24. Hynes, R.O. The extracellular matrix: not just pretty fibrils. *Science*, 326, 1216, 2009.
25. Cosgrove, B.D., Sacco, A., Gilbert, P.M., Blau, H.M. A home away from home: challenges and opportunities in engineering *in vitro* muscle satellite cell niches. *Differentiation*, 78, 185, 2010.
26. Douet, V., Kerever, A., Arikawa-Hirasawa, E., Mercier, F. Fractone-heparan sulphates mediate FGF-2 stimulation of cell proliferation in the adult subventricular zone. *Cell Prolif.*, 46, 137, 2013.
27. Urciuolo, A., Quarta, M., Morbidoni, V., Gattazzo, F., Molon, S., Grumati, P., Montemurro, F., Tedesco, F.S., Blaauw, B., Cossu, G., Vozzi, G., Rando, T.A., Bonaldo, P. Collagen VI regulates satellite cell self-renewal and muscle regeneration. *Nat Commun.*, 4, 1964, 2013.
28. Ingber, D.E. The mechanochemical basis of cell and tissue regulation. *Mech. Chem. Biosyst.*, 1, 53, 2004.
29. Estes, B.T., Gimble, J.M., Guilak, F. Mechanical signals as regulators of stem cell fate. *Curr. Top. Dev. Biol.* 60, 91, 2004.
30. Engler, A.J., Sen, S., Sweeney, H.L., Discher, D.E. Matrix elasticity directs stem cell lineage specification. *Cell*, 126, 677, 2006.
31. Trappmann, B., Gautrot, J.E., Connelly, J.T., Strange, D.G.T., Li, Y., Oyen, M.L., Stuart, M.A.C., Boehm, H., Li, B., Vogel, V., Spatz, J.P., Watt, F.M., Huck, W.T.S. Extracellular-matrix tethering regulates stem-cell fate. *Nat. Mater.*, 11, 642, 2012.
32. Wen, J.H., Vincent, L.G., Fuhrmann, A., Choi, Y.S., Hribar, K.C., Taylor-Weiner, H., Chen, S., Engler, A.J. Interplay of matrix stiffness and protein tethering in stem cell differentiation. *Nat. Mat.*, 13, 979, 2014.
33. Thomson, J.A., Itskovitz-Eldor, J., Shapiro, S.S., Waknitz, M.A., Swiergiel, J.J., Marshall, V.S., Jones, J.M. Embryonic stem cell lines derived from human blastocysts. *Science*, 282, 1145, 1998.
34. Takahashi, K., Tanabe, K., Ohnuki, M., Narita, M., Ichisaka, T., Tomoda, K., Yamanaka, S. Induction of pluripotent stem cells from adult human fibroblasts by defined factors. *Cell*, 131, 861, 2007.
35. Vazin, T., Chen, J., Lee, C.T., Amable, R., Freed, W.J. Assessment of stromal derived inducing activity in the generation of dopaminergic neurons from human embryonic stem cells. *Stem Cells*, 26, 1517, 2008.
36. Xu, C., Inokuma, M.S., Denham, J., Golds, K., Kundu, P., Gold, J.D., Carpenter, M.K. Feeder-free growth of undifferentiated human embryonic stem cells. *Nat Biotechnol.*, 19, 971, 2001.
37. Battista, S., Guarnieri, D., Borselli, C., Zeppetelli, S., Borzacchiello, A., Mayol, L., Gerbasio, D., Keene, D.R., Ambrosio, L., and Netti, P.A. The effect of matrix composition of 3D constructs on embryonic stem cell differentiation. *Biomaterials*, 26, 6194, 2005.
38. Gagner, J.E., Kim, W., Chaikof, E.L. Designing protein-based biomaterials for medical applications. *Acta Biomater.*, 10, 1542, 2014.

39. Ananthanarayanan, B., Little, L., Schaffer, D.V., Healy, K.E., Tirrel, M. Neural stem cell adhesion and proliferation on phospholipid bilayers functionalized with RGD peptides. *Biomaterials*, 31, 8706, 2010.
40. Dingal, P.C.D.P., Discher, D.E. Combining insoluble and soluble factors to steer stem cell fate. *Nat Mat.*, 13, 532, 2014.
41. Kosta, J., Langer, R. Controlled release of bioactive agents. *Cell*, 2, 47, 1984.
42. Solorio, L.D., Fu, A.S., Hernández-Irizarry, R., Alsberg, E. Chondrogenic differentiation of human mesenchymal stem cell aggregates via controlled release of TGF-β1 from incorporated polymer microspheres. *J. Biomed. Mater. Res. A.*, 92, 1139, 2010.
43. Pompe, T., Salchert, K., Alberti, K., Zandstra, P., Werner, C. Immobilization of growth factors on solid supports for the modulation of stem cell fate. *Nat. Protoc.*, 5, 1042, 2010.
44. Masters, K.S. Covalent growth factor immobilization strategies for tissue repair and regeneration. *Macromol. Biosci.*, 11, 1149, 2011.
45. Joddar, B., Hoshiba, T., Chen, G., Ito, Y. Stem cell culture using cell-derived substrates. *Biomater. Sci.*, 2, 1595, 2014.
46. Griffin, M.F., Butler, P.E., Seifalian, A.M., Kalaskar, D.M. Control of stem cell fate by engineering their micro and nanoenvironment. *World J. Stem Cells*, 7, 37, 2015.
47. Discher, D.E., Mooney, D.J., Zandstra, P.W., Growth factors, matrices, and forces combine and control stem cells. *Science*, 324, 1673, 2009.
48. Scadden, D.T., The stem-cell niche as an entity of action. *Nature*, 441, 1075, 2006.
49. Walker, M.R., Patel, K.K., Stappenbeck, T.S., The stem cell niche. *J. Pathol.*, 217, 169, 2009.
50. Ghosh, K., Ingber, D.E., Micromechanical control of cell and tissue development: Implications for tissue engineering. *Adv. Drug Deliv. Rev.*, 59, 1306, 2007.
51. Guilak, F., Cohen, D.M., Estes, B.T., Gimble, J.M., Liedtke, W., Chen, C.S., Control of stem cell fate by physical interactions with the extracellular matrix. *Cell Stem Cell*, 5, 17, 2009.
52. Saha, K., Keung, A.J., Irwin, E.F., Li, Y., Little, L., Schaffer, D.V., Healy, K.E., Substrate modulus directs neural stem cell behavior. *Biophys. J.*, 95, 4426, 2008.
53. Yang, C., Tibbitt, M.W., Basta, L., Anseth, K.S., Mechanical memory and dosing influence stem cell fate. *Nat. Mater.*, 13, 645, 2014.
54. Sacco, A., Doyonnas, R., Kraft, P., Vitorovic, S., Blau, H.M., Self-renewal and expansion of single transplanted muscle stem cells. *Nature*, 456, 502, 2008.
55. Burdick, J.A., Vunjak-Novakovic, G., Engineered microenvironments for controlled stem cell differentiation. *Tiss. Eng. Part A*, 15, 205, 2009.
56. Kwon, K.W., Park, H., Song, K.H., Choi, J.C., Ahn, H., Park, M.J., Suh, K.Y., Doh, J., Nanotopography-guided migration of t cells. *J. Immunol.*, 189, 2266, 2012.

57. Kim, D.H., Provenzano, P.P., Smith, C.L., Levchenko, A., Matrix nanotopography as a regulator of cell function. *J. Cell Biol.*, 197, 351, 2012.
58. Dalby, M.J., Gadegaard, N., Oreffo, R.O.C., Harnessing nanotopography and integrin-matrix interactions to influence stem cell fate. *Nat. Mater.*, 13, 558, 2014.
59. Mills, C.A., Fernandez, J.G., Martinez, E., Funes, M., Engel, E., Errachid, A., Planell, J., Samitier, J., Directional alignment of mg63 cells on polymer surfaces containing point microstructures. *Small*, 3, 871, 2007.
60. Stevens, M.M., George, J.H., Exploring and engineering the cell surface interface. *Science*, 310, 1135, 2005.
61. Caballero, D., Comelles, J., Piel, M., Voituriez, R., Riveline, D., Ratchetaxis: Directed cell migration by local cues. *Trends Cell Biol.*, 25, 815, 2015.
62. Caballero, D., Voituriez, R., Riveline, D., Protrusion fluctuations direct cell motion. *Biophys J*, 107, 34, 2014.
63. Caballero, D., Voituriez, R., Riveline, D., The cell ratchet: Interplay between efficient protrusions and adhesion determines cell motion. *Cell Adh. Mig.*, 9, 327, 2015.
64. Flemming, R.G., Murphy, C.J., Abrams, G.A., Goodman, S.L., Nealey, P.F., Effects of synthetic micro- and nano-structured surfaces on cell behavior. *Biomaterials*, 20, 573, 1999.
65. Dalby, M.J., Riehle, M.O., Yarwood, S.J., Wilkinson, C.D.W., Curtis, A.S.G., Nucleus alignment and cell signaling in fibroblasts: Response to a microgrooved topography. *Exp. Cell Res.*, 284, 272, 2003.
66. Andersson, A.S., Bäckhed, F., von Euler, A., Richter-Dahlfors, A., Sutherland, D., Kasemo, B., Nanoscale features influence epithelial cell morphology and cytokine production. *Biomaterials*, 24, 3427, 2003.
67. Estévez, M., Martínez, E., Yarwood, S.J., Dalby, M.J., Samitier, J., Adhesion and migration of cells responding to microtopography. *J. Biomed. Mater. Res. A*, 103, 1659, 2015.
68. Elias, C.N., Oshida, Y., Lima, J.H.C., Muller, C.A., Relationship between surface properties (roughness, wettability and morphology) of titanium and dental implant removal torque. *J. Mech. Behav. Biomed. Mater.*, 1, 234, 2008.
69. Faia-Torres, A.B., Guimond-Lischer, S., Rottmar, M., Charnley, M., Goren, T., Maniura-Weber, K., Spencer, N.D., Reis, R.L., Textor, M., Neves, N.M., Differential regulation of osteogenic differentiation of stem cells on surface roughness gradients. *Biomaterials*, 35, 9023, 2014.
70. Faia-Torres, A.B., Charnley, M., Goren, T., Guimond-Lischer, S., Rottmar, M., Maniura-Weber, K., Spencer, N.D., Reis, R.L., Textor, M., Neves, N.M., Osteogenic differentiation of human mesenchymal stem cells in the absence of osteogenic supplements: A surface-roughness gradient study. *Acta Biomater.*, 28, 64, 2015.
71. Deligianni, D.D., Katsala, N.D., Koutsoukos, P.G., Missirlis, Y.F., Effect of surface roughness of hydroxyapatite on human bone marrow cell adhesion, proliferation, differentiation and detachment strength. *Biomaterials*, 22, 87, 2000.

72. D'Elia, N.L., Mathieu, C., Hoemann, C.D., Laiuppa, J.A., Santillan, G.E., Messina, P.V., Bone-repair properties of biodegradable hydroxyapatite nanorod superstructures. *Nanoscale*, 2015.
73. Dalby, M.J., Gadegaard, N., Tare, R., Andar, A., Riehle, M.O., Herzyk, P., Wilkinson, C.D.W., Oreffo, R.O.C., The control of human mesenchymal cell differentiation using nanoscale symmetry and disorder. *Nat. Mater.*, 6, 997, 2007.
74. Kilian, K.A., Bugarija, B., Lahn, B.T., Mrksich, M., Geometric cues for directing the differentiation of mesenchymal stem cells. *Proc. Natl. Acad. Sci. U.S.A*, 107, 4872, 2010.
75. Jin, S., Yao, H., Krisanarungson, P., Haukas, A., Ye, K., Porous membrane substrates offer better niches to enhance the wnt signaling and promote human embryonic stem cell growth and differentiation. *Tiss. Eng. Part A*, 18, 1419, 2012.
76. Singhvi, R., Stephanopoulos, G., Wang, D.I.C., Effects of substratum morphology on cell physiology. *Biotechnol. Bioeng.*, 43, 764, 1994.
77. Reubinoff, B.E., Pera, M.F., Fong, C.Y., Trounson, A., Bongso, A., Embryonic stem cell lines from human blastocysts: Somatic differentiation *in vitro*. *Nat. Biotech.*, 18, 399, 2000.
78. Lyu, Z., Wang, H., Wang, Y., Ding, K., Liu, H., Yuan, L., Shi, X., Wang, M., Wang, Y., Chen, H., Maintaining the pluripotency of mouse embryonic stem cells on gold nanoparticle layers with nanoscale but not microscale surface roughness. *Nanoscale*, 6, 6959, 2014.
79. Bae, D., Moon, S.H., Park, B.G., Park, S.-J., Jung, T., Kim, J.S., Lee, K.B., Chung, H.M., Nanotopographical control for maintaining undifferentiated human embryonic stem cell colonies in feeder free conditions. *Biomaterials*, 35, 916, 2014.
80. Carey, B.W., Finley, L.W.S., Cross, J.R., Allis, C.D., Thompson, C.B., Intracellular α-ketoglutarate maintains the pluripotency of embryonic stem cells. *Nature*, 518, 413, 2015.
81. Jaggy, M., Zhang, P., Greiner, A.M., Autenrieth, T.J., Nedashkivska, V., Efremov, A.N., Blattner, C., Bastmeyer, M., Levkin, P.A., Hierarchical micronano surface topography promotes long-term maintenance of undifferentiated mouse embryonic stem cells. *Nano Lett.*, 15, 7146, 2015.
82. Chen, W., Villa-Diaz, L.G., Sun, Y., Weng, S., Kim, J.K., Lam, R.H.W., Han, L., Fan, R., Krebsbach, P.H., Fu, J., Nanotopography influences adhesion, spreading, and self-renewal of human embryonic stem cells. *ACS Nano*, 6, 4094, 2012.
83. Jeon, K., Oh, H.J., Lim, H., Kim, J.H., Lee, D.H., Lee, E.R., Park, B.H., Cho, S.G., Self-renewal of embryonic stem cells through culture on nanopattern polydimethylsiloxane substrate. *Biomaterials*, 33, 5206, 2012.
84. Tsimbouri, P.M., McMurray, R.J., Burgess, K.V., Alakpa, E.V., Reynolds, P.M., Murawski, K., Kingham, E., Oreffo, R.O.C., Gadegaard, N., Dalby, M.J., Using nanotopography and metabolomics to identify biochemical effectors of multipotency. *ACS Nano*, 6, 10239, 2012.

85. Turner, L.A., Dalby, M.J., Nanotopography – potential relevance in the stem cell niche. *Biomater. Sci.*, 2, 1574, 2014.
86. Abagnale, G., Steger, M., Nguyen, V.H., Hersch, N., Sechi, A., Joussen, S., Denecke, B., Merkel, R., Hoffmann, B., Dreser, A., Schnakenberg, U., Gillner, A., Wagner, W., Surface topography enhances differentiation of mesenchymal stem cells towards osteogenic and adipogenic lineages. *Biomaterials*, 61, 316, 2015.
87. Martínez, E., Lagunas, A., Mills, C.A., Rodríguez-Seguí, S., Estévez, M., Oberhansl, S., Comelles, J., Samitier, J., Stem cell differentiation by functionalized micro- and nanostructured surfaces. *Nanomedicine*, 4, 65, 2008.
88. Lee, M.R., Kwon, K.W., Jung, H., Kim, H.N., Suh, K.Y., Kim, K., Kim, K.-S., Direct differentiation of human embryonic stem cells into selective neurons on nanoscale ridge/groove pattern arrays. *Biomaterials*, 31, 4360, 2010.
89. Ferrari, A., Faraci, P., Cecchini, M., Beltram, F., The effect of alternative neuronal differentiation pathways on pc12 cell adhesion and neurite alignment to nanogratings. *Biomaterials*, 31, 2565, 2010.
90. Moe, A.A.K., Suryana, M., Marcy, G., Lim, S.K., Ankam, S., Goh, J.Z.W., Jin, J., Teo, B.K.K., Law, J.B.K., Low, H.Y., Goh, E.L.K., Sheetz, M.P., Yim, E.K.F., Microarray with micro- and nano-topographies enables identification of the optimal topography for directing the differentiation of primary murine neural progenitor cells. *Small*, 8, 3050, 2012.
91. Alapan, Y., Icoz, K., Gurkan, U.A., Micro- and nanodevices integrated with biomolecular probes. *Biotechnol. Adv.*, 2015.
92. Bean, A.C., Tuan, R.S. Stem cells and nanotechnology in tissue engineering and regenerative medicine. *Micro and nanotechnologies in engineering stem cells and tissues*, p. 1, John Wiley & Sons, Inc., 2013.
93. Kingham, E., White, K., Gadegaard, N., Dalby, M.J., Oreffo, R.O.C., Nanotopographical cues augment mesenchymal differentiation of human embryonic stem cells. *Small*, 9, 2140, 2013.
94. Martínez, E., Engel, E., López-Iglesias, C., Mills, C.A., Planell, J.A., Samitier, J., Focused ion beam/scanning electron microscopy characterization of cell behavior on polymer micro-/nanopatterned substrates: A study of cell–substrate interactions. *Micron*, 39, 111, 2008.
95. Caballero, D., Villanueva, G., Plaza, J., Mills, C.A., Samitier, J., Errachid, A., Sharp high-aspect-ratio afm tips fabricated by a combination of deep reactive ion etching and focused ion beam techniques. *J. Nanosci. Nanotechnol.*, 10, 497, 2010.
96. López-Bosque, M.J., Tejeda-Montes, E., Cazorla, M., Linacero, J., Atienza, Y., Smith, K.H., Lladó, A., Colombelli, J., Engel, E., Mata, A., Fabrication of hierarchical micro–nanotopographies for cell attachment studies. *Nanotechnology*, 24, 255305, 2013.
97. Yin, Z., Cheng, E., Zou, H., A novel hybrid patterning technique for micro and nanochannel fabrication by integrating hot embossing and inverse UV photolithography. *Lab Chip*, 14, 1614, 2014.

98. Chen, J., Zhou, Y., Wang, D., He, F., Rotello, V.M., Carter, K.R., Watkins, J.J., Nugen, S.R., Uv-nanoimprint lithography as a tool to develop flexible microfluidic devices for electrochemical detection. *Lab Chip*, 15, 3086, 2015.
99. Verhulsel, M., Vignes, M., Descroix, S., Malaquin, L., Vignjevic, D.M., Viovy, J.L., A review of microfabrication and hydrogel engineering for microorgans on chips. *Biomaterials*, 35, 1816, 2014.
100. Engel, E., Martínez, E., Mills, C.A., Funes, M., Planell, J.A., Samitier, J., Mesenchymal stem cell differentiation on microstructured poly (methyl methacrylate) substrates. *Ann. Anat.*, 191, 136, 2009.
101. Yang, Y., Leong, K.W., Nanoscale surfacing for regenerative medicine. *Wiley Interdiscip. Rev. Nanomed. Nanobiotechnol.*, 2, 478, 2010.
102. Zanchetta, E., Guidi, E., Della Giustina, G., Sorgato, M., Krampera, M., Bassi, G., Di Liddo, R., Lucchetta, G., Conconi, M.T., Brusatin, G., Injection molded polymeric micropatterns for bone regeneration study. *ACS Appl. Mater. Interfaces*, 7, 7273, 2015.
103. Yang, K., Park, E., Lee, J.S., Kim, I.S., Hong, K., Park, K.I., Cho, S.W., Yang, H.S., Biodegradable nanotopography combined with neurotrophic signals enhances contact guidance and neuronal differentiation of human neural stem cells. *Macromol. Biosci.*, 15, 1348, 2015.
104. Kim, T.H., Shah, S., Yang, L., Yin, P.T., Hossain, M.K., Conley, B., Choi, J.W., Lee, K.B., Controlling differentiation of adipose-derived stem cells using combinatorial graphene hybrid-pattern arrays. *ACS Nano*, 9, 3780, 2015.
105. Pan, F., Zhang, M., Wu, G., Lai, Y., Greber, B., Schöler, H.R., Chi, L., Topographic effect on human induced pluripotent stem cells differentiation towards neuronal lineage. *Biomaterials*, 34, 8131, 2013.
106. Qi, L., Li, N., Huang, R., Song, Q., Wang, L., Zhang, Q., Su, R., Kong, T., Tang, M., Cheng, G., The effects of topographical patterns and sizes on neural stem cell behavior. *PLoS One*, 8, e59022, 2013.
107. Ghassemi, S., Meacci, G., Liu, S., Gondarenko, A.A., Mathur, A., Roca-Cusachs, P., Sheetz, M.P., Hone, J., Cells test substrate rigidity by local contractions on submicrometer pillars. *Proc. Natl. Acad. Sci. USA*, 109, 5328, 2012.
108. Skardal, A., Mack, D., Atala, A., Soker, S., Substrate elasticity controls cell proliferation, surface marker expression and motile phenotype in amniotic fluid-derived stem cells. *J. Mech. Behav. Biomed. Mater.*, 17, 307, 2013.
109. Comelles, J., Caballero, D., Voituriez, R., Hortigüela, V., Wollrab, V., Godeau, A., Samitier, J., Martinez, E., Riveline, D., Cells as active particles in asymmetric potentials: Motility under external gradients. *Biophys. J.*, 107, 1513, 2014.
110. Unadkat, H.V., Hulsman, M., Cornelissen, K., Papenburg, B.J., Truckenmüller, R.K., Carpenter, A.E., Wessling, M., Post, G.F., Uetz, M., Reinders, M.J.T., Stamatialis, D., van Blitterswijk, C.A., de Boer, J., An algorithm-based topographical biomaterials library to instruct cell fate. *Proc. Natl. Acad. Sci. USA*, 108, 16565, 2011.

111. Tong, Z., Solanki, A., Hamilos, A., Levy, O., Wen, K., Yin, X., Karp, J.M., Application of biomaterials to advance induced pluripotent stem cell research and therapy. *EMBO J.*, 34, 987, 2015.
112. Chen, A., Lieu, D.K., Freschauf, L., Lew, V., Sharma, H., Wang, J., Nguyen, D., Karakikes, I., Hajjar, R.J., Gopinathan, A., Botvinick, E., Fowlkes, C.C., Li, R.A., Khine, M., Shrink-film configurable multiscale wrinkles for functional alignment of human embryonic stem cells and their cardiac derivatives. *Adv. Mater.*, 23, 5785, 2011.
113. Wang, P.Y., Clements, L.R., Thissen, H., Jane, A., Tsai, W.B., Voelcker, N.H., Screening mesenchymal stem cell attachment and differentiation on porous silicon gradients. *Adv. Funct. Mater.*, 22, 3414, 2012.
114. Park, S.Y., Choi, D.S., Jin, H.J., Park, J., Byun, K.-E., Lee, K.B., Hong, S., Polarization-controlled differentiation of human neural stem cells using synergistic cues from the patterns of carbon nanotube monolayer coating. *ACS Nano*, 5, 4704, 2011.
115. Benazeraf, B., Francois, P., Baker, R.E., Denans, N., Little, C.D., Pourquie, O., A random cell motility gradient downstream of fgf controls elongation of an amniote embryo. *Nature*, 466, 248, 2010.
116. Schumacher, J.A., Hashiguchi, M., Nguyen, V.H., Mullins, M.C., An intermediate level of bmp signaling directly specifies cranial neural crest progenitor cells in zebrafish. *PLoS One*, 6, e27403, 2011.
117. Zhu, J., Liang, L., Jiao, Y., Liu, L., Enhanced invasion of metastatic cancer cells via extracellular matrix interface. *PLoS One*, 10, e0118058, 2015.
118. Comelles, J., Hortigüela, V., Samitier, J., Martínez, E., Versatile gradients of covalently bound proteins on microstructured substrates. *Langmuir*, 28, 13688, 2012.
119. Lagunas, A., Comelles, J., Oberhansl, S., Hortigüela, V., Martínez, E., Samitier, J., Continuous bone morphogenetic protein-2 gradients for concentration effect studies on c2c12 osteogenic fate. *Nanomedicine: NBM*, 9, 694, 2013.
120. Lagunas, A., Martínez, E., Samitier, J., Surface-bound molecular gradients for the high throughput screening of cell responses. *Front. Bioeng. Biotechnol.*, 3, 132, 2015.
121. Miserez, A., Schneberk, T., Sun, C., Zok, F.W., Waite, J.H., The transition from stiff to compliant materials in squid beaks. *Science*, 319, 1816, 2008.
122. Karageorgiou, V., Kaplan, D., Porosity of 3d biomaterial scaffolds and osteogenesis. *Biomaterials*, 26, 5474, 2005.
123. Ostrovidov, S., Seidi, A., Ahadian, S., Ramalingam, M., Khademhosseini, A. Micro- and nanoengineering approaches to developing gradient biomaterials suitable for interface tissue engineering. *Micro and nanotechnologies in engineering stem cells and tissues*, p. 52, John Wiley & Sons, Inc., 2013.
124. Wang, P.Y., Tsai, W.B. Stem cell responses to surface nanotopographies. *Stem cell nanoengineering*, p. 185, John Wiley & Sons, Inc, 2015.
125. Singh, M., Berkland, C., Detamore, M.S., Strategies and applications for incorporating physical and chemical signal gradients in tissue engineering. *Tissue Eng. Part B Rev.*, 14, 341, 2008.

126. Sant, S., Hancock, M.J., Donnelly, J.P., Iyer, D., Khademhosseini, A., Biomimetic gradient hydrogels for tissue engineering. *Can. J. Chem. Eng.*, 88, 899, 2010.
127. Wang, L., Li, Y., Huang, G., Zhang, X., Pingguan-Murphy, B., Gao, B., Lu, T.J., Xu, F., Hydrogel-based methods for engineering cellular microenvironment with spatiotemporal gradients. *Crit. Rev. Biotechnol.*, 1, 2015.
128. Oh, S.H., Park, I.K., Kim, J.M., Lee, J.H., In vitro and in vivo characteristics of pcl scaffolds with pore size gradient fabricated by a centrifugation method. *Biomaterials*, 28, 1664, 2007.
129. Schwarz, K., Epple, M., Hierarchically structured polyglycolide – a biomaterial mimicking natural bone. *Macromol. Rapid Commun.*, 19, 613, 1998.
130. Harley, B.A., Hastings, A.Z., Yannas, I.V., Sannino, A., Fabricating tubular scaffolds with a radial pore size gradient by a spinning technique. *Biomaterials*, 27, 866, 2006.
131. Van Vlierberghe, S., Cnudde, V., Dubruel, P., Masschaele, B., Cosijns, A., De Paepe, I., Jacobs, P.J.S., Van Hoorebeke, L., Remon, J.P., Schacht, E., Porous gelatin hydrogels: 1. Cryogenic formation and structure analysis. *Biomacromolecules*, 8, 331, 2007.
132. Miot, S., Woodfield, T., Daniels, A.U., Suetterlin, R., Peterschmitt, I., Heberer, M., van Blitterswijk, C.A., Riesle, J., Martin, I., Effects of scaffold composition and architecture on human nasal chondrocyte redifferentiation and cartilaginous matrix deposition. *Biomaterials*, 26, 2479, 2005.
133. Kennedy, S.B., Washburn, N.R., Simon Jr, C.G., Amis, E.J., Combinatorial screen of the effect of surface energy on fibronectin-mediated osteoblast adhesion, spreading and proliferation. *Biomaterials*, 27, 3817, 2006.
134. Spijker, H.T., Bos, R., Busscher, H.J., van Kooten, T.G., van Oeveren, W., Platelet adhesion and activation on a shielded plasma gradient prepared on polyethylene. *Biomaterials*, 23, 757, 2002.
135. Kim, T.H., An, D.B., Oh, S.H., Kang, M.K., Song, H.H., Lee, J.H., Creating stiffness gradient polyvinyl alcohol hydrogel using a simple gradual freezing–thawing method to investigate stem cell differentiation behaviors. *Biomaterials*, 40, 51, 2015.
136. Lo, C.T., Throckmorton, D.J., Singh, A.K., Herr, A.E., Photopolymerized diffusion-defined polyacrylamide gradient gels for on-chip protein sizing. *Lab Chip*, 8, 1273, 2008.
137. Hopp, I., Michelmore, A., Smith, L.E., Robinson, D.E., Bachhuka, A., Mierczynska, A., Vasilev, K., The influence of substrate stiffness gradients on primary human dermal fibroblasts. *Biomaterials*, 34, 5070, 2013.
138. Wong, J.Y., Velasco, A., Rajagopalan, P., Pham, Q., Directed movement of vascular smooth muscle cells on gradient-compliant hydrogels. *Langmuir*, 19, 1908, 2003.
139. Du, Y., Hancock, M.J., He, J., Villa-Uribe, J.L., Wang, B., Cropek, D.M., Khademhosseini, A., Convection-driven generation of long-range material gradients. *Biomaterials*, 31, 2686, 2010.

140. Nemir, S., Hayenga, H.N., West, J.L., Pegda hydrogels with patterned elasticity: Novel tools for the study of cell response to substrate rigidity. *Biotechnol. Bioeng.*, 105, 636, 2010.
141. Smith, A.P., Douglas, J.F., Meredith, J.C., Amis, E.J., Karim, A., High-throughput characterization of pattern formation in symmetric diblock copolymer films. *J. Polym. Sci. B Polym. Phys.*, 39, 2141, 2001.
142. Ramalingam, M., Young, M.F., Thomas, V., Sun, L., Chow, L.C., Tison, C.K., Chatterjee, K., Miles, W.C., Simon, C.G., Nanofiber scaffold gradients for interfacial tissue engineering. *J. Biomater. Appl.*, 27, 695, 2013.
143. Kloxin, A.M., Benton, J.A., Anseth, K.S., In situ elasticity modulation with dynamic substrates to direct cell phenotype. *Biomaterials*, 31, 1, 2010.
144. Wang, P.Y., Clements, L.R., Thissen, H., Tsai, W.B., Voelcker, N.H., High-throughput characterisation of osteogenic differentiation of human mesenchymal stem cells using pore size gradients on porous alumina. *Biomater. Sci.*, 1, 924, 2013.
145. Tse, J.R., Engler, A.J., Stiffness gradients mimicking *in vivo* tissue variation regulate mesenchymal stem cell fate. *PLoS One*, 6, e15978, 2011.
146. Oh, S.H., Kim, T.H., Im, G.I., Lee, J.H., Investigation of pore size effect on chondrogenic differentiation of adipose stem cells using a pore size gradient scaffold. *Biomacromolecules*, 11, 1948, 2010.
147. Dickinson, L.E., Kusuma, S., Gerecht, S., Reconstructing the differentiation niche of embryonic stem cells using biomaterials. *Macromol. Biosci.*, 11, 36, 2011.
148. Hazeltine, L.B., Selekman, J.A., Palecek, S.P., Engineering the human pluripotent stem cell microenvironment to direct cell fate. *Biotechnol. Adv.*, 31, 1002, 2013.
149. Anderson, D.G., Putnam, D., Lavik, E.B., Mahmood, T.A., Langer, R. Biomaterial microarrays: rapid, microscale screening of polymer–cell interaction. *Biomaterials*, 26, 4892, 2005.
150. Papp, K., Szittner, A., Prechl, J. Life on a microarray: assessing live cell functions in a microarray format. *Cell. Mol. Life Sci.*, 69, 2717, 2012.
151. Peerani, R., Rao, B.M., Bauwens, C., Yin, T., Wood, G.A., Nagy, A., Kumacheva, E., Zandstra, P.W. Niche-mediated control of human embryonic stem cell self-renewal and differentiation. *EMBO J.*, 26, 4744, 2007.
152. Kusuma, S., Smith, Q., Facklam, A., Gerecht, S. Micropattern size-dependent endothelial differentiation from a human induced pluripotent stem cell line. *J. Tissue Eng. Regen. Med.*, 1985.
153. Hwang, Y.S., Chung, B.G., Ortmann, D., Hattori, N., Moeller, H.C., Khademhosseini, A. Microwell-mediated control of embryoid body size regulates embryonic stem cell fate via differential expression of WNT5a and WNT11. *Proc. Natl. Acad. Sci. USA*, 106, 16978, 2009.
154. Sasaki, M., Shimizu, T., Masuda, S., Kobayashi, J., Itoga, K., Tsuda, Y., Yamashita, J.K., Yamato, M., Okano, T. Mass preparation of size-controlled mouse embryonic stem cell aggregates and induction of cardiac differentiation by cell patterning method. *Biomaterials*, 30, 4384, 2009.

155. Lee, L.H., Peerani, R., Ungrin, M., Joshi, C., Kumacheva, E., Zandstra, P.W. Micropatterning of human embryonic stem cells dissects the mesoderm and endoderm lineages. *Stem Cell Res.*, 2, 155, 2009.
156. Warmflash, A., Sorre, B., Etoc, F., Siggia, E.D., Brivanlou, A.H. A method to recapitulate early embryonic patterning in human embryonic stem cells. *Nat. Methods*, 11, 847, 2014.
157. Soen, Y., Mori, A., Palmer, T.D., Brown, P.O. Exploring the regulation of human neural precursor cell differentiation using arrays of signaling microoenvironments. *Mol. Syst. Biol.*, 2, 37, 2006.
158. Kane, R.S., Takayama, S., Ostuni, E., Ingber, D.E., Whitesides, G.M. Patterning proteins and cells using soft lithography. *Biomaterials*, 20, 2363, 1999.
159. McBeath, R., Pirone, D.M., Nelson, C.M., Bhadriraju, K., Chen, C.S. Cell shape, cytoskeletal tension, and RhoA regulate stem cell lineage commitment. *Dev. Cell*, 6, 483, 2004.
160. Peng, R., Yao, X., Ding, J. Effect of cell anisotropy on differentiation of stem cells on micropatterned surfaces through the controlled single cell adhesion. *Biomaterials*, 32, 8048, 2011.
161. Song, W., Wang, X., Lu, H., Kawazoe, N., Chen, G., Exploring adipogenic differentiation of a single stem cell on poly(acrylic acid) and polystyrene micropatterns. *Soft Matter*, 8, 8429, 2012.
162. Jiang, F., Horber, H., Howard, J., Muller, D.J. Assembly of collagen into microribbons: effects of pH and electrolytes. *J. Struct. Biol.*, 148, 268, 2004.
163. Smith, M., Gourdon, D., Little, W.C., Kubow, K.E., Eguiluz, R.A., Luna-Morris, S., Vogel V. Force-induced unfolding of fibronectin in the extracellular matrix of living cells. *PLoS Biol.*, 5, e268, 2007.
164. Little, W.C., Smith, M.L., Ebneter, U., Vogel, V. Assay to mechanically tune and optically probe fibrillary fibronectin conformations from fully relaxed to breakage. *Matrix Biol.*, 27, 451, 2008.
165. Spatz, J.P., Mossmer, S., Hartmann, C., Möller, M., Herzog, T., Krieger, M., Boyen, H.G., Ziemann, P., Kabius, B. Ordered deposition of inorganic clusters from micellar block copolymer films *Langmuir*, 16, 407, 2000.
166. Xiong, J.P., Stehle, T., Zhang, R.G., Joachimiak, A., Frech, M., Goodman, S.L., Arnaout, M.A. Crystal structure of the extracellular segment of integrin $\alpha V\beta 3$ in complex with an Arg-Gly-Asp ligand. *Science*, 296, 151, 2002.
167. Arnold, M., Cavalcanti-Adam, E.A., Glass, R., Blümmel, J., Eck, W., Kantlehner, M., Kessler, H., Spatz, J. P. Activation of integrin function by nanopatterned adhesive interfaces. *Chem. Phys. Chem.*, 5, 383, 2004.
168. Cavalcanti-Adam, E.A., Micoulet, A., Blummel, J., Auernheimer, J., Kessler, H., Spatz, J.P. Lateral spacing of integrin ligands influences cell spreading and focal adhesion assembly. *Eur. J. Cell. Biol.* 85, 219, 2006.
169. Arnold, M., Schwieder, M., Blümmel, J., Cavalcanti-Adam, E.A., López-Garcia, M., Kessler, H., Geiger, B., Spatz, J.P. Cell interactions with hierarchically structured nano-patterned adhesive surface. *Soft Matter*, 5, 72, 2009.

170. Cavalcanti-Adam, E.A., Volberg, T., Micoulet, A., Kessler, H., Geiger, B., Spatz, J. P. Cell spreading and focal adhesion dynamics are regulated by spacing of integrin ligands. *Biophys. J.*, 92, 2964, 2007.
171. Hon, Y.H., Wang, J.Y., Pan, Y.N. Composition/Phase structure and properties of titanium-niobium alloys. *Mater. Trans.*, 44, 2384, 2003.
172. Medda, R., Helth, A., Herre, P., Pohl, D., Rellinghaus, B., Perschmann, N., Neubauer, S., Kessler, H., Oswald, S., Eckert, J., Spatz, J.P., Gebert, A., Cavalcanti-Adam, E.A. Investigation of early cell-surface interactions of human mesenchymal stem cells on nanopatterned β-type titanium-niobium alloy surfaces. *Interface Focus*, 4, 20130046, 2014.
173. Yamazaki, S., Iwama, A., Takayanagi, S., Eto, K., Ema, H., Nakauchi H. TGF-beta as a candidate bone marrow niche signal to induce hematopoietic stem cell hibernation. *Blood*, 113, 1250e6, 2009.
174. Altrock, E., Muth, C.A., Klein, G., Spatz, J.P., Lee-Thedieck, C. The significance of integrin ligand nanopatterning on lipid raft clustering in hematopoietic stem cells. *Biomaterials*, 33, 3107, 2012.
175. Wang, X., Yan, C., Ye, K., He, Y., Li, Z., Ding, J. Effect of RGD nanospacing on differentiation of stem cells. *Biomaterials*, 34, 2865, 2013.
176. Wang, X., Ye, K., Li, Z., Yan, C., Ding, J. Adhesion, proliferation, and differentiation of mesenchymal stem cells on RGD nanopatterns of varied nanospacings. *Organogenesis*, 9, 280, 2013.
177. Sundelacruz, S., Kaplan, D.L. Stem cell- and scaffold-based tissue engineering approaches to osteochondral regenerative medicine. *Semin. Cell Dev. Biol.*, 20, 646, 2009.
178. Singh, P., Schwarzbauer, J.E. Fibronectin and stem cell differentiation – lessons from chondrogenesis. *J. Cell Sci.*, 125, 3703, 2012.
179. Lim, Y.B., Kang, S.S., Park, T.K., Lee, Y.S., Chun, J.S., Sonn, J.K. Disruption of actin cytoskeleton induces chondrogenesis of mesenchymal cells by activating protein kinase C-α signaling. *Biochem. Biophys. Res. Commun.*, 273, 609, 2000.
180. Lim, Y.B., Kang, S.S., An, W.G., Lee, Y.S., Chun, J.S., Sonn, J.K. Chondrogenesis induced by actin cytoskeleton disruption is regulated via protein kinase C-dependent p38 mitogen-activated protein kinase signaling. *J. Cell. Biochem.*, 88, 713, 2003.
181. Li, Z., Cao, B., Wang, X., Ye, K., Li, S., Ding, J. Effects of RGD nanospacing on chondrogenic differentiation of mesenchymal stem cells. *J. Mater. Chem. B*, 3, 5197, 2015.
182. Choi, C.K.K., Xu, Y.J., Wang, B., Zhu, M., Zhang, L., Bian, L. Substrate coupling strength of integrin-binding ligands modulates adhesion, spreading, and differentiation of human mesenchymal stem cells. *Nano Lett.*, 15, 6592, 2015.
183. Wang, X., Li, S., Yan, C., Liu, P., Ding, J. Fabrication of RGD micro/nanopattern and corresponding study of stem cell differentiation. *Nano Lett.*, 15, 1457, 2015.

184. Gurdon, J.B., Bourillot, P.Y. Morphogen gradient interpretation. *Nature*, 413, 797, 2001.
185. Tasoglu, S., Demirci, U. Bioprinting for stem cell research. *Trends Biotechnol.*, 31, 10, 2013.
186. Ilkhanizadeh, S., Teixeira, A.I., Hermanson, O. Inkjet printing of macromolecules on hydrogels to steer neural stem cell differentiation. *Biomaterials*, 28, 3936, 2007.
187. Phillippi, J.A., Miller, E., Weiss, L., Huard, J., Waggoner, A., Campbell, P. Microenvironments engineered by inkjet bioprinting spatially direct adult stem cells toward muscle- and bone-like subpopulations. *Stem Cells*, 26, 127, 2008.
188. Ker, E.D.F., Chu, B., Phillippi, J.A., Gharaibeh, B., Huard, J., Weiss, L.E., Campbell, P.G. Engineering spatial control of multiple differentiation fates within a stem cell population. *Biomaterials*, 32, 3413, 2011.
189. Krampera, M., Pasini, A., Rigo, A., Scupoli, M.T., Tecchio, C., Malpeli, G., Scarpa, A., Dazzi, F., Pizzolo, G., Vinante, F. HB-EGF/HER-1 signaling in bone marrow mesenchymal stem cells: inducing cell expansion and reversibly preventing multilineage differentiation. *Blood*, 106, 59, 2005.
190. Miller, E.D., Li, K., Kanade, T., Weiss, L.E., Walker, L.M., Campbell, P.G. Spatially directed guidance of stem cell population migration by immobilized patterns of growth factors. *Biomaterials*, 32, 2775, 2011.
191. Wittle, J.D., Barton, D., Alexander, M.R., Short, R.D. A method for the deposition of controllable chemical gradients. *Chem. Commun.*, 1766, 2003.
192. Wells, N., Baxter, M.A., Turnbull, J.E., Murray, P., Edgar, D., Parry, K.L., Steele, D.A., Short, R.D., The geometric control of E14 and R1 mouse embryonic stem cell pluripotency by plasma polymer surface chemical gradients. *Biomaterials*, 30, 1066, 2009.
193. Aplin, A.E., Juliano R.L. Regulation of nucleocytoplasmic trafficking by cell adhesion receptors and the cytoskeleton. *J. Cell Biol.*, 155, 187, 2001.
194. Harding, F.J., Clements, L.R., Short, R.D., Thissen, H., Voelkerm N.H. Assessing embryonic stem cell response to surface chemistry using plasma polymer gradients. *Acta Biomater.*, 8, 1739, 2012.
195. Delatat, B., Goreham, R.V., Vasilev, K., Harding, F., Voelcker, N.H. Subtle changes in surface chemistry affect embryoid body cell differentiation: Lessons learnt from surface-bound amine density gradients. *Tissue Eng. Part A*, 20, 1715, 2014.
196. Arima, Y., Iwata, H. Effect of wettability and surface functional groups on protein adsorption and cell adhesion using well-defined mixed self-assembled monolayers. *Biomaterials*, 28, 3074, 2007.
197. Purpura, K.A., Aubin, J.E., Zandstra, P.W. Sustained *in vitro* expansion of bone progenitors is cell density dependent. *Stem Cells*, 22, 39, 2004.
198. Wang, P.Y., Clements, L.R., Thissen, H., Tsai, W.B., Voelker, N. Screening rat mesenchymal stem cell attachment and differentiation on surface chemistries using plasma polymer gradients. *Acta Biomater.*, 11, 58, 2015.

199. Delatat, B., Mierczynska, A., Ghaemi, S.R., Cavallaro, A., Harding, F.J., Vasilev, K., Voelcker, N.H. Materials displaying neural growth factor gradients and applications in neural differentiation of embryoid body cells. *Adv. Funct. Mater.*, 25, 2737, 2015.
200. Lagunas, A., Comelles, J., Martínez, E., Prats-Alfonso, E., Acosta, G. A., Albericio, F., Samitier, J. Cell adhesion and focal contact formation on linear RGD molecular gradients: study of non-linear concentration dependence effects. *Nanomedicine: NBM*, 8, 432, 2012.
201. Liu, Z., Xiao, L., Xu, B., Zhang, Y., Mak, A.F.T., Li, Y., Man, W.Y., Yang, M. Covalently immobilized biomolecule gradient on hydrogel surface using a gradient generating microfluidic device for a quantitative mesenchymal stem cell study. *Biomicrofluidics*, 6, 024111, 2012.
202. Lund, A.W., Yener, B., Stegemann, J.P., Plopper, G.E. The natural and engineered 3D microenvironment as a regulatory cue during stem cell fate determination. *Tissue Eng. Part B*, 15, 371, 2009.
203. Cukierman, E., Pankov, R., Stevens, D.R., Yamada, K.M., Taking cell-matrix adhesions to the third dimension. *Science*, 294, 1708, 2001.
204. Cukierman, E., Pankov, R., Yamada, K.M., Cell interactions with three-dimensional matrices. *Curr. Op. Cell Biol.*, 14, 633, 2002.
205. Ghibaudo, M., Di Meglio, J.M., Hersen, P., Ladoux, B., Mechanics of cell spreading within 3d-micropatterned environments. *Lab Chip*, 11, 805, 2010.
206. Greiner, A.M., Richter, B., Bastmeyer, M., Micro-engineered 3d scaffolds for cell culture studies. *Macromol. Biosci.*, 12, 1301, 2012.
207. Pampaloni, F., Reynaud, E.G., Stelzer, E.H.K., The third dimension bridges the gap between cell culture and live tissue. *Nat. Rev. Mol. Cell Biol.*, 8, 839, 2007.
208. Baker, B.M., Chen, C.S., Deconstructing the third dimension: How 3d culture microenvironments alter cellular cues. *J. Cell Sci.*, 125, 3015, 2012.
209. Legant, W.R., Miller, J.S., Blakely, B.L., Cohen, D.M., Genin, G.M., Chen, C.S., Measurement of mechanical tractions exerted by cells in three-dimensional matrices. *Nat. Meth.*, 7, 969, 2010.
210. Griffith, L.G., Swartz, M.A., Capturing complex 3d tissue physiology *in vitro*. *Nat. Rev. Mol. Cell Biol.*, 7, 211, 2006.
211. Discher, D.E., Janmey, P., Wang, Y.L., Tissue cells feel and respond to the stiffness of their substrate. *Science*, 310, 1139, 2005.
212. Her, G.J., Wu, H.-C., Chen, M.-H., Chen, M.-Y., Chang, S.-C., Wang, T.-W., Control of three-dimensional substrate stiffness to manipulate mesenchymal stem cell fate toward neuronal or glial lineages. *Acta Biomater.*, 9, 5170, 2013.
213. Ghanian, M.H., Farzaneh, Z., Barzin, J., Zandi, M., Kazemi-Ashtiani, M., Alikhani, M., Ehsani, M., Baharvand, H., Nanotopographical control of human embryonic stem cell differentiation into definitive endoderm. *J. Biomed. Mater. Res. A*, 103, 3539, 2015.

214. Smith, L.A., Liu, X., Hu, J., Wang, P., Ma, P.X., Enhancing osteogenic differentiation of mouse embryonic stem cells by nanofibers. *Tiss. Eng. Part A*, 15, 1855, 2009.
215. Li, X., Liang, H., Sun, J., Zhuang, Y., Xu, B., Dai, J., Electrospun collagen fibers with spatial patterning of sdf1α for the guidance of neural stem cells. *Adv. Health. Mater.*, 4, 1869, 2015.
216. Madhurakkat Perikamana, S.K., Lee, J., Ahmad, T., Jeong, Y., Kim, D.-G., Kim, K., Shin, H., Effects of immobilized bmp-2 and nanofiber morphology on *in vitro* osteogenic differentiation of hmscs and *in vivo* collagen assembly of regenerated bone. *ACS Appl. Mater. Interfaces*, 7, 8798, 2015.
217. Wang, X., Ding, B., Li, B., Biomimetic electrospun nanofibrous structures for tissue engineering. *Mater. Today*, 16, 229, 2013.
218. Xu, T., Miszuk, J.M., Zhao, Y., Sun, H., Fong, H., Electrospun polycaprolactone 3d nanofibrous scaffold with interconnected and hierarchically structured pores for bone tissue engineering. *Adv. Health. Mater.*, 4, 2238, 2015.
219. Shalumon, K.T., Lai, G.J., Chen, C.H., Chen, J.P., Modulation of bone-specific tissue regeneration by incorporating bone morphogenetic protein and controlling the shell thickness of silk fibroin/chitosan/nanohydroxyapatite core–shell nanofibrous membranes. *ACS Appl. Mater. Interfaces*, 7, 21170, 2015.
220. Ma, P.X., Zhang, R., Synthetic nano-scale fibrous extracellular matrix. *J. Biomed. Mater. Res.*, 46, 60, 1999.
221. Tysseling-Mattiace, V.M., Sahni, V., Niece, K.L., Birch, D., Czeisler, C., Fehlings, M.G., Stupp, S.I., Kessler, J.A., Self-assembling nanofibers inhibit glial scar formation and promote axon elongation after spinal cord injury. *J. Neurosci.*, 28, 3814, 2008.
222. Xie, J., Peng, C., Zhao, Q., Wang, X., Yuan, H., Yang, L., Li, K., Lou, X., Zhang, Y., Osteogenic differentiation and bone regeneration of the ipsc-mscs supported by a biomimetic nanofibrous scaffold. *Acta Biomater.*, 29, 365, 2016.
223. Gholizadeh-Ghaleh aziz, S., Gholizadeh-Ghaleh aziz, S., Akbarzadeh, A., The potential of nanofibers in tissue engineering and stem cell therapy. *Artif. Cells Nanomed. Biotechnol.*, 29, 365, 2016.
224. Ahmed, E.M., Hydrogel: Preparation, characterization, and applications: A review. *J. Adv. Res.*, 6, 105, 2015.
225. El-Sherbiny, I.M., Yacoub, M.H., Hydrogel scaffolds for tissue engineering: Progress and challenges. *Glob. Cardiol. Sci. Pract.*, 2013, 316, 2013.
226. Neal, D., Sakar, M.S., Ong, L.L.S., Harry Asada, H., Formation of elongated fascicle-inspired 3d tissues consisting of high-density, aligned cells using sacrificial outer molding. *Lab Chip*, 14, 1907, 2014.
227. Li, H., Wijekoon, A., Leipzig, N.D., 3D differentiation of neural stem cells in macroporous photopolymerizable hydrogel scaffolds. *PLoS One*, 7, e48824, 2012.
228. Sun, A.X., Lin, H., Beck, A.M., Kilroy, E.J., Tuan, R.S., Projection stereolithographic fabrication of human adipose stem cell-incorporated biodegradable

scaffolds for cartilage tissue engineering. *Front Bioeng. Biotechnol.*, 3, 115, 2015.
229. Jeon, O., Alsberg, E., Regulation of stem cell fate in a three-dimensional micropatterned dual-cross-linked hydrogel system. *Adv. Funct. Mater.*, 23, 4765, 2013.
230. Khetan, S., Guvendiren, M., Legant, W.R., Cohen, D.M., Chen, C.S., Burdick, J.A., Degradation-mediated cellular traction directs stem cell fate in covalently cross-linked three-dimensional hydrogels. *Nat. Mater.*, 12, 458, 2013.
231. Lei, Y., Schaffer, D.V., A fully defined and scalable 3d culture system for human pluripotent stem cell expansion and differentiation. *Proc. Natl. Acad. Sci. U.S.A*, 110, E5039, 2013.
232. Kumar, D., Gerges, I., Tamplenizza, M., Lenardi, C., Forsyth, N.R., Liu, Y., Three-dimensional hypoxic culture of human mesenchymal stem cells encapsulated in a photocurable, biodegradable polymer hydrogel: A potential injectable cellular product for nucleus pulposus regeneration. *Acta Biomater.*, 10, 3463, 2014.
233. Murphy, S.V., Atala, A., 3D bioprinting of tissues and organs. *Nat. Biotech.*, 32, 773, 2014.
234. Stanton, M.M., Samitier, J., Sanchez, S., Bioprinting of 3d hydrogels. *Lab Chip*, 15, 3111, 2015.
235. Gao, G., Cui, X., Three-dimensional bioprinting in tissue engineering and regenerative medicine. *Biotechnol. Lett.*, 1, 203, 2015.
236. Chia, H., WU, B., Recent advances in 3D printing of biomaterials. *J. Biol. Eng.*, 9, 1, 2015.
237. Collins, S.F., Bioprinting is changing regenerative medicine forever. *Stem Cells Dev.*, 23, 79, 2014.
238. Bajaj, P., Schweller, R.M., Khademhosseini, A., West, J.L., Bashir, R., 3d biofabrication strategies for tissue engineering and regenerative medicine. *Ann. Rev. Biomed. Eng.*, 16, 247, 2014.
239. Bertassoni, L.E., Cecconi, M., Manoharan, V., Nikkhah, M., Hjortnaes, J., Cristino, A.L., Barabaschi, G., Demarchi, D., Dokmeci, M.R., Yang, Y., Khademhosseini, A., Hydrogel bioprinted microchannel networks for vascularization of tissue engineering constructs. *Lab Chip*, 14, 2202, 2014.
240. Hsieh, F.Y., Lin, H.H., Hsu, S.H., 3d bioprinting of neural stem cell-laden thermoresponsive biodegradable polyurethane hydrogel and potential in central nervous system repair. *Biomaterials*, 71, 48, 2015.
241. Faulkner-Jones, A., Fyfe, C., Cornelissen, D.J., Gardner, J., King, J., Courtney, A., Shu, W., Bioprinting of human pluripotent stem cells and their directed differentiation into hepatocyte-like cells for the generation of mini-livers in 3D. *Biofabrication*, 7, 044102, 2015.
242. Liliang, O., Rui, Y., Shuangshuang, M., Xi, C., Jie, N., Wei, S., Three-dimensional bioprinting of embryonic stem cells directs highly uniform embryoid body formation. *Biofabrication*, 7, 044101, 2015.

243. Parrag, I.C., Zandstra, P.W., Woodhouse, K.A., Fiber alignment and coculture with fibroblasts improves the differentiated phenotype of murine embryonic stem cell-derived cardiomyocytes for cardiac tissue engineering. *Biotechnol. Bioeng.*, 109, 813, 2012.
244. Ker, E.D., Nain, A.S., Weiss, L.E., Wang, J., Suhan, J., Amon, C.H., Campbell, P.G. Bioprinting of growth factors onto aligned sub-micron fibrous scaffolds for simultaneous control of cell differentiation and alignment. *Biomaterials*, 32, 8097, 2011.
245. Peppas, N.A., Hilt, J.Z, Khademhosseini, A., Langer, R. Hydrogels in biology and medicine: from molecular principles to bionanotechnology. *Adv. Mater.*, 18, 1345, 2006.
246. Cushing, M.C., Anseth, K.S. Hydrogel cell cultures. *Science*, 316, 1133, 2007.
247. Tibbitt, M.W., Anseth, K.S. Hydrogels as extracellular matrix mimics for 3D cell culture. *Biotechnol. Bioeng.*, 103, 655, 2009.
248. Lutolf, M.P., Gilbert, P.M., Blau, H.M. Designing materials to direct stem cell fate. Designing materials to direct stem cell fate. *Nature*, 462, 433, 2009.
249. Wosnick, J.H., Shoichet, M. S. Three-dimensional chemical patterning of transparent hydrogels. *Chem. Mater.*, 20, 55, 2008.
250. Angot, E., Loulier, K., Nguyen-Ba-Charvet, K.T., Gadeau, A.P., Ruat, M., Traiffort, E. Chemoattractive activity of sonic hedgehog in the adult subventricular zone modulates the number of neural precursors reaching the olfactory bulb. *Stem Cells*, 26, 2311, 2008.
251. Wylie, R.G., Ahsan, S., Aizawa, Y., Maxwell, K.L., Morshead, C.M., Shoichet, M.S. Spatially controlled simultaneous patterning of multiple growth factors in three-dimensional hydrogels. *Nat. Mater.*, 10, 799, 2011.
252. Khetan, S., Burdick, J.A. Patterning hydrogels in three dimensions towards controlling cellular Interactions. *Soft Matter*, 7, 830, 2011.
253. DeForest, C.A, Anseth. K.S. Cytocompatible click-based hydrogels with dynamically tunable properties through orthogonal photoconjugation and photocleavage reactions. *Nat. Chem.*, 3, 925, 2011.
254. Alberts, B., Johnson, A., Lewis, J., Raff, M., Roberts, K and Walter, P., *Molecular Biology of the Cell*, p. 1616. Garland Science, 2002.
255. DeForest, C.A., Anseth, K.S. Photoreversible patterning of biomolecules within click-based hydrogels. *Angew. Chem. Int. Ed.*, 51, 1816, 2012.
256. Kloxin, A.M., Kasko, A.M., Salinas, C.N., Anseth, K.S. Photodegradable hydrogels for dynamic tuning of physical and chemical properties. *Science*, 324, 59, 2009.

6
Recent Advances in Nanostructured Polymeric Surface: Challenges and Frontiers in Stem Cells

Ilaria Armentano[1]*, Samantha Mattioli[1], Francesco Morena[2], Chiara Argentati[2], Sabata Martino[2], Luigi Torre[1] and Josè Maria Kenny[1]

[1]*Department of Civil and Environmental Engineering, UdR INSTM, University of Perugia, Terni, Italy*
[2]*Department of Chemistry, Biology and Biotechnology, Biochemistry and Molecular Biology Unit, University of Perugia, Perugia, Italy*

Abstract

This chapter describes recent advancing in nanostructured polymeric surfaces by analyzing how they can affect stem cell behavior. Surface is the first part that comes in contact with cells, and surface properties play a critical role in modulating stem cell behavior, guiding the shape and structure of developing tissues, providing mechanical stability, and offering opportunities to deliver inductive molecules to transplanted or migrating cells. In this regard, nanostructured polymeric surface by offering informative microenvironments allowing cells to interpret the biomaterial instructions could provide a fine structure and tunable surface in nanoscales to help the cell adhesion and promote the cell growth and differentiation. The chapter starts with a comprehensive description of the nanostructured polymeric surface by analyzing in detail the main surface modification methods and focusing on the physicochemical properties and then the main mechanisms in the stem cell/surface interactions. A concise examination of microscopic techniques used in estimating the stem cell/nanostructured surface interaction will be described, and in particular fluorescence, electron, and atomic force microscopies will be mainly investigated and compared. The synergism of stem cell biology and biomaterial technology promises to have a profound impact on stem-cell-based clinical applications for tissue regeneration. Thus, the design of novel biomaterials surface is trying to recapitulate the molecular events involved in the production,

---

*Corresponding author: ilaria.armentano@unipg.it

clearance, and interaction of molecules within tissue in pathologic conditions and regeneration of tissue/organs.

Keywords: Nanostructured polymers, surface modifications, stem cells, stem cells and nanostructures, microscopy

6.1 Introduction

In the past decades, modern medicine has been challenged with complex problems, which have led to technological advancements in the area of healthcare, in terms of stem cell culture, new biomaterial developments, and control the culture conditions. The overall goal of nanomedicine is to achieve accurate and early diagnosis, effective treatment with minimal or no side effects and rapid and noninvasive monitoring of treatment efficacy. The development of nanotechnology-based theranostic tests involving cellular-, proteomic-, and genomic-level testing platforms such as microchips represents a paradigm shift in patient care. The main aim is to provide unique, individualized medications for each patient, being more targeted and cost-effective. Based on unique capabilities, nanoscale science probes cells and biomolecules in the physiological states at forces, displacement resolutions and concentrations at the piconewton, nanometer, and picomolar scales, respectively. However, many complex problems still remain unresolved.

Tissue engineering (TE) is a multidisciplinary field combining principles of biology, medicine, and engineering that aims at replacing damaged, injured, or missing organs and tissue with a functional artificial substitute [1, 2]. This substitute can be a combination of both a scaffold and cells [3]. Nowadays, regenerative medicine represents the hold promise for the cure of diseases where tissues are degenerated as consequence of genetic mutation, tumor, or mechanical crash [4, 5]. The paradigm of regenerative medicine is based on the combinations of the most innovative therapeutic tools: adult stem cells, gene therapy, and biomaterials. The aim is the regeneration of the damaged tissues by mimicking the molecular signaling that in physiological condition maintain the tissues homeostasis. Tissue engineering aims to repair or regenerate tissues that have been damaged or lost, which can act as a potential alternative to organ transplantation. The strategy of tissue engineering is to replace a defective organ with an artificial scaffold composed of cells, growth factors, and biomaterials that function as an extracellular matrix (ECM). The success of tissue engineering mainly depends on two essential elements: cells and biomaterials [6].

Therefore, stem cells could represent the source for the generation of new healthy cells within the degenerated tissue. After implantation in the patients, they start to replace the affected cells with new functional cells thank to their biological properties of self-renewal and multipotency that are maintained even in a recipient host (see next section for details). Thus, implanted stem cells engraft within the tissue, respond specifically to the host tissue and differentiate *in loco* in order to repair the damage [7].

In many cases, especially for genetic degenerated diseases, it is necessary also the substitution of the altered gene with its function copy. Therefore, gene transfer technology, by using different of viral vector as gene delivery system, allows the production of engineered stem cells that could be source of new cells genetically corrected [8].

Finally, due to the three-dimensional structure together with the complex tissues architecture, the use of specific designed biomaterial in combination with stem cells became essential. Therefore, different types of biomaterials have been developed in order to respond to the different characteristic of the tissues that must be regenerated [9]. It is a general concern that stem cells recognize biomaterial characteristics and respond with a specific function (e.g. specific cell differentiation) activating a selected signaling cascade call as "mechanotransduction" [9].

TE scaffolds should provide mechanical support and physicochemical cues for the growth of cells to replace tissue [10, 11]. Materials for such scaffolds should possess appropriate mechanical properties, chemical, and biological compatibility, and ideally should degrade in an appropriate time frame [11, 12]. There are three distinct size scales to consider when designing cellular scaffolds: the functional tissue level (>100 μm), the cellular level (1–100 μm), and the subcellular (nanosctructural <1 μm) level. Biocompatibility, biodegradability, providing strength and structure if needed, enabling cell attachment, proliferation, and sometimes even differentiation are just some of the possible necessities. In some other cases including catheters and stents, prevention of cell attachment and adsorption of proteins is required. So, the main aim of the biomaterial scientists is the control of the material synthesis and the comprehension of the mechanism between the material properties and synthesis and cell behavior.

Most of the demands and problems involve the reaction to and interaction with the surrounding tissue of the implanted material once it is implanted in the body. In this respect, the surface of the material plays a key role. It is very difficult to find a material that meets all the requirements. One strategy is to use composite materials that combine the properties of its components. Another way is to use a material that has the required bulk properties (biodegradability, strength, etc.) and to perform a surface treatment to modify the surface characteristics. Biomedical polymers, such as polylactic acid (PLA), poly-ε-caprolactone (PCL), poly(lactic acid-co-glycolic acid) (PLGA), and poly(hydroxybutyrate), are materials with good bulk properties for biomedical applications [13]. They are biocompatible, in some cases also biodegradable, and have good mechanical and structural properties. However, their surface properties are unsuitable to attract cells and a surface treatment is often required. In the past decades, surface treatment of polymers with nonthermal plasmas has been extensively studied, and it has become evident that also for biomedical polymers this is a promising approach. Surface modification of biomedical polymers gives the opportunity to change the surface characteristics of polymeric implants to achieve a better biocompatibility without altering the bulk properties. Due to the versatility of the technology, it can be useful in many different applications [14, 15]. Furthermore, to analyze in a correct and universal way the cell–material interaction is a challenge. Different microscopic technique are used and explored.

This chapter gives an overview of the use of surface modification of biomedical polymers and the influence on cell–material interactions.

The chapter starts with a comprehensive description of the nanostructured polymeric surface, by analyzing in detail the main surface modification methods and then the main mechanisms in the stem cell/surface interactions. A concise examination of microscopic techniques used in estimating the stem cell/nanostructured surface interaction will be described, and in particular fluorescence, electron, and atomic force microscopies will be mainly investigated and compared. Because these types of studies are relatively new, many questions still remain, and the response is related to the cell types.

6.2 Nanostructured Surface

This chapter highlights several techniques that are used to create nanostructured polymeric surface. Nanostructured surface increases cell–material interactions. Nanofeatures promote the adhesion of surrounding cells to scaffolds and the infiltration of processes. This chapter focuses specifically on the effect of highly ordered nano- and micro-topography on stem cell behavior for therapeutic applications. The use of well-controlled methods has been chosen because they are essential for systematically studying the effect of individual variables on cell behavior. In addition, several groups have shown that feature regularity and symmetry may also regulate cell behavior [16–20]. Furthermore, in order to understand the role of the surface in the adhesion of cells to nanostructured materials, it is necessary to understand how protein adsorption is influenced by the polymer surface physical and chemical properties.

Polymers are commonly used in biomedical application because of their excellent bulk properties, such as strength and processability. However, their surface properties are for most application inadequate due to their low surface energy. A surface modification is often needed.

Different techniques are now used to modify the surface and to develop topographies. They can be categorized as top-down or bottom-up approaches [21]. Top-down processing techniques are based on removing molecules from the starting substrate to produce topography, while bottom-up methods involve the addition of material onto the substrate. Plasma approach permits to develop nanotopographical polymeric surface by using both approaches.

Plasma surface modification is a very suitable and versatile technique that does not change the bulk properties of the samples, it can be used to uniformly treat complex-shaped surfaces, and it is a solvent-free technology. Plasma is often referred to as the fourth state of matter. It is a mixture of charged and neutral particles, such as atoms, molecules, ions, electrons, radicals, and photons. There are two main categories: thermal and nonthermal plasmas. Thermal plasmas cannot be used for the surface treatment of polymers because of their high gas temperature. Nonthermal plasmas, however, have a much lower gas temperature but relatively high electron temperature. They do not cause any thermal damage to the surface of heat-sensitive materials, although the reactive species in a nonthermal plasma can cause chemical and physical modifications to the surface [22].

Furthermore, in combination with mask, plasma treatment permits to develop specific surface topography with micrometric spatial distribution and nanometer height. By using a radiofrequency plasma treatment in our group, we developed nanogrooves on polystyrene surface with different height by controlling the time and/or bias parameters [20].

Plasma polymerization has become a well-established method for the surface modification of biomaterials [23]. Applications include tuning the properties of biomaterial surfaces and scaffolds [24, 25], the transfer of cells and the formation of cell sheets, patterning surfaces [23, 26] and commercial culture-ware coatings. The success of plasma treatment in polymer surface modification is due to the fact that only affects the outermost atomic layers of a material. Moreover, it is a fast, versatile, and environmentally friendly technology. The plasma systems can be work in air or in vacuum; however, only in vacuum the plasmas offer a good control over the plasma chemistry.

It was shown that the plasma source was capable of reducing the water contact angle, thus increasing the wettability [13, 20, 22, 27]. It was demonstrated that the at least a part of the treatment is permanent, in fact although the samples are subjected to ageing, the contact angle after 14 days of ageing stays well below the value of the untreated samples. From photoelectron spectroscopy (XPS), it was found that plasma treatments in air, helium, and argon result in surfaces with a similar chemical composition: mainly oxygen containing functionalities are incorporated after plasma treatment. Cell culture tests showed both quantitatively and qualitatively that plasma modification of biodegradable polymeric films increased the initial cell attachment [28, 29]. After 1 day, cells on plasma-treated PLA showed a superior cell morphology in comparison with unmodified PLA samples. After 7 days of culture, this cell morphology is not observed, suggesting that plasma treatment does not affect cell proliferation. From cell culture tests, it was also found that there are not important differences between the different discharge gases when looking at the improved cell–material interactions. Nevertheless, from economical point of view, treatments in air are the best choice [28, 29].

Instead of a uniform surface modification, several authors investigate the creation of gradient or patterned surfaces on biodegradable polyesters by plasma modification techniques, in order to obtain materials with continuously changing properties and functions, for use in challenging applications. On a gradient surface, the chemical composition or topography (or another surface characteristic such as roughness and wettability) gradually varies along one dimension and permits to develop in a whole substrate material with different properties [30].

6.3 Stem Cell

Stem cells are unique cells characterized by self-renewal and pluri-/multipotential properties. Stem cell self-renewal allows maintaining a stem cells reservoir for the subject life span. This property is the consequence of a peculiar cell division. With the asymmetric division, stem cells generate a daughter cell that remains stem cell and a daughter cell that became progenitor cell. With the symmetric division stem cell divides in two identical stem cell daughters [31–35]. Of note, by asymmetric division stem cells provide the correct replacement of stem cells inside the niche (a peculiar tissue area where stem cell reside) and the replacement of the committed cells that give rise differentiated progenies upon selected signals [31–37]. The capability of stem cells to generate a differentiated progeny is called pluri-/multipotential properties, depending on the production of one selected cell lineage or several/all cell lineages present in an animal [3, 34, 35, 37].

Naturally, stem cells exist as embryonic stem cells (ESCs) and adult stem cells (ASCs). ESCs are cells isolated from the inner mass of blastocysts, have self-renewal capability, and are able to generate all cell types. These cells could be used as substitute to germinal stem cells for the generation of animal models [38]. ASCs have similar self-renewal capacity to ESCs but have a more restricted differentiation lineage potential because of they produce a specialized tissue-specific cell type depending on the ASC type [39–42]. Noteworthy, ASCs persist within the niche of adult tissues and organs [43–47] and replace cells within the tissue under physiological and pathological conditions [48–51].

Thank to Takahashi and Yamanaka [52], now exists a new type of stem cells, called induced pluripotent stem cells (iPSCs). These stem cells were generated *in vitro* from somatic differentiated cells engineered to express Oct3/4, Sox2, c-Myc, and Klf4 genes. iPSCs have self-renewal activity and are pluripotent as ESCs. Currently, this technique has been used to generated iPSCs from several degenerated diseases (e.g. Alzheimer, Parkinson) representing an innovative tool for developing a patient-specific stem cell type [3, 53].

6.4 Stem Cell/Surface Interaction

The interaction of the biomedical material with the surrounding tissue is a key factor in the final success of the implant. The response of a cell in contact with the surface and the adhesion of cells to the material play an important role in the biocompatibility of the implant. It is thus important to understand how cells interact with their environment. Cells sense their surroundings through so-called protrusions. Cell behavior is influenced by a combination of soluble factors, direct cell–cell interactions, the insoluble ECM, mechanical forces, and electrical stimuli [3, 4, 54]. The complex interaction of cells with their microenvironment regulates the loss, acquisition, and maintenance of specific cellular functions in physiological and pathological conditions. The microenvironment of a cell consists of other cells of the same or different type, ECM components, fluids, and various molecular factors. The feedback loops of a cell and its surrounding are an indispensable element in cell homeostasis and the control of processes that require a remodeling of existing structures or the generation of new ones, for instance during morphogenesis. It is increasingly recognized that the micromechanics of the environment determine cell fate and function as much as soluble molecular factors do [55–59].

Indeed, the study of molecular mechanisms leading the maintenance of stem cell status and the control of stem cell fate has been used to establish the basic parameters for tissue-engineering applications where stem cells and biomaterials are combined under selected conditions [60]. In this regard, the effect of different surface characteristics, in terms of roughness, smoothness, and nanotopography, has been extensively investigated [60]. Thus, stem cells respond the surface properties by activating the focal adhesion complex and organizing the cytoskeleton fibres as consequence. F-actin, microtubules, and microfilaments change their architecture, the activity of the LINC complex protein, and in turn the chromatin conformation and the gene expression. The overall mechanisms are also known as mechanotransduction [60–64]. For instance, stem cells on grid nanotopography respond re-organizing the cytoskeleton to mimic a square structure, while stem cells on groove assume elongate cytoskeleton morphology [3, 18, 60, 63]. Moreover, the substrate topology could be involved either in stem cell differentiation mechanisms [65] or reprogramming activity [66].

6.5 Microscopic Techniques Used in Estimating Stem Cell/Surface

Various microscopy technologies have been employed to visualize the cell surface at the micro- and nanometer scale resolutions. In this chapter, we focus on the three main microscopies that permit to analyze the cell in different condition and resolution: fluorescence microscopy, electron microscopy, and atomic force microscopy.

6.5.1 Fluorescence Microscopy

Compared to the light microscopy, the fluorescence microscopy allows to monitor cell morphology together with the localization of proteins inside the cell [67].

The principle is that fluorescent molecules that absorb light at one selected wavelength (*excitation*) and emits light at a specific and longer wavelength (*emission*) could be used in order to specifically target cell proteins.

Current fluorescent microscopy is provides by a light from a multiwavelength source that moves through an excitation filter, which allows the selection of the exciting wavelength. The latter is then reflected downward by a dichroic mirror and focused through the objective lens toward the sample. At this stage, in the sample, fluorescent molecules are excited at a specific and longer wavelength to emit fluoresces which is focused by the objective lens. A series of barrier filters blocks all nonspecific wavelengths in order to guarantee an effective signal.

Biological applications of fluorescent microscopy are obtained by coupling antibodies with fluorescent molecules that emit red light (e.g. rhodamine and Texas red), orange light (Cy3), and green light (fluorescein). This technique is immunofluorescence and allows the universal staining of proteins in cells as well as in section of tissues either after fixing treatment with cross-linking agents or in live cells. In this case, a camera could record the fluorescent signal. To this aim, using the gene delivery technology, cells are transduced with vector encoding for the naturally fluorescent protein green fluorescent protein (GFP). In some cases, a fusion protein consisting in the target protein and GFP is generated in order to monitor the processing and localization of that protein in a given cells [68].

An important improvement of this technique is the fluorescent confocal microscopy that combines the characteristic of the classical fluorescence microscopy with the analysis of electronic image of the samples. This is obtained by using a laser that generates a small point of light that is focused to the specimen at a selected depth. Emitted fluorescence then passes through a pinhole aperture (called a confocal aperture). Thus, only light emitted from the plane of focus is able to reach a detector that collect emitted fluorescence. If the procedure is repeated several times at different depths, it is possible to generate a three-dimensional image of the sample.

For the focus of this issue, immunofluorescence is a useful tool of investigation. It allows the staining the cytoskeleton proteins (actin and tubulin) revealing the cellular architecture on a specific surface together with the staining of adhesion proteins that allows the interaction of the cell with that surface.

Fluorescence microscopy allows to make visible the interaction of stem cells with the different material surface and the consequent morphology. Representative images showing the F-actin organization in adult mesenchymal stem cells (MSCs) growth on a glass coverslips are reported (Figure 6.1).

6.5.2 Electron Microscopy

An electron microscope is a type of microscope that uses a beam of accelerated electrons as a source of illumination, in order to create an image

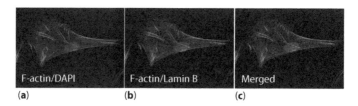

Figure 6.1 Representative fluorescent images of adult bone marrow MSCs cultured on a glass coverslip (a–c). F-actin (staining by falloidin-Alexa-488 [Green]) (Invitrogen); nucleus: DNA (DAPI); nucleus: lamin B (staining by anti-lamin B antibody [Santa-Cruz Biotechnology). (a) F-actin and DAPI, (b) F-actin and lamin B, and (c) merge of A+B images. Magnification: 20×. (Martino S., Unpublished data).

of the specimen. The original form of electron microscope is the transmission electron microscope (TEM) that uses a high-voltage electron beam to create an image. The electron beam is focused by electrostatic and electromagnetic lenses and transmitted through the specimen that is in part transparent to electrons and in part scatters them out of the beam. When it emerges from the specimen, the electron beam carries information about the structure of the specimen that is magnified by the objective lens system of the microscope. Resolution of the TEM is limited primarily by spherical aberration, hardware correction of spherical aberration for the high-resolution transmission electron microscopy (HRTEM) has allowed the production of images with resolution below 50 picometers [69] and magnifications above 50 million times.

The scanning electron microscope (SEM) produces images by probing the specimen with a focused electron beam that is scanned across a rectangular area of the specimen (raster scanning). When the electron beam interacts with the specimen, it loses energy by a variety of mechanisms. The lost of energy is converted into alternative forms such as heat, emission of low-energy secondary electrons and high-energy backscattered electrons, light emission, or X-ray emission, all of which provide signals carrying information about the properties of the specimen surface, such as its topography and composition. The SEM image was constructed from signals produced by a secondary electron detector, in some instruments are shown different types of detectors, these differ from the position inside the chamber and thus able to analyze scattered electrons with different angles in order to highlight diverse features of the sample. Figure 6.2 shows the schematic view of the SEM (a), two SEM images on the same sample with stem cell adherent on PCL films developed by solvent casting process, but acquired by using the InLens detector (b) and the SE2 detector, respectively. The detector-denominated InLens analyzes

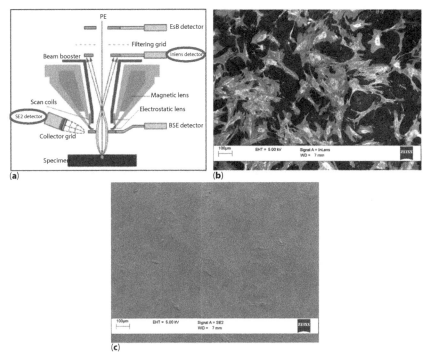

Figure 6.2 Schematic view of terminal part of SEM column with detectors position (a), micrograph of stem cell onto PCL substrate acquired with InLens detector (b) and SE2 detector (c).

the electrons with higher reflection angle, and this implies that the image highlights the structure of the sample or rather the different elements present. In Figure 6.2b are well evident the substrate (PCL film), in dark, and the cells seeded over. The detector called SE2 analyzes the electrons with smaller reflection angle and highlights mainly sample morphological and three-dimensional characteristics. In Figure 6.2c, which represents the same sample shown in the micrograph Figure 6.2b, but analyzed by SE2, the cells are highlighted by a slight variation in color, but it is clear that they are rather flat; the flat cell structure in this image is the result of issue of the methodology of sample preparation was left to dry in the air and that will be discussed shortly. However, because the SEM image relies on surface processes rather than transmission, it is able to image bulk samples up to many centimeters in size and (depending on instrument design and settings) has a great depth of field, and so can produce images that are good representations of the three-dimensional shape of the sample.

In the course of the past decades, scanning electron microscopy has gained access to many scientific and industrial sectors. Today, SEMs are used for a wide range of applications in traditional materials and damage analysis, the electric and semiconductor industries, and even in biology, chemistry, and the life sciences. Thanks to the easy control of today's instruments, many laboratories can hardly imagine work without a SEM for the quick and precise solution of their application problems. However, SEMs also present many other advantages over traditional light microscopy and other analysis methods:

- improved resolution capacities,
- high depth of focus,
- less preparative work,
- simplified interpretation of the images due to the 3D impression,
- use of different contrast mechanisms for the creation of images, and
- relatively easy adaptation and control of additional measuring devices for micro-range analyses.

All these features make the SEM a precious tool for the analysis of the most different materials. Another advantage of SEM is its variety called environmental scanning electron microscope (ESEM) that represented a very attractive innovation for many researchers because it allows imagery of biological as well as material samples, in wet mode, with no prior specimen preparation as for the conventional SEM. This greatly facilitates imaging biological samples that are unstable in the high vacuum of conventional electron microscopes [70]. One of the latest technologies developed for microscopy is the use of focused ion beam (FIB): a particular scanning helium ion microscope that operates in a manner similar to the SEM using a finely focused beam of ions. In scanning helium ion microscope, the detectors provide not only images but can also offer other information about topographic, material, crystallographic, and electrical properties of the sample [71]. Anyway, the combination of different microscopy techniques is a very promising approach to further understanding biological samples and their three-dimensional organization [72].

One of the major problems in the use of electron microscopy or biological material is the sample preparatory. Different methods and protocols have been developed in order to prepare biological samples for electron microscopy analysis, without altering their morphology. Chemical fixation is the commonly used method for preserving them, glutaraldehyde or a combination of glutaraldehyde and formaldehyde are the most frequently used fixative [73]. Since the analysis in the electron microscope entails that the samples are subject to high vacuum levels normally, these must be dehydrated. Critical point drying (CPD) is the main approach to dehydrate the biological samples in such a way that they maintain intact their three-dimensional structure. It is based on the concept that at a certain temperature and pressure, the vapor and liquid phases of carbon dioxide (CO_2) become indistinguishable. Liquid CO_2 from a siphon-tube tank is introduced into the chamber and used to replace the 100% ethanol in the specimen. After the ethanol has been totally replaced by the CO_2, the CPD chamber is raised above the critical point. The temperature is kept above the critical point, while the gaseous CO_2 is vented from the chamber. The process is finished when the CPD is returned to atmospheric pressure. The samples can be dried also in air if the substrate does not allow the use of CPD, or freeze drying. After the dehydrated procedure, the specimen should be totally dry and is ready to be introduced into the vacuum system of the sputter coater and SEM. Depositing a conductive layer on the sample reduces thermal damage, inhibits charging, and improves the secondary electron signal required for SEM analysis. Sputter coating is a technique for depositing a metal coating on specimen surfaces to be examined. The dried sample is attached to support stubs with a variety of materials (colloidal silver, colloidal carbon, double sided tape, or conductive carbon tape, among others) prior to coating with precious metals such as gold–palladium to ensure the electrical conductivity of the specimen surface. Thus, even using a more simple procedure for SEM preparation, the sample requires a few hours before reaching the electron microscopy phase.

6.5.3 Atomic Force Microscopy

The atomic force microscope (AFM), widely used in materials science, has found increasing applications in physical and biological fields [74–76]. AFM is perhaps the most versatile and powerful microscopy technology for studying samples at nanoscale. It is versatile because an AFM can not only image in three-dimensional topography, but it also provides various types of surface measurements to the needs of scientists and engineers. It is powerful because an AFM can generate images at atomic resolution with angstrom scale resolution height information, with minimum sample preparation. The number of research publications making use of AFM to investigate biological processes has also increased, though not at the same rate as in the physical sciences. The number of publications using AFM in vision science has seen only a very slight increase in the past 15 years, suggesting an unrealized potential of AFM in this field.

AFM was first developed to probe nanoscale features of solid materials using its high sensitivity to intermolecular forces (pN) and spatial resolution (nm). AFM has found applications in biology to measure features of cells, such as cellular elasticity [77]. Furthermore, it is a very powerful technique for the biological sciences, allowing samples to be imaged in situ in physiological conditions [78, 79]. Hence, AFM does not require a detailed preparation process as explained before in the case of electron microscopy. The AFM has several advantages over electron microscopy in the study of biological materials, including the ability to image in liquid with minimal sample preparation (no labeling, fixing, or coating). The AFM also allows the topographic characterization of surfaces at resolutions not achievable by optical microscopy. An optical microscope is limited by the diffraction limit of light, while the electron microscope is limited by the electron diffraction limit; the achievable lateral resolution of the AFM is limited by the tip size and shape and is typically on the order of a few nanometers. Furthermore, atomic force microscopy permits to analyze the topography of the sample, by analyzing the substrate and the cell adherent on the substrate.

6.5.3.1 *Instrument*

The height (z) resolution of the AFM is approximately 1 Å, limited only by electronic and thermal noise in the system. AFM provides a 3D profile of the surface on a nanoscale, by measuring *forces* between a sharp probe (<10 nm) and surface at very short distance (0.2–10 nm probe–sample separation). The probe is supported on a flexible cantilever. An AFM uses a cantilever with a very sharp tip to scan over a sample surface.

Atomic force microscopy senses tip–sample interaction forces (electrostatic, van der Waals, bonding, frictional, capillary, and magnetic forces) which occur between a probe tip (attached to a cantilever spring) and the substrate. AFM can work mainly in:

- Contact mode,
- Tapping mode.

In the contact mode, a probing tip has a soft physical contact with the sample. As the scanner gently traces the tip across the sample, contact forces cause the cantilever deflection, following changes in topography. The cantilever deflection is detected by a laser beam that bounces off the back of the cantilever onto a positive sensitive detector. As the cantilever bends, the position of the laser beam on the detector shifts. The photodetector itself can measure the displacements of light as small as 10 Å. As result, the system can detect sub-Angstrom vertical movements of the cantilever tip. Typical forces between the tip and the sample range from 10^{-12} to 10^{-7} N. Per imaging of soft materials like biological samples, the tapping mode AFM is frequently used since the probing tip is only transiently in contact with the surface resulting in a minor destructive interaction with the sample. In this mode, the cantilever is oscillating at its resonance frequency (around 100 KHz) and positioned on the surface for a very small fraction of its oscillating period.

Proper choice of both cantilever stiffness and geometry of the incorporated tip is necessary for successful application of the AFM. To avoid surface damage of delicate biological samples, imaging in contact mode, in either air or fluid, requires cantilevers with low spring constants. A low-spring-constant cantilever is also used for tapping mode in fluid, whereas a stiffer cantilever is needed for tapping mode in air to reduce noise. Low-spring-constant cantilevers also provide better deflection resolution, with a larger deflection for a given force, necessary for measuring intermolecular interactions, which typically range from 10^{-12} to 10^{-7} N. The tip can be functionalized with molecules, proteins, or cells, and the interactions of these tips with a surface of interest can be obtained. Several reviews are available describing these types of experiments [80].

The AFM can be used to study a wide variety of samples (i.e. plastic, metals, glasses, semiconductors, and biological samples such as the walls of cells and bacteria). Unlike scanning electron microscopy, it does not require a conductive sample. However, there are limitations in achieving atomic resolution. The physical probe used in AFM imaging is not ideally sharp. As a consequence, an AFM image does not reflect the true sample topography but rather represents the interaction of the probe with the sample surface, this phenomenon is called tip convolution. AFM permits to analyze the material in different medium:

- Air,
- Controlled atmosphere,
- Vacuum,
- Liquid.

For living cell imaging, the AFM analysis will be conducted in liquid medium, while in the case of fixed cells, the operator can decide the most appropriate medium.

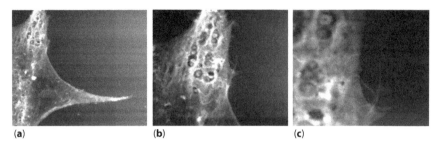

Figure 6.3 AFM images of fibroblasts deposited on glass coverslip and analyzed by atomic force microscopy at different resolutions: 50 × 50 μm (a), 25 × 25 μm (b), and 10 × 10 μm (c), after drying in air. (Armentano I., unpublished data).

Figure 6.3 shows AFM images of fibroblasts deposited on glass coverslip and analyzed after drying in air in tapping mode at different resolutions: 50 × 50 μm (a), 25 × 25 μm (b), and 10 × 10 μm (c). AFM at low resolution permits to analyze the adhesion of cell on the substrates, while at high resolution, the cell nanofeatures are visible.

6.5.3.2 Cell Nanomechanical Motion

The analysis of living cell by AFM permits to study by high-resolution microscopy of the cell migration process. The migration of cell and in particular of MSCs plays a key role in different mechanism and in particular in tumor-targeted delivery vehicles and tumor-related stroma formation [81]. MSCs have the ability to self-renew and can be isolated and expanded from bone marrow and other tissues. Therefore, the common hypothesis is that MSCs possess migratory activity. Indeed, it has been reported that cytokines can stimulate MSC migration and support the potential of MSCs as tumor-targeted delivery vehicles for therapeutic agents by making use of these cytokines from tumors. In addition, a number of studies have found that MSCs can migrate to regions of gastric carcinoma, melanoma, multiple myeloma, and breast cancer. Understanding the precise mechanism underlying the tropism of MSCs induced by tumors has been an area of active study in recent years. Cells are dynamic structures that display nanometer to micrometer scale motions at their cell membranes. The AFM can investigate the nanomechanical motion of the cell surface ranging from yeast [82] to cardiomyocytes [83, 84].

6.5.3.3 Mechanical Properties

AFM permits to analyze the mechanical properties of the substrates at the nanolevel. Mechanical measurements acquired using AFM rely on measuring the force as the tip is pushed toward, indented into, and retracted from the sample or cell surface in this case. The cantilever is mounted on the end of a piezoelectric tube scanner which is used to bring the tip into contact with the surface. The force is measured by recording the deflection (vertical bending) of the cantilever. This deflection is directly proportional to the force. Force–displacement curves are obtaining by monitoring the deflection of the cantilever, and the system allows us to probe the local Young's modulus (E) or "stiffness" of living cells, performs force spectroscopy measurements with piconewton resolution, and provides a sensor to record *in vivo* measurements of the cell wall at sub-nanometer resolution. In particular, AFM is a key tool in acquiring kinetic information and real-time signals of living cells, and is capable of offering *in vivo* single cell diagnostics [84].

Force curves measure the amount of force felt by the cantilever as the probe tip is brought close to – and even indented into – a sample surface and then pulled away. In a force curve analysis, the probe is repeatedly brought toward the surface and then retracted. Force curve analyses can be used to determine chemical and mechanical properties such as adhesion, elasticity, hardness, and rupture bond lengths. The slope of the deflection provides information on the hardness of a sample. The adhesion provides information on the interaction between the probe and sample surface as the probe is trying to break free. Direct measurements of the interactions between molecules and molecular assemblies can be achieved by functionalizing probes with molecules of interest [85].

Liu *et al.* apply AFM in the study of cardiomyocytes. They quantitatively measured its mechanical phenol types, including the contractile force, beat rate, and beat duration by avoiding fluidic disturbances that hampered previous attempts to study cardiomyocytes by AFM. An important problem in the development of new cardiac agents is to determine whether a compound has inotropic (i.e. affecting force generation) or chronotropic (i.e. affecting rate) effects on the cardiomyocytes. Their AFM-based method quantitates these effects separately and thus could facilitate preclinical studies of candidate drugs. Their approach combines measurement of cellular elasticity, beat force, and rate, and it reveals additional parameters than could be seen by imaging positional changes of surface beads or by video microscopy of the cell edges. This approach could be used to analyze cells from patients with cardiomyopathy to understand the potential for gene therapy in these diseases [77].

Docheva *et al.* used AFM to search for differences in cell shape, volume, and elasticity of hMSCs (RS cells and FC cells), hOBs, and osteosarcoma cells (MG63 cell line), investigating if the morphometric and elastic properties of the cells are influenced by the substrates on which they are cultured (polystyrene, glass, and collagen I). Finally, they looked for a link between the above characteristics, the actin organization and the adhesive profile of the chosen cells. To characterize the elasticity of the cells, mean Young's modules were calculated. The only values taken were these calculated from force curves that were recorded in the nuclear and peri-nuclear region [86].

6.6 Conclusions and Future Perspectives

The integration of disciplines from physics (experimental and modeling) to cell biology and developmental biology is important to address in new ways how patterns arise in living organisms, based on the fundamental properties of cell surfaces. In the past, researchers have been limited to topographical patterning at the microscale; however, due to substantial technological advancements, topographical patterning with nanometer resolution has now become a reality. Though the use of nanofabrication techniques is essential for cell types that require nanoscale topography, they are not necessarily the best choice for patterning microscale features due to practical concerns. Fabrication methods with decreasing minimum resolution are generally associated with new and innovative microscopic tools that permit the analysis at the nanolevel the stem cell biomaterial interaction.

Nanostructured polymeric surface can be fabricated to take on a variety of morphologies and properties to more precisely modulate cell behavior and to enhance cell–material interfacial interactions. A better understanding of cell biology, biomaterial science, and cell–material interaction has led to some hopeful advances.

References

1. Tabata, Y., Recent progress in tissue engineering. *Drug Discov. Today* 6, 483, 2001.
2. Vasita, R., Shanmugam, K., Katti, D., Improved biomaterials for tissue engineering applications: surface modification of polymers. *Curr. Top. Med. Chem.* 8, 341, 2008.
3. Martino, S., D'Angelo, F., Armentano, I., Kenny, J.M., Orlacchio, A., Stem cell-biomaterial interactions for regenerative medicine. *Biotechnol. Adv.* 30, 338, 2012.
4. Atala, A., Murphy, S., Regenerative medicine. *JAMA* 313, 1414, 2015.
5. Hutmacher, D.W., Schantz, J.T., Lam, C.X., Tan, K.C., Lim, T.C., State of the art and future directions of scaffold-based bone engineering from a biomaterials perspective. *J. Tissue Eng. Regen. Med.* 1, 260, 2007.

6. Pham, Q.P., Sharma, U.,Mikos. A.G., Electrospinning of polymeric nanofibers for tissue engineering applications: a review. *Tissue Eng.* 12, 1197, 2006.
7. Orlacchio, A., Bernardi, G., Orlacchio, A., Martino, S., Stem cells and neurological diseases. *Discov. Med.* 9, 546, 2010a.
8. Biffi, A., Montini E., Lorioli L., Cesani M., Fumagalli F., Plati T., Baldoli C., Martino S., Calabria A., Canale S., Benedicenti F., Vallanti G., Biasco L., Leo S., Kabbara N., Zanetti G., Rizzo W.B., Mehta N.A., Cicalese M.P., Casiraghi M., Boelens J.J., Del Carro U., Dow D.J., Schmidt M., Assanelli A., Neduva V., Di Serio C., Stupka E., Gardner J., von Kalle C., Bordignon C., Ciceri F., Rovelli A., Roncarolo M.G., Aiuti A., Sessa M., Naldini L. Lentiviral hematopoietic stem cell gene therapy benefits metachromatic leukodystrophy. *Science* 341, 1233158, 2013.
9. Webber, M.J., Appel E.A., Meijer E.W., Langer R. Supramolecular biomaterials. *Nat. Mater.* 15, 13, 2015.
10. Khorshidi S., Solouk A., Mirzadeh H., Mazinani S., Lagaron J. M., Sharifi S., and Ramakrishna S., A review of key challenges of electrospun scaffolds for tissue-engineering applications. *J. Tissue Eng. Regen. Med.* 2005.
11. Chen, G.P., Ushida, T., Tateishi, T. Scaffold design for tissue engineering. *Macromol. Biosci.* 2, 67, 2002.
12. Sabir, M.I., Xu, X., Li, L., A review on biodegradable polymeric materials for bone tissue engineering applications. *J. Mater. Sci.* 44, 5713, 2009.
13. Armentano, I., Dottori, M., Fortunati, E., Mattioli, S., Kenny. J.M., Biodegradable polymer matrix nanocomposites for tissue engineering: a review. *Pol. Deg. Stabil.* 95, 2126, 2010.
14. Morent, R., De Geyter, N., Desmet, T., Dubruel, P., Leys, C., Plasma surface modification of biodegradable polymers: a review. *Plasma Process. Polym.* 8, 171, 2011.
15. Desmet, T., Morent, R., De Geyter, N., Leys, C., Schacht, E., Dubruel, P. Nonthermal plasma technology as a versatile strategy for polymeric biomaterials surface modification: a review. *Biomacromolecules* 10, 2351, 2009.
16. D'Angelo, F., Armentano, I., Cacciotti, I., Tiribuzi, R., Quattrocelli, M., Del Gaudio, C., Fortunati, E., Saino, E., Caraffa, A., Cerulli, G.G., Visai, L., Kenny, J.M., Sampaolesi, M., Bianco, A., Martino, S., Orlacchio, A., Tuning multi/pluri-potent stem cell fate by electrospun poly(L-lactic acid)-calcium-deficient hydroxyapatite nanocomposite mats. *Biomacromolecules* 13, 1350, 2012.
17. D'Angelo, F., Tiribuzi, R., Armentano, I., Kenny, J.M., Martino, S., Orlacchio, A., Mechanotransduction: tuning stem cells fate. *J. Funct. Biomater.* 2, 67, 2011.
18. D'Angelo, F., Armentano, I., Mattioli, S., Crispoltoni, L., Tiribuzi, R., Cerulli, G.G., Palmerini, C.A., Kenny, J.M., Martino, S., Orlacchio, A., Micropatterned hydrogenated amorphous carbon guides mesenchymal stem cells towards neuronal differentiation. *Eur. Cell Mater* 20, 231, 2010.
19. Martino, S., D'Angelo, F., Armentano, I., Tiribuzi, R., Pennacchi, M., Dottori, M., Mattioli, S., Caraffa, A., Cerulli, G.G., Kenny, J.M., Orlacchio, A., Hydrogenated amorphous carbon nanopatterned film designs drive human

bone marrow mesenchymal stem cell cytoskeleton architecture. *Tissue Eng. Part A* 15, 3139, 2009.
20. Mattioli, S., Martino, S., D'Angelo, F., Emiliani, C., Kenny, J.M., Armentano, I., Nanostructured polystyrene films engineered by plasma processes: surface characterization and stem cell interaction. *J. Appl. Pol. Sci.* 14, 131, 2014.
21. Deok-Ho, K., Levchenko A., Suh K.Y. Engineered surface nanotopography for controlling cell–substrate interactions. In: Khademhosseini A., Borenstein J., Toner M., Takayama S., editors. *Micro and Nanoengineering of the Cell Microenvironment: Technologies and Applications.* Norwood, MA: Artech House Publishers; 2008. pp. 185–208.
22. Armentano, I., Ciapetti, G., Pennacchi, M., Dottori, M., Devescovi, V., Granchi, D., Baldini, N., Olalde, B., Jurado, M.J., Alava, J.M., Kenny, J.M., Role of PLLA plasma surface modification in the interaction with human marrow stromal cells. *J. Appl. Pol. Sci.* 114, 3602, 2009.
23. Siow, K.S., Britcher, L., Kumar, S., Griesser, H.J., Plasma methods for the generation of chemically reactive surfaces for biomolecule immobilization and cell colonization. A review. *Plasma Process. Polym.* 3, 392, 2006.
24. Colley, H.E., Mishra, G., Scutt, A.M., McArthur, S.L., Plasma polymer coatings to support mesenchymal stem cell adhesion, growth and differentiation on variable stiffness silicone elastomers. *Plasma Process. Polym.* 6, 831, 2009.
25. Wells, N., Baxter, M.A., Turnbull, J.E., Murray, P., Edgar, D., Parry, K.L., Steele, D.A., Short, R.D., The geometric control of E14 and R1 mouse embryonic stem cell pluripotency by plasma polymer surface chemical gradients. *Biomaterials* 30, 1066, 2009.
26. Sardella, E., Favia, P., Gristina, R., Nardulli, M., d'Agostino, R., Plasma-aided micro- and nanopatterning processes for biomedical applications. *Plasma Process. Polym.* 3, 456, 2006.
27. Fortunati, E., Mattioli, S., Visai, L., Imbriani, M., Fierro, J.L.G., Kenny, J.M., Armentano, I. Combined effects of Ag nanoparticles and oxygen plasma treatments on PLGA morphological, chemical, and antibacterial properties. *Biomacromolecules* 14, 626, 2013.
28. Jacobs, T., Declercq, H., De Geyter, N., Cornelissen, R., Dubruel, P., Leys, C., Beaurain, A., Payen, E., Morent, R., Plasma surface modification of polylactic acid to promote interaction with fibroblasts, *J Mater Sci Mater Med* 24, 469, 2013.
29. Jacobs, T., Morent, R., De Geyter, N., Dubruel, P., Leys, C., Plasma surface modification of biomedical polymers: influence on cell-material interaction. *Plasma Chem. Plasma Process.* 32, 1039, 2012.
30. Engler, A.J., Sen, S., Sweeney, H.L., Discher, D.L. Matrix elasticity directs stem cell lineage specification. *Cell* 2006, 126, 677, 2006.
31. Tulina, N., Matunis, E., Control of stem cell self-renewal in *Drosophila* spermatogenesis by JAK-STAT signaling. *Science* 294, 2546, 2001.
32. Yamashita, Y.M., Fuller, M.T., Jones, D.L., Signaling in stem cell niches: lessons from the Drosophila germline. *J. Cell Sci.* 118, 665, 2005.

33. Rusan, N.M., Peifer, M., A role for a novel centrosome cycle in asymmetric cell division. *J. Cell Biol.* 177, 13, 2007.
34. Yamashita, Y.M., Fuller, M.T., Asymmetric centrosome behavior and the mechanisms of stem cell division. *J. Cell Biol.* 180, 261, 2008.
35. Yamashita, Y.M., The centrosome and asymmetric cell division. *Prion* 3, 84, 2009.
36. Cheng, J., Türkel, N., Hemati, N., Fuller, M.T., Hunt, A.J., Yamashita, Y.M., Centrosome misorientation reduces stem cell division during ageing. *Nature* 456, 599, 2008.
37. Orlacchio, A., Bernardi, G., Orlacchio, A., Martino, S., Stem cells: an overview of the current status of therapies for central and peripheral nervous system diseases. *Curr. Med. Chem.* 17, 595, 2010b.
38. Richards, M., Fong, C.Y., Tan, S., Chan, W.K., & Bongso, A., An efficient and safe xeno-free cryopreservation method for the storage of human embryonic stem cells. *Stem Cells*, 22, 779, 2004.
39. Cossu, G., Bianco, P., Mesoangioblasts vascular progenitors for extravascular mesodermal tissues. *Curr Opin. Genet. Dev.* 13, 537, 2003.
40. McKay, R.D., Stem cell biology and neurodegenerative disease. *Philos. Trans. R Soc. Lond. B Biol. Sci.* 359, 851, 2004.
41. Martino, S., di Girolamo, I., Orlacchio, A., Datti, A., Orlacchio, A., MicroRNA implications across neurodevelopment and neuropathology. *J. Biomed. Biotechnol.* 2009, 654346, 2009b.
42. Martino, S., di Girolamo, I., Tiribuzi, R., D'Angelo, F., Datti, A., Orlacchio, A., Efficient siRNA delivery by the cationic liposome DOTAP in human hematopoietic stem cells differentiating into dendritic cells. *J. Biomed. Biotechnol.* 2009, 410260, 2009c.
43. Lindvall, O., Kokaia, Z., Martinez-Serrano, A., Stem cell therapy for human neurodegenerative disorders – how to make it work. *Nat. Med.* 10(Suppl), S42, 2004.
44. Gritti, A., Galli, R., Vescovi, A.L., Clonal analyses and cryopreservation of neural stem cell cultures. *Methods Mol. Biol.* 438, 173, 2008.
45. Uccelli, A., Moretta, L., Pistoia, V., Mesenchymal stem cells in health and disease. *Nat. Rev. Immunol.* 8, 726, 2008.
46. Ma, D.K., Bonaguidi, M.A., Ming, G.L., Song, H., Adult neural stem cells in the mammalian central nervous system. *Cell Res.* 19, 672, 2009.
47. Lee, P.H., Park, H.J., Bone marrow-derived mesenchymal stem cell therapy as a candidate disease-modifying strategy in Parkinson's disease and multiple system atrophy. *J. Clin. Neurol.* 5, 1, 2009.
48. Brittan, M., Wright, N.A., Gastrointestinal stem cells. *J. Pathol.* 197, 492, 2002.
49. Tumbar, T., Guasch, G., Greco, V., Blanpain, C., Lowry, W.E., Rendl, M., et al., Defining the epithelial stem cell niche in skin. *Science* 303, 359, 2004.
50. Kim, C.F., Jackson, E.L., Woolfenden, A.E., Lawrence, S., Babar, I., Vogel, S., et al., Identification of bronchioalveolar stem cells in normal lung and lung cancer. *Cell* 20121, 823, 2005.
51. Bonner-Weir, S., Weir, G.C., New sources of pancreatic beta-cells. *Nat. Biotechnol.* 23, 857, 2005.

52. Takahashi, K., Yamanaka, S., Induction of pluripotent stem cells from mouse embryonic and adult fibroblast cultures by defined factors. *Cell* 126, 663, 2006.
53. Martino, S., Morena, F., Barola, C., Bicchi, I., Emiliani, C., Proteomics and epigenetic mechanisms in stem cells. *Curr. Proteom.* 11, 193, 2014.
54. McHugh, K.J., Saint-Geniez, M., Tao, S.L. Topographical control of ocular cell types for tissue engineering. *J. Biomed. Mater. Res. Part B* 101B, 1571, 2013.
55. Chen, C.S., Mrksich, M., Huang, S., Whitesides, G.M., Ingber, D.E. Geometric control of cell life and death. *Science* 276, 1425, 1997.
56. Discher, D.E., Janmey, P., Wang, Y.L. Tissue cells feel and respond to the stiffness of their substrate. *Science* 310, 1139, 2005.
57. Engler, A.J., Griffin, M.A., Sen, S., Bonnemann, C.G., Sweeney, H.L., Discher, D.E. Myotubes differentiate optimally on substrates with tissue-like stiffness: pathological implications for soft or stiff microenvironments. *J. Cell Biol.* 166, 877, 2004.
58. Huang. S., Ingber, D.E. The structural and mechanical complexity of cell-growth control. *Nat. Cell Biol.* 1, E131–E138, 1999.
59. Lecuit, T., Lenne, P.F. Cell surface mechanics and the control of cell shape, tissue patterns and morphogenesis. *Nat. Rev. Mol. Cell Biol.* 8, 633, 2007.
60. Chen, W., Shao, Y., Li, X., Zhao, G., Fu, J., Nanotopographical surfaces for stem cell fate control: engineering mechanobiology from the bottom. *Nano Today* 9, 759, 2014.
61. Salmasi, S., Kalaskar, D.M., Yoon, W.W., Blunn, G.W., Seifalian, A.M. Role of nanotopography in the development of tissue engineered 3D organs and tissues using mesenchymal stem cells. *World J. Stem. Cells* 7, 266, 2015.
62. Dalby, M.J., Gadegaard, N., Oreffo, R.O. Harnessing nanotopography and integrin-matrix interactions to influence stem cell fate. *Nat. Mater.* 13, 58, 2014.
63. Connelly, J.T., Gautrot, J.E., Trappmann, B., Tan, D.W., Donati, G., Huck, W.T., Watt, F.M., Actin and serum response factor transduce physical cues from the microenvironment to regulate epidermal stem cell fate decisions. *Nat. Cell Biol.* 12, 711, 2010.
64. Jaalouk, D.E., Lammerding J. Mechanotransduction gone awry. *Nat. Rev. Mol. Cell Biol.* 10, 63, 2009.
65. Teo, B.K., Ankam, S., Chan, L.Y., Yim, E.K., Nanotopography/mechanical induction of stem-cell differentiation. *Methods Cell Biol.* 98, 241, 2010.
66. Downing T.L., Soto J., Morez, C., Houssin, T., Fritz, A., Yuan, F., Chu, J., Patel, S., Schaffer, D.V., Li, S. Biophysical regulation of epigenetic state and cell reprogramming. *Nat. Mater* 12, 1154, 2013,
67. Sluder, G., Nordberg, J.J. Microscope basics. *Methods Cell Biol.* 114, 1, 2013.
68. Yuste, R., Fluorescence microscopy today. *Nat. Methods* 2, 902, 2005.
69. Rolf, E., Rossell, M.D., Kisielowski, C., Dahmen, U., Atomic-resolution imaging with a sub-50-pm electron probe. *Phys. Rev. Lett.* 102 (9), 096101, 2009.
70. Muscariello, L., Rosso, F., Marino, G., Giordano, A., Barbarisi, M., Cafiero, G., Barbarisi, A., A critical review of ESEM applications in the biological field. *J. Cell Physiol.* 205, 328, 2005.

71. Iberi, V., Vlassiouk, I., Zhang, X.G., Matola, B., Linn, A., Joy, D. C., Rondinone, A.J., Maskless lithography and *in situ* visualization of conductivity of graphene using helium ion microscopy. *Sci. Rep.* 5, 11952, 2015.
72. Bidlack, Felicitas B., Huynh, Chuong, Marshman, Jeffrey, Goetze, Bernhard, Helium ion microscopy of enamel crystallites and extracellular tooth enamel matrix. *Front. Physiol.* 5, 395, 2014.
73. Leser, V., Drobne, D., Pipan, Z., Milani, M., Tatti, F. Comparison of different preparation methods of biological samples. *J. Microsc.* 233, 309, 2009.
74. Last, J.A., Russell, P., Nealey, P.F., Murphy, C.J., The applications of atomic force microscopy to vision science. *Invest. Ophthalmol. Vis. Sci.* 51, 6083, 2010.
75. Muller, D.J., Dufrene, Y.F., Atomic force microscopy as a multifunctional molecular toolbox in nanobiotechnology. *Nat. Nanotechnol.* 3, 26, 2008.
76. Parot, P., Dufrene, Y.F., Hinterdorfer, P., et al. Past, present and future of atomic force microscopy in life sciences and medicine. *J. Mol. Recognition* 20, 418–431, 2007.
77. Jianwei Liu, Ning Sun, Marc A. Bruce, Joseph C. Wu, Manish J., Butte atomic force mechanobiology of pluripotent stem cell-derived cardiomyocytes. *PLoS One* 7, 37559, 2012.
78. Muller, D.J., Dufrene, Y.F., Atomic force microscopy as a multifunctional molecular toolbox in nanobiotechnology. *Nat. Nanotechnol.* 3, 261, 2008.
79. Parot, P., Dufrene, Y.F., Hinterdorfer, P., et al. Past, present and future of atomic force microscopy in life sciences and medicine. *J. Mol. Recognition* 20, 418, 2007.
80. Last, J.A., Russell, P., Nealey, P.F., Murphy, C.J., The applications of atomic force microscopy to vision science. *Invest. Ophthalmol. Vis. Sci.* 51, 6083, 2010.
81. Changhong Ke, Jianan Chen, Yajun Guo, ZhengW. Chen, Jiye Cai, Migration mechanism of mesenchymal stem cells studied by QD/NSOM. *Biochim. Biophys. Acta* 1848, 859, 2015.
82. Pelling, A.E., Sehati, S., Gralla, E.B., Valentine, J.S., Gimzewski, J.K., Local nanomechanical motion of the cell wall of *Saccharomyces cerevisiae*. *Science* 305, 1147, 2004.
83. Domke, J., Parak, W.J., George, M., Gaub, H.E., Radmacher, M. Mapping the mechanical pulse of single cardiomyocytes with the atomic force microscope. *Eur. Biophys. J* 28, 179–186, 1999.
84. Dufrene, *Life at the Nanoscale. Atomic Force Microscopy of Living Cells*. Edited by Dufrene. Pan Stanford Publishing.
85. Wilson, R.A., Bullen, H.A. Introduction to Scanning Probe Microscopy (SPM): Basic Theory Atomic Force Microscopy (AFM). Creative Commons Attribution-Noncommercial-Share Alike 2, 2006.
86. Docheva, D., Padula, D., Popov, C., Mutschler, W., Clausen-Schaumann, H., Schieker, M., Researching into the cellular shape, volume and elasticity of mesenchymal stem cells, osteoblasts and osteosarcoma cells by atomic force microscopy. *J. Cell. Mol. Med.* 12, 537, 2008.

7
Laser Surface Modification Techniques and Stem Cells Applications

Çağrı Kaan Akkan

Institute of Biomedical Engineering, Boğaziçi University, Istanbul, Turkey

Abstract

While designing a medical implant chemical inertness, mechanical stability and biocompatibility of the selected material must be favorable. Due to the knowledge and experiences behind the production process, today, most of the implants show high success when they meet with the body. However, although their main properties support the fundamental requirements, in some cases, further modifications on their surfaces may be needed for their fast acceptance by the body. It is well known that the surface properties of materials, such as chemistry, roughness, and wetting, play a critical role when these materials are implanted into the bodies. Success of an implant is lying behind these factors since they are the controller parameters of the protein adhesion to the implant surface, thus the cell–surface interaction. Therefore, introducing new chemical groups over surfaces accompanying the topographical modifications on their surfaces (or chemical modification alone) may be concluded with better cell adhesion and proliferation. In these circumstances, lasers may be the only well-suited device for surface modifications. Surface modification of implants by laser ended with precise and repeatable results while supporting fast processing speed in comparison to the other methods. Additionally, possibility to control the laser parameters, such as wavelength, pulse duration, frequency, and output energies, brings the opportunity to modify any kind of materials in nano- and/or microscale while preventing the material surface from chemical changes. On the other hand, it is also possible to modify the chemistry of implant surfaces with and without topographical changes, which is especially important for the stem cell researches since the environmental conditions are leading the differentiation of stem cells to a cell line, dominantly. This chapter presents the interaction of

Corresponding author: kaan.akkan@boun.edu.tr; akkankaanis@gmail.com

Ashutosh Tiwari, Bora Garipcan and Lokman Uzun (eds.) Advanced Surfaces for Stem Cell Research, (167–198) © 2017 Scrivener Publishing LLC

laser with materials and several techniques applied by lasers for chemical and physical surface modifications, which have the possibility to apply for stem cell researches or which have already been applied.

Keywords: Laser, surface, stem cell, chemistry, topography

7.1 Introduction

It is well known that the surface properties of materials, such as chemistry, roughness and wetting, play a critical role when these materials are implanted into the bodies. Success of an implant is lying behind these factors since they are the controller parameters of the protein adhesion to the implant surface, thus the cell–surface interaction. Therefore, introducing new chemical groups over surfaces accompanying the topographical modifications on their surfaces (or chemical modification alone) may be concluded with better cell adhesion and proliferation. In these circumstances, lasers may be the only well-suited device for surface modifications since the produced surface structures/features are as follows:

- Repeatable,
- Precise,
- Fast,
- Clean from process related surface contaminations.

Additionally, possibility to control the laser parameters, such as wavelength, pulse duration, frequency, and output energies, brings the opportunity to modify any kind of materials in nano- and/or microscale while preventing the material surface from chemical changes. On the other hand, it is also possible to modify the chemistry of implant surfaces with and without topographical changes, which are especially important for the stem cell researches since the environmental conditions are leading the differentiation of stem cells to a cell line, dominantly.

7.2 Fundamental Laser Optics for Surface Structuring

Since success of an implant, which is produced and/or altered by using lasers, is lying at the used laser parameters, knowing what happens when laser beam strikes the material surface is the start point to understand the mechanism behind the surface modification of an implant material. In this case, we can talk about several basic facts:

- Absorptivity and reflectivity of the laser beam by the material surface,
- Effect of the incoming laser light polarization,
- Operation mode of the laser,
- Beam quality factor,
- Laser pulse energy/power.

These parameters define the final surface structure that one can obtain.

7.2.1 Definitive Facts for Laser Surface Structuring

7.2.1.1 Absorptivity and Reflectivity of the Laser Beam by the Material Surface

Absorptivity determines what percentage of the light will penetrate in to the material. Therefore, this parameter plays a major factor for the surface structuring by using lasers. Absorptivity (A) of a surface can be calculated as follows:

$$A = 1 - R - T \qquad (7.1)$$

for transparent materials where R and T are reflectivity and transmissivity, respectively. In the case of an opaque material, absorptivity becomes as follows:

$$A = 1 - R \qquad (7.2)$$

Since absorptivity is related to the reflectivity of the material, calculating the reflectivity becomes crucial. However, reflectivity changes for any material depending on the used laser wavelength. For the normal angle of laser incidence, reflectivity can be found via

$$R = \frac{(1-n)^2 + k^2}{(1+n)^2 + k^2} \qquad (7.3)$$

where n is the real part and k is the imaginary part of the refractive index. Real and imaginary values of the refractive index of titanium (Ti) including the reflectance percentages for selected wavelengths can be seen in Table 7.1.

Table 7.1 Real and imaginary parts of titanium and its reflectance for selected lasers and wavelengths.

Wavelength (nm) (Laser)	n	k	~R (%)
193 (ArF)	1.1800	1.6467	36.8
248 (KrF)	1.2717	1.82	39.9
266 (Nd:YAG-4th)	1.27	2.0357	45.4
355 (Nd:YAG-3th)	1.8257	2.8721	55.0
532 (Nd:YAG-2nd)	2.4793	3.3511	57.5
1064 (Nd:YAG-1th)	3.4654	4.0085	61.5

7.2.1.2 Effect of the Incoming Laser Light Polarization

Depending on the chosen application, laser light may be faced with 0° of incidence angle. However, incidence angle may vary for some kind of applications, such as direct laser interference structuring [1, 2]. For such conditions, reflectivity of the surface changes proportional to the incidence angle and polarization of the laser light becomes more important in the mean of surface absorbance. In nature, light can be polarized to three different states defined as linear polarization, circular polarization, and elliptical polarization. However, most of the lasers are used at linear polarization state. On the other hand, linear polarization is classified inside into two different states that are "s-polarization" and "p-polarization". Since the light is an electromagnetic wave, it has electric and magnetic field vectors propagating in space/medium in relation with the propagation direction. When the laser light is polarized to "s-"state, one can consider that the oscillation of electric field vector of the laser beam is parallel to the plane of incidence (Figure 7.1a). In this approach, "p-"polarization state will have the electric field oscillation perpendicular to the plane of incidence [3] (Figure 7.1b).

According to this information, reflectance of a surface for a laser light that incidence with an angle of "α" can be found for "s-" and "p-"polarization states are

$$R_s = \frac{(n-\cos\alpha)^2 + k^2}{(n+\cos\alpha)^2 + k^2} \tag{7.4}$$

$$R_p = \frac{(n-1/\cos\alpha)^2 + k^2}{(n+1/\cos\alpha)^2 + k^2} \tag{7.5}$$

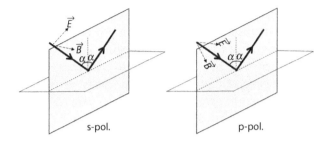

Figure 7.1 s- and p-Polarization states of laser light.

7.2.1.3 Operation Mode of the Laser

Lasers are divided into two groups depending on their working regimes which are defined as continuous wave (CW) or pulsed lasers. CW lasers emit continuous laser output like the laser pointers that we use in our daily life. Although these pointers are used to indicate some points in space, there are also high-power CW lasers for material processing as well. When the photon of the laser strikes at the substrate surface, lattice starts to heat up by vibrations which is called "lattice heating time" and stays for few picoseconds [4, 5]. Due to the short time scale of lattice vibration, most of the used lasers operating cause heat formation at the surface. On the other hand, this extended interaction time may be used to break the weaker bonds of the material to let it evaporate at lower temperatures than its melting point [5].

Due to their working principles, CW lasers cause expanded heat-affected zone over the work piece/substrates because of their continuous interaction with the substrate where the material heat conductivity is also in charge of this process. This can be a benefit for some cases for the substrate/implant preparations for the sintering of ceramic implant powders [6]. However, if the substrate surface wants to be structuring to a specific value and/or topography precisely, pulsed lasers are the best option. Since the pulsed lasers interact with the substrate for a short period of time, heat diffusion to the inner regions of the material will be reduced or eliminated. Here, pulse duration (pulse length) of the laser has high importance.

Pulsed lasers are also divided into groups in their selves:

- Modulated pulsing (gated pulsing): System of this type of pulsing is based on the switch on/off of a CW laser. This can be supported by an electronic circuit or by a mechanical shutter. Due to the simple working principle, output power of these types of lasers is limited to the output of the CW laser in use.

- Super pulsing: Super pulsing differs in a narrow field than modulated pulsing where the output power of the used CW laser operating at the modulation regime can be rising up to 2 or 3 times more by using electronic configurations.
- Hyper pulsing: Defines the pulsed lasers which are specially designed to give high-peak power outputs with short pulse durations. Lasers designed with Q-Switch systems and Mode-Locking systems are a kind of examples to this mechanism. In this mechanism, laser pulse is emitted only for short period of time to interact with the material which can go down to femtosecond (fs) time scales and gives an output power within the ranges of gigawatts per pulse [5, 7–10].

Although they cause heat generation at the surface, μs and ns lasers are used to structure the surfaces with fine features with ultra-less heat formation and less molten material at rest. In the case of usage of μs and ns laser pulses, due to the high-peak powers, plasma formation over the work piece has to be avoided, which may cause extra heat load to the substrate. On the other hand, laser beam may be blocked by the generated plasma, which seize the absorption of the laser by the material and changes the expected results, which is called "plasma shielding" [4, 8, 11, 12].

For ultra-short laser pulses, heat generation around the laser-processed area is not expected since the pulse duration of the laser stays well below the lattice heating time for most of the cases (0.5–50 ps), which ends up with highly precise surface structures [4, 5, 9].

7.2.1.4 Beam Quality Factor

Beam quality factor, also called as beam propagation factor, which measures and determines the quality of a laser beam. It is indicated by "M^2" (factor) and let us calculate how tight a laser beam can be focused in comparison to its diffraction limited version, which is a Gaussian beam. M^2 factor can be calculated by measuring the half-angle divergence "θ" of the laser beam propagating freely in space or in an optical medium:

$$M^2 = \frac{\theta \pi w_0}{\lambda} \qquad (7.6)$$

where λ is the wavelength and w_0 is the beam waist at the focal plane.

For a Gaussian beam, M^2 factor is given as "1" which is a theoretical value and for real laser beams, and M^2 value is larger than 1.

On the other hand, M² is not only a parameter which determines how the intensity distribution of the laser beam is. Intensity distribution of a laser can be defined by the cavity modes which are directly related to the laser cavity design. For a laser cavity, two different modes can be discussed which are longitudinal and transverse modes. Longitudinal modes define which wavelength can be obtained at the outside which are close to the central wavelength. This fact only effects the linewidth of the laser. Besides, transverse modes define the beam intensity profile mainly where the intensity profile is given by an indication of "TEM_{nm}". TEM is the abbreviation of transverse electromagnetic modes while "m" and "n" are used to indicate the number of the dark patterns at the intensity profile, horizontally and vertically. For a Gaussian laser output, output mode is indicated as "TEM_{00}". However, lasers do not operate at perfect Gaussian mode in real. Multimode output of a laser is usual obtained by the sum of different high-order modes where the optical intensity profile is divided by dark patterns depending on the orders "n" and "m" vertically and horizontally, respectively. However, therefore multimode lasers emit high-energy outputs [5, 10, 13, 14].

7.2.1.5 Laser Pulse Energy/Power

In order to structure a substrate surface, laser pulse energy/power plays a critical role. As mentioned in Section 7.2.3, lasers are defined into two groups as CW or pulsed lasers. Due to their working regimes, output powers differ. It is suitable to talk about power unit only for CW lasers, while "peak powers" are one of the leading parameters for the pulsed lasers which can be calculated by dividing the output energy "E" of the laser to its pulse duration "τ". Additionally, it is also possible to mention about an average power for pulsed lasers as well, which can be found by multiplying the output energy of the laser with its working frequency "f" [9]. When the pulse energy of the laser is increased, its peak power increases proportionally where the structuring will happen by ablation process for short laser pulses. In Figure 7.2, effect of the pulse energy on polyetheretherketone (PEEK) polymer at constant pulse length (10 ns) can be seen. It is obvious that the structure depth and width are increasing with the increased pulse energy [9].

When we concentrated on creating surface structures, pulsed lasers can be taken into account, as mentioned. In most of the cases, pulsed lasers operating at μs pulse durations and lower are used to create the required structures where the material removal regime is mostly defined by ablation process [9, 15].

Figure 7.2 Effect of laser pulse energy/power on PEEK polymer. Pulse length is fixed to 10 ns, while the pulse energy used as 1.35, 2.2, and 4.5 mJ from left to right, respectively. Corresponding peak powers are calculated as 135, 220, and 450 kW, respectively [9].

7.2.2 Ablation by Laser Pulses

Ablation process is the material removal via vaporization of the material or braking of the molecular bonds at the interested area. In order to create a structured area, focusing of the laser beam is crucial since the ablated area is mostly related to the spot size of the laser at the focal plane and the laser power where they both pointed out the laser intensity. Therefore, focusing ability of the optical system in use is the critical starting point of material ablation.

7.2.2.1 Focusing the Laser Beam

Collimated laser light can be focused to a tight spot by positive lenses with specific focal lengths. In order to obtain a diffraction limited spot size, aspherical lenses are the preferred optics. However, due to their relatively high costs, cheap but effective spherical lenses can be used. In order to obtain a well-focused laser light, clear aperture of the lens must be filled by the laser light. Since there are two mainly used positive spherical lenses are available in the market, attention must be paid which surface of the lens will be directed to the laser source. In the case of a plano-convex lens, which has a flat surface on one side, curved surface of the lens must be directed to the laser for safety issues. Besides of that, for most of the plano-convex lenses, that kind of usage will help to reduce the optical aberrations as well.

The performance of the focusing is not only related to the properties of the selected lens, but also to the properties of the laser beam. For a Gaussian beam profile, beam can be focused to a diffraction limited value which can be calculated as given in Equation 7.7:

$$2\omega_0 = \frac{4\lambda F}{\pi D} \tag{7.7}$$

where ω_0 is the beam radius at the focal plane, λ is the wavelength of the laser beam, F is the focal length of the lens, and D is the diameter of the laser beam before focusing. Depth of focus (DOF) for such a laser beam can also be calculated as follows:

$$DOF = 2 * X_R = \frac{2\pi\omega_0^2}{\lambda} = \frac{8\lambda F^2}{\pi D^2} \qquad (7.8)$$

where X_R is the Rayleigh length. Rayleigh length is defined as the distance from the beam waist where the beam Radius is increased by a factor of $2^{1/2}$.

7.2.2.2 Ablation Regime

When light get in touch with the material, no heat observes at the beginning although the process will be finalized with the production of heat. The parameters like ambient conditions and absorption properties of the material and the properties of the light source give the final result. For high-energy laser applications, although the process is thermal or non-thermal, due to the high light intensity at the focal area (or interested area), high excitation densities exceed 10^{22} species per cm^3.

After the first engage of light with the material, electron–electron collision starts the heating procedure which usually takes time from 10^{-14} to 10^{-12} s for metals. Reflecting the gained energy to the lattice (electron–phonon interaction) causes the vibration of lattice which is called "lattice heating time" and for its relaxation time, period of 10^{-12} to 10^{-10} s is needed. For nonmetals, time needed for the relaxation time is much longer which stays from 10^{-12} to 10^{-6} s [14].

In the case of a pulsed laser, ablation of a material from the surface occurs via pulse to pulse material removing regime which may happens through material vaporization (photothermal ablation) or through bond braking via high energetic photons (photochemical).

7.2.2.2.1 Photothermal Ablation

Photothermal ablation occurs when the pulse length of the used laser is longer than the thermal relaxation time of the material lattice [4, 9]. With every single laser pulses, identical amount of the material can be removed from the surface. After the removal of the material, resultant depth of the structure, which is the ablation depth, can be calculated by the Arrhenius equation for photothermal process. The depth from a single-pulse ablation is given by

$$d_{thermal} = Ae^{-E/RT}$$

where E, R, T, and A are the activation energy, ideal gas constant, temperature at the laser focal area, and effective frequency factor (μm/pulse), respectively. Arrhenius equation can be simplified to

$$d_{thermal} = Ae^{-B/\phi} \quad (7.9)$$

when the temperature rise at the localized are is linear. Here, B is the constant for the process and ϕ is the laser fluence [16, 17].

7.2.2.2.2 Photochemical Ablation

Photothermal ablation occurs when the pulse length of the used laser is shorter than the thermal relaxation time of the material lattice [4, 9]. Another approach for the observation of photochemical ablation is the factor of the used laser's wavelength. If the photon energy is sufficient enough to break the chemical bond, heat production does not observed [5]. In this case, UV lasers are the frequently used light sources for photochemical processes [5, 9]. The resultant structure depth from the photochemical ablation process is defined by Beer's law as given:

$$d_{photochemical} = \alpha^{-1} \ln(\phi/\phi_{th}) \quad (7.10)$$

where ϕ_{th} is the threshold fluence of laser in order to ablate material and α is the absorption coefficient of the material. Although photochemical process mostly occurs when the UV lasers are in use, in the lack of the exact wavelength matching for the bond to break, multiphoton ablation can be used for photochemical process. Ultrashort pulse lasers are the address for these kinds of applications [18].

7.3 Methods for Laser Surface Structuring

Laser surface structuring does not cover only the physical modification of the surface but also the chemical modifications. Following sections will explain the methods that one can use for the modification of implant surfaces via physical and chemical modifications by using lasers.

7.3.1 Physical Surface Modifications by Lasers

In the literature, lots of methods for the physical surface modification of materials, especially the implant materials, can be found. In this chapter, only the three main and the most rapid and effective physical surface modification methods will be introduced.

7.3.1.1 Direct Structuring

Direct laser structuring is based on the focusing of the laser light by using a lens or lens group. By using this technique, it is possible to generate single surface structure which will have the shape of the incoming laser beam. Therefore, scanning the laser beam over the sample by galvo scanners or moving the substrate by using translating stages is necessary (Figure 7.3). In the case of galvo scanner usage, special F-Theta lens is used which focuses the laser beam to a fixed circular spot over the substrate regardless of the incidence angle.

Today, most of the manufacturers support the information of a selected lens. However, it is necessary to calculate the focal length of lens combination which corrects many different optical aberrations and therefore improve the optical quality, such as: diffraction limited focusing [19]. Focal length of lens combination can be calculated for two different lenses as follows:

$$f_{com} = \left(\frac{1}{f_1} + \frac{1}{f_2} - \frac{d}{f_1 f_2} \right)^{-1} \qquad (7.11)$$

where f_1, f_2, and d are the focal length of the lens-1, lens-2, and the distance between the lenses, respectively. On the other hand, distance between the

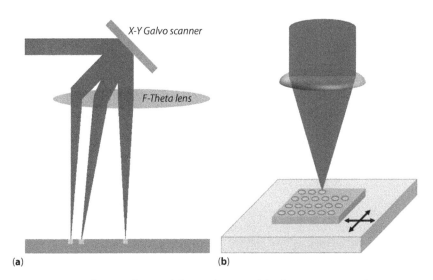

Figure 7.3 Direct laser surface modification by using (a) galvo scanner in order to move the laser beam over the substrate surface and (b) via scanning the substrate by using translation stages mainly in x- and y-axes.

focal plane and the second surface of the second lens can be calculated as follows:

$$x = \frac{f_2(f_1 - d)}{f_2 + f_1 - d} \tag{7.12}$$

In this kind of structuring approach, focusing ability of the lens plays a critical role for the generated spot size (radius) on the substrate surface. Spot size of a focused laser beam which has Gaussian intensity distribution can be calculated as follows:

$$2r = \frac{4\lambda f}{\pi D} \tag{7.13}$$

where the f is the focal length, λ is the wavelength of the laser, and D is the diameter of the laser beam entering the lens [5].

In real, lasers cannot have perfect Gaussian intensity distribution and tend to diverge in space. Therefore, M^2 term is used to indicate the beam quality (divergence) where these kinds of lasers are called multimode laser. Although it is possible to reach small M^2 values close to theoretical Gaussian, spot size term must be corrected for real beam (r_R) by using quality term:

$$2r_R = \frac{4M^2 \lambda f}{\pi D} \tag{7.14}$$

DOF is the change of the spot size at the focal plane. There are two common acceptances for the definition of DOF which are either the 5% of increasing or $\sqrt{2}$ times enlargement of spot size at the focal plane. DOF can be calculated from the beam waist equation at a distance z for a Gaussian beam as given in Equation 7.15:

$$r(z) = r\sqrt{1 + \left(\frac{\lambda \Delta z}{\pi r^2}\right)^2} \tag{7.15}$$

where $r(z)$ is the radius of the laser spot at the distance of z. For $\sqrt{2}$ times enlargement of spot size, Δz can be found as follows:

$$\Delta z = \pm 4 \frac{\lambda}{\pi} \left(\frac{f}{D}\right)^2 \tag{7.16}$$

and DOF will be

$$\text{DOF} = 8\frac{\lambda}{\pi}\left(\frac{f}{D}\right)^2$$

For multimode lasers, quality factor M^2 must be added to the beam waist equation:

$$r_R(Z) = r\sqrt{1 + \left(\frac{M^2 \lambda \Delta z}{\pi r_R^2}\right)^2} \qquad (7.17)$$

Solving Equation 7.17 via using 7.14, Δz_R and DOF_R can be found for real laser beam as follows:

$$\Delta z_R = \pm 4\frac{\lambda}{\pi}\left(\frac{f}{D}\right)^2 M^2 \qquad (7.18)$$

and DOF_R will be

$$\text{DOF} = 8\frac{\lambda}{\pi}\left(\frac{f}{D}\right)^2 M^2$$

7.3.1.2 Beam Shaping Optics

In order to create special structures, it is necessary to reshape the incoming laser beam. In this case, a mechanical mask, microlens array, and its optical setups can be used.

7.3.1.2.1 Mask Shaping
In order to ablate the substrate with special structure, one mechanical mask can be used by placing the mask to the optical path. By using this technique, any kind of shape can be transferred to the substrate surface (Figure 7.4).

In order to fulfill the mask, laser beam may need to be expanded by using a beam expander where simply two lenses are used to diverge and re-collimate the beam. Masked beam can be focused afterward by using a positive lens to form smaller features over the substrate surface.

7.3.1.3.2 Microlens Array
Microlens array is a single optical element which has multiple lenses over, in an order. These lenses can be formed at any shape as desired to form

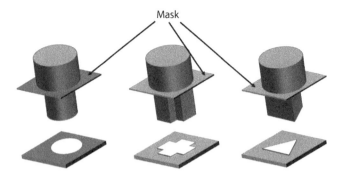

Figure 7.4 Mask shaping technique to ablate specific shape of material from the substrate.

different shapes at the focal plane, such as spherical, square, and cylindrical. Besides, these lenses can be created over the glass in different distribution instead of linear distribution. Orthogonal distribution is one kind of example.

Structuring with microlens arrays has basically an adaptation of direct structuring. It differs in the application methods with the direct structuring where it is possible to create over hundreds of identical structured over substrate surface just by using single laser pulse. Amount of the structures can rise up to few thousands which is limited with the design of the microlens array. Additionally, it is possible to combine this optical setup with x- and y-axes translation stage in order to create more structures by shifting the substrate.

Although structuring with microlens array is simple and fast in comparison to the methods mentioned before, attention must be paid for keeping the lens surface cleaning. Since this array has short focal distances, ablated materials can be deposited over the backside of the lens surface where working with metals and ceramics may be problematic in this way. In order to prevent this problem, longer focal length microlens arrays can be used or some further beam shaping optical setups can be applied which are nonimaging and imaging beam homogenization in order to create new patterns over the substrate surface with longer working distance from a short focal length microlens array [9].

7.3.1.3.2.1 Nonimaging Beam Homogenization
Beam homogenization systems originally used to transform the Gaussian intensity distribution of a laser beam into flat-top intensity profile. However, since they generated an array of light at the image plane, they can be further used to structure the substrate surface by the generated bright spots where their intensities are mainly identical.

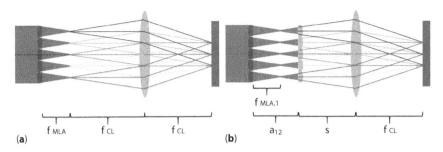

Figure 7.5 (a) Nonimaging and (b) imaging laser beam homogenization setups for pattern generation.

Nonimaging beam homogenization setup consists of simple one microlens array and one collecting lens as can be seen in Figure 7.5a. For both imaging and nonimaging beam homogenizations, Fresnel number (FN) plays an important role for the understanding of what type of homogenization setup is needed to be used. For nonimaging beam homogenization setup, FN bigger than 10 or better bigger than 100 is necessary for homogeneous intensity distribution at the generated pattern over the focal plane. FN can be calculated from Equation 7.19 [20].

$$FN = \frac{P_{MLA} D_{IP}}{4\lambda f_{CL}} \quad (7.19)$$

where P_{MLA}, D_{IP}, λ, and f_{CL} are the period of the microlens array, dimension of the illuminated region at the image plane, wavelength, and the focal length of the collection lens, respectively. D_{IP} can be calculated as follows:

$$D_{IP} = \frac{P_{MLA} f_{CL}}{f_{MLA}} \quad (7.20)$$

Since the microlens in the homogenization systems acts like a diffraction grating, there is a pattern occurs at the image plane where the focal length of the collecting lens is located. Period of the generated patterns can be calculated by the following equation:

$$P = \frac{f_{CL} \lambda}{P_{MLA}} \quad (7.21)$$

7.3.1.3.2.2 Imaging Beam Homogenization
In imaging beam homogenization, optical setup is suggested to use if the FN given by Equation 7.19 is smaller 10. In this kind of optical setup, two

microlens arrays are being used where both of the microlens arrays are suggested to be identical. In imaging beam homogenization system, obtained diffraction patterns at the image plane are generated more homogeneous in their intensity profile. Dimension of the obtained illuminated region at the image plane for this kind of optical setup is given as follows:

$$D_{IP} = \frac{P_{MLA.1} f_{CL} (f_{MLA.1} + f_{MLA.2} - a_{12})}{f_{MLA.1} f_{MLA.2}} \qquad (7.22)$$

Period of the generated diffraction pattern for imaging beam homogenization setup can be found by using the Equation 7.21 [20, 21]. Schematic illustration of imaging beam homogenization setup is given in Figure 7.5b.

7.3.1.3 Direct Laser Interference Patterning

Direct laser interference patterning (DLIP) is based on the interfering of minimum two identical laser beams over the substrate surface in order to produce periodical surface structures where the features size can be from tens of nanometers to micrometer scale [14, 22]. Its basic is the laser interference lithography (LIL) where the interfering laser beams are used to create periodical structures without need of a mask. However, LIL still needs for development of the photoresist where DLIP does not need [23, 24]. By using DLIP, it is possible to create periodical structures on biocompatible polymers, metals, and ceramics with an easy and fast method where these kind of structures can be used to guide and align the cells over [1, 2, 22]. High pulse energy and long coherent length of the laser are the key parameters for the generation of the well-separated interference fringes in high resolution in this kind of application [1, 2, 22, 25]. Periodicity of the generated patterns is defined by the periodicity of the interference fringes which can be calculated for two interfering identical laser beam by

$$P = \frac{\lambda}{2n \sin \alpha} \qquad (7.23)$$

where λ, n, and α are the wavelength of the used laser, refractive index of the medium that the laser propagates in and the half angle between the interfering laser beams, respectively. In the case of the interfering of three identical laser beam, periodicity of the generated patterns can be calculated as follows:

$$P = \frac{\lambda}{\sqrt{3} n \sin \alpha} \qquad (7.24)$$

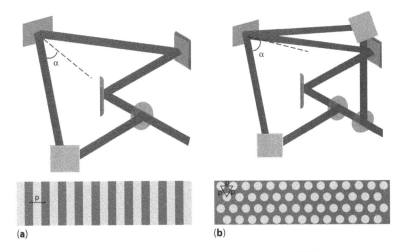

Figure 7.6 Schematic illustration of DLIP of (a) double beam and (b) triple beam. Illustration of the results from the corresponding processes is given under the setups.

Optical setups and corresponding illustrations of the periodical structures can be seen in Figure 7.6.

7.3.2 Chemical Surface Modification by Lasers

7.3.2.1 Pulsed Laser Deposition

Pulsed laser deposition (PLD) is a deposition technique where the target material is evaporated by laser pulses to form a thin film over the substrate. By using PLD, many different materials can be deposited. Especially, deposition of complex alloys with the exact stoichiometry makes PLD a valuable tool [26]. Although PLD was started to be applied in 1965, its actual discovery was in 1980 [27].

The reason behind the being attractive of PLD for surface chemical modification is lie behind its unique advantages which are

- Stoichiometric deposition materials, especially the complex alloys;
- Can be applied under vacuum or reactive gas atmosphere to form a new stoichiometry;
- Enhanced adhesion between the deposited thin film and the substrate surface;
- Suitable for multilayer depositions from different materials;
- Controlled film growth via the number of the applied laser pulses (10^{-3} to 1 Å per pulse) [9, 12, 27–31].

184 Advanced Surfaces for Stem Cell Research

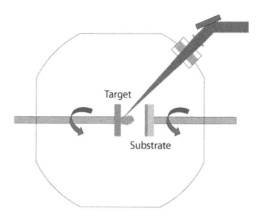

Figure 7.7 Schematic illustration of basic PLD setup.

Basic PLD setup is illustrated in Figure 7.7. Simply, laser beam, which enters to the vacuum chamber through a highly transmitting viewport, gets focused over the target material for the fast evaporation. Selected viewport has to be made of high transmitting material for a broad range of the electromagnetic spectrum. This range must go down to deep-ultraviolet (DUV) region since most of the material shows better absorbance at these wavelengths. Preferably, incidence angle of the laser beam to the target is 45°; however, it can be tilted more or less gently, as required. Rotation of the target by a rotation system is necessary if the laser beam is stationary in order to prevent evaporation from the same location continuously. Otherwise, there may be a change at the plume direction which may cause inhomogeneity at the deposited film [9, 14, 32]. Additionally, extra rotation system can be added to the system for the rotation of the sample holder to get homogeneous deposits over the surface. For a balanced deposition, distance between the target and the sample has to be between 3 and 8 cm. On the other hand, this range can be increased depending on the application [9, 14].

To obtain a uniform and high-quality deposit at the substrate surface, choosing the right laser parameters plays important role. Primarily, pulse length of the laser beam has to be short, preferably in the nanosecond region, to realize fast evaporation of the target material which is necessary for the evaporation of complex alloys. Lasers used in PLD are mostly operated in UV region as mentioned above. Excimer lasers (193 nm, 248 nm) and third or fourth harmonic of Nd:YAG lasers (266 nm, 355 nm) are suitable to use in PLD. Besides, it has been known that IR lasers had found place to work for PLD applications, too [12, 26].

7.3.2.1.1 Plasma Plume Formation

When the laser pulse shone on target material, some parts of its intensity will be absorbed by the material and transformed to the heat energy by electronic excitations. Here, reflectivity of the material for the used laser wavelength plays a critical role which can be calculated from Equations 7.4 and/or 7.5. On the other hand, reflectivity may change with the porosity of the surface (Bäuerle D 1996). When light is absorbed, the temperature at the laser focal rises well above the evaporation temperature of the target and generate heat can penetrate to a certain depth of "heat penetration depth" or "heat diffusion length" "I_{th}". Heat penetration depth for conducting materials is

$$I_{th} = 2\sqrt{D\tau_L} \quad (7.25)$$

where τ_L is the pulse length of the laser and D is the thermal diffusivity.

$$D = \frac{k}{\rho c_p} \quad (7.26)$$

where κ is the thermal conductivity, "ρ" is the mass density, and "c_p" is the specific heat at constant pressure.

For insulating materials such as polymers and ceramics, heat penetration depth limited by the attenuation depth (optical penetration depth) which is calculated as follows:

$$I_{opt} = 1/\alpha \quad (7.27)$$

where α is the absorption coefficient [9, 12, 27–29].

Time needed for the evaporation of the material can be calculated by using Equation 7.28.

$$t_v = \frac{\pi}{4D}\left(\frac{k\Delta T \tau_L}{F(1-R)}\right)^2 \quad (7.28)$$

where F is the laser fluence and ΔT is the temperature difference [14].

When the temperature at the laser spot area reaches to the evaporation temperature, some of the material gets ablated and a plasma layer forms over the target surface which is so-called "Knudsen layer" which has a thickness around 100 µm and temperature around 10^4 K. Since the time needed to reach evaporation temperature is well below the pulse length of the laser, rest of the incoming pulse gets absorbed by the formed layer and expends adiabatically toward the substrate surface because of the generated reverse pressure [9, 12, 26, 31].

7.3.2.1.2 Overcoming the Main Disadvantage: Increasing the Film Homogeneity

Although PLD has several advantageous over the most common deposition techniques, it suffers from a disadvantage which is the limitation of the deposition area. In PLD, due to the used single laser beam which strikes the target material mostly from the same position over the target, deposited films have usually a thickness distribution which is thicker in the center and decrease to the tails. In order to overcome this problem, several approaches were applied, such as scanning the laser beam over the target surface for the changing of the deposition position, rotating the substrate holder and so on [33–41]. Kreutz et al. obtained a homogeneous deposit at an area of 10 mm × 40 mm via translating the substrate in single axis [33]. Pryds et al. rotated the substrate holder additional to the scanning of the laser beam on target surface which causes deposition from different regions and resultant with a increased homogeneity to diameter of 90–125 mm [34, 35]. Börner et al. combined the translation and rotation of the substrate to obtain 60 mm diameter of homogeneous deposition [37]. An innovative approach for homogeneous coating at an increased area was applied by Schey et al. They used in their setup a rotation cylindrical target which is under strike of a linearly focused laser beam to deposit from a linear range. By using a mask to limit the deposition area and via scanning the substrate in single direction, they achieved a homogeneous area of 7 cm × 20 cm [40]. Another simple and cheap but effective approach was applied by Akkan et al. In their optical setup, they used the nonimaging beam homogenization system in order to generate several diffraction patterns over the target material which causes instantaneous deposition from several locations, therefore, increases the area of the deposit together with the homogeneity up to 50 mm in diameter [41].

7.3.2.2 Laser Surface Alloying

Laser surface alloying (LSA) aims to create new alloy via mixing the added material into the molten region of the bulk material melted by the laser source. Main interest of LSA is to create wear and corrosion-resistive surfaces on bulk materials. Modified region at the surface is usually stays around 1–2000 μm without affecting the properties of bulk material. By using LSA, it is possible to create new biocompatible surfaces with increased corrosion and wear resistance, such as silica with $TiAl_6V_4$, molybdenum with nitinol and hydroxyapatite (HA)–titanium oxide composite with 316L stainless steel (SS) [42–44].

LSA starts as the laser melting of both adding material and the base material. Laser melting realizes in a short time with the transformation of the laser light to the heat energy within the short time scale of the laser–material interaction time. This process causes high temperature gradient between the melt surface layer and the bulk material. Intensive mixing of materials is obtained during the laser melting due to the convection movements caused by temperature differences between the top surface and the bottom of the melt pool. Here, shielding gas, which is usually inert, increases the temperature difference between the layers. To obtain a uniform mixture, time to keep the surface layer as melt has to be sufficiently long. Therefore, laser output parameters and the scanning speed of the laser beam or the substrates have to be controlled precisely. Usually, scanning velocity for LSA process is around 10 cm/s. Schematic illustration of a LSA setup is given in Figure 7.8.

As well as the heating process of the material, cooling cycle has also importance on the quality of the produced alloy. In order to prevent segregation of the materials, high cooling rates are suggested. On the other hand, rapid solidification causes the formation of fine columnar or dendritic structures where the realized internal microstructures are depending on the chemical composition of the alloy, phase transformation, and the cooling of the re-solidified region. Regular cooling rates are about 10^6 K/s when CW lasers are used. This rate rises up to 10^{10}–10^{12} K/s if the laser is operating at the pulsed regime (Q-Switched). By using these kinds of lasers, it is possible to obtain and/or produce glassy alloys which have thickness resolution in the nanoscale [5, 14, 45, 46].

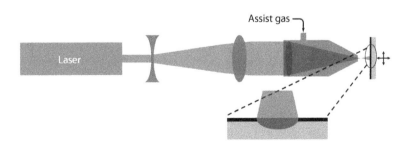

Figure 7.8 Schematic illustration of LSA. Material wanted to be mixed with the base material is introduced to the surface and melted together with the base material by laser source to form homogeneous alloy.

7.3.2.3 Laser Surface Oxidation and Nitriding

Surface oxidation is one of the methods that can be used in the biomaterial science. It has been known that there is a high correlation between the cells and the surface oxygen level for the *in vitro* experiment. Therefore, controlling the oxygen rate over the surface can affect the cell–surface interaction, too [47, 48].

Surfaces of many bare materials such as Al, Ti, Si, Cr, and Zn immediately react with oxygen in air, even at room temperature to form a thin native oxide layer. Thickness of the native oxide layer of these materials usually stays around 10–100 Å. Formed native oxide layer protects the material from further oxidation; however, it may be needed to increase the oxide layer thickness [14]. Besides, it may be needed to increase the surface oxygen of the material's surfaces [49, 50].

Although there are different ways to increase the level of surface oxygen, such as oxygen RF plasma, lasers can be used to create or increase the oxygen level at the substrate surface [9, 50, 51]. Oxidation process with lasers is mainly performed under air or O_2 atmospheres. Basic setup for surface oxidation can be seen in Figure 7.9.

Laser surface oxidation can be investigated in two forms: pyrolytic (photothermal) and photolytic (photochemical). During the laser surface oxidation with photothermal process, thickness of the native oxide layer over the substrate material has high importance. Temperature rise at the laser spot is highly related to this thickness where the surface native oxide layer influences the optical and the thermal properties of the bulk material. The increased temperature at the laser-heated area, diffusion and reaction properties of the molecules increase to form new oxide layers. On the other hand, multiple layers with different oxygen levels can be supported by laser surface oxidizing method, as well, such as CuO and Cu_2O from copper, Fe_2O_3, Fe_3O_4, and FeO from iron [14].

Photochemical oxidation can be observed only if the laser photon energy (wavelength) matches to the energy needed for the selective excitation of the material surface species. By applying photochemical oxidation process, it is possible to increase the oxygen level of the surface by this technique

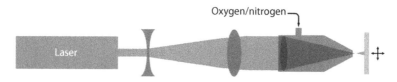

Figure 7.9 Schematic illustration of laser surface oxidation and nitriding.

without increasing the surface temperature to the melting temperature of the material which is suitable for materials with low melting temperature. Far-UV laser, e.g. ArF laser with 193 nm wavelength, can produce gas-phase atomic oxygen and ozone which can react with the surfaces. On the other hand, most lasers operating in the UV region can be used to increase the oxygen containing species over the material surface which can increase the hydrophilicity of the material for enhanced cell–surface interaction [9, 48, 49].

Similar to surface oxidation, surface nitriding can be applied to several materials such as Al, Ti, Fe, and steel to increase the hardness and wear resistance of the materials, while the main properties of the material remains unchanged [14, 46, 52]. Besides, nitride surface finds place in biomedical field [53, 54]. For laser surface nitriding applications, different lasers sources can be used, e.g. excimer laser, CO_2, and Nd:YAG lasers. These lasers may operate either CW or pulse mode. Main system is based on melting the material surface and simultaneously introduced the molten material with N_2 gas or gas mixture of N_2 and Ar. Additionally, it has been known that the thickness of the nitride layer is depending on the nitrogen gas purity at the environment, as well as the laser power [52]. By using pulsed lasers, it is possible to obtain various thick nitride layers on the top of Ti from nm to μm scale. Schematic illustration of laser surface nitriding setup is given in Figure 7.9 together with surface oxidation setup.

7.4 Stem Cells and Laser-modified Surfaces

It has been known that the surface properties, such as wettability and roughness, influence the interaction between the implant material and the cells [9, 55]. Therefore, controlling these parameters help to figure out the true potential usage of the materials for implantations. Besides, formation of the cell type primarily over the implant surface also determines the future of the implant if it fails or not [56]. In order to overcome or control these parameters, there are several different approaches; however, lasers are found to be one of the best devices since they promise to produce identical surface properties for every repeated processes, quickly and precisely [1, 9, 57–59]. Therefore, usage potential of lasers for the production and/or modification of implant materials is increasing.

Although there are numerous studies available for the investigation of different cell lines on laser-structured surface, it is rare for stem cell applications. The common surface treatment methods support surface features in both micro- and nanoscales may be, however, in a random distribution.

On the other hand, it has been known that the specifically organized surface features support differentiation of stem cell to different cell types. In this case, surface topography detected by the focal contacts forces the nucleus to change the gene expression, which supports the differentiation of the stem cell to a specific cell line [60–63]. Here, cell morphology imaged by the common imaging techniques may help to estimate the differentiating form of the stem cell. It has been reported as the observed cell with an elongated or star-like cell morphology points out the osteogenic differentiation, while round cell morphology points out the adipogenic differentiation [64, 65]. Dumas *et al.* investigated the effect of surface features in micro- and nanoscales created by femtosecond pulsed laser on Ti-6Al-4V titanium alloy with mouse pluripotent mesenchymal stem cells (MSCs). They used a scanning system to create 20 and 60 μm width microstructures over the material surface while the self-organized nanofeatures which are 600 nm at width were created parallel and perpendicular to the microstructures depending on the polarization state of the laser beam. They observed that the MSCs are getting oriented with the direction of the nanogrooves where the best cell orientation with 90% and 95% were observed at 60 μm parallel and 20 μm parallel structures. Besides, they did not observe a significant difference at the cell proliferation for the compared surface structures. On the other hand, they were pointed out in their work that nanofeatures support the adhesion of MSCs in comparison to polished surfaces where there was more focal adhesion observed on these surfaces [59]. Parameters and results are summarized in Table 7.2.

Another study by using femtosecond laser-structured surfaces was performed by Oya *et al.* where they investigated the morphological changes of the MSC. They created nanostructures on Grade 2 Ti substrate around 500 nm pitch and 100 nm by using femtosecond laser operating at 190 fs of pulse width with 780 nm of wavelength. They used human synovium-derived MSCs for the *in vitro* experiments. They observed in their study that the created nanofeatures are increasing the surface roughness to 2–3 times higher value in comparison to as-prepared Ti surfaces, and these structures are forcing the cell to change their body shape and get oriented in the direction of the nanogrooves. Directional growth of the MSCs was observed at the first hour of seeding. This effect becomes more clear 6 h of culturing. They observed that cell seeded on laser structures surface shows significantly enhanced spreading and adhesion in comparison to the flat Ti surface even after 1 h of culturing [66]. Although their results show no significant differences after 24 h, it has been known that the early adhesion of cell on surfaces effects the future of the implant [58]. Parameters and results are summarized in Table 7.2.

Table 7.2 List of applied surface modifications techniques on various materials with chosen stem cell type and their cellular results.

Surface modification	Laser	Material	Cell type	Cell results	Ref.
Periodic nanostructures inside microgrooves	Femtosecond, 120 fs, 800 nm, 5 kHz	Ti-6Al-4V	Mouse pluripotent MSC	• Cell orientation • Enhanced adhesion	59
Periodic nanostructures	Femtosecond, 190 fs, 780 nm, 1 kHz	GR-2 Ti	Human synovium-derived MSC	• Shaped cell body • Cell orientation • Enhanced spreading and adhesion	66
Laser microwelding (effect of dendritic structures)	CW fiber laser, 1091 nm	NiTi	MSC	• Increased cell adhesion at melt zone on dendritic structures	67 68
Periodic nanostructures inside microspots	Femtosecond, <130 fs, 780 nm, 1 kHz	316L SS	Bone marrow MSC	• Enhanced adhesion • Enhanced Ca^{2+} deposition	58
Periodic nanostructures	Nd:YAG, 5ns, 266nm, 10Hz	PET	Bone marrow MSC	• Enhanced proliferation • Adipogenic differentiation (dif.)	69
Periodic nanostructures	Nd:YAG 38 ns, 355 nm	Polyimide	MSC	• Adipogenic dif. @ 15 μm/20 μm • Osteogenic dif. @ 2 μm/20 μm	70
TiO_2 coating by PLD	Picosecond, 10 ps, 1 MHz	Glass	Human bone marrow MSC	• Enhanced proliferation • Fibroblast-like cell morphology	71
Combinatorial PLD (Ag-HA/β-TCP)	KrF; 25 ns, 248 nm, 10 Hz	–	MSC	• Nontoxic effect up to 0.6% of Ag containing	72

LASER SURFACE MODIFICATION TECHNIQUES 191

Another interesting study was performed by Waugh et al. where they investigated the behavior of MSCs on laser melted nitinol (NiTi) since NiTi finds application fields for orthopedic implants and meanwhile the need of laser microwelding for the production of miniaturized medical implants. They observed that the MSCs have tended to adhere to the melt zone (resolidified locations), which may be due to the presence of dendritic structure. Here, increased surface roughness of this interested zone may be the reason of their result. Additionally, it was reported by the authors that the surface chemistry at the region where the MSCs are mainly interested in is rich for titanium dioxide (TiO_2) [67, 68]. Parameters and results are summarized in Table 7.2.

Although Ti and its alloys are favorite for the production of implants, SS is another material can be used for implants. Especially, the 316L SS with relatively high strength, corrosion resistance, biocompatibility, and reasonable costs makes this alloy one of the interesting materials for the implant production. In this manner, Kenar et al. investigated the effect of surface structures created in micro- and nanoscales by femtosecond lasers on 316L SS. In their work, they applied identical surface structures at different densities over the surface and test them with human bone marrow (hBM) MSCs in the presence of osteogenic differentiation medium. They observed that the adhered cells on laser-treated surface are much more in comparison to the nontreated SS surface. Additionally, when they investigated the formation of Ca^{2+} ions on the comparable surface, they observed that the laser-modified surfaces show higher deposition rates, although the obtained results are nonsignificant [58]. Parameters and results are summarized in Table 7.2.

As well as femtosecond lasers, nanosecond lasers can be used to create periodic surface structures over materials where the reflection of the laser beam from the base material which is coated with another material or direct interfering of two or more identical laser beams can be used for direct ablation [69, 70]. Reboller et al. investigated the effect of nanosecond pulsed UV laser-created nanostructures over poly(ethylene terephthalate) (PET) on adhesion of MSCs. They illuminate the PET-coated silicon wafer with 266 nm of laser beam, which operates at 5 ns pulse duration, where the reflected light and the rest incoming light to the PET surface interfere to form the nanostructures. Depending on the polarization of the incoming laser beam, they observed either linear and parallel or circular like nanostructures. They observed at the cell culture experiments laser-structured surface has shown enhanced cell proliferation, regardless of the structure type. Additionally, increased b-catenin and actin regions point out that the cells are communicating with each other and are going to differentiate to adipocytes where the detected catenin in the cell nucleus would be the indication [69]. Abagnale *et al.* investigated the morphology, proliferation, focal adhesions, and *in vitro* differentiation of MSCs on laser-modified polyimide structures by DLIP technique. They used a linearly polarized nanosecond pulse laser operates at 355 nm wavelength and 38 ns pulse duration to form different scales of surface structures. They observed that the surface containing 15 μm of ridge at 20 μm of periodic structures have shown higher adipogenic differentiation, while this amount is much more less when the ridge size set to 2 μm (at 20 μm period). On the other hand, when the same surfaces are examined for osteogenic differentiation, surface with 2 μm ridge has shown much better results [70]. Parameters and results are summarized in Table 7.2.

Surface modification by laser is not limited only with topographical modification, as mentioned before. Chemical modification by using PLD method is another way to modify the surface chemistry, thus the surface wettability, while relatively protecting the predetermined surface topography. Details of PLD can be seen in Section 7.3.2.1. Similar to topographic surface modification, there is not so much work available for the stem cell application on chemically modified surfaces by using PLD, while several can be found for regular cell lines. Kaitainen *et al.* investigated the morphology and growth of the human MSCs on deposited thin TiO_2 films, which were deposited by ultrashort PLD (UPLD) under oxygen atmosphere. They observed that the proliferation rate of the hMSCs is enhanced on deposited surface, while the cells are exhibiting fibroblast-like morphology on all surface types [71]. Here, negligible effect of PLD on surface roughness may be the reason why the hMSCs showed similar behavior on all surfaces. Socol *et al.* applied dual-coating technique, called combinatorial PLD (CPLD) to deposit HA and $β$-tricalcium phosphate layers containing silver (Ag) inside its matrix in order to investigate the toxic effect of Ag on MSCs. They observed that the region which contains more Ag particles causes a decrease at the roughness which follows a path from higher to lower roughness value. They also identify their results by using electron dispersive X-ray spectrometer. When they check the toxicity of the Ag particles on MSCs, they observed that content up to 0.6% of Ag shows non-toxic behavior [72]. Parameters and results are summarized in Table 7.2.

7.5 Conclusions

It is clear that the laser surface modification has several advantages on common surface modifications techniques with their fast, precise, and repeatable results. On the other hand, there is a lack of knowledge on the interaction of stem cell with modified surface either chemically or physically by using laser where it has to be investigated more to apply their benefits to this field of science.

References

1. Lee, J., Schwarz, L.K., Akkan, C.K., Miró, M.M., Torrents, O., Schäfer, K.-H., Veith, M., Aktas, C., Guidance of glial cells and neurites from dorsal root ganglia by laser induced periodic patterning of biphasic core/shell nanowires. *Phys. Status Solidi*, 210, 952, 2013.
2. Lee, J., Miró, M.M., Akkan, C.K.C., Haidar, A., Metzger, W., Schwarz, L.K., Zaporojtchenko, V., Schäfer, K.H., Abdul-Khaliq, H., Veith, M., Aktas, C., Miro, M., Development of multi scale structured Al/Al2O3 nanowires for controlled cell guidance. *J. Biomed. Nanotechnol.*, 9, 295, 2013.
3. Hecht, E. (Ed.), *Optics*, Addison Wesley, San Francisco, 2001.
4. Knowles, M.R.H., Rutterford, G., Karnakis, D., Ferguson, A., Micromachining of metals, ceramics and polymers using nanosecond lasers. *Int. J. Adv. Manuf. Technol.*, 33, 95, 2007.
5. Steen, W.M., J., M., and Mazumder, J. (Ed.), *Laser Material Processing*, Springer London, London, 2010.
6. Qian, B., Shen, Z., Laser sintering of ceramics. *J. Asian Ceram. Soc.*, 1, 315, 2013.
7. Ghany, K.A., Newishy, M., Cutting of 1.2 mm thick austenitic stainless steel sheet using pulsed and CW Nd·YAG laser. *J. Mater. Process. Technol.*, 168, 438, 2005.
8. Kannatey-Asibu, E. (Ed.), *Principles of Laser Materials Processing*, John Wiley & Sons, Inc., Hoboken, NJ, USA, 2009.
9. Akkan, C.K., Micro/Nano Modification of PEEK Surface for Possible Medical Use., 2015.
10. Thyagarajan, K., Ghatak, A. (Ed.), *Lasers*, Springer US, Boston, MA, 2011.
11. Mutlu, M., Kacar, E., Akman, E., Akkan, C.K., Demir, P., Demir, A., Effects of the laser wavelength on drilling process of ceramic using Nd: YAG laser. *J. Laser Micro/Nanoengineering*, 4, 84, 2009.
12. Schou, J, Laser beam-solid interactions: fundamental aspects, in: *Materials Surface Processing by Directed Energy Techniques*, Pauleau, Y. (Ed.), pp. 35–66, Elsevier, Oxford, 2006.
13. Silfvast, W.T. (Ed.), *Laser Fundamentals*, Cambridge University Press, Cambridge, UK.
14. Bäuerle, D. (Ed.), *Laser Processing and Chemistry*, Springer Berlin Heidelberg, Berlin, Heidelberg, 2011.
15. Romoli, L., Fischer, F., Kling, R., A study on UV laser drilling of PEEK reinforced with carbon fibers. *Opt. Lasers Eng.*, 50, 449, 2012.

16. Babu, S. V., D'Couto, G.C., Egitto, F.D., Excimer laser induced ablation of polyetheretherketone, polyimide, and polytetrafluoroethylene. *J. Appl. Phys.*, 72, 692, 1992.
17. Tsunekawa, M., Nishio, S., Sato, H., Laser ablation of polymethylmethacrylate and polystyrene at 308 nm: Demonstration of thermal and photothermal mechanisms by a time-of-flight mass spectroscopic study. *J. Appl. Phys.*, 76, 5598, 1994.
18. Hancock, G., Laser studies of gas-phase kinetics and photochemistry. *J. Chem. Soc. Faraday Trans. 2*, 84, 429, 1988.
19. Lens Tutorial, http://www.thorlabs.de/newgrouppage9.cfm?objectgroup_id=8790, 2016
20. Suss Microoptics, SMO TechInfo Sheet 10 - Beam Homogenizing., 2008.
21. Völkel, R., Weible, K.J., Laser beam homogenizing: limitations and constraints. *Proc. SPIE*, 7102, 71020J, 2008.
22. Kiefer, K., Lee, J., Haidar, A., Miró, M.M., Akkan, C.K., Veith, M., Aktas, O.C., Abdul-Khaliq, H., Alignment of human cardiomyocytes on laser patterned biphasic core/shell nanowire assemblies. *Nanotechnology*, 25, 495101, 2014.
23. Zhu, X., Xu, Y., Yang, S., Distortion of 3D SU8 photonic structures fabricated by four-beam holographic lithography with umbrella configuration. *Opt. Express*, 15, 16546, 2007.
24. Misawa, H., Kondo, T., Juodkazis, S., Mizeikis, V., Matsuo, S., Holographic lithography of periodic two- and three-dimensional microstructures in photoresist SU-8. *Opt. Express*, 14, 7943, 2006.
25. Lasagni, F.A., and Lasagni, A.F. (Eds.), *Fabrication and Characterization in the Micro-Nano Range*, Springer Berlin Heidelberg, Berlin, Heidelberg, 2011.
26. Norton, D.P., Pulsed laser deposition of complex materials: progress toward applications, *Pulsed Laser Deposition of Thin Films*, Eason, R. (Ed.), pp. 1–31, John Wiley & Sons, Inc., Hoboken, NJ, USA, 2006.
27. Eason, R. (Ed.), *Pulsed Laser Deposition of Thin Films*, John Wiley & Sons, Inc., Hoboken, NJ, USA, 2006.
28. Svendsen, W., Ellegaard, O., Schou, J., Laser ablation deposition measurements from silver and nickel. *Appl. Phys. A Mater. Sci. Process.*, 63, 247, 1996.
29. Willmott, P.R., Huber, J.R., Pulsed laser vaporization and deposition. *Rev. Mod. Phys.*, 72, 315, 2000.
30. Gottmann, J., Kreutz, E.W., Pulsed laser deposition of alumina and zirconia thin films on polymers and glass as optical and protective coatings. *Surf. Coatings Technol.*, 116-119, 1189, 1999.
31. Carrado, A., Pelletier, H., and Rol, T., Nanocrystalline Thin Ceramic Films Synthesised by Pulsed Laser Deposition and Magnetron Sputtering on Metal Substrates for Medical Applications, in: *Biomedical Engineering – From Theory to Applications*, Reza Fazel-R. (Ed.), InTech, Croatia, 2011.
32. Doughty, C., Findikoglu, A.T., Venkatesan, T., Steady state pulsed laser deposition target scanning for improved plume stability and reduced particle density. *Appl. Phys. Lett.*, 66, 1276, 1995.

33. Kreutz, E., Backes, G., Mertin, M., Large area pulsed laser deposition of ceramic films. *Surf. Coatings Technol.*, 97, 435, 1997.
34. Pryds, N., Toftmann, B., Bilde-Sørensen, J.B., Schou, J., Linderoth, S., Thickness determination of large-area films of yttria-stabilized zirconia produced by pulsed laser deposition. *Appl. Surf. Sci.*, 252, 4882, 2006.
35. Pryds, N., Schou, J., Linderoth, S., Large-area production of yttria-stabilized zirconia by pulsed laser deposition. *J. Phys. Conf. Ser.*, 59, 140, 2007.
36. Boughaba, S., Islam, M., McCaffrey, J.P., Sproule, G.I., Graham, M.J., Ultrathin Ta2O5 films produced by large-area pulsed laser deposition. *Thin Solid Films*, 371, 119, 2000.
37. Börner, H., Hochmuth, H., Schurig, T., Quan, Z., Lorenz, M., Optimization of large area pulsed laser deposition of YBaCuO thin films by SNMS depth profiling and Rutherford backscattering. *Fresenius. J. Anal. Chem.*, 353, 619, 1995.
38. Bouazaoui, M., Capoen, B., Caricato, A.P., Chiasera, A., Fazzi, A., Ferrari, M., Leggieri, G., Martino, M., Mattarelli, M., Montagna, M., Romano, F., Tunno, T., Turrel, S., Vishnubhatla, K., Pulsed laser deposition of Er doped tellurite films on large area. *J. Phys. Conf. Ser.*, 59, 475, 2007.
39. Pignolet, A., Welke, S., Curran, C., Alexe, M., Senz, T., Hesse, D., Large area pulsed laser deposition of aurivilliustype layered perovskite thin films. *Ferroelectrics*, 202, 285, 1997.
40. Schey, B., Biegel, W., Kuhn, M., Stritzker, B., Large-area pulsed laser deposition of YBCO thin films: homogeneity and surface. *Appl. Phys. A Mater. Sci. Process.*, 69, S419, 1999.
41. Akkan, C.K., May, A., Hammadeh, M., Abdul-Khaliq, H., Aktas, O.C., Matrix shaped pulsed laser deposition: New approach to large area and homogeneous deposition. *Appl. Surf. Sci.*, 302, 149, 2014.
42. Fasasi, A.Y., Roy, S.K., Galerie, A., Pons, M., Caillet, M., Laser surface alloying of Ti-6Al-4V with silicon for improved hardness and high-temperature oxidation resistance. *Mater. Lett.*, 13, 204, 1992.
43. Man, H.C., Ho, K.L., Cui, Z.D., Laser surface alloying of NiTi shape memory alloy with Mo for hardness improvement and reduction of Ni2+ ion release. *Surf. Coatings Technol.*, 200, 4612, 2006.
44. Ghaith, E.S., Hodgson, S., Sharp, M., Laser surface alloying of 316L stainless steel coated with a bioactive hydroxyapatite–titanium oxide composite. *J. Mater. Sci. Mater. Med.*, 26, 83, 2015.
45. Draper, C.W., Ewing, C.A., Laser surface alloying: a bibliography. *J. Mater. Sci.*, 19, 3815, 1984.
46. Major, B. et al., Laser processing for surface modification by remelting and alloying of metallic systems, in: *Materials Surface Processing by Directed Energy Techniques*, Pauleau, Y. (Ed.), pp. 241–274, Elsevier, Oxford, 2006.
47. Ratner, B.D., Hoffman, A.S., Schoen, F.J., and Lemons, J.E. (Ed.), *Biomaterials science: an introduction to materials in medicine*, Academic Press, California, 1996.

48. Hao, L., and Lawrence, J. (Ed.), *Laser Surface Treatment of Bio-Implant Materials*, John Wiley & Sons, Ltd, Chichester, UK, 2005.
49. Bremus-Koebberling, E.A., Meier-Mahlo, U., Henkenjohann, O., Beckemper, S., Gillner, A., Laser structuring and modification of polymer surfaces for chemical and medical microcomponents. *SPIE Proceedings*, 274, 2004.
50. Riveiro, A., Soto, R., Comesaña, R., Boutinguiza, M., del Val, J., Quintero, F., Lusquiños, F., Pou, J., Laser surface modification of PEEK. *Appl. Surf. Sci.*, 258, 9437, 2012.
51. Ha, S.W., Hauert, R., Ernst, K.H., Wintermantel, E., Surface analysis of chemically-etched and plasma-treated polyetheretherketone (PEEK) for biomedical applications. *Surf. Coatings Technol.*, 96, 293, 1997.
52. Raaif, M., El-Hossary, F.M., Negm, N.Z., Khalil, S.M., Kolitsch, A., Höche, D., Kaspar, J., Mändl, S., Schaaf, P., CO_2 laser nitriding of titanium. *J. Phys. D. Appl. Phys.*, 41, 085208, 2008.
53. Wang, M., Ning, Y., Zou, H., Chen, S., Bai, Y., Wang, A., Xia, H., Effect of Nd:YAG laser-nitriding-treated titanium nitride surface over Ti6Al4V substrate on the activity of MC3T3-E1 cells. *Biomed Mater Eng.*, 24, 643, 2014.
54. May, A., Agarwal, N., Lee, J., Lambert, M., Akkan, C.K., Nothdurft, F.P., Aktas, O.C., Laser induced anisotropic wetting on Ti – 6Al – 4V surfaces. *Mater. Lett.*, 138, 21, 2015.
55. Veith, M., Lee, J., Miró, M.M., Akkan, C.K.K., Dufloux, C., Aktas, O.C.C., Martinez Miró, M., Akkan, C.K.K., Dufloux, C., Aktas, O.C.C., Bi-phasic nanostructures for functional applications. *Chem. Soc. Rev.*, 41, 5117, 2012.
56. Takagi, M., Bone-implant interface biology. Foreign body reaction and periprosthetic osteolysis in artificial hip joints. *J. Clin. Exp. Hematop.*, 41, 81, 2001.
57. Oberringer, M., Akman, E., Lee, J., Metzger, W., Akkan, C.K.C.K., Kacar, E., Demir, A., Abdul-Khaliq, H., Pütz, N., Wennemuth, G., Pohlemann, T., Veith, M., Aktas, C., Reduced myofibroblast differentiation on femtosecond laser treated 316LS stainless steel. *Mater. Sci. Eng. C*, 33, 901, 2013.
58. Kenar, H., Akman, E., Kacar, E., Demir, A., Park, H., Abdul-Khaliq, H., Aktas, C., Karaoz, E., Femtosecond laser treatment of 316L improves its surface nanoroughness and carbon content and promotes osseointegration: an in vitro evaluation. *Colloids Surfaces B Biointerfaces*, 108, 305, 2013.
59. Dumas, V., Rattner, A., Vico, L., Audouard, E., Dumas, J.C., Naisson, P., Bertrand, P., Multiscale grooved titanium processed with femtosecond laser influences mesenchymal stem cell morphology, adhesion, and matrix organization. *J. Biomed. Mater. Res. Part A*, 100A, 3108, 2012.
60. Tang, J., Peng, R., Ding, J., The regulation of stem cell differentiation by cell-cell contact on micropatterned material surfaces. *Biomaterials*, 31, 2470, 2010.
61. Wei Song, Hongxu Lu, Kawazoe, N., Guoping Chen, Gradient patterning and differentiation of mesenchymal stem cells on micropatterned polymer surface. *J. Bioact. Compat. Polym.*, 26, 242, 2011.

62. Davidson, P.M., Özçelik, H., Hasirci, V., Reiter, G., Anselme, K., Microstructured surfaces cause severe but non-detrimental deformation of the cell nucleus. *Adv. Mater.*, 21, 3586, 2009.
63. Chalut, K.J., Kulangara, K., Giacomelli, M.G., Wax, A., Leong, K.W., Deformation of stem cell nuclei by nanotopographical cues. *Soft Matter*, 6, 1675, 2010.
64. Kilian, K.A., Bugarija, B., Lahn, B.T., Mrksich, M., Geometric cues for directing the differentiation of mesenchymal stem cells. *Proc. Natl. Acad. Sci.*, 107, 4872, 2010.
65. Martínez, E., Engel, E., Planell, J.A., Samitier, J., Effects of artificial micro- and nano-structured surfaces on cell behaviour. *Ann. Anat. – Anat. Anzeiger*, 191, 126, 2009.
66. Oya, K., Aoki, S., Shimomura, K., Sugita, N., Suzuki, K., Nakamura, N., Fujie, H., Morphological observations of mesenchymal stem cell adhesion to a nanoperiodic-structured titanium surface patterned using femtosecond laser processing. *Jpn. J. Appl. Phys.*, 51, 125203, 2012.
67. Waugh, D.G., Lawrence, J., Chan, C.W., Hussain, I., Man, H.C., Laser melting of NiTi and its effects on *in vitro* mesenchymal stem cell responses. *Laser Surf. Eng.*, 653, 2015.
68. Chan, C.W., Hussain, I., Waugh, D.G., Lawrence, J., Man, H.C., Effect of laser treatment on the attachment and viability of mesenchymal stem cell responses on shape memory NiTi alloy. *Mater. Sci. Eng. C*, 42, 254, 2014.
69. Rebollar, E., Pérez, S., Hernández, M., Domingo, C., Martín, M., Ezquerra, T.A., García-Ruiz, J.P., Castillejo, M., Physicochemical modifications accompanying UV laser induced surface structures on poly(ethylene terephthalate) and their effect on adhesion of mesenchymal cells. *Phys. Chem. Chem. Phys.*, 16, 17551, 2014.
70. Abagnale, G., Steger, M., Nguyen, V.H., Hersch, N., Sechi, A., Joussen, S., Denecke, B., Merkel, R., Hoffmann, B., Dreser, A., Schnakenberg, U., Gillner, A., Wagner, W., Surface topography enhances differentiation of mesenchymal stem cells towards osteogenic and adipogenic lineages. *Biomaterials*, 61, 316, 2015.
71. Kaitainen, S., Mähönen, A.J., Lappalainen, R., Kröger, H., Lammi, M.J., Qu, C., TiO2 coating promotes human mesenchymal stem cell proliferation without the loss of their capacity for chondrogenic differentiation. *Biofabrication*, 5, 025009, 2013.
72. Socol, G., Socol, M., Sima, L., Petrescu, S., Enculescu, M., Miroiu, M., Stefan, N., Cristescu, R., Mihailescu, C.N., Stanculescu, A., Sutan, C., Mihailescu, I.N., Phosphates, C., Combinatorial Pulsed Laser Deposition of Ag-Containing Calcium Phosphate Coatings. *Dig. J. Nanomater. Biostruct.*, 7, 563, 2012.

8
Plasma Polymer Deposition: A Versatile Tool for Stem Cell Research

M. N. Macgregor-Ramiasa* and K. Vasilev

Mawson Institute, University of South Australia, Adelaide, Australia

Abstract

Stem cell-based therapies are emerging as a potent new approach to treat a variety of diseases. This chapter is focused on the application and potential of plasma-assisted material modification in stem cells culture substrates and therapies. The effect of substrate chemical and physical properties achieved using plasma-coating methods on the fate of stem cells is reviewed. In particular, plasma-assisted methods to produce tailored surface chemistries, controlled topography and stiffness, surface gradient substrates, and modify 3D structures are explained together with their significance in addressing challenges faced in conventional stem cell culture.

Keywords: Plasma modification, surface chemistry, nanoroughness, nanoengineering, oxazoline, surface stiffness, gradients substrates

8.1 Introduction

Stem cells are considered as one of the greatest promises of modern medicine [1]. They have an unequalled therapeutic potential with future prospect spanning over the fields of regenerative medicine [2], tissue engineering [3], cancer cure [4], diabetes [5], and autoimmune disease [6] treatment and much more. In all cases, stem cell research and therapy success rely on the controlled and effective culture of this utmost delicate cell type. However, to this day, progress in this research field is still limited due

---

*Corresponding author: melanie.ramiasa@unisa.edu.au

Ashutosh Tiwari, Bora Garipcan and Lokman Uzun (eds.) *Advanced Surfaces for Stem Cell Research*, (199–232) © 2017 Scrivener Publishing LLC

to the lack of appropriate *in vitro* culture systems for important stem cell lineages. Challenges lie in establishing model culture materials allowing to control the differentiation and growth of pluripotent cells [7]. The importance of biomaterials in overcoming this obstacle is no longer disputed and relentless effort are being undertaken worldwide to develop ideal culture substrates. Recent works investigating cell–material interaction have demonstrated that a variety of factors influence the fate of stem cells [8]. For instance, it is well accepted that the chemistry, topography [9], stiffness, and three-dimensional (3D) shape of the growth support play an important role in directing cell adhesion and spreading, and eventually also differentiation pathways. Most often, several of these surface properties need to be carefully combined in order to achieve the targeted effect [10].

A novel approach allowing to control the surface properties relevant to the fate of stem cells is plasma-assisted surface modification [11]. Plasma-assisted deposition is a gaseous coating technique which uses plasmas produced from volatile organic precursors to form thin polymeric films on surfaces. This technique developed in the early 1900s has the advantages to be substrate independent, highly reproducible, and environmentally friendly, in the sense that it does not require substrates pre-treatment or organic solvents and thus generates virtually no waste. This one step and easily scalable coating technique can be used to modify the chemistry, wettability, roughness, and adhesion properties of materials. For all these reasons, plasma deposition has been widely used in a variety of research and applied field of material science. In particular, plasma deposition has recently become a tool of choice for the production of novel functional biomaterials [12].

In the specific field of stem cell research, plasma-deposited polymers are attracting a growing interest because they can be used to tailor both the chemical and mechanical properties of the stem cell culture substrates. This chapter presents the state of the art on the use of plasma deposition-based substrate in the field of stem cell research. The question of controlling substrate chemistry and biocompatibility, by means of carefully choosing an appropriate organic precursor, is addressed first. Ways to functionalize the polymer substrates with growth factors and other biomolecules are also described [13]. Furthermore, we describe how the combined use of plasma deposition and other techniques can help controlling the topographical patterns of the substrates and investigating the associated cell response [14]. Finally, we show how this versatile technique can also be used to investigate the effect of substrate stiffness [15] and 3D landscape on the fate of pluripotent cell [16], before demonstrating that systematic studies can also been conducted using plasma polymer gradient on a variety of different substrates [17, 18]. An outlook on the future of plasma-based methods for the generation of new biomaterials is presented at the end of this chapter.

8.2 The Principle and Physics of Plasma Methods for Surface Modification

Plasma-modified substrates have been at the forefront of the cell culture industry for over half a century. Cell culture was historically conducted in glass, hence the common saying *"in vitro"*. The rapid development of polymers in the 19th century first brought plastic ware into biology labs. However, untreated plastic substrates generally do not support cell adhesion and growth. A revolutionary plasma-assisted technique was developed by Michael Erchak in 1964 to render polymer substrates more hydrophilic [19]. Since then, the oxygen plasma treatment of polystyrene has been behind all modern cell culture plastic ware.

Plasma is the 4th state of matter and consists of neutral and ionized molecules, atoms, and radicals in an excited gas phase. It is typically generated using an electrical or radio frequency discharge. The excited species present in plasma react with virtually any material surface, reactive or inert, that is presence. As a result, a material substrate undergoing plasma treatment will bear changes in its physicochemical properties such as wettability and elemental composition. In the case of common plastic culture ware, for example, the oxygen plasma treatment applied to the polystyrene produces hydroxyl surface groups which increase the wettability of this otherwise rather bio-hostile substrate. The benefits of plasma-based techniques for substrate modification are multiple. These techniques have the advantage to be reproducible, relatively inexpensive, and versatile, in the sense that sample geometry and bulk properties are no limitations to the success of the process. As such, plasma coatings bind strongly a variety of substrates (polymer, metals, ceramics, etc.) and form pinhole free layers. Also, last but not least, plasma techniques produce essentially zero organic waste and are competitively scalable which make them an attractive alternative for industrial applications where eco-friendliness and cost effectiveness are sought after attributes. Plasma-based surface modifications encompass a range of techniques which were describe in quite some detail by Chu *et al.* [12]. In the following sections, we briefly describe the principle of plasma-based modifications directly applicable to the engineering of biomaterials for cell culture purposes. These techniques are separated in two categories based on the nature of the plasma source, namely simple gases or organic precursors.

8.2.1 Plasma Sputtering, Etching an Implantation

The plasma state of inert (Ar, Ne, He, etc.) or reactive (O_2, N_2, NH_3, CO_2, etc.) gases are often used to modify the bulk properties of materials. Gaseous plasmas are typically generated by continuous or non-uniform electric field using radio or microwave frequency or laser pulses. Whether the plasma is generated at low or atmospheric pressures, several competing process are at play when the plasma is in contact with a solid substrate. When the excited species present in the plasma phase collide with the solid substrate, they transfer energy to atoms in the surface layer of the materials. If sufficient energy is acquired, elements of the materials happen to leave the surface (Figure 8.1a). The degradation of the outmost layer of the bulk material is typically referred to as sputtering, while the further loss of exposed surface material is called etching (Figure 8.1b). This process is commonly used in the biomedical field for surface cleaning and sterilization, but it can also, at its extreme, be used for surface roughening and patterning [11]. It is also possible for the gaseous species to directly react with the substrate and induce surface modification. This second interaction with plasmas is particularly important for polymer substrates. While polymer undergoing plasma etching generally have surface properties very similar to their bulk phase, plasma modification process can result in the alteration of the

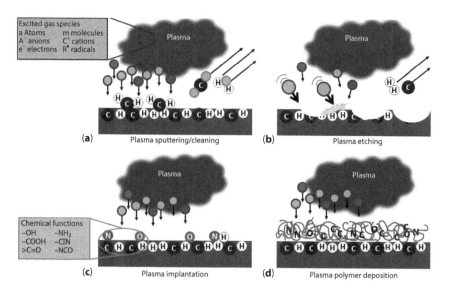

Figure 8.1 Schematic of the different plasma-based method used for the surface modification of biomaterials: (a) plasma sputtering/cleaning, (b) plasma etching, (c) plasma implantation, and (d) plasma polymer deposition.

polymer structure itself (change in cross-link density, broken bonds, etc.) and also to the implantation of new chemical functions. For this reason, this second mechanism is often referred to as plasma implantation (Figure 8.1c). Using adequate vapor-phase precursors and substrates, it is also possible for reactive species to be "stored" within the polymeric material. These reactive species can later react with molecules at the surface such as proteins present in culture media [20]. As mentioned earlier, the surface modification of polymers with oxygen plasma is readily used to increase the hydrophilicity of otherwise hydrophobic plastics [21]. Other gases such as carbon dioxide or nitric oxide can also be used for this purpose. Alternatively, fluorinated gases can be used to increase the hydrophobicity of substrates [22].

However, it has been brought to light that glow discharge plasma modification of surfaces suffer from recovery effect. In other word, the functionalities obtained by these methods, that merely modify the base material itself, tend to fade over extended period of time as the polymer chains undergo unavoidable reorientation [23, 24].

8.2.2 Plasma Polymer Deposition

Plasma deposition differs from plasma surface modification in the fact that a thin organic coating is formed over the surface of the original material. As a result, it is possible to generate surfaces with lasting properties that completely differ from those of the bulk materials (Figure 8.1d). For plasma deposition, small and volatile organic molecules are used as precursors for the generation of excited species in the plasma state. To be suitable for use in plasma polymer deposition, the precursor molecules need to have low molecular weight and relatively high vapor pressure in order to be easily excited and fractionated. The precursors are then converted into high-molecular-weight molecules via complex recombination reaction mechanisms between the ions, electrons, and radicals present in the plasma phase. Plasma-deposited polymer stands apart from polymer generated via conventional ionic or radical polymerization techniques. Generally, they are not constituted of repeating units of monomer but rather of random arrangement of precursors fractions with various degree of cross-linking depending on the deposition conditions used. Among the many parameter directing the final chemical and physical properties of the films, ignition power, precursor flow rate, and deposition time are the most influencial [25]. It is well accepted that high powers and low flow rates tend to facilitate greater fragmentation and cross-linking. As a result, the film generated is typically more stable than those generated at low powers and high flow rates. Low flow rate and high powers, on the other hand, promote the retention of the original precursor functionality. These films are consequently relatively more reactive. A careful balance between precursor flow rate and power is crucial to promote film growth, as opposed to substrate etching [25]. In the film growth regime increasing deposition time increases film thickness. Several studies have investigated the growth and adhesion properties of plasma-deposited polymer films [26]. The general consensus of this investigation is that excellent adhesion can be achieved on a variety of substrates and that beyond 5 nm in thickness, the nature of the underlying materials does not affect the properties of the polymer films [27]. This characteristic is particularly relevant for stem cell culture and implant coating, both biomedical fields where critical mechanical properties of the bulk material (resilience, elasticity, etc.) need to be retained and supplemented with biocompatible surface properties.

Overall, in view of the variety of precursor available and possible deposition parameters combination, the range of plasma-deposited films is close to infinite, and so are their properties. In the following sections, we describe how the versatile plasma-based modification options have been used to serve the field of cell growth support biomaterials, with a particular focus on the specific case of stem cell culture.

8.3 Surface Properties Influencing Stem Cell Fate

Driven by the rapid emergence of cell-based therapies, research focusing on the development of improved biomaterial is thriving [7]. Although temperature, pH, and the availability of appropriate nutriments and growth factors are essential for the survival of any type of cells in culture, research in the field of cell culture systems has demonstrated that phenomena occurring at the interface between cells and their culture support also greatly influence cellular behaviors. For instance, the regulation of biomolecules, cell adhesion, division, and communication pathways all depend on the growth substrate properties. In particular, substrates that mimic properties of the natural extra cellular matrix environment can greatly enhance the viability of particular cell types [28]. It has been found that human mesenchymal stem cell (hMSC) preferentially differentiate into neurones on surfaces with nanogroove patterns [29] and also on soft supports while stiffer extra cellular matrix promote differentiation into myogenic lineages [30]. Nonetheless, the culture of stem cells for cell therapies faces challenges beyond the ones generally encountered by conventional cell lines culture. Due to their pluripotent nature, stem cells tend to differentiate into heterogeneous cells type depending on the external stimuli they are subjected to. In other words, the physicochemical cues from the extra cellular matrix and culture environment determine finely stem cells proliferation and differentiation pathways. The growth material may, for example, be chosen for its ability to either keep the cells in a undifferentiated state for as long as possible (for cell therapy) or to direct differentiation into a desired cell type (for tissue engineering) [31]. In both cases, the success of stem cell culture relies on understanding which factors influence the fate of stem cell and how to control them. Reservations exist, however, on the clinical use of stem cells cultured *in vitro* due to the routine use of animal derived product in common cell culture practice. The associated risk of disease transmission raises safety concerns. Furthermore, stem cell typically undergoes phenotypic and functional changes when cultured for extended periods of time. Thus, new methods allowing xenogenic-free upscalable culture of stem cells are needed. Most effort is currently focusing on growth support capable to bind extracellular matrix component RGD adhesion ligand and/or to deliver appropriate growth factors.

The wettability [32], chemistry, and reactivity [33] of the artificial culture support are prominent factor in directing the adhesion, growth and differentiation of stem cells just like its topography [34], elasticity and also its dimension. The versatility of plasma-assisted surface modification permits the preparation of substrates with controlled properties on all these fronts. The following sections demonstrate how plasma-based growth substrates can be tailored for the need of a particular cell culture system and also be used to scan a range of properties for research purposes in order to determine the optimum properties required for any particular need.

8.3.1 Plasma Methods for Tailored Surface Chemistry

The first critical step of *in vitro* culture is for the cells to attach to the artificial extracellular environment. It is well accepted that cell adhesion relies on the initial formation of a layer of adsorbed proteins [35]. Protein adsorption occurs faster than cell settlement in culture conditions. Cells surface receptors, such as integrins, then recognize favorable active binding sites in the protein film formed on the culture substrate. The binding of integrin receptors (fibronectin, laminin, etc.) to surface adsorbed biomolecules present in the media (fibrinogen, vitronectin, etc.) is controlled by peptide signaling pathways. This is the onset of the cell adhesion process. It is therefore critical for biomaterials to facilitate protein adsorption in an appropriate conformation.

A substrate's wettability, charge, and polarity greatly influence its protein adsorption capability [36, 37]. For example, substrates with moderate hydrophilicity (contact angle between 60° and 100°) have been found to generally promote the adsorption of proteins such as immunoglobulin, albumin, fibrinogen, and fibronectin [38–40]. However, a surface ability to adsorb proteins is not directly proportional to its performance in terms of cell adhesion. It is indeed well known that adsorbed albumin molecules, for example, tend to decrease cell adhesion. In general, for the cells to recognize surface bound proteins and adhere, the later need to have retained their bioactive conformation and adopted accessible orientation [41, 42]. Since protein adsorbed on rather hydrophobic surfaces tend to be denatured, substrates with contact angles between 40° and 70° are generally more suitable for cell adhesion [43]. Surface charge also affects proteins adsorption [44, 45]. This is mainly due to the fact that biomolecules are in an amphoteric state in physiological conditions. At pH 7.3, carboxylic acid groups present on peptides, enzymes, and proteins are negatively charged while the amine groups are protonated. As a result, charged and polar substrates engage in ionic and hydrogen bonds with biomolecules. Although protein layers bound ionically to the substrate may be sufficient to support rapid cell adhesion, these kinds of protein film fluctuate overtime due to continual competing adsorption–desorption processes [46]. For the culture of stem cells over extended period of time, this may not be ideal. The intrinsic chemistry of a substrate governs both its wettability and its surface charge but can also provide functional groups capable to bind biomolecules irreversibly via covalent bonds. Therefore, tailoring the surface chemistry of growth supports is a potent way to increase stem cell culture performance.

Among the many available methods for modifying a substrate chemical functionality, plasma polymer deposition is particularly resourceful due to the variety of organic precursor that can be used to produce films with tailored chemistry facilitating protein binding [47]. Oxygen-rich and nitrogen-rich plasma-based surfaces have been particularly well studied for their application as biomaterials [48] and are described in more details below. A summary of the most common plasma polymer deposition precursors used for protein adhesion and cell culture studies is provided in Table 8.1.

8.3.1.1 Oxygen-rich Surfaces

Oxygen-rich chemistry tends to favor cellular attachment. Groups such as hydroxyls, carboxyls, and carbonyls are polar in culture condition and enter ionic interaction with cell adhesion-mediating molecules.

Table 8.1 List of the organic precursors used to prepare plasma-deposited film, summarizing their chemical formula, the nature of the surface functionality formed, nature of the chemical bond formed with biomolecules, class of biomolecules and cell type tested, and corresponding literature references.

Precursor	Formula	Surface functionality	Reaction with biomolecule	Biomolecule tested	Cells tested	Refs
Allylalcohol	⌁⌁OH	Hydroxyl	Ionic and hydrogen binding		Murine ovary cells Endothelial cells	49–53
Acrylic acid	⌁⌁COOH	Carboxyl	Ionic, Hydrogen and acid–base binding Carbodiimine mediated amide covalent bond	Collagen Heparin Hyaluronic acid	Smooth muscle Fibroblasts, osteoblasts Keranocytes, Endothelial cells	53–59
Ally glycidyl ether		Epoxy	Covalent amine bond Covalent amine bond	Lysosome Protein g Protein A IgG antibodies		60, 61
Glycidyl methacrylate						
Propanal		Aldehyde	Covalent imine bond	Albumin Streptavidin	lymphocytes	47, 62–64

(Continued)

Table 8.1 Cont.

Precursor	Formula	Surface functionality	Reaction with biomolecule	Biomolecule tested	Cells tested	Refs
Allylamine	$\diagup\!\!\diagdown\!\!\diagup\!\text{NH}_2$	Amine, amide	Ionic, Hydrogen and acid-base binding Carbodiimine mediated amide covalent bond	Heparin Epidermal growth factor	Osteoblasts, fibroblasts, endothelial cells, neural cells, MSC	65–74
Ethylene diamine	$\text{H}_2\text{N}\diagup\!\!\diagdown\!\text{NH}_2$	Amine, amide	Ionic, hydrogen and acid-base binding	Glucose isomerase		75
Alkylamine	$\diagup\!\!\diagdown\!\!\diagup\!\text{NH}_2$	Amine, amide	Carbodiimine-mediated amide covalent bond			76–78
Alkyloxazoline		Oxazoline, amine, amide	Covalent amide bond Ionic and hydrogen binding	Albumin, FGF, Streptavidin, Fibronectin, Anti-EpCAM	Fibroblasts, Macrophages, Kidney stem cells, hMSC	79, 80

8.3.1.1.1 Alcohol

Surfaces with alcohol functionality can be prepared by plasma deposition of allyl alcohol [49–51] and also via the water plasma treatment of polymer surfaces [52]. Hydroxyl-rich polymer substrates typically have increased hydrophilicity compared to their unmodified counterpart [81]. Following hydroxyl functionalization, Lee *et al.* reported a decrease in the contact angle of a variety of polymer by 20–50° which had a favorable effect on the adhesion, spreading, and growth of ovary cells.

8.3.1.1.2 Acrylate

Acrylate surfaces groups can be conjugated to proteins, such as fibronectin, vitronectin, lamilin, and sialoproteins. The resulting substrata are particularly favorable for stem cell culture, and peptide acrylate-conjugated substrates are readily available from Sigma-Aldrich. Although there is no report about their use for the specific culture of cell lines or stem cells, acrylate-rich surfaces have been prepared via plasma polymerisation [82].

8.3.1.1.3 Carboxylic Acid

Small and volatile carboxylic acid such as acrylic acid can be plasma polymerized to produce stable films rich in carboxylic acid functionalities [54]. Rossini *et al.* investigated plasma deposition conditions for these films and demonstrated that film deposited at lower plasma ignition powers retained more –COOH functionality [83]. In direct correlation, films deposited at low powers were more prone to hydrogen and acid–base bonding, and increased protein adsorption occurred as the plasma ignition power decreased. Collagen was immobilized on similar films by Gusta *et al.* [55] The resulting substrate proves to be excellent for smooth muscle cell adhesion and rapid growth. Acrylic acid containing plasma polymers also promotes fibroblasts [56], osteoblasts [57], and human keranocytes [58] attachment. In particular, Detomaso *et al.* found that carboxyl surface density of 2.3×10^{-9} mol/cm^2 is sufficient to produce beneficial effects on fibroblasts attachment and growth [54]. Oxygen plasma treatment can also lead to the formation of surface carboxylic acid functionalities [59].

Coupling agent such as carbodiimine and *N*-hydroxysuccinimide (NHS) can then be used to enable the formation of a covalent bond between carboxylic acids and the amine groups present on the biomolecules [84]. Heparin and hyaluronic acid have been covalently coupled to the carboxylic acid functionality of plasma-treated polyethylene [59]. A drawback of carbodiimine-mediated amide bond formation is the side reactions that occur within proteins themselves when carboxylic acid group present on the biomolecules react with their amine groups.

8.3.1.1.4 Esters, Anhydrides, and Aldehydes

It is also possible to deposit plasma polymer films containing oxygen-rich reactive functions that will spontaneously form covalent bonds with biomolecules without the use of superfluous linking agents. Although it can be challenging to retain sufficient functionality through the plasma polymerization process, by using gentle deposition conditions films with reactive esters [85] and anhydrides [86] surface chemistries have successfully been generated. In both case, amide bonds are spontaneously formed via nucleophilic substitution from primary amine groups. Peptide, proteins, and antibodies were effectively immobilized on these types of substrates in a gentle and effective manner. Other classes of chemical functionality allowing direct covalent binding with proteins are aldehyde and epoxy compound. Epoxides react with nucleophile amine groups present on biomolecule to form amine bonds. Allyl glycidyl ether and glycidyl methacrylate precursors were both used to form epoxide-rich films via pulsed plasma polymerization and used to covalently bind proteins and DNA [60, 61]. Aldehyde also reacts with biomolecules through their nucleophilic primary amine groups, this time to form imine bonds [47]. Although the imine bond formation reaction is inherently reversible via hydrolysis, each biomolecule binding is essentially irreversible because multiple conjugation sites engage bonds with the substrate aldehyde groups. Stable aldehyde-rich films generated from the plasma deposition of propanal have been used for the covalent binding of proteins such as albumin and streptadivin [62]. Retention of streptavidin bioactivity was confirmed using biotinylated gold nanoparticle binding assay [63] and the culture of lymphocytes on these materials confirmed their biocompatibility [64].

Nevertheless, cell responses to oxygen-rich surface functionality appear to vary from one cell type to another. For example, a comparative study used the versatility of plasma polymer deposition to investigate the influence of hydroxyl, carboxyl, and carbonyl surface density on the growth of bovine aortic endothelial cells. Interestingly, cell growth was not influenced by hydroxyl and carboxyl concentration but did increase with increasing carbonyl functions [53]. These findings suggest that for the culture of stem cells, care will need to be taken to optimize the substrate chemistry toward any particular goal.

8.3.1.2 Nitrogen-rich Surfaces

8.3.1.2.1 Amine

Amine-rich substrates are among the most popular chemistries used to develop modern biomaterials [87]. As much as quaternary ammonium is cytotoxic, primary amine is biocompatible and well known for enhancing cell growth. The most common plasma media used to confer amine functionalities to biomaterials are ammonia, for non-film forming precursor, and allylamine [65, 66] for plasma polymer deposition [88]. However, a variety of other volatile amine monomers can be used to generate plasma-deposited films with a varying amount of surface primary amines and amides groups. These include, but are not limited to ethylendiamine [75], propylamine [76], butlyamine [77], and heptylamine [78]. Amine-rich surfaces are typically moderately hydrophilic, with contact angles ranging from 60° to 80°. Protonated amine (NH_2) groups are particularly favorable for protein adhesion because they have a localized positive charge in cell culture condition which engaged with negatively charged biomolecules. The electrostatic adsorption of proteins is thought to confer amine-rich surfaces their biocompatible characteristic. Burns et al. developed substrates with a varying amount of amine functionality and surface charge via the plasma deposition of diaminocyclohexane [89]. They reported that although albumin adsorption was not affected by surface charge variation, lysozyme tends to adsorb preferentially on positively charged surfaces. Plasma-deposited polyallylamine films have also been reported to promote the focal adhesion of osteoblasts [67], the growth of fibroblasts [68, 69], neural and other cells [70]. Furthermore, as mentioned in the previous section, amine groups can form covalent bonds with carboxyl functions through the use of linking agents (carbodiimine, NHS, trifluoroacteic anhydride); therefore, amine-rich surfaces can be used to covalently bind biomolecule via their carboxylic acid group. A variety of substrates (titania [71], stainless steel [72]) modified by allylamine plasma deposition were, in this way, functionalized with heparin. The use of linking reagent is, nonetheless, not ideal due to the complexity of the reaction and the assembly of protein aggregates [84]. As for oxygen-rich substrates, nitrogen-rich substrata capable to covalently bind biomolecule without using linking agent are preferable. Such alternative are described in the following section.

Ammonia glow discharge plasma is another popular plasma-based surface modification technique capable to generate surface bound amine functionalities. Diamond like carbon films were also functionalized with heparin following an ammonia plasma treatment [90]. The cell adhesion molecule expression of human endothelial cells increased on PET surface modified via ammonia glow discharge plasma [91]. Similar substrate was successfully coupled to the carboxylic acid group of epidermal growth factor using carbodiimine and NHS as linking agent. The modified PET surfaces were found to trigger cell growth and decrease apoptosis [92]. This technique was also used to generate amine-rich surface chemistry on nylon and polypropylene substrata [73]. A study of amine-rich growth substrate for MSC culture demonstrated that nitrogen-rich substrata reduced the expression of type X collagen, a marker associated with endochondral ossification [73]. In a follow-up study, Mwale et al. [74] demonstrated the amine enrichment of nylon surface via ammonia glow discharge plasma affects MSC differentiation pathway evidenced by the modulation of RUNX, osteocalcin, sialoprotein, and alkaline phosphatase expression.

8.3.1.2.2 Oxazolines

Recent studies have shown that it is also possible to prepared oxazoline-based plasma polymer films with nitrogen-rich functionality that enable a spontaneous irreversible binding with biomolecules without using reducing agents [79]. Ramiasa et al. have developed a method to produce plasma-deposited polyoxazoline film from methyl and ethyl oxazolines. The films obtained were hydrophilic and stable in physiological conditions. These films bind irreversibly a variety of biomolecules such as albumin, streptavidin, fibronectin, antibodies, and growth factors. The irreversible nature of the bond was challenged by washes with SDS and high-ionic-strength solutions [80]. The unmodified plasma-deposited polyoxazoline films supported the growth of several types of cells such as macrophages, fibroblast, and also embryonic kidney stem cells. Spontaneous binding of fibroblast growth factor polyoxazoline deposited from methyloxazoline promoted the rapid growth of hMSC which remained undifferentiated after 5 days of culture.

8.3.1.3 *Systematic Studies and Copolymers*

Although both oxygen-rich and nitrogen-rich surfaces seem to promote protein adsorption and in turn cell adhesion and spreading, the mechanisms instigating cell growth on these two types of substrates may be different. Steele et al. indeed demonstrated that nitrogen-rich plasma-deposited film supported endothelial cell and fibroblasts adhesion and growth in protein deprived media, while oxygen rich substrates did not [93]. These findings indicate that although protein binding facilitates cell adhesion, it may not be the only determining factor affecting cell adhesion. In the past decade, several groups examined how surface chemistry directs stem cell fate. Lan et al. [94] investigated the effect of hydroxyl, carboxyl, and amine surface functionality on the proliferation and differentiation of muscle-derived stem cells (myocytes). Their results demonstrated that the surface chemistry modulated the fate of stem cells beyond adhesion and spreading stages. In particular, chemistry-dependent variation in integrin binding was found to modulate cell differentiation pathways. Several other groups found that human stem cell lineage, such as hMSC and hASC, was largely sensitive to surface chemistry [95–97]. Shroder *et al.* compared the adhesion, growth, and differentiation of hMSC on plasma polymer deposited from allylamine and acrylic acid precursors [98]. Cells spread significantly faster on the polyallylamine films. Nonetheless, both coating facilitated osteogenic differentiation. Liu *et al.* in particular used plasma polymer films rich in amine, carboxyl, and methyl groups to screen the chemistry-dependent behavior of human adipose stem cell *in vitro* [99]. Interestingly, they reported that while both amine and carboxyl functionality allowed protein binding, cell behavior differed on these two coatings. Amine-rich substrate promoted osteogenic cell differentiation, while carboxyl functionality supported non-differentiated proliferation. Although significant insight has recently been provided on the various surface chemistries capable to support stem cell culture, further works are required in order to gain a comprehensive understanding of stem cell–surface interaction [100]. Combining chemical functionality and varying their surface density could help underpinning these complex mechanisms. We already demonstrated how plasma polymer depositions can be used to screen a range of surface chemistries, but it can also be used to assess the effect of functionality surface density. The method indeed allows the formation of composite substrates prepared from mixture of different precursors with perfect control on the proportion of each surface chemical component. The versatility of plasma surface modification methods can thus be used to screen ranges of varying chemistry and wettability in a very simple way. In Figure 8.2, the wetting properties of common homogeneous plasma polymer films are illustrated. By simply choosing different precursors, the contact angle can be varied from 95° to 50°, and the hydrophilicity of the substrate can be finely tuned by using precursors combinations. Copolymers of allylamine and oxygen-rich [101, 102] or alkyl [103] precursors were produced via plasma polymer deposition and tested for cell culture purposes. The copolymerization with allyl alcohol, for example, resulted in enhanced fibroblast viability.

8.3.2 Plasma for Surface Topography

Beyond the effect of surface chemistry, previous works have shown that physical interactions between biomaterials and stem cells also influence their behaviour [31]. Functionalized nanostructured materials can potentially be used to control the microscale bioregulation that governs

214 Advanced Surfaces for Stem Cell Research

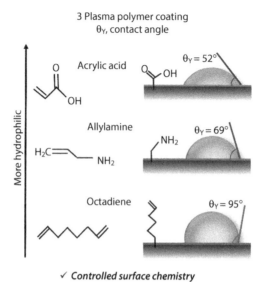

Figure 8.2 Wettability of plasma-deposited polymer films prepared from acrylic acid, allylamine, and octadiene precursors. The young water contact angle of the substrate can be increased from 53° for acrylic acid coating to 95° for octadiene coating.

the fate of stem cells [104]. Indeed, the physical interaction of stem cell with their growth support leads to deformation of the extracellular matrix and is believed to initiate cell differentiation by prompting gene expression and suppression [8]. Nanostructured surfaces can direct the cells focal attachment and induce changes in the cell shape by exerting forces on the cell membranes [105]. More precisely, nanotopological cues are sensed by the cell via their filopodia [9]. This induces changes in the cell cytoskeleton structure by changing the density and/or a number of focal adhesion points. These deformations of the cell cytoskeleton vary depending on the size, shape, and geometry of the nanofeatures [8, 106]. Ensuing mechano-transduction mechanisms result in alteration of gene expression. The surface roughness of titanium substrate [107], for example, is known to influence bone marrow cell behavior and ultimately the properties of the cultured cells. Besides, titanium microroughness promotes the differentiation of MSC into osteoblasts [108, 109]. Possible pathways responsible for this include the modification of collagen synthesis [110]. Also, stem cell culture scaffold consisting of aligned fibers of polylactic acid (PLA) reportedly generates aligned neural cells [111] and other groups have demonstrated similar behaviors on nanogrooves [29, 112]. It is also worth noting that the nanotopography of a substrate directly changes its wettability in quite a radical manner [113, 114]. As a consequence, some of the effect that surface roughness has on cell behavior may be due to factors going beyond the effects of pure physical interaction. For example, moderately hydrophilic surfaces may become more hydrophilic once roughened, what would certainly favor protein and cell adhesion.

Plasma-based approaches are particularly attractive for the production [115] and modification of surfaces with nanoscale topography. When experimental conditions are carefully controlled, plasma etching and implantation processes can be used to produce micro and nanofeatures [116]. Selected research group have perfected plasma-assisted nanoscale self-assembly and reported on the deterministic creation of a variety of nanostructures [117]. These include nanomaterials consisting of nanoparticles [118], nanotubes [119], nanocones [120], and nanofibers [121]. The control over the physical properties of the nanofeatures is unique to the plasma-assisted methods in that the building blocks generated in the plasma are of nanoscale dimensions. In order to understand the effect of surface topography on stem cell fate, it is important to screen a variety of complex nanoscale architectures. The range of nanomaterials produced via plasma-assisted method, where defect size, shape, and geometry can be tailored constitutes an ideal selection of model substrates for such investigations.

The development of patterning technique has helped our understanding of cell behavior at the micron and nanoscale. Techniques such as electron beam lithography allow the making of close to any type of topographical features in a spatially controlled manner. These features can be used to mimic the natural cell environment and explore cell–substrate interaction. However, with these methods, there is often a need to optimize surface chemistry in a subsequent step.

Regardless of the mode of production of the micro or nanorough substrates, plasma implantation and plasma deposition of polymer films are effective ways to perform the final surface modification in order to achieve a suitable surface chemistry for cellular behavior investigations [122]. Nanocones generated by colloidal lithography were, for example, coated with plasma-deposited polyacrylic acid and POE-like coatings in order to investigate the intricate role of surface topography, chemistry, and wettability on human keranocytes [123]. Thin plasma polymer layers were also coated over surface bound nanoparticles ranging from 16 to 68 nm in diameter and preserved the nanotopographical landscape of the underlying substrate (Figure 8.3) [124].

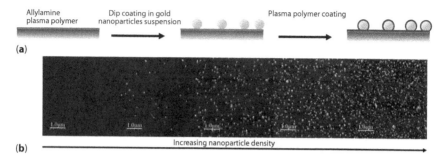

Figure 8.3 (a) Schematic of the generation of nanotopography gradients using plasma polymer deposition methods. (b) AFM images of 16 nm gold nanoparticle gradient overcoated with a plasma polymer layer.

8.3.3 Plasma for Surface Stiffness

In vivo, cells interact with a variety of tissue whose stiffness and elastic properties vary greatly. Bone tissue, for example, is hard, with young modulus of 15 MPa, while the very soft brain tissue stiffness is only 1 kPa. A number of studies have reported on the behavior of different cell types on substrates with varying stiffness. Neurons appear to grow better on soft substrate with elastic modulus below 0.5 kPa, while fibroblast "prefers" harder surfaces (10 kPa). In view of these observations, it does not come as a surprise that stem cells too are susceptible to the stiffness and elasticity of their growth substrates [30]. Evans *et al.*, for example, reported increased spreading, growth, and gene expression of embryonic stem cells on increasing substrate stiffness from 0.04 to 2.7 MPa [125]. Besides promoting cell proliferation, substrates stiffness has also been shown to induce selective differentiation of mesenchymal stem cells [30]. Typically as the substrate stiffness increases, neurogenic, myogenic, and ultimately osteogenic phenotype are promoted. Interestingly, cells have historically been cultured on materials such as glass and polystyrene which are several orders of magnitude stiffer than any biological tissue (50 and 3 GPa, respectively). In order to mimic the elasticity of natural ECM, new biomaterials are being investigated.

Polydimethylsiloxane (PDMS)-based elastomers, for example, are a biocompatible type of material whose roughness and elastic modulus can be controlled and tailored for specific applications. This unique set of properties is particularly interesting for the culture of stem cell whose differentiation can be guided by surface stiffness. However, the high hydrophobicity of PDMS makes it hostile to cell adhesion. Several surface modification strategies have been used to try and enhance cell adhesion on PDMS. The methods include protein physisorption [126], and also oxygen [127], or argon [128] plasma modification to initiate the subsequent deposition of polymers. These non-covalent modification techniques, however, tend not to sustain the long incubation period necessary to study stem cell differentiation [23, 129]. Mcarthur et al. [15] have shown that plasma polymer deposition, on the contrary, was a valid way to modify the surface of soft elastomer for the study of mesenchymal stem cell differentiation. Polyacrylic acid plasma-deposited thin films were used to study the adhesion, growth, and differentiation of stem cell on substrata with stiffness varying from 1 KPa to 1 MPa. The nanometer thin plasma polymer layer enhanced mesenchymal stem cell growth by increasing the wettability of the surface and allowing protein adsorption while preserving the elasticity of the underlying substrates. Overall, greater osteogenic differentiation was observed on the stiffest substrates. This type of chemical modification was reported to enhance the attachment and proliferation of several cell types, including hMSC [130].

Other soft biomaterials have been produced using layer-by-layer polymer deposition techniques [131, 132]. In the work of Hopp et al. [132], the stiffness of the polymer layer was varied in a gradient fashion in order to screen stiffnesses ranging from 0.5 to 110 MPa. In order to cancel out any residual chemical variation resulting from the polymer deposition process, uniform plasma-deposited coatings of allylamine or acrylic acid were used to confer the stiffness gradient a homogeneous outer chemistry. Hopp et al. demonstrated that fibroblast adhesion, growth, and proliferation were greater on the stiffer end of the gradient. The use of two homogeneous plasma-deposited films with different chemistries further allows them to conclude that the substrate stiffness had a greater impact on the cell behavior than the substrate chemistry or wettability. We describe other application of plasma methods to produce gradients substrata in the following section.

8.3.4 Plasma for Gradient Substrata

In nature, biomolecules gradient governs a variety of biological processes. Eventually, biomaterial capable to mimic these naturally occurring gradients could help optimizing and controlling stem cells *in vitro* microenvironments [133, 134]. In the meantime, gradients substrates are a powerful screening tool to investigate the effect of gradually changing surface properties on cell behavior. Plasma polymer deposition methods have been used to generate gradients of surface chemistry [17, 135], wettability [136], and surface charge [137] as well as gradient of surface topography [138, 139] or stiffness [132]. Plasma-based surface chemistry gradient were used to generate gradients of surface bound proteins such as fibrinogen [140] and streptavidin [62]. The same plasma polymer gradient method was used to interrogate the optimal growth factor density necessary to induce neural differentiation in mouse embryonic stem cells [141]. In this study, plasma polymer gradient of hydroxyl to aldehyde functionalities was used to bind increasing amount of nerve growth factor (NGF). A critical value of 52.9 ng cm^{-2} NGF density was identified, above which cell attachment and proliferation did no longer increase. The effect of surface wettability on the pluripotency of mouse embryonic stem cells was tested on plasma-deposited gradients of octadiene and acrylic acid [142]. The water contact angle over the gradient decreased from 88° on the alkyl side to just under 50° on the carboxyl side of the gradient. Pluripotency was retained when the cell spreading was less than 120 μm^2 which occurred over parts of the gradient with poor substrate adhesivity. This initial study highlighted the potential of gradient substrata to identify threshold experimental conditions [18]. Surface chemistry gradient ranging from 100% polyallylamine to 100% octadiene were then used to investigate the fate of several stem cell lineage [143–145]. These gradient substrates further helped identifying culture conditions favoring pluripotency retention for mouse embryonic stem cells: increased cell adhesion occurred toward high nitrogen content, while stem cell marker expression was greater on the octadiene side of the gradient [143]. This result was attributed to a combined and complex effect of both colony size and cell–substrate interaction. For mouse kidney stem cell on the other hand, it appeared that the difference in chemistry dictated the cell differentiation into different lineages: differentiation into podocyte occurred at high nitrogen content, while the cell preferentially differentiated into proximal tubule cells on the octadiene-rich side of the gradients [145]. A very recent study on the fate of human adipose stem cells on the same gradient surfaces was able to evidence osteogenic differentiation toward the allylamine-rich side of the gradients [144]. The authors were also able to identify ERK1/2 as an important signaling pathway protein.

Plasma polymer deposition methods have also been used to prepare and/or coat nanotopography gradients with uniform outermost surface chemistries. One way to generate nanotopography gradients is to dip a uniformly positively charged substrate (e.g. plasma-deposited polyallylamine) into a suspension of negatively charged nanoparticles (e.g. carboxyl-functionalized gold nanoparticles). The chemistry of the resulting nanoroughness gradients is not uniform: it consists of the underlying plasma polymer mounted with gold islands. In order to assess the effect of nanotopography alone, a conformal plasma polymer layer of uniform chemistry may be deposited over the nanoparticles [138]. This method is represented in a schematic in Figure 8.3a. Since pinhole-free films as thin as 5 nm can be achieved, the topographical features of the gradient are retained. The size of the nanoparticle used can be easily varied and a variety of overcoating used to investigate dual mechanisms. Gradients substrates of gold nanoparticles 16, 38, and 68 nm in diameter were used to interrogate the effect of nanotopography on cell adhesion, spreading, and proliferation. Typical AFM images of a nanoparticle density gradient overcoated with a plasma polymer layer are shown in Figure 8.3b. Fibroblasts were found to adhere better and spread well on all nanoparticles sizes. Osteoblasts, on the other hands adhere preferentially on 16nm gradients. When the nanodefect size was increased to 38 and 68nm, osteoblast cell adhered in lower number than on flat substrate with the same chemistry [124].

Nanotopography gradient were also coated with allylamine and octadiene copolymer in order to investigate further the differentiation behavior of mouse kidney stem cells. In Figure 8.4, representative micrographs of mKSC immunostained in green for WT1 podocyte markers are presented. The cells growing on the section of the gradient with increased nanoparticle density tend to differentiate into podocytes. Surface chemistry also affects cell differentiation, as for a comparable nanoparticle density, increased podocyte differentiation is observed on the allylamine rich substrates.

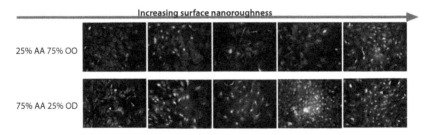

Figure 8.4 Mouse kidney stem cell differentiation over 16 nm nanotopography gradient with allylamine and octadiene copolymer overcoating. 25% allylamine, 75% octadiene, top row. 75% allylamine, 25% octadiene bottom row. The nanoparticle gradient density increases toward the right. Wils tumor podocyte marker is stain with alexa fluor 488 green, actin red is used as counter stain. Stem cell differentiation into podocytes is more pronounced on the denser part of the nanotopography gradient, and on the nitrogen rich surface chemistry.

8.3.5 Plasma and 3D Scaffolds

Natural stem cell niches are 3D and present cues in various directions. Three-dimensional cell culture scaffolds have major advantages over classic 2D substrates. In 3D scaffolds, cells can adopt their natural morphology and contact the surrounding ECM and other cells in all directions. Supplementing the physical effect of 3D scaffold with biochemical cues (growth factor delivery, oxygen level control [146], etc.) allows to further control the fate of stem cells which are particularly responsive to their microenvironment [147].

Hydrogels have been used for decades as stem cells culture media. They can be tailored to closely resemble the natural environment of stem cells in the human body and present the advantages to have high elasticity and water content and to allow the diffusion of growth factors and gases. These biocompatible scaffolds can, in combination with appropriate extracellular matrix molecules, maintain cell undifferentiated or facilitate their differentiation into a variety of cell lineage, such as cardiac [148], hepatic [149], and endothelial cells [150–154]. One of the limitations of hydrogels is that their stiffness is limited. As such, following the point discussed in the previous section on surface stiffness, hydrogel typically cannot be used for the differentiation of stem cells into lineage that require hard substrate (e.g. bones, cartilages). New alternative are being developed to close this gap. Three-dimensional scaffolds for the culture of bone cell lineage, for example, need to have similar properties to the stiff and porous mineral bone phase. Electrospun polymer nanofibrous scaffolds have been used to induce the proliferation and differentiation of mSC into various lineage, such as oestogenic, adipogenic, and chondrogenic [155] and 3D carbon structures used for hard tissue engineering [156]. Hybrid materials of synthetic 3D-printed calcium phosphate scaffolds also share several bulk properties with the natural ECM of osteoblast (stiffness, mechanical integrity, interconnected porosity); however, their surface chemistry does not favor cell adhesion [157]. Very recently, the cellularization of calcium phosphate scaffold modified by ammonia glow discharge plasma and plasma-deposited allylamine has been explored [158]. Another very recent piece of work from Favia *et al.* depicts advances made in this field so far and report on the plasma modification of 3D polycaprolactone and polylactic acid scaffolds with ethylene:nitrogene, ethylene:hydrogen, and ethylene:acrylic acid precursor mix [159]. The plasma modification of the scaffolds improved their cytocompatibility and their osteo-integration. Allylamine plasma polymers have also been successfully deposited onto other 3D structures such as PLA scaffold [160] and titanium alloys [161].

8.4 New Trends and Outlook

New ways of stimulating stem cell differentiation are being investigated. Films of polypyrrole doped with large electrolytes have been used to induce nerve cell differentiation upon electrical stimulation [162]. Electrically conductive 3D scaffolds have been fabricated using carbon nanotubes and used to induce stem cell differentiation into neural and cardiac lineage with electric signals [163, 164]. In view of these new developments, one can expect conductive coating to also attract interest for *in vitro* stem cell culture. Several groups have reported on the making of conductive plasma polymers coating with interesting chemical properties [165].

Another developing field of plasma science is the so-called "plasma medicine" which consists in targeting tissues or cells directly with plasma jets [166]. Because this branch of plasma science does not involve the engineering of advanced materials, it was not developed here. Nevertheless, it has been shown that spatially targeted plasma agitation could induce the differentiation of neuronal stem cells [167].

Finally, plasma polymers can also be engineered in order to contain nanocavity or porosity used to store and deliver small compound such as drugs or nanoparticulates [168]. This emerging application of plasma polymers could become of interest for the making of advanced stem cell culture systems. In such systems, growth factor or gases could be slowly made available to the stem cells in culture in order to direct differentiation pathway over an extended period of time.

8.5 Conclusions

The behaviors of stem cells in culture, their growth, proliferation, and differentiation are guided by both chemical and physical cues. The properties of the material on which the stem cells are grown are of paramount importance in both understanding and guiding the cell differentiation.

Plasma-assisted substrate modification methods can be used to selectively adjust biomaterials properties in order to guide stem cell behavior. This can be achieved by customizing the surface chemistry or tailoring its physical properties. The deposition of nanometer thin plasma polymer films with hydrophilic properties and reactive chemical functionality can enhance protein adsorption and in turn cell adhesion. Additionally, functionalization of chemically reactive plasma-deposited polymer films with ECM component or growth factors is straightforward. This approach can be used to further guide the fate of stem cells, whether it is to maintain stemness or to direct differentiation. Plasma-based techniques offer the advantage to be substrate independent. In consequences, biocompatible and functional plasma polymer films can be deposited onto any type of substrates. As such plasma deposition is a versatile tool capable to facilitate and study cell adhesion, growth, and spreading on surfaces with original nanotopography, on substrates with various elasticity and even on and within 3D scaffolds. Plasma processes and treatments encompass a large variety of methods, and so the range of cell culture substrates that can potentially be generated is as unlimited as the diversity of their properties. Researchers therefore are only limited by their imagination as to how, that is by which means and in which way, they should mimic the natural extra cellular environment of stem cells for better end results.

References

1. Tabar, V. & Studer, L. Pluripotent stem cells in regenerative medicine: challenges and recent progress. *Nat Rev. Genetics* 15, 82–92, 2014.
2. Mizuno, H., Tobita, M. & Uysal, A.C. Concise review: adipose-derived stem cells as a novel tool for future regenerative medicine. *Stem Cells* 30, 804–810, 2012.
3. Jaklenec, A., Stamp, A., Deweerd, E., Sherwin, A. & Langer, R. Progress in the tissue engineering and stem cell industry "are we there yet?". *Tissue Eng B: Rev* 18, 155–166, 2012.
4. Clarke, M.F. *et al.* Cancer stem cells—perspectives on current status and future directions: AACR Workshop on cancer stem cells. *Cancer Res.* 66, 9339–9344, 2006.
5. Rezania, A. *et al.* Maturation of human embryonic stem cell–derived pancreatic progenitors into functional islets capable of treating pre-existing diabetes in mice. *Diabetes* 61, 2016–2029, 2012.
6. Tyndall, A. Successes and failures of stem cell transplantation in autoimmune diseases. *ASH Education Program Book* 2011, 280–284, 2011.
7. Lutolf, M.P., Gilbert, P.M. & Blau, H.M. Designing materials to direct stem-cell fate. *Nature* 462, 433–441, 2009.
8. Guilak, F. *et al.* Control of stem cell fate by physical interactions with the extracellular matrix. *Cell Stem Cell* 5, 17–26, 2009.
9. Dalby, M.J., Gadegaard, N. & Oreffo, R.O.C. Harnessing nanotopography and integrin-matrix interactions to influence stem cell fate. *Nat. Mater.* 13, 558–569, 2014.
10. Discher, D.E., Mooney, D.J. & Zandstra, P.W. Growth factors, matrices, and forces combine and control stem cells. *Science* 324, 1673–1677, 2009.

11. Borghi, F.F. et al. Emerging stem cell controls: nanomaterials and plasma effects. *J. Nanomater.* 2013, 15, 2013.
12. Chu, P.K., Chen, J.Y., Wang, L.P. & Huang, N. Plasma-surface modification of biomaterials. *Mater. Sci. Eng. R Rep.* 36, 143–206, 2002.
13. Meade, K.A. et al. Immobilization of heparan sulfate on electrospun meshes to support embryonic stem cell culture and differentiation. *J. Biol. Chem.* 288, 5530–5538, 2013.
14. Girard-Lauriault, P.-L. et al. Atmospheric pressure deposition of micropatterned nitrogen-rich plasma-polymer films for tissue engineering. *Plasma Processes Polym.* 2, 263–270, 2005.
15. Colley, H.E., Mishra, G., Scutt, A.M. & McArthur, S.L. Plasma polymer coatings to support mesenchymal stem cell adhesion, growth and differentiation on variable stiffness silicone elastomers. *Plasma Processes Polym.* 6, 831–839, 2009.
16. Barry, J.J.A., Howard, D., Shakesheff, K.M., Howdle, S.M. & Alexander, M.R. Using a core–sheath distribution of surface chemistry through 3D tissue engineering scaffolds to control cell ingress. *Adv. Mater.* 18, 1406–1410, 2006.
17. Zelzer, M. et al. Investigation of cell–surface interactions using chemical gradients formed from plasma polymers. *Biomaterials* 29, 172–184, 2008.
18. Harding, F.J., Clements, L.R., Short, R.D., Thissen, H. & Voelcker, N.H. Assessing embryonic stem cell response to surface chemistry using plasma polymer gradients. *Acta Biomater.* 8, 1739–1748, 2012.
19. Erchak, J.M. (Google Patents, 1968).
20. Bilek, M.M. & McKenzie, D.R. Plasma modified surfaces for covalent immobilization of functional biomolecules in the absence of chemical linkers: towards better biosensors and a new generation of medical implants. *Biophys. Rev.* 2, 55–65, 2010.
21. Wiedemair, J. et al. In-situ AFM studies of the phase-transition behavior of single thermoresponsive hydrogel particles. *Langmuir* 23, 130–137, 2006.
22. Woodward, I., Schofield, W., Roucoules, V. & Badyal, J. Super-hydrophobic surfaces produced by plasma fluorination of polybutadiene films. *Langmuir* 19, 3432–3438, 2003.
23. Guimond, S. & Wertheimer, M.R. Surface degradation and hydrophobic recovery of polyolefins treated by air corona and nitrogen atmospheric pressure glow discharge. *J. Appl. Polym. Sci.* 94, 1291–1303, 2004.
24. Chatelier, R.C., Griesser, H.J., Steele, J.G. & Johnson, G. (Google Patents, 1995).
25. Yasuda, H. Plasma polymerization. Academic press, 2012.
26. Michelmore, A., Martinek, P., Sah, V., Short, R.D. & Vasilev, K. Surface morphology in the early stages of plasma polymer film growth from amine-containing monomers. *Plasma Processes Polym.* 8, 367–372, 2011.
27. Vasilev, K., Michelmore, A., Griesser, H.J. & Short, R.D. Substrate influence on the initial growth phase of plasma-deposited polymer films. *Chem. Commun.* 3600–3602, 2009.

28. Cha, C., Liechty, W.B., Khademhosseini, A. & Peppas, N.A. Designing biomaterials to direct stem cell fate. *ACS Nano* 6, 9353–9358, 2012.
29. Yim, E.K.F., Pang, S.W. & Leong, K.W. Synthetic nanostructures inducing differentiation of human mesenchymal stem cells into neuronal lineage. *Exp. Cell Res.* 313, 1820–1829, 2007.
30. Engler, A.J., Sen, S., Sweeney, H.L. & Discher, D.E. Matrix elasticity directs stem cell lineage specification. *Cell* 126, 677–689, 2006.
31. Saha, K., Pollock, J.F., Schaffer, D.V. & Healy, K.E. Designing synthetic materials to control stem cell phenotype. *Curr. Opin. Chem. Biol.* 11, 381–387, 2007.
32. Ayala, R. et al. Engineering the cell–material interface for controlling stem cell adhesion, migration, and differentiation. *Biomaterials* 32, 3700–3711, 2011.
33. Lee, J.H., Jung, H.W., Kang, I.K. & Lee, H.B. Cell behaviour on polymer surfaces with different functional groups. *Biomaterials* 15, 705–711, 1994.
34. Lampin, M., Warocquier-Clérout, R., Legris, C., Degrange, M. & Sigot-Luizard, M. Correlation between substratum roughness and wettability, cell adhesion, and cell migration. *J. Biomed. Mater. Res.* 36, 99–108, 1997.
35. Ruoslahti, E. & Pierschbacher, M.D. New perspectives in cell adhesion: RGD and integrins. *Science* 238, 491–497, 1987.
36. Arima, Y. & Iwata, H. Effect of wettability and surface functional groups on protein adsorption and cell adhesion using well-defined mixed self-assembled monolayers. *Biomaterials* 28, 3074–3082, 2007.
37. Gray, J.J. The interaction of proteins with solid surfaces. *Curr. Opin. Struct. Biol.* 14, 110–115, 2004.
38. Tamada, Y. & Ikada, Y. Effect of preadsorbed proteins on cell adhesion to polymer surfaces. *J. Colloid Interface Sci.* 155, 334–339, 1993.
39. Sethuraman, A., Han, M., Kane, R.S. & Belfort, G. Effect of surface wettability on the adhesion of proteins. *Langmuir* 20, 7779–7788, 2004.
40. Xu, L.-C. & Siedlecki, C.A. Effects of surface wettability and contact time on protein adhesion to biomaterial surfaces. *Biomaterials* 28, 3273–3283, 2007.
41. Bax, D.V., McKenzie, D.R., Bilek, M.M.M. & Weiss, A.S. Directed cell attachment by tropoelastin on masked plasma immersion ion implantation treated PTFE. *Biomaterials* 32, 6710–6718, 2011.
42. Lhoest, J.B., Detrait, E., Van Den Bosch De Aguilar, P. & Bertrand, P. Fibronectin adsorption, conformation, and orientation on polystyrene substrates studied by radiolabeling, XPS, and ToF SIMS. *J. Biomed. Mater. Res.* 41, 95–103, 1998.
43. Van Wachem, P. et al. Interaction of cultured human endothelial cells with polymeric surfaces of different wettabilities. *Biomaterials* 6, 403–408, 1985.
44. Pasche, S., Vörös, J., Griesser, H.J., Spencer, N.D. & Textor, M. Effects of ionic strength and surface charge on protein adsorption at PEGylated surfaces. *J. Phys. Chem. B* 109, 17545–17552, 2005.

45. Hallab, N., Bundy, K., O'connor, K., Clark, R. & Moses, R. Cell adhesion to biomaterials: correlations between surface charge, surface roughness, adsorbed protein, and cell morphology. *J. Long. Term Eff. Med. Implants* 5, 209–231, 1994.
46. Vroman, L., Adams, A., Fischer, G. & Munoz, P. Interaction of high molecular weight kininogen, factor XII, and fibrinogen. *Blood* 55, 156–159, 1980.
47. Coad, B.R., Jasieniak, M., Griesser, S.S. & Griesser, H.J. Controlled covalent surface immobilisation of proteins and peptides using plasma methods. *Surf. Coat. Technol.* 233, 169–177, 2013.
48. Ratner, B.D. Plasma deposition for biomedical applications: a brief review. *J. Biomater. Sci. Polym. Ed.* 4, 3–11, 1993.
49. Ameen, A.P., Short, R.D. & Ward, R. The formation of high surface concentrations of hydroxyl groups in the plasma polymerization of allyl alcohol. *Polymer* 35, 4382–4391, 1994.
50. Gancarz, I., Bryjak, J., Bryjak, M., Poźniak, G. & Tylus, W. Plasma modified polymers as a support for enzyme immobilization 1.: Allyl alcohol plasma. *Eur. Polym. J.* 39, 1615–1622, 2003.
51. Fally, F., Virlet, I., Riga, J. & Verbist, J.J. Detailed multitechnique spectroscopic surface and bulk characterization of plasma polymers deposited from 1-propanol, allyl alcohol, and propargyl alcohol. *J. Appl. Polym. Sci.* 59, 1569–1584, 1996.
52. Lee, J.H., Park, J.W. & Lee, H.B. Cell adhesion and growth on polymer surfaces with hydroxyl groups prepared by water vapour plasma treatment. *Biomaterials* 12, 443–448, 1991.
53. Ertel, S.I., Chilkoti, A., Horbetti, T.A. & Ratner, B.D. Endothelial cell growth on oxygen-containing films deposited by radio-frequency plasmas: the role of surface carbonyl groups. *J. Biomater. Sci. Polym. Ed.* 3, 163–183, 1992.
54. Detomaso, L., Gristina, R., Senesi, G.S., d'Agostino, R. & Favia, P. Stable plasma-deposited acrylic acid surfaces for cell culture applications. *Biomaterials* 26, 3831–3841, 2005.
55. Gupta, B., Plummer, C., Bisson, I., Frey, P. & Hilborn, J. Plasma-induced graft polymerization of acrylic acid onto poly(ethylene terephthalate) films: Characterization and human smooth muscle cell growth on grafted films. *Biomaterials* 23, 863–871, 2002.
56. Detomaso, L., Gristina, R., d'Agostino, R., Senesi, G.S. & Favia, P. Plasma deposited acrylic acid coatings: Surface characterization and attachment of 3T3 murine fibroblast cell lines. *Surf. Coat. Technol.* 200, 1022–1025, 2005.
57. Daw, R. *et al.* Plasma copolymer surfaces of acrylic acid/1,7 octadiene: Surface characterisation and the attachment of ROS 17/2.8 osteoblast-like cells. *Biomaterials* 19, 1717–1725, 1998.
58. France, R., Short, R. & Dawson, R. Attachment of human keratinocytes to plasma co-polymers of acrylic acid/octa-1,7-diene and allyl amine/octa-1,7-diene. *J. Mater. Chem.* 8, 37–42, 1998.

59. Favia, P. et al. Immobilization of heparin and highly-sulphated hyaluronic acid onto plasma-treated polyethylene. *Plasmas Polym.* 3, 77–96, 1998.
60. Thierry, B., Jasieniak, M., de Smet, L.C., Vasilev, K. & Griesser, H.J. Reactive epoxy-functionalized thin films by a pulsed plasma polymerization process. *Langmuir* 24, 10187–10195, 2008.
61. Harris, L., Schofield, W. & Badyal, J. Multifunctional molecular scratchcards. *Chem. Mater.* 19, 1546–1551, 2007.
62. Coad, B.R. et al. Immobilized streptavidin gradients as bioconjugation platforms. *Langmuir* 28, 2710–2717, 2012.
63. Coad, B.R. et al. Functionality of proteins bound to plasma polymer surfaces. *ACS Appl. Mater. Interfaces* 4, 2455–2463, 2012.
64. Christo, S.N. et al. Individual and population quantitative analyses of calcium flux in T-cells activated on functionalized material surfaces. *Aust. J. Chem.* 65, 45–49, 2012.
65. Shard, A.G. et al. A NEXAFS examination of unsaturation in plasma polymers of allylamine and propylamine. *J. Phys. Chem. B* 108, 12472–12480, 2004.
66. Hook, A.L., Thissen, H., Quinton, J. & Voelcker, N.H. Comparison of the binding mode of plasmid DNA to allylamine plasma polymer and poly (ethylene glycol) surfaces. *Surf. Sci.* 602, 1883–1891, 2008.
67. Finke, B. et al. The effect of positively charged plasma polymerization on initial osteoblastic focal adhesion on titanium surfaces. *Biomaterials* 28, 4521–4534, 2007.
68. Ren, T.B., Weigel, T., Groth, T. & Lendlein, A. Microwave plasma surface modification of silicone elastomer with allylamine for improvement of biocompatibility. *J. Biomed. Mater. Res. A* 86A, 209–219, 2008.
69. Hamerli, P., Weigel, T., Groth, T. & Paul, D. Surface properties of and cell adhesion onto allylamine-plasma-coated polyethylenterephtalat membranes. *Biomaterials* 24, 3989–3999, 2003.
70. Harsch, A., Calderon, J., Timmons, R. & Gross, G. Pulsed plasma deposition of allylamine on polysiloxane: a stable surface for neuronal cell adhesion. *J. Neurosci. Methods* 98, 135–144, 2000.
71. Yang, Z. et al. Improved hemocompatibility guided by pulsed plasma tailoring the surface amino functionalities of TiO_2 coating for covalent immobilization of heparin. *Plasma Processes Polym.* 8, 850–858, 2011.
72. Yang, Z. et al. The covalent immobilization of heparin to pulsed-plasma polymeric allylamine films on 316L stainless steel and the resulting effects on hemocompatibility. *Biomaterials* 31, 2072–2083, 2010.
73. Nelea, V. et al. Selective inhibition of type X collagen expression in human mesenchymal stem cell differentiation on polymer substrates surface-modified by glow discharge plasma. *J. Biomed. Mater. Res. A* 75A, 216–223, 2005.
74. Mwale, F. et al. The effect of glow discharge plasma surface modification of polymers on the osteogenic differentiation of committed human mesenchymal stem cells. *Biomaterials* 27, 2258–2264, 2006.

75. Choukourov, A. et al. Properties of amine-containing coatings prepared by plasma polymerization. *J. Appl. Polym. Sci.* 92, 979–990, 2004.
76. Fally, F., Doneux, C., Riga, J. & Verbist, J. Quantification of the functional groups present at the surface of plasma polymers deposited from propylamine, allylamine, and propargylamine. *J. Appl. Polym. Sci.* 56, 597–614, 1995.
77. Gancarz, I., Bryjak, J., Poźniak, G. & Tylus, W. Plasma modified polymers as a support for enzyme immobilization II. Amines plasma. *Eur. Polym. J.* 39, 2217–2224, 2003.
78. Gengenbach, T.R., Chatelier, R.C. & Griesser, H.J. Characterization of the ageing of plasma-deposited polymer films: global analysis of X-ray photoelectron spectroscopy data. *Surf. Interface Anal.* 24, 271–281, 1996.
79. Ramiasa, M. et al. Plasma polymerised polyoxazoline thin films for biomedical applications. *Chem. Commun.* 51, 20, 4279–4282, (2015).
80. Macgregor-Ramiasa, M.N., Cavallaro, A.A. & Vasilev, K. Properties and reactivity of polyoxazoline plasma polymer films. *J. Mater.* 3, 30, 6327–6337, 2015.
81. Rinsch, C.L. et al. Pulsed radio frequency plasma polymerization of allyl alcohol: controlled deposition of surface hydroxyl groups. *Langmuir* 12, 2995–3002, 1996.
82. Francesch, L., Garreta, E., Balcells, M., Edelman, E.R. & Borrós, S. Fabrication of bioactive surfaces by plasma polymerization techniques using a novel acrylate-derived monomer. *Plasma Processes Polym.* 2, 605–611, 2005.
83. Rossini, P., Colpo, P., Ceccone, G., Jandt, K.D. & Rossi, F. Surfaces engineering of polymeric films for biomedical applications. *Mater. Sci. Eng. C* 23, 353–358, 2003.
84. Strola, S. et al. Comparison of surface activation processes for protein immobilization on plasma-polymerized acrylic acid films. *Surf. Interface Anal.* 42, 1311–1315, 2010.
85. Duque, L., Menges, B., Borros, S. & Förch, R. Immobilization of biomolecules to plasma polymerized pentafluorophenyl methacrylate. *Biomacromolecules* 11, 2818–2823, 2010.
86. Mishra, G., Easton, C.D., Fowler, G.J. & McArthur, S.L. Spontaneously reactive plasma polymer micropatterns. *Polymer* 52, 1882–1890, 2011.
87. Aziz, G., Geyter, N.D. & Morent, R. Incorporation of primary amines via plasma technology on biomaterials. 2015.
88. Griesser, H.J., Chatelier, R.C., Gengenbach, T.R., Johnson, G. & Steele, J.G. Growth of human cells on plasma polymers: putative role of amine and amide groups. *J. Biomater. Sci. Polym. Ed.* 5, 531–554, 1994.
89. Burns, N.L., Holmberg, K. & Brink, C. Influence of surface charge on protein adsorption at an amphoteric surface: effects of varying acid to base ratio. *J. Colloid Interface Sci.* 178, 116–122, 1996.
90. Steffen, H.J., Schmidt, J. & Gonzalez-Elipe, A. Biocompatible surfaces by immobilization of heparin on diamond-like carbon films deposited on various substrates. *Surf. Interface Anal.* 29, 386–391, 2000.

91. Pu, F.R., Williams, R.L., Markkula, T.K. & Hunt, J.A. Effects of plasma treated PET and PTFE on expression of adhesion molecules by human endothelial cells *in vitro*. *Biomaterials* 23, 2411–2428, 2002.
92. Charbonneau, C. *et al.* Chondroitin sulfate and epidermal growth factor immobilization after plasma polymerization: a versatile anti-apoptotic coating to promote healing around stent grafts. *Macromol. Biosci.* 12, 812–821, 2012.
93. Steele, J.G. *et al.* Roles of serum vitronectin and fibronectin in initial attachment of human vein endothelial cells and dermal fibroblasts on oxygen- and nitrogen-containing surfaces made by radiofrequency plasmas. *J. Biomater. Sci. Polym. Ed.* 6, 511–532, 1995.
94. Lan, M.A., Gersbach, C.A., Michael, K.E., Keselowsky, B.G. & García, A.J. Myoblast proliferation and differentiation on fibronectin-coated self assembled monolayers presenting different surface chemistries. *Biomaterials* 26, 4523–4531, 2005.
95. Glennon-Alty, L., Williams, R., Dixon, S. & Murray, P. Induction of mesenchymal stem cell chondrogenesis by polyacrylate substrates. *Acta Biomater.* 9, 6041–6051, 2013.
96. Liu, X. *et al.* Adipose stem cells controlled by surface chemistry. *J. Tissue Eng. Regen. Med.* 7, 112–117, 2013.
97. Phillips, J.E., Petrie, T.A., Creighton, F.P. & García, A.J. Human mesenchymal stem cell differentiation on self-assembled monolayers presenting different surface chemistries. *Acta Biomater.* 6, 12–20, 2010.
98. Schröder, K. *et al.* Capability of differently charged plasma polymer coatings for control of tissue interactions with titanium surfaces. *J. Adhes. Sci. Technol.* 24, 1191–1205, 2010.
99. Liu, X., Feng, Q., Bachhuka, A. & Vasilev, K. Surface chemical functionalities affect the behavior of human adipose-derived stem cells *in vitro*. *Appl. Surf. Sci.* 270, 473–479, 2013.
100. Hwang, Y., Phadke, A. & Varghese, S. Engineered microenvironments for self-renewal and musculoskeletal differentiation of stem cells. *Regen. Med.* 6, 505–524, 2011.
101. Hawker, M.J., Pegalajar-Jurado, A., Hicks, K.I., Shearer, J.C. & Fisher, E.R. Allylamine and allyl alcohol plasma copolymerization: synthesis of customizable biologically-reactive three-dimensional scaffolds. *Plasma Processes Polym.* 2015.
102. Beck, A.J. *et al.* Plasma co-polymerisation of two strongly interacting monomers: acrylic acid and allylamine. *Plasma Processes Polym.* 2, 641–649, 2005.
103. Zuber, A., Robinson, D., Short, R., Steele, D. & Whittle, J. Development of a surface to increase retinal pigment epithelial cell (ARPE-19) proliferation under reduced serum conditions. *J. Mater. Sci. Mater. Med.* 25, 1367–1373, 2014.
104. Yim, E.K., Darling, E.M., Kulangara, K., Guilak, F. & Leong, K.W. Nanotopography-induced changes in focal adhesions, cytoskeletal

organization, and mechanical properties of human mesenchymal stem cells. *Biomaterials* 31, 1299–1306, 2010.
105. McNamara, L.E. *et al.* Nanotopographical control of stem cell differentiation. *J. Tissue Eng.* 1 120623, 13 pages, 2010.
106. Dalby, M.J. *et al.* The control of human mesenchymal cell differentiation using nanoscale symmetry and disorder. *Nat. Mater.* 6, 997-1003, 2007.
107. Deligianni, D.D. *et al.* Effect of surface roughness of the titanium alloy Ti-6Al-4V on human bone marrow cell response and on protein adsorption. *Biomaterials* 22, 1241–1251, 2001.
108. Wall, I., Donos, N., Carlqvist, K., Jones, F. & Brett, P. Modified titanium surfaces promote accelerated osteogenic differentiation of mesenchymal stromal cells *in vitro. Bone* 45, 17–26, 2009.
109. Olivares-Navarrete, R. *et al.* Direct and indirect effects of microstructured titanium substrates on the induction of mesenchymal stem cell differentiation towards the osteoblast lineage. *Biomaterials* 31, 2728–2735, 2010.
110. Martin, J.Y. *et al.* Effect of titanium surface roughness on proliferation, differentiation, and protein synthesis of human osteoblast-like cells (MG63). *J. Biomed. Mater. Res.* 29, 389–401, 1995.
111. Yang, F., Murugan, R., Wang, S. & Ramakrishna, S. Electrospinning of nano/micro scale poly(l-lactic acid) aligned fibers and their potential in neural tissue engineering. *Biomaterials* 26, 2603–2610, 2005.
112. Lee, M.R. *et al.* Direct differentiation of human embryonic stem cells into selective neurons on nanoscale ridge/groove pattern arrays. *Biomaterials* 31, 4360–4366, 2010.
113. Ramiasa, M., Ralston, J., Fetzer, R. & Sedev, R. The influence of topography on dynamic wetting. *Adv. Colloid Interface Sci.* 206, 275–293, 2014.
114. Quéré, D. Wetting and roughness. *Ann. Rev. Mater. Res.* 38, 71–99, 2008.
115. Ostrikov, K., Neyts, E. & Meyyappan, M. Plasma nanoscience: from nanosolids in plasmas to nano-plasmas in solids. *Adv. Phys.* 62, 113–224, 2013.
116. Meyyappan, M. A review of plasma enhanced chemical vapour deposition of carbon nanotubes. *J. Phys. D: Appl. Phys.* 42, 213001, 2009.
117. Mariotti, D., Švrček, V. & Kim, D.-G. Self-organized nanostructures on atmospheric microplasma exposed surfaces. *Appl. Phys. Lett.* 91, 183111, 2007.
118. Huang, X.Z. *et al.* Plasmonic Ag nanoparticles via environment-benign atmospheric microplasma electrochemistry. *Nanotechnology* 24, 095604, 2013.
119. Kumar, S., Levchenko, I., Ostrikov, K.K. & McLaughlin, J.A. Plasma-enabled, catalyst-free growth of carbon nanotubes on mechanically-written Si features with arbitrary shape. *Carbon* 50, 325–329, 2012.
120. Kumar, S. *et al.* Copper-capped carbon nanocones on silicon: plasma-enabled growth control. *ACS Appl. Mater. Interfaces* 4, 6021–6029, 2012.
121. Ostrikov, K.K., Levchenko, I., Cvelbar, U., Sunkara, M. & Mozetic, M. From nucleation to nanowires: a single-step process in reactive plasmas. *Nanoscale* 2, 2012–2027, 2010.

122. Antonini, V. et al. Combinatorial plasma polymerization approach to produce thin films for testing cell proliferation. *Colloids Surf. B. Biointerfaces* 113, 320–329, 2014.
123. Sardella, E. et al. Nano-structured cell-adhesive and cell-repulsive plasma-deposited coatings: chemical and topographical effects on keratinocyte adhesion. *Plasma Processes Polym.* 5, 540–551, 2008.
124. Goreham, R.V., Mierczynska, A., Smith, L.E., Sedev, R. & Vasilev, K. Small surface nanotopography encourages fibroblast and osteoblast cell adhesion. *RSC Adv.* 3, 10309–10317, 2013.
125. Evans, N.D. et al. Substrate stiffness affects early differentiation events in embryonic stem cells. *Eur. Cells Mater.* 18, 1–13; discussion 13–14, 2009.
126. Gray, D.S., Tien, J. & Chen, C.S. Repositioning of cells by mechanotaxis on surfaces with micropatterned Young's modulus. *J. Biomed. Mater. Res. A* 66A, 605–614, 2003.
127. Prichard, H.L., Reichert, W.M. & Klitzman, B. Adult adipose-derived stem cell attachment to biomaterials. *Biomaterials* 28, 936–946, 2007.
128. Hsiue, G.H., Lee, S.D., Chuen-Thuen Chang, P. & Kao, C.Y. Surface characterization and biological properties study of silicone rubber membrane grafted with phospholipid as biomaterial via plasma induced graft copolymerization. *J. Biomed. Mater. Res.* 42, 134–147, 1998.
129. Lee, J.N., Jiang, X., Ryan, D. & Whitesides, G.M. Compatibility of mammalian cells on surfaces of poly(dimethylsiloxane). *Langmuir* 20, 11684–11691, 2004.
130. Lanniel, M. et al. Substrate induced differentiation of human mesenchymal stem cells on hydrogels with modified surface chemistry and controlled modulus. *Soft Matter* 7, 6501–6514, 2011.
131. Ai, H. et al. Biocompatibility of layer-by-layer self-assembled nanofilm on silicone rubber for neurons. *J. Neurosci. Methods* 128, 1–8, 2003.
132. Hopp, I. et al. The influence of substrate stiffness gradients on primary human dermal fibroblasts. *Biomaterials* 34, 5070–5077, 2013.
133. Liu, Z. et al. Covalently immobilized biomolecule gradient on hydrogel surface using a gradient generating microfluidic device for a quantitative mesenchymal stem cell study. *Biomicrofluidics* 6, 24111–2411112, 2012.
134. Rivera, A. & Baskaran, H. The effect of biomolecular gradients on mesenchymal stem cell chondrogenesis under shear stress. *Micromachines* 6, 330, 2015.
135. Whittle, J.D., Barton, D., Alexander, M.R. & Short, R.D. A method for the deposition of controllable chemical gradients. *Chem. Commun.* (14), 1766–1767, 2003, DOI: 10.1039/B305445B.
136. Mierczynska, A. et al. pH-tunable gradients of wettability and surface potential. *Soft Matter* 8, 8399–8404, 2012.
137. Lee, J.H., Lee, J.W., Khang, G. & Lee, H.B. Interaction of cells on chargeable functional group gradient surfaces. *Biomaterials* 18, 351–358, 1997.
138. Goreham, R.V. et al. A substrate independent approach for generation of surface gradients. *Thin Solid Films* 528, 106–110, 2013.

139. Goreham, R.V., Short, R.D. & Vasilev, K. Method for the generation of surface-bound nanoparticle density gradients. *J. Phys. Chem. C* 115, 3429–3433, 2011.
140. Vasilev, K. et al. Creating gradients of two proteins by differential passive adsorption onto a PEG-density gradient. *Biomaterials* 31, 392–397, 2010.
141. Delalat, B. et al. Materials displaying neural growth factor gradients and applications in neural differentiation of embryoid body cells. *Adv. Funct. Mater.* 25, 2737–2744, 2015.
142. Wells, N. et al. The geometric control of E14 and R1 mouse embryonic stem cell pluripotency by plasma polymer surface chemical gradients. *Biomaterials* 30, 1066–1070, 2009.
143. Harding, F., Goreham, R., Short, R., Vasilev, K. & Voelcker, N.H. Surface bound amine functional group density influences embryonic stem cell maintenance. *Adv. Healthcare Mater.* 2, 585–590, 2013.
144. Liu, X. et al. Surface chemical gradient affects the differentiation of human adipose-derived stem cells via ERK1/2 signaling pathway. *ACS Appl. Mater. Interfaces* 7, 18473–18482, 2015.
145. Murray, P. et al. The potential of small chemical functional groups for directing the differentiation of kidney stem cells. *Biochem. Soc. Trans.* 38, 1062–1066, 2010.
146. Grayson, W.L., Zhao, F., Izadpanah, R., Bunnell, B. & Ma, T. Effects of hypoxia on human mesenchymal stem cell expansion and plasticity in 3D constructs. *J. Cell. Physiol.* 207, 331–339, 2006.
147. Burdick, J.A. & Vunjak-Novakovic, G. Engineered microenvironments for controlled stem cell differentiation. *Tissue Eng. A* 15, 205–219, 2009.
148. Ventura, C. et al. Butyric and retinoic mixed ester of hyaluronan A novel differentiating glycoconjugate affording a high throughput of cardiogenesis in embryonic stem cells. *J. Biol. Chem.* 279, 23574–23579, 2004.
149. Maguire, T. et al. Control of hepatic differentiation via cellular aggregation in an alginate microenvironment. *Biotechnol. Bioeng.* 98, 631–644, 2007.
150. Xu, C. et al. Feeder-free growth of undifferentiated human embryonic stem cells. *Nat. Biotechnol.* 19, 971–974, 2001.
151. Battista, S. et al. The effect of matrix composition of 3D constructs on embryonic stem cell differentiation. *Biomaterials* 26, 6194–6207, 2005.
152. Chang, C.F. et al. Three-dimensional collagen fiber remodeling by mesenchymal stem cells requires the integrin–matrix interaction. *J. Biomed. Mater. Res. A* 80, 466–474, 2007.
153. Ponticiello, M.S., Schinagl, R.M., Kadiyala, S. & Barry, F.P. Gelatin-based resorbable sponge as a carrier matrix for human mesenchymal stem cells in cartilage regeneration therapy. *J. Biomed. Mater. Res.* 52, 246–255, 2000.
154. Gerecht, S. et al. Hyaluronic acid hydrogel for controlled self-renewal and differentiation of human embryonic stem cells. *Proc. Nat. Acad. Sci.* 104, 11298–11303, 2007.
155. Li, W.J., Tuli, R., Huang, X., Laquerriere, P. & Tuan, R.S. Multilineage differentiation of human mesenchymal stem cells in a three-dimensional nanofibrous scaffold. *Biomaterials* 26, 5158–5166, 2005.

156. Jun Han, Z. *et al.* Carbon nanostructures for hard tissue engineering. *RSC Adv.* 3, 11058–11072, 2013.
157. Rezwan, K., Chen, Q.Z., Blaker, J.J. & Boccaccini, A.R. Biodegradable and bioactive porous polymer/inorganic composite scaffolds for bone tissue engineering. *Biomaterials* 27, 3413–3431, 2006.
158. Bergemann, C. *et al.* Continuous cellularization of calcium phosphate hybrid scaffolds induced by plasma polymer activation. *Mater. Sci. Eng. C* 59, 514–523, 2016.
159. Intranuovo, F. *et al.* Plasma processing of scaffolds for tissue engineering and regenerative medicine. *Plasma Chem. Plasma Process.* 36, 1–12, 2015.
160. Barry, J.J.A., Silva, M.M.C.G., Shakesheff, K.M., Howdle, S.M. & Alexander, M.R. Using plasma deposits to promote cell population of the porous interior of three-dimensional poly(D,L-lactic acid) tissue-engineering scaffolds. *Adv. Funct. Mater.* 15, 1134–1140, 2005.
161. Wu, S. & Chu, P.K. Biomimetic interfaces of plasma- modified titanium alloy in *Biomimetic Architectures by Plasma Processing: Fabrication and Applications*, edited by Surojit Chattopadhyay, 181–226, Pan Stanford Publishing Pte. Ltd., 2014, ISBN 978-981-4463-94-2.
162. Xiao, L., Kerry, J.G., Simon, E.M. & Gordon, G.W. Electrical stimulation promotes nerve cell differentiation on polypyrrole/poly (2-methoxy-5 aniline sulfonic acid) composites. *J. Neural Eng.* 6, 065002, 2009.
163. Kam, N.W.S., Jan, E. & Kotov, N.A. Electrical stimulation of neural stem cells mediated by humanized carbon nanotube composite made with extracellular matrix protein. *Nano Lett.* 9, 273–278, 2008.
164. Shin, S.R. *et al.* Carbon-nanotube-embedded hydrogel sheets for engineering cardiac constructs and bioactuators. *ACS Nano* 7, 2369–2380, 2013.
165. Robert, B. *et al.* Effect of oxidant on the performance of conductive polymer films prepared by vacuum vapor phase polymerization for smart window applications. *Smart Mater. Struct.* 24, 035016, 2015.
166. Fridman, G. *et al.* Applied plasma medicine. *Plasma Processes Polym.* 5, 503–533, 2008.
167. Xiong, Z. *et al.* Selective neuronal differentiation of neural stem cells induced by nanosecond microplasma agitation. *Stem Cell Res.* 12, 387–399, 2014.
168. Vasilev, K. Nanoengineered plasma polymer films for biomaterial applications. *Plasma Chem. Plasma Process.* 1–14, 2013.

9
Three-dimensional Printing Approaches for the Treatment of Critical-sized Bone Defects

Sara Salehi[1†], Bilal A. Naved[1†] and Warren L. Grayson[1,2,3]*

[1]*Translational Tissue Engineering Center, Johns Hopkins University School of Medicine, Baltimore, MD, USA*
[2]*Department of Biomedical Engineering, Johns Hopkins University School of Medicine, Baltimore, MD, USA*
[3]*Department of Material Sciences & Engineering, Johns Hopkins University School of Engineering, Baltimore, MD, USA*

Abstract

Autografts are the current gold standards for treating critical-sized bone defects. However, advances in three-dimensional printing (3DP) technologies have lead to their increased use in the treatment of bone defects. In this chapter, we outline three general categories of 3DP: laser-based, extrusion-based, and ink-based strategies, which are used in developing surgical guides for reconstructing bone, nonbiodegradable implants or porous, biodegradable scaffolds that can be used with or without stem cells to induce the regeneration of new bone, and bioprinting of stem cells. We discuss the advantages and disadvantages of the various 3DP technologies for treating critical-sized defects (CSDs). Additionally, we present a meta-analysis of published studies, categorizing the research according to the methods being used to address bone defects. This chapter provides a comprehensive overview of the advantages and disadvantages of different 3DP technologies for bone repair and regeneration.

Keywords: Three Dimensional Printing, 3DP, critical sized bone defect, CSD, SLS, SLM, EBM, bioprinting, inkjet, bone, tissue engineering, custom-designed

---

*Corresponding author: wgrayson@jhmi.edu
†These authors contributed equally to this work.

9.1 Background

9.1.1 Treatment Approaches for Critical-sized Bone Defects

It is estimated that over one million bone fractures requiring hard tissue transplantation occur each year in the USA at an economic burden of $3 billion USD per year [1]. As the population ages, it is expected that this number will increase substantially. Transplants are required for the treatment of critical-sized defects (CSDs), in which the bone never fully regenerates [2]. The gold standard for bone grafting is autologous reconstruction because of its histocompatibility and nonimmunogenicity. While autografts possess the essential components to achieve osteoinduction, osteogenesis, and osteoconduction, the harvesting procedure may result in significant donor site injury and morbidity, deformity, scarring, and the surgical risks of bleeding, inflammation, infection, and chronic pain [3, 4]. This approach may also suffer from the lack of high-quality bone available for harvesting. The second most common treatment for bone defects is the allograft; this involves transplanting donor bone tissue, often from a cadaver. In comparison to autografts, allogenic grafts are associated with risks of immune rejection and the transmission of infection. They have reduced osteoinductive properties and no cellular component because the donor grafts are processed via irradiation or freeze-drying. There is also a shortage in allograft bone material [5]. To overcome the limitations of autografts and allografts, other approaches to substitute lost bone have included the surgical implantation of alloplastic materials (such as titanium or cobalt-chromium alloys) into defect sites. However, alloplastic materials pose a risk of material failure and breakage within the patient. Regeneration aims to provide a solution to the problems that come with reconstruction and replacement by enhancing repair mechanisms to create new living tissue in defect sites. While successful regenerative approaches for healing CSDs are limited, progress in the field of tissue engineering continues to grow.

Regenerative approaches to heal CSDs involve the use of biological bone grafts, synthetic grafts, and the delivery of growth factors. However, no regenerative approach currently provides all the qualities of autologous cancellous bone. Different categories of grafting materials are available, but whether one is using a scaffold or hydrogel for bone regeneration, a ceramic or metal for bone replacement, or a mold to guide surgical reconstruction, these approaches for treating CSDs involve the use of some material (natural or synthetic) to help regenerate, replace, or rebuild the defect.

Porous bone scaffolds can be made by a variety of methods. Chemical/gas foaming [6] solvent casting, particle/salt leaching [7, 8], freeze-drying [9], thermally-induced phase separation [10], and foam–gel [11] are some that have been used extensively. However, important parameters like pore size, shape, and interconnectivity are not fully controlled in these approaches [8, 10, 11]. In addition, bone, especially in the craniomaxillofacial region, often features complex anatomical shapes that vary from patient-to-patient. Successful approaches for bone engineering would require a manufacturing process that is detailed, precise, and customizable.

9.1.2 History of the Application of 3D Printing to Medicine and Biology

In 1984, a US patent was filed by 3D Systems Corporation to sequentially add layers of photopolymerized materials in custom configurations with UV light [12]. Soon, other methods took advantage of metal sintering, material phases, chemical binders, lasers, melting, or even photon polymerization to manufacture materials by adding custom layers to each other. Figure 9.1 briefly represents important landmarks of additive manufacturing (AM) technology development and their application in tissue engineering. These processes of AM to create three-dimensional (3D) objects had tremendous potential for the medical field. In 1999, a group of engineers and medical researchers in Singapore were the first group to report the manufacture of bioresorbable scaffolds for tissue-engineering applications using fused deposition modeling (FDM) technology [13]. In the same year, cells were printed by a laser-directed technique. This group went on to found Osteopore and received Food and Drug Administration (FDA)

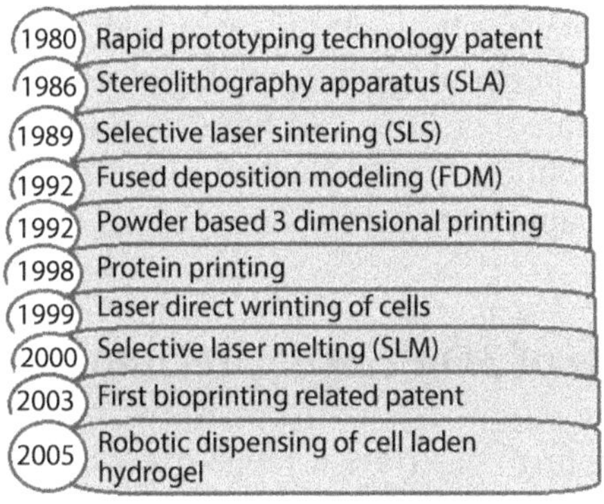

1980	Rapid prototyping technology patent
1986	Stereolithography apparatus (SLA)
1989	Selective laser sintering (SLS)
1992	Fused deposition modeling (FDM)
1992	Powder based 3 dimensional printing
1998	Protein printing
1999	Laser direct writing of cells
2000	Selective laser melting (SLM)
2003	First bioprinting related patent
2005	Robotic dispensing of cell laden hydrogel

Figure 9.1 Historical overview of AM in tissue engineering.

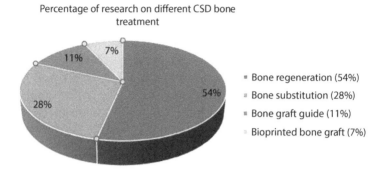

Figure 9.2 Percentage of 196 published articles related to particular aspect of treatment of critical-sized bone defects.

approval for 3D-printed bioresorbable polycaprolactone (PCL) polymer implants for neurosurgical, orthopedic, and maxillofacial surgical use. In 2000, a Michigan team demonstrated the manufacture of craniofacial hydroxyapatite (HA) bioceramic scaffolds using stereolithography (SLA/STL) technology [14]. The scaffolds were implanted into a Yucatan mini-pig mandible defect model. These approaches of printing porous scaffolds to support tissue growth have applied clinically for a 3D-printed PCL windpipe [15], mandible [16], and pelvis [17]. In 2003, a patent was filed for the use of inkjet printers to directly 3D-print cells in a 'bioink' [18]. This technology has been developed and used to created human blood vessels [19], ears [20], livers [21], bone, and skeletal muscle [22], though none of these approaches have been sufficiently developed for use in human patients to date.

Three-dimensional printing (3DP) approaches intersect with the treatment of CSDs in four primary ways: (1) development of surgical guides; (2) engineering of nondegradable implants; (3) creation of porous, biodegradable scaffolds that can be used with or without stem cells to induce the regeneration of new bone; and (4) bioprinting of stem cells with or without scaffolds to regenerate new bone. We have included a meta-analysis that describes the distribution of published studies reporting how 3DP technologies have been applied to the treatment of CSDs (Figure 9.2).

9.2 Overview of 3D Printing Technologies

The term "3D printing" describes a range of technologies that build 3D objects by adding layer upon layer of material. These materials may be plastics, metals, ceramics, or even cells encapsulated within a bioink. 3DP allows for detailed, precise, and customized builds making it highly suitable for manufacturing materials for the reconstruction, replacement, or regeneration of bone. Some of the most common considerations for using different methodologies in bone tissue engineering include the geometrical complexity that may be produced, the properties of the materials that may be used, and whether or not biomolecules can be included. While there exist numerous AM methodologies that are applicable to produce materials for the four aforementioned approaches to treating CSDs, they may be broadly categorized into three groups: laser-based, extrusion-based, or ink-based. Laser-based technologies use laser stimulation to bond material powders or fluid medium into desired layers. Extrusion-based technologies extrude molten thermoplastic materials that either cool and physically bond or are further solidified via UV stimulation to form object layers. Ink-based technologies print liquid or aerosol chemical binders that chemically bond material powders together to form layers of the desired object. We briefly review some of the commercially available AM techniques commonly used for the treatment of bone CSDs.

9.2.1 Laser-based Technologies

9.2.1.1 Stereolithography

SLA/STL features a reservoir of fluid medium capable of solidifying in response to a prescribed stimulation, i.e. UV light. The stimulation is applied as a graphic pattern at the specified working surface of the reservoir to form thin, solid, individual layers of the build material (likely a photopolymer). Each layer represents a cross section of the 3D object to be produced. Superposition of successive adjacent layers is automatically accomplished as each layer solidifies. To facilitate this, a platform bed, to which the first layer is secured, is moved away from the working surface in a programmed manner allowing fresh liquid to flow into the working surface position. This new liquid layer is then converted to solid material by the programmed UV light and immediately adheres to the preceding solid layer. The process iterates according to programmed instructions from computer-aided designs (CADs) until the entire 3D object is formed. Afterward, parts are removed, cleaned, and post-cured [12].

SLA, by nature of its methodology, can obtain complex internal features. Using surface modification strategies or common cell-seeding methods, SLA is also suitable for incorporating biomolecules like growth factors and proteins as well as cells [23]. However, because SLA involves using a laser-driven free radical polymerization reaction, it is limited to only photopolymeric materials. Materials used for bone engineering with SLA include poly(propylene) fumarate (PPF), PPF/diethyl fumarate (DEF), PPF/DEF-HA, poly(D,L) lactide (PDLLA)/HA, or beta-tricalcium phosphate (β-TCP). A limitation of SLA is that it may require support structures to prevent the built object from collapsing under hydrostatic pressure. These support structures can be difficult to remove and may interfere with desired physical and/or chemical material properties. Advantages of SLA for bone engineering include its ability to build large parts, to obtain complex internal features, to bioprint, and to obtain good accuracy. SLA has a dimensional resolution of about 0.1 mm [24].

9.2.1.2 Selective Laser Sintering

Selective laser sintering (SLS) features a reservoir of polymer powder that is spread over the surface of a bed. The fabrication chamber is maintained at a temperature just below the melting point of the powder. Laser stimulation is applied as a graphic pattern on the working surface of the bed. By slightly elevating the temperature such that the powder melts and fuses together, a thin layer is built representing a single cross section of the 3D object to be produced. Then, the part bed is programmed to move down an increment equivalent to the height of one lamina. A roller then spreads another layer of powder over the working surface and the process is iterated as layers are sequentially built upon each other according to programmed instructions from CADs [25].

While SLS uses powder instead of a fluid medium like SLA, the unbonded regions are filled with powder and support the parts of the built object that would otherwise collapse under hydrostatic pressure if made with SLA. This means that SLS has no need for support structures. Also, SLS may be used with more materials since the bonding of powder is a physical process rather than a chemical one requiring photopolymers. Materials suitable for bone engineering with SLS include PCL, Nano HA, Calcium Phosphate (CaP), poly-hydroxybutarate-cohydroxyvalerate (PHBV), carbonated hydroxyapatite (CHA), CHA/poly(L-lactic acid) (PLLA), β-TCP, and PHBV. As SLS does not require postprocessing and produces scaffolds with good mechanical properties, is considered an economically viable solution for applications in bone regeneration.

9.2.1.3 Selective Laser Melting

Selective laser melting (SLM) is similar to SLS but utilizes laser stimulation to melt metal rather than polymer powders together on a print bed to 3D print according to CAD instructions [26]. One of the superior methods for printing metallic materials used in bone engineering is SLM. SLM is able to print with materials such as alumina, ceramic, cobalt–chromium (Co–Cr) alloy, copper-based alloys, gold, molybdenum, nickel-based alloys (IN718), platinum, silver, stainless steel, and titanium (Ti6Al4V). The mechanical properties of objects made by SLM are often superior to those of objects made using other 3DP techniques and may actually be too strong for proper integration with surrounding host tissue due to stress shielding (see the three dimensionally printed implants for bone substitutions section). However, SLM is expensive and slow and may result in nonuniform distributions of heat throughout the object, which might result in warping and inconsistent mechanical properties within the build part.

9.2.1.4 Electron Beam Melting

Electron beam melting (EBM) is similar to SLM and utilizes an electron beam to melt metal powders together. EBM requires support structures to reduce stresses and to prevent warping of the heated metal [27]. Materials suitable for printing with EBM include Co–Cr, IN718, and Ti6Al4V. However, EBM is also slow and expensive with a more limited range of materials for printing.

9.2.1.5 Two-photon Polymerization

To fabricate 3D structures of small resolution, two-photon polymerization (2PP/TPP) may be used. 2PP utilizes the unique properties associated with the simultaneous absorption of two photons relative to single-photon-mediated absorption processes like STL, SLS, or SLM. This technique uses a femtosecond (fs)-pulsed laser focused on a volume of photopolymer solution. The technology works one voxel at a time with polymerization occurring at the focal point of the beam. Biocompatible polymer networks can be produced with a spatial resolution significantly below 1 μm and may have resolutions down to the 100 nm level [28]. TPP is the only printing methodology that can manufacture at the 100 nm level. It is ideal for integrating nano- and microscale features into materials. This may be particularly useful to pattern tissue engineering constructs with surface features or biomolecules to enhance bone tissue regeneration. It may utilize photopolymers or hydrogels. However, while it may print at nano- and microscales, it is not useful for printing materials of sizes relevant to CSDs. Thus, TPP is often used in conjunction with other 3DP methodologies to further improve the properties of materials for bone engineering.

9.2.2 Extrusion-based Technologies

9.2.2.1 Fused Deposition Modeling

FDM takes use of extruding molten plastic. A plastic filament that provides material is fed to an extrusion nozzle. The nozzle is heated to melt the plastic and has a mechanism which allows the flow of melted plastic to be turned on and off. The nozzle is mounted and can be moved in both horizontal and vertical directions so that as it is moved over a table it deposits plastic that cools and bonds in the desired shape. Between each layer, the distance between the nozzle and the table is increased so that the next layer may be extruded. The entire process is repeated sequentially until the desired 3D object is completed [29].

FDM requires no supports but is restricted in the materials it can use due to the need of a molten phase. Also because of the print method, it may not be suitable for printing most proteins and cells. Biomolecules would need to be functionalized onto the scaffolds after manufacturing. However, there are modified versions of FDM that use a lower temperature and thereby allow the extrusion of biomolecules. Material choices suitable for low-temperature extrusion are even more restricted. Materials used for bone engineering by FDM include TCP, TCP/propylene (PP), alumina, PCL, and TCP/PCL. Parts produced by FDM have good mechanical properties but anisotropy in the z-direction. In addition, it is important to note that objects built by FDM may be susceptible to warping while cooling if there is nonuniform airflow. FDM is highly reproducible, with a relatively moderate speed, which enables control over the major physical characteristics of the resulting scaffold, such as mechanical properties, porosity, and pore shape [13].

9.2.2.2 Material Jetting

Material jetting is similar to inkjet printing but instead of jetting an ink-based chemical binder onto a bed of powder, material jet printing extrudes liquid photopolymer onto a build tray. UV light is then used to cure the layers as they build up. Post-curing is not required [30]. Material jetting is limited to printing with photopolymers, poly(methyl methacrylate), and wax. It is mainly used in the production of molds and surgical guides due to its material restriction of printing with select wax-like substances. Material jetting features a dimensional resolution of about 0.016 mm [31–33].

9.2.3 Ink-based Technologies

9.2.3.1 Inkjet 3D Printing

Inkjet 3DP dispenses droplets of ink that bind powders together. Inkjet print heads are positioned over a bed of powder (plastic, metal, etc.). The print head dispenses ink in a programmed shape and causes the powder to bond together. Then, the distance between the print head and the stage is increased incrementally, and a fresh layer of powder is rolled onto the stage. The inkjet printer then traces another desired shape with the chemical binder and the second layer bonds to the previously printed layer. Each layer is sequentially printed by repeating the process until the desired 3D object is complete. The parts are allowed to dry within the powder bed and are then generally strengthened with cyanoacrylate or epoxy [34]. Inkjet printing has a mild process that is suitable for printing biomolecules. However, constructs built from it are often fragile and may need postprocessing to strengthen. Inkjet is particularly suited for printing hydrogels to use in bioprinting approaches to regenerate bone. Other materials utilized by inkjet printing for bone engineering include PCL, HA, bioactive glasses, mesoporous bioactives, glass/alginate composites, poly(lactic acid) (PLA), PLA/polyethylene glycol (PEG), PLA/PEG/G5 glass, and poly(hydroxymethylglycolide co-e-caprolactone) (PHMGCL). Inkjet printing is fast and cheap and can print in colors, which is useful for depicting various tissue layers when creating surgical guides. It has a print resolution of 20–25 microns and prints with inks of viscosity lower than 20 centiPoise (cP).

9.2.3.2 Aerosol Jet Printing

Aerosol jet printing is similar to inkjet printing; however, it uses an atomizer to turn the inks into a mist of droplets that are 2–5 microns in diameter. The aerosol mist is then delivered to the deposition head where it is focused through a profiled nozzle. This high-velocity stream travels from the nozzle to the substrate and may be driven via STL data to pattern the ink in 2D or 3D layers to form objects [35]. Aerosol printing is similar to inkjet printing; however, it has a higher resolution (5–10 microns) than inkjet. Also, aerosol printing may use a greater range of materials with significantly lower viscosities than inkjet printing. The viscosity requirement for aerosol inks ranges from 0.5 to 2000 cP. Aerosol printing may utilize all of the materials of inkjet printing and also allows for printing on nonsmooth surfaces due to the ability to print at varying distances. Inkjet printing must print at set distances per layer. However, because of the necessity to atomize the inks, aerosol printing may not be suitable for printing biomolecules, and since it is a relatively nascent technology, it is more expensive than other 3DP methodologies.

9.3 Surgical Guides and Models for Bone Reconstruction

Surgical simulation models aid in preparation for direct reconstruction of bone defects by reproducing the visual, auditory, and haptic aspects of the clinical procedure to reduce medical errors. All three categories of 3DP are available for use in manufacturing surgical guide materials/models.

9.3.1 Laser-based Surgical Guides

Laser-based printing methods for creating surgical guides include STL and SLS. Laser-based printing methods feature high precision, accuracy, and strength when sustaining the shear or cutting forces from drills, oscillating saws, or bone knives. They also feature the greatest range of production times. However, they require expensive equipment. SLS and STL were found to have a dimensional error of 1.79% and 0.63%, respectively when creating models of a patient's mandible for surgical reconstruction [31]. Dimensional precision is important when using 3D-printed models in the planning of complex surgeries [36–39]. The cost of SLS and STL for producing a preoperative model of a patient's mandible was reported to be $250 and $500, respectively [31]. A recent study assessed the preoperative preparation time to 3D print surgical guides for use in treating bone cancer patients who needed surgical resection. Guides printed by STL and SLS took 5.2 +/− 1.3 h and 51.7 +/− 12.9 h [37], respectively.

9.3.2 Extrusion-based Surgical Guides

Extrusion-based printing requires less expensive equipment but has low accuracy for printing larger-sized guide plates. Extrusion-based methods may utilize multiple-head deposition systems (MHDS) allowing for the simulation of various tissue types when using various material ratios. Material jetting was found to have a dimensional error of 2.14% when creating a preoperative model of a patient's mandible. The cost of material jetting for producing this model was $350 [31]. In addition, models created by material jetting do not require finishing, only a jet of pressurized water to remove support structures and to provide smoothness. The preoperative preparation time to 3D print a surgical guide via FDM was 19.3 +/− 6.5 h [37]. Of the nine surgical guides created by FDM in this study, one of them fractured before use [37].

Another clinical case study utilized extrusion-based material jetting to create a temporal bone model individualized for the patient for preoperative simulation of a complicated surgery [40]. It was felt that a "preoperative simulation" would be ideal for investigating the utility of 3D-printed surgical guides. The utility of MHDS was demonstrated in the creation of a multi-material temporal bone model using a material jetter (PolyJet, Stratasys). Materials with varying ratios of different polymers achieved unique biomechanical properties for the different anatomical structures. The 3D-printed model of the child's temporal bone is shown with landmark features identified in Figure 9.3.

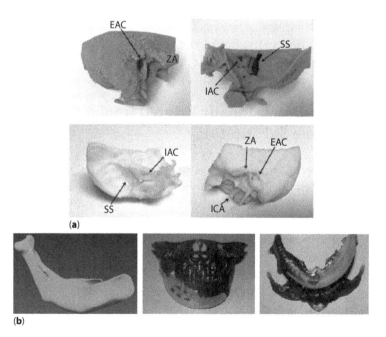

Figure 9.3 Three-dimensional printed surgical guides. Lateral and medial views of 3D temporal bone models. (a) Bone model based on clinical CT scan and micro-CT data. Adapted from [40]. (b) Three-dimensional printed surgical guide and implant used for surgical simulation and bone substitution, respectively. Adapted from [116].

9.3.3 Ink-based Surgical Guides

Inkjet printing allows for high-precision printing of various colors, which is favorable for simulating multiple materials. However, the equipment for it is expensive and the resulting material is relatively brittle [41]. Inkjet-based printing methods may also utilize MHDS allowing for the simulation of multiple tissues types when using various material ratios. Inkjet printing was found to have a dimensional error of 3.14% when creating a pre-operative model of a patient's mandible [31]. The cost of inkjet printing for producing this model was $150. The preoperative preparation time to 3D print a surgical guide via inkjet printing was 8.6 +/− 1.9 h [37]. Of the five surgical guides created by inkjet printing, two guide plates fractured before use [37].

In summary, 3D-printed surgical guides are useful tools for simulating bony anatomy, soft tissue anatomy, drilling cortical bone, drilling trabecular bone, and cadaver work. In metrics of utility, surgical guides were rated to have superior ease of use and safety with an average of 4.5 on a 5-point scale overall value to surgical simulation [37]. However, supply of autograft and allograft material is still limited resulting in the need for alternative approaches to treat CSDs. Nondegradable implants offer an alternative material to substitute native bone in cases where there is insufficient material to treat large defects with autograft or allograft reconstruction.

9.4 Three-dimensionally Printed Implants for Bone Substitution

Alloplastic materials as alternatives to autografts and allografts offer the advantages of osteoconductive bone substitutes that avoid problems of donor site morbidity or insufficient supply of donor tissue. Metallic materials make up a major portion of alloplastic bone implants. In load-bearing applications metals are favored over ceramics or polymers due to their combination of high mechanical strength and fracture toughness [42]. While the strength of titanium implants makes them attractive, it may be detrimental due to the effect of stress-shielding. Hence building lighter implants with comparable mechanical properties to surrounding bone is a critical consideration to reduce the stress-shielding effect and increase implant longevity. Additive manufacturing techniques are being exploited to manufacture porous metallic implants with reduced mechanical strength and increased bioactivity. Table 9.1 presents a number of bone implants that have been built by AM technologies.

Table 9.1 Selection of bone implant fabricated by AM techniques.

AM technologies	Material	Feature	Refs.
Extrusion-based technologies	Titanium alloy	• Completely interconnected pores • Highly permeable due to increased pore size and interconnecting pore size • CaP coating increase metal bioactivity	[49]
Ink-based technologies	β-TCP-BGH	• Layer thickness between 50 and 75 μm; high thickness makes implant brittle, small layer thickness caused direct sheering of the layers during the printing process • Homogeneous distribution of the two components is important for a reliable binding reaction and homogeneous properties of the fabricated implant	[117]
	Mg_3PO_4	• Enhanced mechanical performance by post-hardening regime • The compressive strength to values of 10 MPa for struvite and 35 MPa for newberyite • Higher solubility of the magnesium phosphates than calcium phosphate cements	[52]
	Ti/HA	• Close compressive strength to bone • Increased bioactivity • Irregular pores and no dominating crystal	[53]
Laser-based technologies SLM EBM	TiAl6V4 CoCrMo TiAl6V4	• Higher hardness and favorable corrosion behavior • Easier process control, higher part density, and lower roughness • Higher tensile strength than cast material and nominal wrought • Higher elongation and yield stress • Controlled porosity and varied Young Modulus • Influenced mechanical properties by dimension of the solid structures and pore sizes	[118] [119] [27, 44] [120]

9.4.1 Laser-based Technologies for Metallic Bone Implants

It is challenging to produce metal alloy implants with open cellular structures using conventional microcasting or sintering technologies. Laser-based 3DP technologies provide a unique manufacturing avenue to create patient-specific complex structures for any alloy precursor powders. Bone Implant fabrication requires open-cellular structures with preselected elastic moduli [43], EBM and SLM have been used to produce implants. EBM is energy efficient and fast which can decrease the cost of fabrication [44]. However, there are options to optimize the process like beam focus selection and melt scan beams, which are currently dependent on the powder properties [43]. Pore size and porous volume of around 50–400 μm and 30–40%, respectively, for porous metallic implants have been claimed as optimal parameters to achieve both mechanical properties and bone ingrowth by several researchers [45].

Porous titanium implants fabricated by EBM can have controlled porosity. Microstructure–property relationships generally determine material performance. In Figure 9.4, a Co–29Cr–6Mo alloy femoral knee implant prototype is shown; it can be varied as necessary for adequate bone ingrowth, strength, and density, due to the density influence on effective stiffness. The incorporation of porous surfaces for cell ingrowth may provide some advantages relative to surface coating techniques with respect

Figure 9.4 Co–29Cr–6Mo alloy femoral knee implant prototype fabricated by EBM. The insert illustrates the porous inner surface zone (p) composing the monolith. Adapted from [27].

to implant integration [43]. EBM can also be used to fabricate diamond-structured porous Ti–6Al–4V implants with a wide range of Young's moduli ranging from 0.2 to 12.2 GPa by controlling the pore size and the size of cellular structures. The processing parameters for EBM influence the mechanical properties of the implants. Layer-by-layer generation of the structures leads to anisotropic properties. Hence, the loading direction should be considered prior to printing [46]. The cooling rate for Ti–6Al–4V builds influences the residual dislocation density, and leads to variations in micro-indentation hardness [27]. It has been observed that higher-energy input make surface features smooth [47].

An improved EBM technology named SEBM (selective electron beam melting) has also been used to produce porous implants by selective melting of discrete powder layers by an electron beam under vacuum. This improvement makes generating a huge variety of cellular metal structures with different densities and cell morphologies feasible. Another limitation that has been addressed by this technology is working under vacuum, since atmospheric gas absorption might decrease the ductility of titanium. Heinl *et al.* produced the cellular Ti–6Al–4V structures with interconnected porosity by SEBM. It has been suggested that the implant having mechanical properties similar to those of human bone, might minimize stress-shielding effects [46]. Therefore, AM technologies have shown to be a promising way of porous implant fabrication that can lead to reduced shielding effects and weight.

9.4.2 Extrusion-based Technologies for Bone Implants

Extrusion-based techniques have also been used to produce bone implants and have the capacity to produce materials with gradients of structures. For example, in FDM approaches, each layer may have a different fiber diameter, thickness, fiber spacing, and fiber orientation. Hence parametric analyses may be used to determine how scaffolds perform as a function of their physical characteristics [48]. Li *et al.* reported the first fabrication of porous titanium implants with fully interconnected pores by FDM in 2006. Various FDM process' parameters and implant properties have been studied. Dispensing pressure, feeding rate, and initial height between nozzle and platform are able to influence implant structure. By changing the space between the fibers, scaffolds with highly uniform internal honeycomb-like structures, controllable pore morphology and complete pore interconnectivity can be obtained. *In vitro* studies of Ti6Al4V implants showed that increases in the porosity and pore size (and hence permeability) of the 3D fiber deposition Ti6Al4V implants causes a positive effect on the amount of new bone growth [49, 50].

9.4.3 Ink-based Technologies for Bone Implants

Powder printing is an attractive technology due to its rapid and inexpensive fabrication ability in medical applications. This technique is commonly performed with polymer modified gypsum or starch-based materials. Powder used in powder printing must allow the formation of thin powder layers of 100–200 μm thickness without grooves in the surface and harden with the binder solution during printing. However, these constructs suffer from low mechanical performance, and hence, they need to be infiltrated with epoxy polymer after printing and drying for reinforcement [51]. Vorndran *et al.* applied another solidification mechanism during 3DP based on a hydraulic setting reaction between the powder and the binding liquid. $Mg_3(PO_4)_2$ powder was used and resulted in high printing accuracy and sufficient mechanical strength to remove the printed samples from the powder bed. The post-hardening process increased the compressive strength [52].

Functionally graded materials enhance bioactivity of implants further and provide ideal bone substitute materials for reconstruction and restoration. Qian *et al.* 3D-printed a titanium/hydroxyapatite (Ti/HA) composite implant with gradients of pore size and particle spacing. They varied the pore size for cell growth by changing the layer thicknesses. In addition, reducing interparticle spacing between adjacent Ti/HA particles improved its mechanical properties. Functionally graded implants fabricated by 3DP is currently an intense area of new research and parameters such as layer thickness, powder scale and sintering temperature need to be optimized for bone implants [53].

9.5 Scaffolds for Bone Regeneration

Bone tissue engineering aims to regenerate bone to restore structure and function in CSDs. It is shown in Figure 9.2 that scaffold-based approaches constitute the largest portion of ongoing research for treating CSDs. Scaffolds are temporary support structures to assist cells in forming tissues of the proper architecture. Cell survival, colonization, migration, proliferation, and differentiation are critically dependent on the scaffold's internal architectural design (pore shape, size, and interconnectivity), physical and material composition, and physical properties (mechanical strength and degradation rate) [54, 55]. Successful scaffolds for bone tissue engineering must also promote bone growth through osteoinduction and/or osteoconduction, resorb in concert with bone growth, and prevent soft tissue generation at the bone/scaffold interface. Besides necessary physical and mechanical properties, scaffolds should also be available to surgeons on short notice and adaptable to irregular wound sites [56].

9.5.1 Laser-based Printing for Regenerative Scaffolds

Scaffolds should have bioactive properties allowing them to attract and induce cells to form bone. Inorganic materials like HA and β-TCP are present in natural bone and have osteoinductive properties [57, 58]. They interact favorably with bone cells and promote direct integration between implant and bone and are often incorporated into extruded materials to improve bioactivity. Several groups have successfully demonstrated the printing of bioactive minerals or extracellular matrix via STL [59, 60]. However, SLS-printed HA/PCL composite scaffolds were shown to have reduced surface hydrophobicity, which promotes cell adhesion. STL was shown to produce scaffolds made of bioglass that exhibited high biaxial bending strength in complex shapes. However, STL demonstrates limitations when creating internal features. When used to produce HA scaffolds, the resulting constructs had low porosity and the polymer solvent residuals remained within potentially interfering with cellular processes for regeneration.

SLS-printed PLG/HA scaffolds demonstrated decreased layer thickness and reduced interlayer bond strength. Laser-based printing methods often create scaffolds of limited mechanical strength and require postprocessing techniques like sintering for strengthening (see section titled *9.5.4.2 Postprocessing: Sintering*). However, sintering exposes the scaffold to temperatures of up to 1300 °C limiting the ability to functionalize the scaffold with growth factors or cells unless additional postprocessing steps are taken (see section titled *9.5.4.3 Postprocessing: Functionalization*). Another strategy to improve strength is to change the composition of the material. When SLS was used to produce a PCL/HA composite the resulting scaffold exhibited higher compressive strength. A CHAp/PLLA composite printed via SLS demonstrated higher compressive strength than both polymer counterparts.

The use of SLS to print PCL scaffolds with stiffnesses of ~15–300 MPa has been reported [61]. These values fall within the reported range for human trabecular bone. However, there was a lack of interconnected pores. Thirteen months post-implantation, the patient's scaffold was exposed and had failed to generate tissue. Histological analysis indicated a prevalence of connective tissue and little bone regeneration, suggesting that the lack of interconnected pores was a critical limitation.

9.5.2 Extrusion-based Printing for Regenerative Scaffolds

Extrusion-based printing methods are suitable for incorporating bioactive components to bone scaffolds. FDM was used to incorporate TCP or bone extracellular matrix particles into PCL. The bioactive composites were shown to have enhanced osteoinductive properties [59, 62]. While inorganic materials may be extruded to create bioactive scaffolds, the ceramic material is brittle. To solve this, composite materials may be extruded. FDM produced scaffolds made of a PCL/TCP composite were demonstrated to have increased stiffness, a compressive strength close to bone, a fully interconnected network of pores, and osteoconductive properties.

Shim et al. used an FDM printer with multiple print heads (an MHDS) to compare the effectiveness of PCL/PLGA polymeric scaffolds to PCL/PLGA/TCP composite scaffolds for regenerating a CSD in the calvarium of the rabbit [63, 64]. They report that when β-TCP was added via the MHDS a significant amount of new bone formation was observed. Using 3D-printed PCL/PLGA scaffolds, 11.85 +/− 5.61% ($p < 0.05$) of the defect was filled with bone after 8 weeks. However, the 3D-printed PCL/PLGA/TCP composite scaffold regenerated 31.93 +/− 10.80% ($p < 0.05\%$) of the defect after eight weeks (Figure 9.5) [65].

Extrusion-based methods provide precision of both gross and internal architectures. Gross architecture is important to attaining the exact anatomical shapes necessary for clinical implementation. Temple et al. illustrate the capabilities of 3DP to produce anatomically shaped, vascularized bone grafts using PCL Figure 9.6 [66]. They first optimized scaffold porosity for uniform cell seeding of adipose-derived stem cells (ASCs) and also determined the ideal melt temperature to ensure sufficient accuracy and speed of printing. After determining that a porosity of 60% would allow most uniform cell seeding and a melt temperature of 80°C would ensure optimal accuracy and speed of printing, they implanted the cell-laden scaffolds into a mouse model and demonstrated the formation of vasculature. Finally, the 3DP technology was applied to clinically relevant geometries like the mandible and maxilla. The manufactured scaffolds successfully replicated features as complex as the temporomandibular joint condyle and the maxillary opening to the nasal cavity while maintaining the optimized level of porosity for cell seeding [66].

Extrusion techniques like FDM are able to also carefully control internal architecture [67]. They have precise control over pores size [68]. The precise control of extrusion-based methods over pore size is further demonstrated by the creation of scaffolds with a gradient of pore sizes that resulted in improved osteogenesis [69] in comparison to scaffolds with homogenous pore sizes.

As a case study, cylindrical scaffolds of medical-grade PCL incorporating 20% of TCP microparticles were fabricated by FDM creating a fully interconnected scaffold with 70% porosity and 22.2 MPa of elastic modulus. When these scaffolds were functionalized with recombinant human BMP 7 (rhBMP-7) and implanted in 10-cm CSDs in a sheep model. Defect

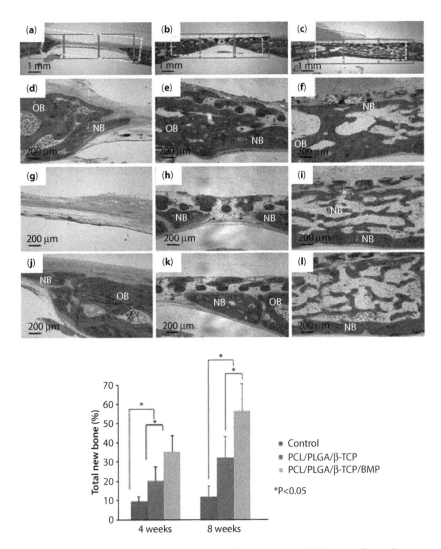

Figure 9.5 *In vivo* study of PCL/PLGA/β-TCP. Total new bone (%) at 4 and 8 weeks post-implantation in each study group [65]. Histologic section at 8 weeks after surgery. (a, d, g, j) Uncovered control group. (b, e, h, k) PCL/PLGA/β-TCP group. New bone formation was observed beneath membranes. (c, f, i, l) PCL/PLGA/β-TCP/BMP group. New bone formed continuously at the inferior borders of defects. Left, center, and right red boxes in (a, b, c) correspond to (d, e, f), (g, h, i), and (j, k, l), respectively. Yellow dotted lines indicate the placement of the 8 mm defect.

bridging was observed within 3 months, and after 12 months, significantly greater bone formation and superior mechanical strength were observed for the 3D-printed scaffolds relative to the gold standard bone autologous graft [70].

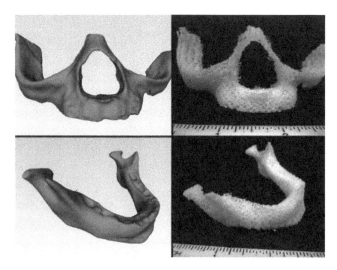

Figure 9.6 Three-dimensional printed anatomically shaped, porous scaffolds. Adapted from [68].

9.5.3 Ink-based Printing for Regenerative Scaffolds

Inkjet printing may also be used to print bioactive inorganics like HA and β-TCP. The printing technique was used to create mineralized structures within calcium phosphate scaffolds to house cells [71]. However, ink-based methods must carefully choose the binder to prevent pH-related damage to the cells. Laser- or extrusion-based techniques that require heat for printing and are not suitable for growth factor printing. However, the incorporation of bioactive growth factors like bone morphogenetic protein (BMP) is an investigated option for inkjet printing since its process occurs at room temperature. However, ink-based approaches are less precise in their pore creation.

The internal architecture of a scaffold consists of its pore geometry, pore shape, and interconnectivity. Proper internal architecture is beneficial for promoting cell adhesion, mechanical integration between host and scaffold via bone ingrowth, and the transport of waste and nutrients [55, 72, 73]. Studies on the effect of designed pore architecture versus a more random architecture resulting from salt-leaching methods found that, despite similar porosity, pore size, and surface area, the designed architecture resulted in higher pore interconnectivity and therefore more uniform cell distribution within the scaffold [74]. Pore size has been demonstrated to affect osteoblast proliferation and migration through collagen–glycosaminoglycan scaffolds. Larger pores of around 300 μm diameter resulted in higher cell numbers throughout the scaffold [75]. It was found that a threshold pore size of ~100 μm in hydrogels [76] was required for vascularization, a critical component of tissue survival and integration. Tarafder *et al.* reports a 3D-printed TCP scaffold with controlled internal architecture and interconnected pores [77]. Porosity obtained was between 42% and 63% with pore sizes between 400 and 500 μm.

While suitable porosities have been achieved with inkjet printing it is at the compromise of mechanical strength. Pore size, percent porosity, and pore shape influence the mechanical properties of bioceramic scaffolds [78–81]. One strategy to improve mechanical strength of inkjet-printed scaffolds is to use multiple print heads to create composites. Inzana *et al.* used inkjet printing to develop composite Ca-P/collagen scaffolds [71]. Type I collagen is a critical component of bone extracellular matrix and plays important roles in bone's strength and toughness. Collagen incorporation into Ca–P cements has been shown to improve cellular attachment, viability, proliferation, activity, and mechanical properties [82–84]. Research groups are developing composite scaffolds of many different types for bone tissue engineering to build the structure with optimized parameters. Table 9.2 presents a selection of different types of scaffolds and indicates the variety of materials and technologies being implemented to fabricate bone scaffold.

9.5.4 Pre- and Post-processing Techniques

Mechanical strength is a scaffold property that is closely related to other properties. As the porosity increases, there is less scaffold material to provide strength to withstand *in vivo* stresses in sites of clinical application. The addition of bioactive components like ECM can also decrease mechanical strength [59, 60, 85]. While selection of printing material holds the greatest significance in determining mechanical strength, there are various pre- and postprocessing techniques that may be used to enhance AM methods of scaffold design.

9.5.4.1 Pre-processing

Proper mechanical strength remains a hurdle despite the ability to strengthen scaffolds with sintering. Three-dimensional printing approaches may further increase the mechanical strength of bioceramic scaffolds by adding trace minerals to the calcium phosphate-based printing material. These dopants are mineral oxides that are 3DP with the calcium-phosphate materials and remain embedded in the manufactured bone scaffolds. The mineral phase of natural bone contains many trace elements such as Na^+, Mg^{2+},

Table 9.2 Selection of bone scaffold fabricated by AM techniques.

AM technologies		Material	Features	Ref.
Extrusion Based Technologies	Direct	Gelatin	Suspends cells in a gel at low temperature; Control over the mechanical properties, porosity, and pore shape	[112]
		Hydrogel	Closely resemble the natural ECM, very low stiffness	[113]
		PCL	Low melting temperature, glass transition temperature, and high thermal stability; Simple to print; highly elastic, reproducible bioresorbable scaffold; fully interconnected pore network	[121]
		PLGA	High glass transition temperature and challenging with a higher extrusion temperature; Hydrophilic and better cell adhesion and proliferation; Brittle, less ductile and higher mechanical strength comparing to PCL	[122]
		CPC	Highly bioactive, no swelling and no crack formation	[123]
		TCP, HA	Osteoinductive, brittle; lower rates of degradation than hydrogels	[124]
		PP/TCP	Decreased tensile strength with a high tensile modulus; Increased stiffness, close compressive strength to bone	[125]
		PCL/TCP	Fully interconnected network of pores and osteoconductive	[126]
		PLA/nHA	Diminished mechanical properties; supported vessel cells growth	[127]
		PCL/TiO$_2$	Controlled degradation and improved bioactivity with superior hardness	[128]
		PCL/BG	Suitable for nonload-bearing applications; PCL/BG is more elastic than PCL/TiO$_2$	
		Hydrogel/Si filler	Promoted viscosity and stiffness; higher expression of BMPs genes	[129]
		PCL/BAG fillers	Increased viscosity, improved compressive modules	[130]
			Improved Osteoinduction and porosity	[65]
		PCL/PLGA/TCP	Increased mechanical strength(compressive strength), highly elastic	[131]

THREE-DIMENSIONAL PRINTING APPROACHES 255

Extrusion Based Technologies		Ce-MCS/PCL	Higher cerium content upregulated biological activities, minor decrease in compressive strength but still in the range of human trabecular bone	[132]
		PHBHHx–MBG	Hierarchical porosity and high compressive strength; Roughened surface Well-ordered pore arrangement, large surface area; Improved bioactivity due to lower contact angle Higher viscosity higher precision and resolution;	[133]
		β-TCP/alginate	Comparable resistance to compression to that of native trabecular bone, and higher Young's modulus	[134]
		PCL/PLGA	Better remodeling process; balancing glass transition temperature; Proper compressive strength and biocompatible; Enhanced compressive modulus	[135]
		PDA/PLA	More uniform degree of hydrophilicity; Enhanced cell attachment and proliferation	[136]
		PLA/PCL	Enhanced hydrophilicity, improved compressive strength without significant difference in porosity; Higher osteogenic gene expression	
		nHA/PEGDA-PEG/ TGF-β1	Increased collagen synthesis due to n HA and TGF-β1 linear correlation between hMSC adhesion and nHA concentration; Incorporating near physiological concentrations of nHA improved cell adhesion; Interconnected large pores	
		L-PLA(PLLA)	High void fraction and improved cell attachment; Few interconnected pores; exhibited no gradients	[137]
		PCL	High Young's modulus and variable, depending on temperature and humidity Resilient under influence of scaffold architecture	[138]
		PLGA	Efficient cell seeding due to architectural designs; Attainable large pore without resolution compensation, improved cell attachment; Favorable to make complex shaped bone scaffold (M Lee et al.)	[139]

(Continued)

Table 9.2 Cont.

AM technologies	Material	Features	Ref.
Ink Based Technologies	TCP	Higher compressive strength comparing to conventional techniques	[140]
		Sintering cause transformation from β- to α-TCP; β-TCP has higher compressive strength than HA; Brittle and slow degradation rate; Highly osteoconductive	[141]
	HA	Low compressive strength; Proper interconnecting channels; Better cell attachment due to inclined layer	[142]
	BCP	High porosity and low compressive strength; rising TCP content increases compressive strength, decrease in modulus of toughness with increasing Ca/P ratio	[143]
	Bioglass Si	Low fracture toughness and strength; Having both microporosity in the struts and a designed macroporosity; increased porosity and decreased mechanical strength	[144]
		Lacks osteoconductivity due to its rapid degradation; Improved mechanical strength and osteogenesis using PVA as binder	[145]
	Bioceramic-HSP	Flexibility in design and printing method; Controllable mechanical properties due to adjustable porosity; Higher compressive strength and modulus comparing to conventionally made bioceramic; fast degradation rate	
	CaP/Collagen I	Improved strength and cellular viability	[71]
		Collagen fiber formation may not occur by powder-based inkjet printing; Reduced viscosity, surface tension and reliable thermal inkjet printing	
	TCP/PCL	Maximum flexural strength and cell viability was improved	[79]
		Rough surface and no change in compressive strength; Delayed degradation and interconnected pores	
	SrO-TCP	High compressive strength by increasing SrO and MgO	[58]

Laser Based Technologies		MgO-TCP	High osteoid, bone and haversian canal formation induced by the presence of SrO and MgO dopants	[146]
		TCP/alginate	Enhanced osteogenesis and vasculogenesis; Significant increase in mechanical strength; reliable printing with continuous micro-porous ceramic matrix	[117]
		CaP/BG	Increased the four-point bending strength; No effect on cement reaction by glassy phase	[147]
		Fe/Mn	High porosity Lower mechanical properties to natural bone and suitable for nonload-bearing bones; increased corrosion current density and corrosion rate	[148] [149]
	SLS	PCL	Larger as-produced pore comparing to FDM or SLA Thermally stable; high strut surface roughness and porosity	[150]
		Bioglass	Minor discrepancy between computational design and fabricated scaffold due to excess material in scaffold Unidirectional pore channels; At a low laser power, necking between particles and at a high laser power, holes and sinking of the powder layer	[151] [152]
	STL	HA/nSiO2 nHA/PCL	Greater mechanical strength due to heat treatment of bioceramics Predesigned, well-ordered macropores and interconnected micropores; Improved surface hydrophilicity; Lower porosity and higher compressive strength; harder sintering	
		PEEK/HA	Good integrity obtained by keeping HA composition less than 40% Laser being able to sinter a high melting point polymer in a much lower temperature environment	[153]
		PVA/HA	Easy sample handling due to PVA particles; Good network of necking	[154]

(Continued)

Table 9.2 Cont.

AM technologies	Material	Features	Ref.
	PLG/HA	Decreased layer thickness; Reduced interlayer bond strength and z resolution; Relatively uniform porosity	[155]
	nCaP/PHBV	Intact structure and good handling stability after sintering; Fully interconnected porous structure; Higher compressive strength and modulus than polymer counterpart; Stronger under compression than CaP, PHBV and not less stiff	[156]
	nHA/PLGA/TGFβ	Excellent corroboration between the predesigned porous scaffold and resultant 3D-printed structure with excellent horizontal and vertical microchannel formation	[157]
	Bioglass® 45S5	Discernable high resolution lines for 40% and 60% in-fill density scaffolds. High biaxial bending strength; high resolution, suitable for complex shaped Low porosity and pore size gradient; presence of polymer solvent residual	[158]
	PLGA,PCL,PLLA Chitosan, alginate	Highest resolution and fast fabrication; no significant effect on mechanical properties of polymeric scaffold; Minimized structural defect due to high pH in the mold removal process	[159]

Zn^{2+}, Si^{4+}, Sr^{2+} [57, 58]. Trace amounts of strontium enhance bone regeneration when incorporated into synthetic bone grafts [86–88]. Tarafder et al. reported that the existence of strontium and magnesium in TCP scaffolds enhances their compressive strength, *in vitro* degradation kinetics, osteoblast interactions, and *in vivo* bone formation [58, 89, 90]. Tarafder et al. investigated the effect of multiscale porosity and SrO and MgO dopants in 3DP TCP scaffolds for *in vivo* bone regeneration in a rabbit femoral defect model. The SrO and MgO dopants resulted in smaller grain size, increased density, decreased pore size, higher compressive strength (that was further improved by microwave sintering), increased bone total formation, and increased vascularization [90].

9.5.4.2 Post-processing: Sintering

Sintering uses heat and pressure to compact the scaffold material, thereby strengthening it. Conventional sintering dissipates heat from the outside to the inside of the material requiring longer sintering times and undesired grain growth [91]. However, microwave sintering causes materials to absorb microwave energy allowing heat to be generated within the scaffold. Microwave sintering allows for improved heating uniformity, shorter sintering time, controlled grain growth, higher densification, and improved mechanical properties [91–93]. Sintering represents one of many 3DP postprocessing techniques that may be used to tailor particular properties of the manufactured scaffolds for the clinical application. When used on bioceramics like TCP for bone tissue engineering, sintering has been shown to increase porosity and interconnectivity (due to the creation of micropores), increase mechanical strength, and increase cell density when implanted into rat femoral defects [77]. Tarafder et al. have reported a significant increase in the mechanical strength of macroporous TCP scaffolds via microwave sintering compared with conventional heat or pressure sintering.

9.5.4.3 Post-processing: Functionalization

When the 3D-printed composite scaffolds were post-processed via immersion in a collagen-encapsulated BMP-2 solution and subsequently functionalized, they were found to generate even more bone. BMPs are a sub-class of the transforming growth factor β family [94] and induce bone formation by triggering the angiogenesis, migration, proliferation, and differentiation of mesenchymal stem cells [95, 96]. BMP-2 and BMP-7 are FDA approved and used to treat craniomaxillofacial and periodontal bone defects [97, 98]. The sustained release of rhBMP-2 from the composite scaffold was observed for up to 28 days and after 8 weeks improved regeneration in the defect to 55.93 +/− 14.04% (p < 0.05%) of the total defect volume [65].

9.6 Bioprinting

Bioprinting involves the precise deposition of biomaterials such as cells or proteins that are encapsulated in hydrogels or bioinks. Sequential layering in predetermined shapes allows for the creation of 3D tissue constructs. Bioprinters output biomaterials at determined positions in the xyz plane by the CADs input by users (often in .STL file format). The hydrogels and bioinks that encapsulate the cells or proteins are meant to support and protect the cells during printing [99]. Laser-based, extrusion-based, and ink-based bioprinters all exist. Laser-assisted bioprinting utilizes laser stimulation to vaporize portions of hydrogel solution and release droplets that fall on a platform below. It has a resolution of 10 μm [100]. Extrusion bioprinting is similar to inkjet printing but uses pneumatic systems, pumps, or screws to extrude hydrogels of cells [100]. Inkjet bioprinting does not require much change from traditional inkjet printing due to its inherently mild conditions. While bioprinters are superior in coordinating spatial location of cells and biomolecules, they suffer from limited materials and mechanical properties. Bioprinters almost exclusively use hydrogels and as a result will often lack the mechanical strength required to bear physiological loads.

Bioprinting approaches have been used to create viable 2D and layered 3D cell-laden constructs using inkjet printers [101, 102]. A major hurdle in the bone tissue engineering is creating a vascular network to allow for transport of nutrients and wastes. Simple diffusion is not enough as there is a limit of 100–200 mm that nutrients and waste will diffuse through. Bioprinters have been shown able to print tubular structures with endothelial cells [103]. Hydrogels provide a support matrix and a hydrated microenvironment to the cells that enhance nutrient and waste diffusion. Work is also being done to design hydrogel systems that can present biochemical and physical stimuli to guide cellular processes like migration, proliferation, and differentiation [104]. The ability to manipulate cellular processes will be helpful in engineering complex features like a vascular network. Table 9.3 provides a summary of bioprinted structures and their features.

Polymer hydrogels are ideal candidates for the development of printable materials for tissue engineering. Hydrogels present remarkable tunability of rheological, mechanical, chemical, and biological properties; high biocompatibility; and similarity to native ECM [105].

Table 9.3 Selection of bone scaffold fabricated by bioprinting.

AM technologies	Material	Cell	Feature	Ref.
INKJET-BASED TECHNOLOGY	PEGDMA/nBG	hMSC	• Highest cell viability due to interaction between hMSC with HA	[160]
	PEGDMA/nHA	hMSC	• Increased compressive modulus of PEGDMA/nHA to PEGDMA/nBG	
	PEG-GelMA	hMSC	• nHA is more effective than BG for osteogenesis • Even distribution of hMSC and high biocompatibility surviving the printing process due to simultaneous assembly of Cells and scaffold in a layer-by layer approach without extra steps for cell encapsulation • PEG improves GelMA application in inkjet bioprinting and mechanical properties of cell laden scaffold improves (modulus of 1–2 MPa) • GelMA promoted early differentiation of hMSC	[114]
EXTRUSION-BASED TECHNOLOGY	PCL/TCP-composite hydrogel(Gelatin, Fibrinogen, HA, Glycerol)	Human AFSCs	• High cell viability, and cells were protected from physical stress during printing • PCL co-printing with hydrogel due to biocompatibility, flexibility and low melting temperature • Cell damage from heat transfer because of PCL and microchannel distance gap between PCL and hydrogel	[22]
	PCL	hMSCs	• Increased porosity results in decreased scaffold stiffness and compressive modulus • Increasing amplitude between filaments improve cell accessibility	[161]

(Continued)

Table 9.3 Cont.

AM technologies	Material	Cell	Feature	Ref.
	MG/CBD-BMP2-collagen microfibers	BMSC hASC	• Even pore size distribution • Mimicking the complexity of the native microenvironment by microextrusion bioprinter	[162]
	nHA/sodium alginate/gelatin	hASC	• HA could promote hASCs 3D constructs to osteogenic differentiation • Well-interconnected porous structure • Poor strength of the materials	[163]

Three-dimensional bioprinting techniques allow for the deposition of cells and bioactive hydrogel matrices into spatially organized layered constructs that have controlled architecture [106]. The approach is to create a construct that mimics native tissue with respect to its anatomical geometry, spatial organization of cells, and the microenvironment of the cells [107]. The idea is that by incorporating various cell types at predetermined locations, specifically osteoprogenitors and endothelial progenitors, vascular beds may be integrated with surrounding bone extracellular matrix to engineer native bone.

Laser bioprinting is less common than inkjet and extrusion bioprinting, but its applications have been increasingly used for tissue regeneration. Laser-assisted methods have been reported to enable printing of various materials with a broad range of viscosities, thus overcoming a common limitation of both extrusion and inkjet printers. Similarly, it is believed that the laser pulse used for the ejection transfer of the bioink has negligible effects in post-printing cell viability with several types of cells [108].

Inkjet 3D printers have been extensively used to print living cells and tissue-engineering constructs. Although most inkjet printers are compatible with high cell viability [109], the shear stresses caused by their extrusion through small orifices of the print head can be a limiting factor. Concerns relative to the material viscosity associated with frequent clogging have also been reported [110]. Therefore, cells should be homogeneously distributed in low-viscosity inks or in the form of liquid slurries to prevent variations in the printing quality [109]. Three-dimensional printing of polymers and hydrogels generally relies on the use of materials with controlled viscosity, which then defines the range of printability of the ink. Polymer inks, which are typically printed in the prepolymer phase, need to be viscous enough to allow for structural support of subsequent printed layers while being fluid enough to prevent nozzle clogging. To address the challenges of developing printable viscous inks, alginate hydrogels have been cross-linked with calcium ions immediately before the ink leaves the printing head or just after [111].

To demonstrate the ink-based technique, an inkjet bioplotter system was used to simultaneously deposit cells and hydrogels into 3D constructs [112]. The deposited cells maintained viability and the ability to differentiate. Another study [113] combined controlled growth factor release with 3D inkjet bioprinting to produce hydrogel constructs with temporal control of biomolecular activity. Their study demonstrated that microparticles could be used to control the delivery of BMP-2 microparticles in bioprinted alginate constructs that also contained mesenchymal stem cells (MSCs). The MSCs maintained their ability to differentiate into osteoblasts *in vitro* and *in vivo*. The release of the BMP-2 from the microparticles had two phases. The first phase featured a burst release of about 30% of the growth factor due to the diffusion from the outer regions of the particle [113]. The second phase featured sustained release due to the enzymatic degradation of the microparticles by collagenases secreted by cells. The BMP-2 was shown to enhance differentiation of the MSCs.

In addition to spatiotemporal control of cell types and growth factors, a successful bioprinting approach would have to manufacture scaffolds that mimic the properties of tissue-specific extracellular matrix [114]. Gao et al. bioprinted PEG-GelMA hybrid scaffolds with a compressive modulus 10 times greater than those reported by other groups. They were able to produce scaffolds with a modulus of 1–2 MPa [115], which, nevertheless, remains much lower than the modulus of bone (10–30 GPa).

Although bioprinted hydrogels present a range of advantages for tissue engineering, such as the ability of exposing cells to highly hydrated 3D microenvironments that closely resemble the natural ECM [105], they generally present very low stiffness (in the kPa range) compared with the majority of load-bearing tissues that constitute the craniofacial region (in the GPa range). Their success for regeneration of load-bearing tissues, which are primarily constituted of proteins and minerals with relatively lower cell density, is yet to be demonstrated. Recent studies demonstrated a 3D bioprinter that can fabricate human-scale tissue constructs of anatomical shapes. Mechanical stability was achieved by printing cell-laden hydrogels with integrated biodegradable polymers. Microchannels were incorporated into the constructs to improve diffusion of nutrients and wastes. With this approach, it was possible to fabricate both the mandible and calvarial bones [22].

9.7 Conclusion

Three-dimensional printing technologies have the potential to significantly improve treatment of critical-sized bone defects. The various approaches include (1) developing surgical guides for reconstructing bone; (2) engineering nondegradable implants; (3) creating porous, biodegradable scaffolds to regenerate new bone; or (4) bioprinting new tissues with stem cells. Three general classes of 3DP are employed: laser-based, extrusion-based, and ink-based strategies. Each type of 3DP presents unique advantages and disadvantages for utilization in different approaches for treating CSDs. Some of the characteristics to be considered when deciding upon the 3DP technology to select are cost, speed, and geometrical complexity that may be produced, properties of the materials that may be used, and whether or not biomolecules can be included. A universal solution for treating all CSDs does not exist, and the advantages or disadvantages of different 3DP technologies depend on the particular treatment approach.

List of Abbreviation

2D	Two dimensional
3D	Three dimensional
3DP	Three-dimensional printing
AM	Additive manufacturing
BMP	Bone morphogenetic protein
β-TCP	Beta-tricalcium phosphate
CAD	Computer-aided design
CaP	Calcium phosphate
CHA	Carbonated hydroxyapatite
Co–Cr	Cobalt–chromium
CSD	Critical-sized defect
CT	Computed tomography
DEF	Diethyl fumarate
EBM	Electron beam melting
ECM	Extracellular matrices
FDA	Food and Drug Administration
FDM	Fused deposition modeling
GPa	Giga Pascal
HA	Hydroxyapatite
MgO	Magnesium oxide
PDLLA	Poly(D,L) lactide
PCL	Poly(caprolactone)
PDMS	Poly(dimethylsiloxane)
PEG	Poly(ethylene glycol)
PEGDMA	Poly(ethylene glycol) dimethacrylate
PHBV	Poly hydroxybutarate-cohydrxyvalerate
PHMGCL	Poly (hydroxymethylglycolide co-e-caprolactone)
PGA	Poly(glycolic acid)
PLA	Poly lactic acid
PLGA	Poly(lactic-co-glycolic) acid
PLLA	Poly lactic acid
PPF	Poly(propylene fumarate)
SLA/STL	Stereolithography
SLM	Selective laser melting
SrO	Strontium oxide
Ti_6Al_4V	Titanium alloy
TPP/2PP	Two-photon polymerization
USD	US dollar
UV	Ultraviolet

References

1. B. M. Desai, "Osteobiologics," *Am. J. Orthop. (Belle Mead. NJ).*, vol. 36, no. 4 Suppl, pp. 8–11.
2. E. Georgios, "Ruiz José Luis Calvo Guirado, and Romanos," *Crit. size defects bone Regen. Exp. Rabbit calvariae Syst. Rev. Qual. Eval. using ARRIVE Guidel. oral Implant. Res.*, 2014.
3. E. Younger and M. Chapman, "Morbidity at bone graft donor site," *J Orthop Trauma*, vol. 3, pp. 192–95, 1989.
4. J. J. Tiedeman, K. L. Garvin, T. A. Kile, and J. F. Connolly, "The role of a composite demineralized bone matrix and bone marrow in the treatment of osseous defects," *Orthopedics*, vol. 18, no. 12, pp. 1153–1158, 1995.
5. A. R. Amini, C. T. Laurencin, and S. P. Nukavarapu, "Bone Tissue Engineering: Recent Advances and Challenges," *Crit. Rev. Biomed. Eng.*, vol. 40, no. 5, pp. 363–408, 2012.
6. M. Kucharska, *Mater. Lett.*, vol. 85, no. 124–127, 2012.
7. M. Stoppato, E. Carletti, V. Sidarovich, A. Quattrone, R. E. Unger, C. J. Kirkpatrick, and A. Motta, Influence of scaffold pore size on collagen I development: a new *in vitro* evaluation perspective. *Journal of Bioactive and Compatible Polymers*, 28(1), 16–32, 2013.
8. H. Cao and N. Kuboyama, "A biodegradable porous composite scaffold of PGA/beta-TCP for bone tissue engineering," *Bone*, vol. 46, no. 2, pp. 386–395.
9. N. Sultana and M. Wang, "Fabrication of HA/PHBV composite scaffolds through the emulsion freezing/freeze-drying process and characterisation of the scaffolds," *J. Mater. Sci. Mater. Med.*, vol. 19, no. 7, pp. 2555–2561.
10. Z. I. N. K. W. T. S. H. T. K. C. Hutmacher DW Schantz T, "Mechanical properties and cell cultural response of polycalrolactone scaffolds designed and fabricated via fused deposition modelling," *J. Biomedial Mater. Res.*, vol. 55, no. September 2015, pp. 203–216, 2001.
11. H. Yoshikawa, N. Tamai, T. Murase, and A. Myoui, "Interconnected porous hydroxyapatite ceramics for bone tissue engineering," *J. R. Soc. Interface*, vol. 6 Suppl 3, pp. S341–8.
12. W. Hull, Charles, "Apparatus for production of three-dimensional objects by stereolithography." U.S. Patent No. 4,575,330. 11 Mar. 1986.
13. D. W. Hutmacher, "Scaffolds in tissue engineering bone and cartilage," *Biomaterials*, vol. 21, no. 24, pp. 2529–2543, Dec. 2000.
14. S. J. Hollister, R. a Levy, T. M. Chu, J. W. Halloran, and S. E. Feinberg, "An image-based approach for designing and manufacturing craniofacial scaffolds," *Int. J. Oral Maxillofac. Surg.*, vol. 29, pp. 67–71, 2000.
15. D. A. Zopf, S. J. Hollister, M. E. Nelson, R. G. Ohye, and G. E. Green, "Bioresorbable airway splint created with a three-dimensional printer," *N. Engl. J. Med.*, vol. 368, no. 21, pp. 2043–2045, May 2013.
16. I. Gibson, D. W. Rosen, and B. Stucker, *Additive Manufacturing Technologies*. Boston, MA: Springer US, 2010.

17. C. Zeng, J. Xiao, Z. Wu, and W. Huang, "Evaluation of three-dimensional printing for internal fixation of unstable pelvic fracture from minimal invasive para-rectus abdominis approach: A preliminary report," *Int. J. Clin. Exp. Med.*, vol. 8, no. 8, pp. 13039–13044, 2015.
18. Boland, Thomas, William Crisp Wilson Jr, and Tao Xu. "Ink-jet printing of viable cells." U.S. Patent No. 7,051,654. 30 May 2006.
19. H. W. Kang, S. J. Lee, I. K. Ko, C. Kengla, J. J. Yoo, and A. Atala, A 3D bioprinting system to produce human-scale tissue constructs with structural integrity. *Nature Biotechnology*, 34(3), 312–319, 2016.
20. J.S. Lee, J. M. Hong, J. W. Jung, J.H. Shim, J.H. Oh, and D.W. Cho, "3D printing of composite tissue with complex shape applied to ear regeneration," *Biofabrication*, vol. 6, no. 2, p. 024103, 2014.
21. Ma, X., Qu, X., Zhu, W., Li, Y.S., Yuan, S., Zhang, H., Liu, J., Wang, P., Lai, C.S.E., Zanella, F. and Feng, G.S. Deterministically patterned biomimetic human iPSC-derived hepatic model via rapid 3D bioprinting. *Proceedings of the National Academy of Sciences*, 113(8), pp. 2206–2211, 2016.
22. H.W. Kang, S. J. Lee, I. K. Ko, C. Kengla, J. J. Yoo, and A. Atala, "A 3D bioprinting system to produce human-scale tissue constructs with structural integrity," *Nat. Biotechnol.*, no. October 2015, vol.34, no.3, pp.312-319, 2016.
23. P. Lan, *J. Mater. Sci.Mater. Med*, vol. 20, pp. 271–279, 2009.
24. A. Cohen, A. Laviv, P. Berman, R. Nashef, and J. Abu-Tair, "Mandibular reconstruction using stereolithographic 3-dimensional printing modeling technology," *Oral Surgery, Oral Med. Oral Pathol. Oral Radiol. Endodontology*, vol. 108, no. 5, pp. 661–666, 2009.
25. L. Lü, J. Y. H. Fuh, and Y. S. Wong, Selective Laser Sintering. *Laser-Induced Materials and Processes for Rapid Prototyping* pp. 89–142. Springer US, 2001.
26. S. Bremen, W. Meiners, and A. Diatlov, Selective laser melting. *Laser Technik Journal*, 9(2), 33–38, 2012.
27. L. Murr, S. Gaytan, D. Ramirez, E. Martinez, J. Hernandez, K. Amato, P. Shindo, F. Medina, and R. Wicker, "Metal fabrication by additive manufacturing using laser and electron beam melting technologies," *J. Mater. Sci. Technol.*, vol. 28, no. 1, pp. 1–14, 2012.
28. B. H. Cumpston, J. E. Ehrlich, S. M. Kuebler, M. Lipson, S. R. Marder, D. McCord-Maughon and M. C. Rumi, Three-dimensional microfabrication using two-photon polymerization. *Micromachining and Microfabrication* pp. 168–168. International Society for Optics and Photonics, 1998.
29. L. Wood, Rapid automated prototyping. Industrial Press, Inc, 1993.
30. M. Hofmann, "3D printing gets a boost and opportunities with polymer materials," *ACS Macro Lett.*, vol. 3, no. 4, pp. 382–386, 2014.
31. D. Ibrahim, T. L. Broilo, C. Heitz, M. G. de Oliveira, H. W. de Oliveira, S. M. W. Nobre, J. H. G. dos Santos Filho, and D. N. Silva, "Dimensional error of selective laser sintering, three-dimensional printing and PolyJet™ models in the reproduction of mandibular anatomy," *J. Cranio-Maxillofacial Surg.*, vol. 37, no. 3, pp. 167–173, 2009.

32. J. Faber, P. M. Berto, and M. Quaresma, "Rapid prototyping as a tool for diagnosis and treatment planning for maxillary canine impaction," *Am. J. Orthod. Dentofac. Orthop.*, vol. 129, no. 4, pp. 583–589, 2006.
33. E. K. Sannomiya, J. V. L. Silva, A. A. Brito, D. M. Saez, F. Angelieri, and G. da Silva Dalben, "Surgical planning for resection of an ameloblastoma and reconstruction of the mandible using a selective laser sintering 3D biomodel," *Oral Surgery, Oral Med. Oral Pathol. Oral Radiol. Endodontology*, vol. 106, no. 1, pp. 36–40, 2008.
34. Calvert, Paul. "Inkjet printing for materials and devices.," *Chem. Mater. 13*, vol. 10, no. 2001 SRC - GoogleScholar FG - 0, pp. 3299–3305.
35. H. Liu and T. J. Webster, "Enhanced biological and mechanical properties of well-dispersed nanophase ceramics in polymer composites: From 2D to 3D printed structures," *Mater. Sci. Eng. C*, vol. 31, no. 2, pp. 77–89, 2011.
36. J. S. Bill, J. F. Reuther, W. Dittmann, N. Kübler, J. L. Meier, H. Pistner, and G. Wittenberg, "Stereolithography in oral and maxillofacial operation planning," *Int. J. Oral Maxillofac. Surg.*, vol. 24, no. 1 PART 2, pp. 98–103, 1995.
37. P. S. D'Urso, R. L. Atkinson, I. J. Bruce, D. J. Effeney, M. W. Lanigan, W. J. Earwaker, A. Holmes, T. M. Barker, and R. G. Thompson, "Stereolithographic (SL) biomodelling in craniofacial surgery," *Br. J. Plast. Surg.*, vol. 51, no. 7, pp. 522–530, 1998.
38. C. Santler, H. Karcher, and C. Ruda, "Indications and limitations of three-dimensional models in cranio-maxillofacial surgery," *J. Cranio-Maxillo-Facial Surg.*, vol. 26, no. 1, pp. 11–16, 1998.
39. D. N. Silva, M. Gerhardt de Oliveira, E. Meurer, M. I. Meurer, J. V. Lopes da Silva, and A. Santa-Bárbara, "Dimensional error in selective laser sintering and 3D-printing of models for craniomaxillary anatomy reconstruction," *J. Cranio-Maxillofacial Surg.*, vol. 36, no. 8, pp. 443–449, 2008.
40. A. S. Rose, J. S. Kimbell, C. E. Webster, O. L. Harrysson, E. J. Formeister, and C. A. Buchman, Multi-material 3D models for temporal bone surgical simulation. *Annals of Otology, Rhinology & Laryngology*, 124(7), 528–536, 2015.
41. J. Fu, Z. Guo, Z. Wang, X. Li, H. Fan, J. Li, Y. Pei, G. Pei, and D. Li, "Use of four kinds of three-dimensional printing guide plate in bone tumor resection and reconstruction operation," *Chinese J. Reparative Reconstr. Surg.*, vol. 28, no. 3, pp. 304–308, 2014.
42. D. A. Puleo and W. W. Huh, "Acute toxicity of metal ions in cultures of osteogenic cells derived from bone marrow stromal cells," *J. Appl. Biomater.*, vol. 6, no. 2, pp. 109–116, 1995.
43. L. E. Murr, S. M. Gaytan, D. A. Ramirez, E. Martinez, J. Hernandez, K. N. Amato, P. W. Shindo, F. R. Medina, and R. B. Wicker, "Metal fabrication by additive manufacturing using laser and electron beam melting technologies," *J. Mater. Sci. Technol.*, vol. 28, no. 1, pp. 1–14, 2012.
44. J. Parthasarathy, B. Starly, S. Raman, and A. Christensen, "Mechanical evaluation of porous titanium (Ti6Al4V) structures with electron beam melting (EBM)," *J. Mech. Behav. Biomed. Mater.*, vol. 3, no. 3, pp. 249–259, 2010.

45. T. Albrektsson, P.-I. Brånemark, H.-A. Hansson, and J. Lindström, "Osseointegrated titanium implants: requirements for ensuring a long-lasting, direct bone-to-implant anchorage in man," *Acta Orthop. Scand.*, vol. 52, no. 2, pp. 155–170, 1981.
46. P. Heinl, L. Müller, C. Körner, R. F. Singer, and F. A. Müller, "Cellular Ti-6Al-4V structures with interconnected macro porosity for bone implants fabricated by selective electron beam melting," *Acta Biomater.*, vol. 4, no. 5, pp. 1536–1544, 2008.
47. N. W. Hrabe, P. Heinl, B. Flinn, C. Körner, and R. K. Bordia, "Compression-compression fatigue of selective electron beam melted cellular titanium (Ti-6Al-4V)," *J. Biomed. Mater. Res. – Part B Appl. Biomater.*, vol. 99 B, no. 2, pp. 313–320, 2011.
48. J. P. Li, S. H. Li, C. a Van Blitterswijk, and K. de Groot, "A novel porous Ti6Al4V: characterization and cell attachment," *J. Biomed. Mater. Res. A*, vol. 73, no. 2, pp. 223–33, 2005.
49. J. P. Li, P. Habibovic, M. van den Doel, C. E. Wilson, J. R. de Wijn, C. A. van Blitterswijk, and K. de Groot, "Bone ingrowth in porous titanium implants produced by 3D fiber deposition," *Biomaterials*, vol. 28, no. 18, pp. 2810–2820, 2007.
50. J. P. Li, J. R. De Wijn, C. A. Van Blitterswijk, and K. De Groot, "Porous Ti6Al4V scaffold directly fabricating by rapid prototyping: Preparation and *in vitro* experiment," *Biomaterials*, vol. 27, no. 8, pp. 1223–1235, 2006.
51. A. Pfister, U. Walz, A. Laib, and R. Mülhaupt, "Polymer ionomers for rapid prototyping and rapid manufacturing by means of 3D printing," *Macromol. Mater. Eng.*, vol. 290, no. 2, pp. 99–113, 2005.
52. E. Vorndran, K. Wunder, C. Moseke, I. Biermann, F. A. Müller, K. Zorn, and U. Gbureck, "Hydraulic setting Mg 3 (PO 4) 2 powders for 3D printing technology," *Adv. Appl. Ceram.*, vol. 110, no. 8, pp. 476–481, 2011.
53. C. Qian, F. Zhang, and J. Sun, "Fabrication of Ti/HA composite and functionally graded implant by three-dimensional printing," *Biomed. Mater. Eng.*, vol. 25, no. 2, pp. 127–136, 2015.
54. A. G. Mikos, Y. Bao, L. G. Cima, D. E. Ingber, J. P. Vacanti, and R. Langer, "Preparation of poly(glycolic acid) bonded fiber structures for cell attachment and transplantation," *J. Biomed. Mater. Res.*, vol. 27, no. 2. pp. 183–189, 1992.
55. S. Yang, K.F. Leong, Z. Du, and C.K. Chua, "The design of scaffolds for use in tissue engineering. Part II. Rapid prototyping techniques," *Tissue Eng.*, vol. 8, no. 1, pp. 1–11, 2002.
56. K. J. Burg, S. Porter, and J. F. Kellam, "Biomaterial developments for bone tissue engineering," *Biomaterials*, vol. 21, no. 23, pp. 2347–2359, 2000.
57. A. Bandyopadhyay, S. Bernard, and W. Xue, "Calcium phosphate-based resorbable ceramics: influence of MgO, ZnO and SiO2 dopants," *Ceram Soc*, vol. 89 SRC-, pp. 2675–2688, 2006.
58. S. S. Banerjee, S. Tarafder, N. M. Davies, A. Bandyopadhyay, and S. Bose, "Understanding the influence of MgO and SrO binary doping on the

mechanical and biological properties of beta-TCP ceramics," *Acta Biomater.*, vol. 6, no. 10, pp. 4167–4174.

59. B. P. Hung, B. A. Naved, E. L. Nyberg, M. Dias, C. A. Holmes, J. H. Elisseeff, A. H. Dorafshar, and W. L. Grayson, "Three-dimensional printing of bone extracellular matrix for craniofacial regeneration," *ACS Biomater. Sci. Eng.*, p. acsbiomaterials.6b00101, May 2016. http://pubs.acs.org/doi/abs/10.1021/acsbiomaterials.6b00101

60. S. Lohfeld, S. Cahill, V. Barron, L. McHugh, L. Durselen, L. Kreja, C. Bausewein, and A. Ignatius, "Fabrication, mechanical and *in vivo* performance of polycaprolacton/tricalcium phosphate composite scaffolds," *Acta Biomater.*, no. 8, pp. 3446–3456, 2012.

61. S. Eshraghi and S. Das, "Mechanical and microstructural properties of polycaprolactone scaffolds with one-dimensional, two-dimensional, and three-dimensional orthogonally oriented porous architectures produced by selective laser sintering," *Acta Biomater.*, vol. 6, no. 7, pp. 2467–2476, 2010.

62. A. Sawyer, S. Song, E. Susanto, P. Chuan, C. Lam, M. Woodruff, D. Hutmacher, and S. Cool, "The stimulation of healing within a rat calvarial defect by mPCL-TCP/collagen scaffolds loaded with rhBMP-2," *Biomaterials*, no. 30, pp. 2479–2488, 2009.

63. J. Shim, "Fabrication of blended polycaprolacton/poly(lactic-co-glycolic acid)/B-tricalcium phosphate thin membrane using solid freeform fabrication technology for guided bone regeneration," *Tissue Eng. Part A*, vol. 19, pp. 317–28, 2013.

64. W. McKay, S. Peckham, and J. Badura, "A comprehensive clinical review of recombinant human bone morphogenetic protein-2," *Int. Orthop.*, vol. 31, pp. 729–34, 2007.

65. J.H. Shim, M.C. Yoon, C.M. Jeong, J. Jang, S.I. Jeong, D.W. Cho, and J.B. Huh, "Efficacy of rhBMP-2 loaded PCL/PLGA/β-TCP guided bone regeneration membrane fabricated by 3D printing technology for reconstruction of calvaria defects in rabbit," *Biomed. Mater.*, vol. 9, no. 6, p. 065006, 2014.

66. J. P. Temple, D. L. Hutton, B. P. Hung, P. Y. Huri, C. A. Cook, R. Kondragunta, X. Jia, and W. L. Grayson, "Engineering anatomically shaped vascularized bone grafts with hASCs and 3D-printed PCL scaffolds," *J. Biomed. Mater. Res. – Part A*, vol. 102, no. 12, pp. 4317–4325, 2014.

67. S. J. Hollister, "Porous scaffold design for tissue engineering," *Nat. Mater.*, vol. 4, no. 7, pp. 518–524, Jul. 2005.

68. J. P. Temple, D. L. Hutton, B. P. Hung, P. Y. Huri, C. A. Cook, R. Kondragunta, X. Jia, and W. L. Grayson, "Engineering anatomically shaped vascularized bone grafts with hASCs and 3D-printed PCL scaffolds," *J. Biomed. Mater. Res. – Part A*, vol. 102, no. 12, pp. 4317–4325, Dec. 2014.

69. A. Di Luca, B. Ostrowska, I. Lorenzo-Moldero, A. Lepedda, W. Swieszkowski, C. Van Blitterswijk, and L. Moroni, "Gradients in pore size enhance the osteogenic differentiation of human mesenchymal stromal cells in three-dimensional scaffolds," *Sci. Rep*, no. 6, p. 22898, 2016.

70. A. Cipitria, J. Reichert, D. Epari, S. Saifzadeh, A. Berner, H. Schell, M. Metha, M. Schuetz, G. Duda, and D. Hutmacher, "Polycaprolactone scaffold and reduced rhBMP-7 dose for the regeneration of critical-sized defects in sheep tibiae," *Biomaterials*, vol. 34, no. 38, pp. 9960–68, 2013.
71. J. Inzana, D. Olvera, S. Fuller, J. Kelly, O. Graeve, E. Schwarz, S. Kates, and H. Awad, "3D printing of composite calcium phosphate and collagen scaffolds for bone regeneration," *Biomaterials*, vol. 35, no. 13, pp. 4026–4034, 2014.
72. H. R. R. Ramay and M. Zhang, "Biphasic calcium phosphate nanocomposite porous scaffolds for load-bearing bone tissue engineering," *Biomaterials*, vol. 25, no. 21, pp. 5171–5180.
73. L. G. Sicchieri, G. E. Crippa, P. T. de Oliveira, M. M. Beloti, and A. L. Rosa, "Pore size regulates cell and tissue interactions with PLGA-CaP scaffolds used for bone engineering," *J. Tissue Eng. Regen. Med.*, vol. 6, no. 2, pp. 155–162.
74. F. P. W. Melchels, J. Feijen, and D. W. Grijpma, "A review on stereolithography and its applications in biomedical engineering," *Biomaterials*, vol. 31, no. 24, pp. 6121–6130, 2010.
75. C. Murphy, M. Haugh, and F. O'Brien, "The effect of mean pore size on cell attachment, proliferation and migration in collagen-glycosaminoglycan scaffolds for bone tissue engineering," *Biomaterials*, no. 31, pp. 461–66, 2010.
76. C. Chiu, M. Cheng, H. Engel, S. Kao, J. Larson, S. Gupta, and E. Brey, "The role of pore size on vascularization and tissue remodeling in PEG hydrogels," *Biomaterials*, no. 32, pp. 6045–51, 2011.
77. S. Tarafder, V. K. Balla, N. M. Davies, A. Bandyopadhyay, and S. Bose, "Microwave-sintered 3D printed tricalcium phosphate scaffolds for bone tissue engineering," *J. Tissue Eng. Regen. Med.*, vol. 7, no. 8, pp. 631–641.
78. K. D. Groot, "Effect of porosity and physicochemical properties on the stability, resorption, and strength of calcium phosphate ceramics," *Ann Sci*, vol. 523, pp. 227–233, 1988.
79. S. Tarafder and S. Bose, "Polycaprolactone-coated 3D printed tricalcium phosphate scaffolds for bone tissue engineering: *in vitro* alendronate release behavior and local delivery effect on *in vivo* osteogenesis Solaiman Tarafder and Susmita bose *," *ACS applied materials & interfaces*, vol. 6, no. 13, pp. 9955–9965. 2014.
80. S. Bose, J. Darsell, M. Kintner, and C. Eng, "Pore size and pore volume effects on alumina and TCP ceramic scaffolds," *Mater. Sci.*, vol. 23 SRC, pp. 479–486, 2003.
81. A. Hattiangadi and A. Bandyopadhyay, "Strength degradation of nonrandom porous ceramic structures under uniaxial compressive loading.," *Ceram Soc*, vol. 83 SRC, pp. 2730–2736, 2004.
82. F. Tamimi, B. Kumarasami, C. Doillon, U. Gbureck, D. Le Nihouannen, E. L. Cabarcos, and J. E. Barralet, "Brushite-collagen composites for bone regeneration," *Acta Biomater.*, vol. 4, no. 5, pp. 1315–1321.

83. J. Moreau, M. Weir, and H. Xu, "Self-setting collagen-calcium phosphate bone cement: mechanical and cellular properties," *J Biomed Mater Res*, vol. 91, no. 2, pp. 605–613, 2009.
84. R. Perez and M. Ginebra, "Injectable collagen/alpha-tricalcium phosphate cement: collagen-mineral phase interactoins and cell response," *J. Mater. Sci. Mater. Med.*, vol. 24, no. 2, pp. 381–93, 2013.
85. S. Wang, C. Wong Po Foo, A. Warrier, M. Poo, S. Heilshorn, and X. Zhang, "Gradient lithography of engineered proteins to fabricate 2D and 3D cell culture microenvironments," *Biomed. Microdevices*, no. 11, pp. 1127–34, 2009.
86. N. Pors, "The biological role of strontium," *Bone*, vol. 35, pp. 583–588, 2004.
87. N. Lakhkar, I. Lee, H. Kim, V. Salih, I. Wall, and J. Knowles, "Bone formation controlled by biologically relevant inorganic ions: role and controlled delivery from phosphate-based glasses," *Adv. Drug Deliv. Rev.*, vol. 65, pp. 405–20, 2013.
88. S. Bose, K.-W. Han, M.-J. Lee, and H. Kim, "Intestinal protective effects of herbal-based formulations in rats against neomycin insult," *Evid. Based. Complement. Alternat. Med.*, vol. 2013, p. 161278, 2013.
89. S. Bose, S. Tarafder, S. S. Banerjee, N. M. Davies, and A. Bandyopadhyay, "Understanding *in vivo* response and mechanical property variation in MgO, SrO and SiO$_2$ doped β-TCP," *Bone*, vol. 48, no. 6, pp. 1282–1290.
90. S. Tarafder, W. S. Dernell, A. Bandyopadhyay, and S. Bose, "SrO- and MgO-doped microwave sintered 3D printed tricalcium phosphate scaffolds: Mechanical properties and *in vivo* osteogenesis in a rabbit model," *J. Biomed. Mater. Res. – Part B Appl. Biomater.*, vol. 103, no. 3, pp. 679–690, 2015.
91. P. Yadoji, R. Peelamedu, D. Agrawal, and B. Eng, "Microwave sintering of Ni-Zn ferrites: comparison with conventional sintering," *Mater Sci*, vol. 98 SRC, pp. 269–278, 2003.
92. A. Chanda, S. Dasgupta, S. Bose, and C. Eng, "Microwave sintering of calcium phosphate ceramics," *Mater. Sci.*, vol. 29 SRC, pp. 1144–1149, 2009.
93. S. Bose, S. Dasgupta, S. Tarafder, and A. Bandyopadhyay, "Microwave-processed nanocrystalline hydroxyapatite: simultaneous enhancement of mechanical and biological properties," *Acta Biomater.*, vol. 6, no. 9, pp. 3782–3790.
94. M. R. Urist, "Bone: formation by autoinduction," *Science*, vol. 150, no. 3698, pp. 893–899.
95. P. C. Bessa, M. Casal, and R. L. Reis, "Bone morphogenetic proteins in tissue engineering: the road from the laboratory to the clinic, part I (basic concepts)," *J. Tissue Eng. Regen. Med.*, vol. 2, no. 1, pp. 1–13.
96. P. C. Bessa, M. Casal, and R. L. Reis, "Bone morphogenetic proteins in tissue engineering: the road from laboratory to clinic, part II (BMP delivery)," *J. Tissue Eng. Regen. Med.*, vol. 2, no. 2–3, pp. 81–96, 2008.

97. A. White, A. Vaccaro, J. Hall, P. Whang, B. Friel, and M. McKee, "P, R, A, G, C and applications of BMP-7/OP-1 in fractures, nonunions and spinal fusion Int," *Orthop*, pp. 31735–41 SRC, 2007.
98. M. J. Danesh-Meyer, "Tissue engineering in periodontics and implantology using rhBMP-2," *Ann. R. Australas. Coll. Dent. Surg.*, vol. 15, pp. 144–149.
99. S. Murphy and A. Atala, "3D bioprinting of tissues and organs," *Nat. Biotechnol.*, vol. 32, no. 8, pp. 773–785, 2014.
100. F. Obregon, C. Vaquette, S. Ivanovski, D. Hutmacher, and L. Bertassoni, "Three-dimensional bioprinting for regenerative dentistry and craniofacial tissue engineering," *Journal of Dental Research*, 94, pp. 1–10, 2015.
101. E. H. Groeneveld and E. H. Burger, "Bone morphogenetic proteins in human bone regeneration," *Eur. J. Endocrinol.*, vol. 142, no. 1, pp. 9–21, 2000.
102. M. Yamamoto, Y. Takahashi, and Y. Tabata, "Controlled release by biodegradable hydrogels enhances the ectopic bone formation of bone morphogenetic protein," *Biomaterials*, vol. 24, no. 24, pp. 4375–4383, 2003.
103. M. T. Poldervaart, H. Gremmels, K. Van Deventer, J. O. Fledderus, F. C. Oner, M. C. Verhaar, W. J. A. Dhert, and J. Alblas, "Prolonged presence of VEGF promotes vascularization in 3D bioprinted scaffolds with defined architecture," *J. Control. Release*, vol. 184, no. 1, pp. 58–66, 2014.
104. J. L. Drury and D. J. Mooney, "Hydrogels for tissue engineering: scaffold design variables and applications," *Biomaterials*, vol. 24, no. 24, pp. 4337–4351, 2003.
105. N. Annabi, A. Tamayol, J. Uquillas, M. Akbari, L. Bertassoni, C. Cha, G. Camci-Unal, M. Dokmeci, N. Peppas, and A. Khademhosseini, "Anniversary article: rational design and applications of hydrogels in regenerative medicine," *Adv Matr.*, vol. 26, no. 1, pp. 85–123, 2014.
106. V. Mironov, T. Boland, T. Trusk, G. Forgacs, and R. Markwald, "Organ printing: computer-aided jet-based 3D tissue engineering," *Trends Biotechnol.*, vol. 21, p. 157, 2003.
107. W. C. Wilson and T. Boland, "Jr. and Cell and organ printing protein and cell printers," *Anat. Rec. Mol. Cell Evol. Biol.* 272, 491, vol. 1 SRC - G, 2003.
108. L. Koch, S. Kuhn, H. Sorg, M. Gruene, and S. Schlie, "Laser printing of skin cells and human stem cells," *Tissue Eng. Part C Methods*, vol. 16, no. 5, pp. 847–854, 2010.
109. T. Xu, J. Jin, C. Gregory, J. Hickmann, and T. Boland, "Inkjet printing of viable mammalian cells," *Biomaterials*, vol. 26, no. 1, pp. 93–99, 2005.
110. P. Bajaj, R. Schweller, A. Khademhosseini, J. West, and R. Bashir, "3D biofabrication strategies for tissue engineering and regenerative medicine," *Annu. Rev. Biomed. Eng.*, no. 16, pp. 247–276, 2014.
111. S. Bakarich, R. I. Gorkin, M. Panhuis, and G. Spinks, "Three-dimensional printing fiber reinforced hydrogel composites," *ACS Appl. Mater. Interfaces*, vol. 6, no. 18, pp. 15998–16006, 2014.
112. N. E. Fedorovich, J. Alblas, J. R. de Wijn, W. E. Hennink, A. J. Verbout, and W. J. A. Dhert, "Hydrogels as extracellular matrices for skeletal tissue

engineering: state-of-the-art and novel application in organ printing," *Tissue Eng.*, vol. 13, no. 8, pp. 1905–1925.

113. M. T. Poldervaart, H. Wang, J. van der Stok, H. Weinans, S. C. Leeuwenburgh, F. C. Öner, ... and J. Alblas, Sustained release of BMP-2 in bioprinted alginate for osteogenicity in mice and rats. PLoS One, 8(8), e72610, 2013.

114. G. Gao, A. F. Schilling, K. Hubbell, T. Yonezawa, D. Truong, Y. Hong, G. Dai, and X. Cui, "Improved properties of bone and cartilage tissue from 3D inkjet-bioprinted human mesenchymal stem cells by simultaneous deposition and photo cross-linking in PEG-GelMA," *Biotechnol. Lett.*, vol. 37, no. 11, pp. 2349–2355, 2015.

115. X. Cui, K. Breitenkamp, M. Lotz, and D. D. Lima, "Synergistic action of fibroblast growth factor-2 and transforming growth factor-beta1 enhances bioprinted human neocartilage formation," *Biotechnol. Bioeng.*, vol. 109, no. 9, pp. 2357–2368, 2012.

116. S. Singare, Q. Lian, W. P. Wang, J. Wang, Y. Liu, D. Li, and B. Lu, "Rapid prototyping assisted surgery planning and custom implant design," *Rapid Prototyp. J.*, vol. 15, no. April 2016, pp. 19–23, 2009.

117. C. Bergmann, M. Lindner, W. Zhang, K. Koczur, A. Kirsten, R. Telle, and H. Fischer, "3D printing of bone substitute implants using calcium phosphate and bioactive glasses," *J. Eur. Ceram. Soc.*, vol. 30, no. 12, pp. 2563–2567, 2010.

118. B. Vandenbroucke and J.P. Kruth, "Selective laser melting of biocompatible metals for rapid manufacturing of medical parts," *Rapid Prototyp. J.*, vol. 13, no. 4, pp. 196–203, 2007.

119. L. E. Murr, S. A. Quinones, S. M. Gaytan, M. I. Lopez, A. Rodela, E. Y. Martinez, D. H. Hernandez, E. Martinez, F. Medina, and R. B. Wicker, "Microstructure and mechanical behavior of Ti-6Al-4V produced by rapid-layer manufacturing, for biomedical applications," *J. Mech. Behav. Biomed. Mater.*, vol. 2, no. 1, pp. 20–32, 2009.

120. L. E. Murr, E. V. Esquivel, S. A. Quinones, S. M. Gaytan, M. I. Lopez, E. Y. Martinez, S. W. Stafford, "Microstructures and mechanical properties of electron beam-rapid manufactured Ti–6Al–4V biomedical prototypes compared to wrought Ti–6Al–4V", *Materials characterization*, vol. 60, no. 2, pp. 96–105, 2009.

121. T. Patrício, M. Domingos, A. Gloria, and P. Bártolo, "Characterisation of PCL and PCL/PLA scaffolds for tissue engineering," *Procedia CIRP*, vol. 5, 2012, pp. 110–114, 2013.

122. S. H. Park, D. S. Park, J. W. Shin, Y. G. Kang, H. K. Kim, T. R. Yoon, J. W. Shin, ''Scaffolds for bone tissue engineering fabricated from two different materials by the rapid prototyping technique: PCL versus PLGA," *J. Mater. Sci. Mater. Med.*, vol. 23, no.11, pp. 2671–2678, 2012

123. A. R. Akkineni, Y. Luo, M. Schumacher, B. Nies, A. Lode, and M. Gelinsky, "3D plotting of growth factor loaded calcium phosphate cement scaffolds," *Acta Biomater.*, vol. 27, pp. 264–274, 2015.

124. J. M. Sobral, S. G. Caridade, R. A. Sousa, J. F. Mano, and R. L. Reis, "Three-dimensional plotted scaffolds with controlled pore size gradients: Effect of scaffold geometry on mechanical performance and cell seeding efficiency," *Acta Biomater.*, vol. 7, no. 3, pp. 1009–1018, 2011.
125. S. J. Kalita, S. Bose, H. L. Hosick, and A. Bandyopadhyay, "Development of controlled porosity polymer-ceramic composite scaffolds via fused deposition modeling," *Mater. Sci. Eng. C*, vol. 23, no. 5, pp. 611–620, 2003.
126. B. Rai, S. H. Teoh, K. H. Ho, D. W. Hutmacher, T. Cao, F. Chen, and K. Yacob, "The effect of rhBMP-2 on canine osteoblasts seeded onto 3D bioactive polycaprolactone scaffolds," *Biomaterials*, vol. 25, no. 24, pp. 5499–5506, 2004.
127. B. Holmes, K. Bulusu, M. Plesniak, and L. G. Zhang, "A synergistic approach to the design, fabrication and evaluation of 3D printed micro and nano featured scaffolds for vascularized bone tissue repair," *Nanotechnology*, vol. 27, no. 6, p. 064001, 2016.
128. A. a Mäkitie, J. Korpela, L. Elomaa, M. Reivonen, A. Kokkari, M. Malin, H. Korhonen, X. Wang, J. Salo, E. Sihvo, M. Salmi, J. Partanen, K.S. Paloheimo, J. Tuomi, T. Närhi, and J. Seppälä, "Novel additive manufactured scaffolds for tissue engineered trachea research," *Acta Otolaryngol.*, vol. 133, no. 4, pp. 412–7, 2013.
129. J. R. Xavier, T. Thakur, P. Desai, M. K. Jaiswal, N. Sears, E. Cosgriff-Hernandez, R. Kaunas, and A. K. Gaharwar, "Bioactive nanoengineered hydrogels for bone tissue engineering: A growth-factor-free approach," *ACS Nano*, vol. 9, no. 3, pp. 3109–3118, 2015.
130. J. Korpela, A. Kokkari, H. Korhonen, M. Malin, T. Narhi, and J. Seppalea, "Biodegradable and bioactive porous scaffold structures prepared using fused deposition modeling," *J. Biomed. Mater. Res. - Part B Appl. Biomater.*, vol. 101, no. 4, pp. 610–619, 2013.
131. M. Zhu, J. Zhang, S. Zhao, and Y. Zhu, "Three-dimensional printing of cerium-incorporated mesoporous calcium-silicate scaffolds for bone repair," *J. Mater. Sci.*, vol. 51, no. 2, pp. 836–844, 2016.
132. S. Yang, J. Wang, L. Tang, H. Ao, H. Tan, T. Tang, and C. Liu, "Mesoporous bioactive glass doped-poly (3-hydroxybutyrate-co-3-hydroxyhexanoate) composite scaffolds with 3-dimensionally hierarchical pore networks for bone regeneration," *Colloids Surfaces B Biointerfaces*, vol. 116, pp. 72–80, 2014.
133. G. S. Diogo, V. M. Gaspar, I. R. Serra, R. Fradique, and I. J. Correia, "Manufacture of β-TCP/alginate scaffolds through a Fab@home model for application in bone tissue engineering," *Biofabrication*, vol. 6, no. 2, p. 025001, 2014.
134. J. Y. Kim and D. W. Cho, "Blended PCL/PLGA scaffold fabrication using multi-head deposition system," *Microelectron. Eng.*, vol. 86, no. 4–6, pp. 1447–1450, 2009.

135. C. T. Kao, C. C. Lin, Y. W. Chen, C. H. Yeh, H. Y. Fang, and M. Y. Shie, "Poly(dopamine) coating of 3D printed poly(lactic acid) scaffolds for bone tissue engineering," *Mater. Sci. Eng. C*, vol. 56, pp. 165–173, 2015.
136. Y. Zhang, L. Xia, D. Zhai, M. Shi, Y. Luo, C. Feng, B. Fang, J. Yin, J. Chang, and C. Wu, "Mesoporous bioactive glass nanolayer-functionalized 3D-printed scaffolds for accelerating osteogenesis and angiogenesis," *Nanoscale*, vol. 7, pp. 19207–19221, 2015.
137. J. Zeltinger, D. Ph, J. K. Sherwood, D. Ph, D. a Graham, R. Müeller, D. Ph, L. G. Griffith, and D. Ph, "Adhesion, proliferation, and matrix deposition," *Tissue Eng.*, vol. 7, no. 5, pp. 557–572, 2001.
138. A.V. Do, B. Khorsand, S. M. Geary, and A. K. Salem, "3D Printing of scaffolds for tissue regeneration applications," *Adv. Healthc. Mater.*, vol. 4, no. 12, 2015.
139. M. Lee, J. C. Y. Dunn, and B. M. Wu, "Scaffold fabrication by indirect three-dimensional printing," *Biomaterials*, vol. 26, no. 20, pp. 4281–4289, 2005.
140. R. Detsch, S. Schaefer, U. Deisinger, G. Ziegler, H. Seitz, and B. Leukers, "*In vitro* -osteoclastic activity studies on surfaces of 3D printed calcium phosphate scaffolds," *J. Biomater. Appl.*, vol. 26, no. 3, pp. 359–380, 2011.
141. K. Igawa, M. Mochizuki, O. Sugimori, K. Shimizu, K. Yamazawa, H. Kawaguchi, K. Nakamura, T. Takato, R. Nishimura, S. Suzuki, M. Anzai, U. Il Chung, and N. Sasaki, "Tailor-made tricalcium phosphate bone implant directly fabricated by a three-dimensional ink-jet printer," *J. Artif. Organs*, vol. 9, no. 4, pp. 234–240, 2006.
142. C. Wu, W. Fan, Y. Zhou, Y. Luo, M. Gelinsky, J. Chang, and Y. Xiao, "3D-printing of highly uniform CaSiO3 ceramic scaffolds: preparation, characterization and *in vivo* osteogenesis," *J. Mater. Chem.*, vol. 22, no. 24, p. 12288, 2012.
143. M. Castilho, C. Moseke, A. Ewald, U. Gbureck, J. Groll, I. Pires, J. Teßmar, and E. Vorndran, "Direct 3D powder printing of biphasic calcium phosphate scaffolds for substitution of complex bone defects," *Biofabrication*, vol. 6, no. 1, p. 015006, 2014.
144. A. Zocca, H. Elsayed, E. Bernardo, C. M. Gomes, M. A. Lopez-Heredia, C. Knabe, P. Colombo, and J. Gunster, "3D-Printed silicate porous bioceramics using a non-sacrificial preceramic polymer binder," *Biofabrication*, vol. 7, no. 2, p. 25008, 2015.
145. Y. Luo, D. Zhai, Z. Huan, H. Zhu, L. Xia, J. Chang, and C. Wu, "Three-dimensional printing of hollow-struts-packed bioceramic scaffolds for bone regeneration," *ACS Appl. Mater. Interfaces*, vol. 7, no. 43, pp. 24377–24383, 2015.
146. M. Castilho, J. Rodrigues, I. Pires, B. Gouveia, M. Pereira, C. Moseke, J. Groll, A. Ewald, and E. Vorndran, "Fabrication of individual alginate-TCP scaffolds for bone tissue engineering by means of powder printing," *Biofabrication*, vol. 7, no. 1, p. 015004, 2015.
147. D. T. Chou, D. Wells, D. Hong, B. Lee, H. Kuhn, and P. N. Kumta, "Novel processing of iron-manganese alloy-based biomaterials by inkjet 3-D printing," *Acta Biomater.*, vol. 9, no. 10, pp. 8593–8603, 2013.

148. J. M. Williams, A. Adewunmi, R. M. Schek, C. L. Flanagan, P. H. Krebsbach, S. E. Feinberg, S. J. Hollister, and S. Das, "Bone tissue engineering using polycaprolactone scaffolds fabricated via selective laser sintering," *Biomaterials*, vol. 26, no. 23, pp. 4817–4827, 2005.
149. S. Van Bael, T. Desmet, Y. C. Chai, G. Pyka, P. Dubruel, J. P. Kruth, and J. Schrooten, "*In vitro* cell-biological performance and structural characterization of selective laser sintered and plasma surface functionalized polycaprolactone scaffolds for bone regeneration," *Mater. Sci. Eng. C*, vol. 33, no. 6, pp. 3404–3412, 2013.
150. E. A. R. Gmeiner, U. Deisinger, J. Schönherr, B. Lechner, R. Detsch, A. R. Boccaccini, J. Stampfl, "Additive manufacturing of bioactive glasses and silicate bioceramics," *J. Ceram. Sci. Technol.*, vol. 6, no. 2, pp. 75–86, 2015.
151. F. H. Liu, "Fabrication of bioceramic bone scaffolds for tissue engineering," *J. Mater. Eng. Perform.*, vol. 23, no. 10, pp. 3762–3769, 2014.
152. Y. Xia, P. Y. Zhou, X. S. Cheng, Y. Xie, C. Liang, C. Li, and S. G. Xu, "Selective laser sintering fabrication of nano-hydroxyapatite/poly-ε-caprolactone scaffolds for bone tissue engineering applications," *Int. J. Nanomed.*, vol. 8, pp. 4197–4213, 2013.
153. K. H. Tan, C. K. Chua, K. F. Leong, C. M. Cheah, P. Cheang, M. S. Abu Bakar, and S. W. Cha, "Scaffold development using selective laser sintering of polyetheretherketone-hydroxyapatite biocomposite blends," *Biomaterials*, vol. 24, no. 18, pp. 3115–3123, 2003.
154. F. E. Wiria, C. K. Chua, K. F. Leong, Z. Y. Quah, M. Chandrasekaran, and M. W. Lee, "Improved biocomposite development of poly(vinyl alcohol) and hydroxyapatite for tissue engineering scaffold fabrication using selective laser sintering," *J. Mater. Sci. Mater. Med.*, vol. 19, no. 3, pp. 989–996, 2008.
155. Y. C. Ho, F. M. Huang, and Y. C. Chang, "Cytotoxicity of formaldehyde on human osteoblastic cells is related to intracellular glutathione levels," *J. Biomed. Mater. Res. B. Appl. Biomater.*, vol. 83, no. 2, pp. 340–344, 2007.
156. B. Duan, M. Wang, W. Y. Zhou, W. L. Cheung, Z. Y. Li, and W. W. Lu, "Three-dimensional nanocomposite scaffolds fabricated via selective laser sintering for bone tissue engineering," *Acta Biomater.*, vol. 6, no. 12, pp. 4495–4505, 2010.
157. N. J. Castro, R. Patel, and L. G. Zhang, "Design of a novel 3D printed bioactive nanocomposite scaffold for improved osteochondral regeneration," *Cell. Mol. Bioeng.*, vol. 8, no. 3, pp. 416–432, 2015.
158. G. Mitteramskogler, R. Gmeiner, R. Felzmann, S. Gruber, C. Hofstetter, J. Stampfl, J. Ebert, W. Wachter, and J. Laubersheimer, "Light curing strategies for lithography-based additive manufacturing of customized ceramics," *Addit. Manuf.*, vol. 1, pp. 110–118, 2014.
159. T. M. G. Chu, J. W. Halloran, S. J. Hollister, and S. E. Feinberg, "Hydroxyapatite implants with designed internal architecture," *J. Mater. Sci. Mater. Med.*, vol. 12, no. 6, pp. 471–478, 2001.

160. G. Gao, A. F. Schilling, T. Yonezawa, J. Wang, G. Dai, and X. Cui, "Bioactive nanoparticles stimulate bone tissue formation in bioprinted three-dimensional scaffold and human mesenchymal stem cells," *Biotechnol. J.*, vol. 9, no. 10, pp. 1304–1311, 2014.
161. M. Domingos, F. Intranuovo, T. Russo, R. De Santis, A. Gloria, L. Ambrosio, J. Ciurana, and P. Bartolo, "The first systematic analysis of 3D rapid prototyped poly(ε-caprolactone) scaffolds manufactured through BioCell printing: the effect of pore size and geometry on compressive mechanical behaviour and *in vitro* hMSC viability," *Biofabrication*, vol. 5, no. 4, p. 045004, 2013.
162. M. Du, B. Chen, Q. Meng, S. Liu, X. Zheng, C. Zhang, H. Wang, H. Li, N. Wang, and J. Dai, "3D Bioprinting of BMSC-laden methacrylamide gelatin scaffolds with CBD-BMP2-collagen microfibers," *Biofabrication*, vol. 7, no. 4, p. 044104, 2015.
163. Y. G. W. and Y. W. Xiao-Fei Wang, Pei-Jun Lu, Yang Song, Yu-Chun Sun, "Nano hydroxyapatite particles promote osteogenesis in a three-dimensional bio-printing construct consisting of alginate/gelatin/hASCs," *RSC Adv.*, vol. 8, no. January, pp. 6832–6842, 2016.

10
Application of Bioreactor Concept and Modeling Techniques to Bone Regeneration and Augmentation Treatments

Oscar A. Deccó and Jésica I. Zuchuat*

Bioengineering, Faculty of Engineering, National University of Entre Ríos, Oro Verde, Entre Ríos, Argentina

Abstract

During almost two decades of research in the field of Tissue Engineering and Regenerative Medicine, a variety of biomaterials and techniques have been developed, and a series of results and clinical evidence have been published to support and improve bone regeneration and augmentation processes. The latest techniques include the use of growth factors and platelet-rich plasma preparations in combination with natural or artificial scaffolds, membranes, and stem cells. However, the application of these concepts depends heavily on a favorable microenvironment for the regeneration, which is provided by the use of the bioreactor. The bioreactor is a system in which a microenvironment can be created to promote bone regeneration by providing favorable biomechanical conditions, protecting against infection, and allowing proper nutrition of the regenerating tissue by vascularization. The latest research in this field is very promising as preliminary studies have shown good results concerning bone quality and architecture. In relation to the bioreactor principle, computational modeling represents a very effective tool to predict bone ingrowth. In addition, based on the obtained results, the use of these data to prevent diseases in the human osseous–articular system is expected in the near future.

Keywords: Bone regeneration, tissue engineering, bioreactor, bioengineering, computational modeling

---

Corresponding author: jzuchuat@bioingenieria.edu.ar

Ashutosh Tiwari, Bora Garipcan and Lokman Uzun (Eds.) Advanced Surfaces for Stem Cell Research, (279–322) © 2017 Scrivener Publishing LLC

10.1 Bone Tissue Regeneration

Two phenomena can occur when bone tissue suffers structural damage: repair or healing in which tissue does not keep its original architecture or function and does not recover its original properties and function or bone regeneration in which a new tissue originates with all of the properties of the original one.

A bone injury is followed by an immediate response, then bone repair and finally bone remodeling, similarly to the healing of a wound. The blood coagulation process and clot melting are followed by the repair of injured tissue, and all of these events are called homeostasis. The numerous cellular protagonists involved in bone homeostasis are responsible for the connections established between bone tissue, the immune system, and the vascular compartment.

After an injury, the platelets activated by the thrombin release factors stimulate vasoconstriction and trigger the coagulation cascade, where the fibrinogen contained in the plasma is converted into fibrin. The blood clot contains chemotactic and mitogenic growth factors (GFs) such as platelet-derived growth factor (PDGF) and the transforming growth factor beta (TGF-β) that provide a temporary matrix in which cells can grow [1].

Clot dissolution occurs through the action of plasmin, a serine protease that circulates as the inactive pro-enzyme plasminogen. Plasmin causes fibrinolysis, replacing the clot with granulation tissue. Plasminogen binds to fibrinogen and fibrin, and they incorporate into the clot. The tissue activator of plasminogen (tPA) and to a minor degree urokinase are other serine proteases released by the cells from the clot that convert the plasminogen into plasmin [1].

The granulation tissue is highly vascularized, being rich in collagen, hyaluronic acid, proteoglycans and fibronectin as well as macrophages, fibroblasts, lymphocytes, and mesenchymal cells from the periosteum's bone marrow, from the circulation and from the surrounding soft tissue. Mesenchymal progenitor cells are attracted by morphogenetic proteins and other molecules that stimulate their mitogenic activity and induce differentiation into their osteochondrogenic lineage [2].

When a bone injury is in close contact with the bone extremities, with a suitable vascular supply and stability, guided bone regeneration (GBR) occurs by intramembranous ossification.

During the different stages of bone repair, the influence of exogenous factors that interfere in the healing process has been observed to improve the final quality of bone. These factors include the following:

10.1.1 Proinflammatory Cytokines

In the late 1990s, research identified a new group of cytokines within the tumoral necrosis factor (TNF) family that are required for the control of bone remodeling [3–5]. The role of these cytokines is, in part, to begin the bone remodeling response.

As consequence of an injury, proinflammatory cytokines (IL-1 and TNF-α) and vasoactive amines are released, which increase the tissue blood flow and modify the endothelial permeability. The cytokines modify the adhesive properties of endothelial cells, which allows the migration of leukocytes to the site of the injury.

Two types of discrete reabsorption can be observed during the repair. The first occurs at the end of the endochondral period, in which mineralized cartilage is removed and primary bone formation occurs. During this period, the levels of macrophage colony-stimulating factor (MCSF), the receptor activator of nuclear kB factor Ligand (RANKL), and osteoprotegerin (OPG) are elevated [6].

Although most of the cytokines have been associated with the process of bone remodeling, they are absent during this period, with the exception of TNF-α, which begins to increase at the end of the period of endochondral resorption and is essential in the progression and transition of this phase in the process of injury repair [7].

In the bone, TNF-α increases the production of osteoclastogenic cytokines such as MCSF and RANKL and acts directly on the osteoclast precursors to enhance RANKL-induced osteoclastogenesis by promoting bone resorption, even in the absence of high levels of this ligand [6].

The second type of resorption occurs during the secondary bone formation, in which the level of cytokines increases and the levels of OPG, RANKL, and MCSF decrease, which shows that the processes of endochondral resorption and bone remodeling are different.

Reducing the resorption or the secondary bone remodeling or improving the bone coupling will lead to quick bone repair or to quick bone strength recovery [8].

10.1.2 Transforming Growth Factor Beta

TGF-β is responsible for the regulation of osteoclastogenesis and the survival of the osteoclasts, in part through stimulating of the activity of OPG, which inhibits the formation and function of osteoclasts [9]. All of the isoforms of TGF-β (β1, β2, and β3) produce the same effect on this protein.

The bone morphogenetic proteins (BMPs) can induce bone formation, improving significantly the process of bone repair. The levels of BMPs increase from the initiation of the injury [10], which suggests that their delivery at different stages of the healing process can promote bone repair.

10.1.3 Angiogenesis in Regeneration

The regeneration of any tissue creates a demand on the surrounding tissue that requires increased blood flow. The dependence the bone quality achieved and the blood flow developed in the process of bone repair has been extensively demonstrated [11].

During the endochondral formation of new bone in fracture healing and bone injury repair, the transition of the cartilaginous callus to the new bone represents a crucial stage in the process of repair. This stage includes four coordinated biological events: the apoptosis of chondrocytes, degradation and removal of the cartilage matrix, new vascularization in the areas of repair, and the recruitment of osteogenic cells, resulting in the differentiation and production of bone matrix.

Angiogenesis is regulated in two ways: one dependent on vascular endothelial growth factor (VEGF) and the other dependent on angiopoietin [12]. It is believed that both processes are functional during the process of repair.

VEGF is a glycoprotein that exclusively acts on endothelial cells. It stimulates the proliferation and migration of these cells and their tubular organization and inhibits their apoptosis. This GF is detected during fracture healing and is expressed throughout the chondrogenic phase, reaching maximal levels during the late stages of calcification of the cartilage tissue, when resorption is initiated.

The family of proteins related to VEGF includes mitogens of endothelial cells and essential mediators of neo-angiogenesis [13]. It has been demonstrated that VEGF plays a central role in neo-angiogenesis and in the formation of endochondral bone. The osteoblasts express elevated quantities of VEGF and serve as primary regulators of angiogenesis in fracture healing and bone repair.

Several studies [14, 15] have demonstrated that the BMPs stimulate the expression of VEGF and their receptors, suggesting an intimate relationship between these two families in the promotion of new bone.

The second way includes angiopoietin-1 and -2 and their receptors. Angiopoietins are vascular morphogenetic proteins associated with the formation of larger blood vessels and the development of collateral branches from existent vessels [11]. The expression of angiopoietin-1 is induced during the first stages of bone repair, which suggests that the initial vascular growth in the periosteum plays an important role in the repair process [16]; angiopoietin-1 plays a critical role in the maturation of vessels and mediates the migration, adhesion, and survival of endothelial cells. Angiopoietin-2 interrupts the connections between the endothelium and the perivascular cells and promotes cellular death and vascular regression; however, in combination with VEGF, it promotes neo-vascularization [17].

The instability of defects and the low oxygen tension in the area favor the differentiation of mesenchymal stem cells (MSCs) to their osteogenic lineage [18]. Then, this tissue is replaced by bone through endochondral ossification, as described above.

The new mineralized tissue that lies between the reticular bone and the preexistent bone, as well as between osteoconductive surfaces and/or biomaterials, is of laminar type.

New bone formation to the level of membranes or an implantable prosthesis is possible due to the conductivity of the material. It is unlikely that existing mature osteoblasts are the only cells responsible for bone formation, so it is believed that, as during bone formation and fracture healing, progenitor cells are attracted to the damage site, where the signals that stimulate their differentiation into osteoblasts are received.

Thus, angiogenesis and intramembranous ossification occur concurrently during bone development, fracture healing, and the osseointegration of dental implants.

It can be concluded that the development of embryonic bone, bone regeneration, and osseointegration all depend on a sufficient number of mesenchymal progenitor cells at the site of development or repair, making migration as well as mitogenic and osteo-chondrogenic differentiation subject to the presence of signaling molecules and GFs at the site of new bone formation. The use of a matrix or a scaffold [19] into which cells can adhere plays a fundamental role in the regeneration process.

10.2 Actual Therapeutic Strategies and Concepts to Obtain an Optimal Bone Quality and Quantity

Current strategies to improve GBR are based on the improvement of the natural healing processes of the body through the use of scaffolds, GFs, signaling molecules, and mesenchymal progenitor cells.

The development of these new strategies is associated with technological progress in the material sciences, bioengineering, biotechnology, biochemistry, and bioinformatics.

10.2.1 Guided Bone Regeneration Based on Cells

One of the major aims of modern medicine and the sciences that support it is to create new technologies and techniques capable of regenerating the human body to its original function, after an illness or trauma.

The advances in tissue engineering and regenerative medicine, especially in the area of stem cells in the past decades, offer a new option and hope to patients. Stem cells are not only entities of biological organization, responsible for the development and regeneration of tissues and organs but also are units in evolution by natural selection. Stem cells are defined as clonogenic cells capable of self-renewal through cellular division and of differentiation into multiple cell lineages [20], two well-defined properties that distinguish them from other cells of the organism. Because of these characteristics, stem cells have generated great enthusiasm in the scientific world and represent a valuable tool with unprecedented potential for biomedical research and its therapeutic applications.

To realize the therapeutic potential of stem cells, the mechanisms through which these cells self-renew, differentiate, and work must be understood.

Most of the actual studies are focused in two types of stem cells:

10.2.1.1 Embryonic Stem Cells

Embryonic stem cells (ESCs) can be isolated from fetal blood, amniotic fluid, and the placenta [21] and are derived from embryos in the blastocyst stages from the ages of 4–5 days. Blastocysts are thin, hollow structures formed by hundreds of cells. The outer layer forms the placenta and other support tissues, and the inner layer gives rise to tissues that forms the organism. ESCs are pluripotent, which means that they have the potential to differentiate into cellular types of all somatic lineages [22]. Although they present many advantages to regenerative medicine and tissue engineering, their use is limited due to ethical issues leading to the prohibition of isolation of these cells from human embryos. In addition to this ethical controversy, a large number of technical problems exist, as ESCs can easily form tumor tissues and they are not immunologically compatible [22].

10.2.1.2 Adult Stem Cells

Adult stem cells (ASCs) are undifferentiated cells that originate from mature tissues and organs, and they are multipotent cells with the ability to differentiate into the specific cellular types of the tissue or organ in which they are placed [22]. The ASCs are found in most of the tissues of the body, and their main function is to maintain homeostasis in the surrounding tissue. These cells are widely used in regenerative medicine because they can be harvested from the patient, cultured *in vitro*, exposed to the appropriate factors, depending on the tissue under regeneration, and finally re-implanted in the patient. Because these cells are autologous, immune response is avoided. Ethical issues are also avoided. The ASCs also represent a lower risk of malignant transformation.

The most studied ASCs that can be extracted from the bone marrow are hematopoietic stem cells (HSCs), which are responsible for the formation of all types of blood cells; endothelial progenitor cells (EPCs); and MSCs.

10.2.1.3 Mesenchymal Stem Cells

MSCs are present in many tissues and give rise to a variety of cells depending on a large number of interacting chemical, biological and mechanical factors [22]. They can be differentiated into osteoblasts, chondrocytes, adipocytes and stromal cells that support blood formation. MSCs are highly useful in clinical application because they can be expanded many times in culture. Medical applications of these cells are being studied for the treatment of illness, including in coronary arteries [23], Parkinson's disease [24], and liver regeneration [25].

The proliferation and differentiation of stem cells cultured *in vitro* is regulated by different signaling pathways, so it is suggested that this behavior is similar in biological media. Chemical, biological, and mechanical factors and the surrounding microenvironment determine the differentiation path and whether the cells remain multipotent or differentiate into a specific cellular type [26]. The extracellular matrix is the biological substrate in which cells adhere, migrate, and grow. It also provides a structural support for the cells, modulates intracellular communication, and regulates the signaling events induced by soluble factors [21].

In a multicellular organism, each MSC has its own microenvironment, consisting of many factors that influence cell proliferation and differentiation state in tissues. Generally, these factors are classified into two categories [27]:

a. Soluble signals: GFs, metabolites and dissolved gases.
b. Insoluble signals: composition, architecture and elasticity of the extracellular matrix (ECM) and the intercellular interactions.

The proliferation and differentiation of cells in culture are extremely important issues for basic research, drug discovery, tissue engineering, and regenerative medicine; thus, the incorporation of these signals in cultures *in vitro* is essential to recreate the microenvironment of these cells.

Glycoproteins from the EMC produce insoluble signals that play an important role in the specification of the cellular fate of MSCs by influencing signaling pathways affected by soluble signals [28].

Cell-based bone regeneration is a process of several stages. Initially, autologous cells with high proliferative and osteogenic potential, such as bone marrow stromal cells (BMCs), are extracted and seeded in osteoconductive scaffolds that can consist of various materials, such as metal alloys [29, 30], hydroxyapatite, collagen, Poly-Lactic Acid (PLA) and Poly-Glycolic Acid (PGA) [31, 32], or corals [33]. To improve cellular adhesion, the scaffolds can be treated with various natural and synthetic materials.

Interaction between the transplanted cells and the scaffold cover is essential for successful formation of bone tissue. Thus, an osteoconductive scaffold induces the differentiation of BMCs into osteoblasts. For therapeutic application, these cells can be cultivated and then implanted in the defect site [34]. Under optimal conditions, transplanted cells will form a mineralized matrix in the scaffold. The matrix and the scaffold will then be replaced by neo-formed bone during the remodeling phase.

10.2.2 Guided Bone Regeneration Based on Platelet-Rich Plasma (PRP) and Growth Factors

A strategy to improve GBR that has been the subject of a number of research studies in the recent past is the use of autologous platelet concentrates to supplement the graft.

Platelets originate from the fragmentation of the megakaryocyte cytoplasm; they have no nucleus and cannot replicate. Their lifespan is from 7 to 10 days [35]. Due to the short lifetime of platelets, their influence decreases gradually over the subsequent stages [36, 37]. The platelet action in the context of bone recovery operates by two mechanisms: during the early stage, through the release of GFs and cytokines; and later, by chemotaxis and macrophage activation [38]. Although the release of protein can last for 1 h, the mean life of a GF is only a few minutes.

It was previously thought that activated blood platelets contributed only to hemostasis at the beginning of the coagulation cascade. It is now known that they also aid in the transport of GFs at the beginning of the healing processes [39]. Platelets may act as natural biological mediators, regulating cellular processes during tissue regeneration, including chemotaxis, extracellular matrix differentiation, and synthesis, by binding to specific cell surface receptors [40].

Platelets are deposited at the site of an injury and can promote the early phase of healing and GBR. They act as carriers of GFs [38] and bioactive molecules [41], controlling the release of the proteins from their own alpha granules [42]. The GFs are inactive within the granules. In response to platelet aggregation or to contact with the connective tissue, platelets are activated, releasing the alpha granules, which then release the GFs by active extrusion through the cellular membrane [43]. Upon release, the GFs are attached to carbohydrate chains and histones [44], completing their structure and activation.

GFs exert pleiotropic effects on the wound healing process, some of which overlap and include angiogenesis, mitogenesis, cellular differentiation, and the regulation of normal cellular functions [44]. While they have been found to positively influence the repair process, the beneficial effects of platelet concentrates in GBR remain controversial [45].

GFs are proteins that will be sent from one cell to another to transmit a concrete signal, which can include migration, differentiation, and activation. Cells that receive the signal may be near or distant. Depending on the interactions, the response of the cells can be as follows:

a. Endocrine, when a GF acts on remote cells.
b. Autocrine, when a GF acts on the cell that produces it.
c. Paracrine, when a GF acts on nearby cells.
d. Juxtacrine, when a GF acts on a cell receptor in close contact with another cell.

Each GF has one or more of these functions, and its specific actions in a cell will depend on its environment. The amount of GF released is not relevant, as GFs only act after their union with specific receptors. Once released, the GF must interact with its corresponding receptor, which must be present in sufficient quantity and functional activity to transmit the appropriate stimuli.

The receptors are proteins that are embedded in the cellular membrane, with dual or adjacent active sites, and consist of three regions [46]:

1. An extracellular domain where the GFs are bound.
2. A transmembrane domain.
3. A cytoplasmic domain, which contains secondary messengers.

This union activates the transmembrane receptors and induces a high-energy phosphate bond to signaling molecules in the target cells' cytoplasm. These signaling molecules (secondary messengers) are proteins.

Once activated, they are released into the nucleus, where they initiate the expression of genes associated a specific function in tissue regeneration, such as controlled cellular replication, capillary proliferation, osteoid production, or collagen synthesis [44].

The most important secondary messengers are as follows:

- Adenylcyclase, an enzyme activated by G protein, which catalyzes the reaction of adenosine triphosphate (ATP) to cyclic adenosine monophosphate (AMPc).
- Phospholipase C (PLC), which activates protein kinase C to initiate the protein phosphorylation.
- Tyrosine kinase (TK).
- Serine–threonine kinase (STK), which is the directly responsible for the phosphorylation of target proteins.

When this process is concluded, the cycle of cellular proliferation begins.

Thus, the GFs control and regulate biosynthesis, and platelets initiate and modulate the wound healing. The PRP is simply a medium to regulate the amount of active GF to accelerate the rate and degree of bone healing, achieving the repair of damaged tissue and/or the augmentation of bone volume during the regeneration process.

Platelet-derived growth factor (PDGF): There are three isomers of PDGF (-AA, -AB, and -BB), which act in different forms and at different time in bone regeneration and wound healing. Their actions are superposed and depend on the cells with which they interact [44]. PDGF is weakly expressed during the inflammatory phase but increases and then remains constant during the repair phase. PDGF has many functions, including the stimulation of the proliferation of stem cells, initiating bone repair and helping in the intramembranous phase, originating the bone callus [1]; its most important biological activities include chemotaxis and mitogenesis, through the stimulation and synthesis of DNA. Its activity is mediated by protein tyrosine kinase signaling.

Vascular endothelial growth factor (VEGF): VEGF is essential for the regulation of both physiological and pathological angiogenesis. Its actions are limited to the endothelial cells, which possess receptors for this GF. It induces various biological responses, including the recruitment of inflammatory cells; the proliferation of endothelial cells and synthesis of basal lamina, resulting in vascular growth into the wound; and the absorption of fatty acids [47]. VEGF is also a vasodilator depending on the dose and increases the vascular permeability [44].

Transforming growth factor beta (TGF-β): TGF-β is a member of a superfamily of molecules. Its three isoforms ($\beta1$, $\beta2$, and $\beta3$) are important in mammalian tissues, with a well-established role in osteogenesis [1]. It stimulates and inhibits the proliferation of different mesenchymal tissues; this approach is manifested synergistically or antagonistically, as a function of its interaction with other GFs. TGF-β acts in the phase of collagen self-assembly during bone remodeling and intervenes in bone metabolism by stimulating OPG and by inhibiting osteoclast formation and the resorptive function of mature osteoclasts [46].

Fibroblast Growth Factor (FGF): There are two types of FGF, FGF-1 (acidic), and FGF-2 (basic). They comprise a large family of polypeptides that regulate cellular functions, such as mitogenesis, cellular differentiation, receptor modulation, and protease production and allow the maintenance of cells [1]. It has been demonstrated [48] that the expression of TGF-β in osteoblastic cells is improved by FGF. These factors play an important role in bone formation through the regulation of osteogenesis.

FGF is essential to the maintenance of bone homeostasis [48]. It is expressed by macrophages and other inflammatory cells, mesenchymal cells and chondrocytes.

Both FGF-1 and FGF-2 are detected in the granulation tissue during the first stages of wound healing [49]. In case of FGF-1, the mitogenic effects seem to affect the chondrocytes. FGF-2 is generally more potent than FGF-1 and has attracted more attention in investigations of wound-healing processes [49]. Its presence is particularly prominent in the soft bone tissue and in the periosteum. In addition to its mitogenic and angiogenic properties, it has been shown to stimulate bone resorption, suggesting the potential to influence many phases of the early repair of post-traumatic events, continuing until the end of the bone remodeling callus.

Insulin-like Growth Factor Types I and II (IGF-I and IGF-II): The IGFs are proteins and are also termed somatomedins or sulfation factors. These GFs are detected in high concentrations in periosteum cells, in healing bone tissue and in the development of ectopic bone tissue induced by demineralized bone. The IGFs produced by bone cells not only act as autocrine and paracrine regulators, but also are incorporated into the bone matrix and released later (during the resorption phase), augmenting the proliferation of osteoblastic precursor cells [1, 48]. IGFs do not only contribute to bone formation but also modulate osteoclastic function, leading to bone remodeling during the bone repair phase.

10.2.2.1 Bone Morphogenetic Proteins

The BMPs are a subfamily of the TGF-β superfamily of polypeptides and represent a group of proteins that play a critical role in ROG, specifically in growth regulation, differentiation, and cellular apoptosis during bone development, involving osteoblasts and chondrocytes [48]. The BMPs have more selective effects in bone than TGF-β, and highly promising results have been shown in bone regeneration in animal models [49].

During bone regeneration, the expression of BMPs rises by 2- to 7-fold. Several investigations [50, 51] have demonstrated that BMPs are expressed during the first phases of bone repair, where small quantities of proteins are susceptible to release from the bone extracellular matrix. During intra-membranous formation, osteoprogenitor cells from the inner layer of the periosteum (cambium layer) respond to this initial low level of release of the extracellular matrix components and begin to differentiate.

During the first 7–14 days after the fracture, the expression of BMP-2 and BMP-4 is maximal in chondroid precursors, while osteoblasts show only moderate levels of expression [48]. It is known that in fracture healing, BMPs act primarily as activators of differentiation in osteoprogenitor and mesenchymal cells destined to become osteoblasts and chondrocytes. Once these primitive cells mature, BMP expression is drastically reduced. BMP expression emerges transiently in chondrocytes and osteoblasts during their respective periods of matrix formation, then returns to lower levels during callus remodeling [48].

Several preclinical studies [50, 52, 53] have assessed the efficacy of recombinant human bone morphogenetic protein (rhBMP) in the healing of critical-sized bone defects and in the acceleration of this process. The use of rhBMP-2 has demonstrated efficacy in the healing of critical-sized bone defects in animal models, including rat, rabbit, sheep, and dog. Preclinical and human studies have demonstrated that rhBMP-2 induces bone formation [50], which integrates with the surrounding tissue by bone remodeling. Although the results of these preclinical studies have been promising, the doses of rhBMPs required to induce bone formation are high (0.75–1.5 mg/ml) which suggests that, to produce a clinically important effect, large quantities of recombinant proteins are needed.

10.2.3 Guided Bone Regeneration Based on Barrier Membranes

The goal of technologies based on barrier membranes is the restoration of supporting tissue, lost as a consequence of inflammatory processes or trauma, loss of dental pieces, or vertical bone loss due to periodontal disease, among other causes.

The purposes of the use of barrier membranes in techniques pursuing the regeneration or augmentation of bone tissue (vertical or horizontal) include the following: to facilitate bone regeneration in critical- and noncritical-sized bone defects, to improve implant healing, to induce complete bone regeneration, to improve the results of bone grafts, and to serve as a container for bone substitutes [54, 55]. This type of membrane, which can be of a resorbable or nonresorbable nature (Figure 10.1), is introduced as a physical obstacle to prevent tissue invasion, such as of epithelial or conjunctive tissue, from interfering with or avoiding with preventing the healing process and/or bone neo-formation.

These membranes act in a similar way to the traditional flap, but with improved characteristics, as the membrane does not cause bone reabsorption, a situation that is observed when the soft tissue is under tension, while performing the flap. The membrane provides stability and protection to the blood clot, simultaneously creating a space for cell migration based on the inductive activity supplied by the GFs present in the blood, in addition to favoring the angiogenesis [56] that is fundamental to bone regeneration.

Preliminary studies have demonstrated that the use of nonresorbable membranes as mechanical barriers resulted in complete healing of bone defects *in vivo* [57], and the use of collagen membranes prevented the apical migration of the epithelium and provided support for the insertion of new connective tissue in bone tissue regeneration [58].

The process of regeneration occurring within the barrier membrane involves angiogenesis and the migration of osteogenic cells from the periphery toward the center to create a well-vascularized granulation tissue. The initial deposition of the blood clot is followed by vascular ingrowth and new bone tissue deposition, subsequent lamellar bone formation, and finally bone remodeling [56].

To achieve the proposed objectives, several studies have been conducted to assess the use of membranes in different configurations, either alone or in combination with GFs, placing or not placing grafts or bone substitutes or forming part of a 'bioreactor' system as described above.

When membranes are used in combination with bone substitutes to promote the filling of defects, it has been demonstrated that the best results are achieved more quickly when the substitutes are covered by a membrane [56, 59]. The use of membranes in combination with bone grafts

Figure 10.1 Three-dimensional membrane model.

has allowed the achievement of bone augmentation in clinical situations in which important bone atrophy has been produced [60].

The biomaterial characteristics, engineering design and physicochemical properties of the membranes can significantly influence the bone formation process or bone volume augmentation at the site of a defect [61, 62].

A biomaterial is defined as a material intended to form an interface with biological systems to evaluate, treat, increase, or replace a tissue, organ, or function of the organism. A biomaterial can also be defined as a synthetic, natural, or modified natural material intended to be in contact with and to interact with a biological system. All biomaterials designed to be introduced into the body must satisfy safety and efficiency requirements, validated by *in vitro* and *in vivo* studies, including biocompatibility, which allows the material to interact properly with the host tissue without triggering adverse reactions.

10.2.4 Guided Bone Regeneration Based on Scaffolds

The transplantation of autologous bone has historically been the 'gold standard' to recover bone loss, as it satisfies all of the characteristics needed to support bone reconstruction: osteogenesis, osteoinduction, and osteoconduction [63]. Autografts have mature osteoblasts and MSCs that can form new bone through a porous structure in the trabecular bone, which allows the growth of blood vessels, resorption during bone remodeling, and release of GFs to stimulate the differentiation of progenitor cells into osteogenic cells to form new bone. However, the main disadvantage of these grafts is their acquisition, which is associated with high morbidity at the donor site, as well as the possibility of infection and an inadequate amount of harvested tissue.

The use of synthetic biomaterials allows the surgeon to supply and improve the GBR. Scaffolds used in bone tissue engineering have been synthesized from synthetic and natural calcium phosphates, as well as a large variety of synthetic polymers (e.g. PLA/PGA) and natural substances (collagen and fibrin, among others). These scaffolds, independently of their composition, must be chosen based on a series of considerations [44, 64]:

- They must have a microstructure capable of supporting cells and their functions, so generally requiring a porous structure.

- The chemical composition of the matrix is important due to its influence on cell adhesion and the phenotypic expression of the infiltrating cells.
- Because the objective is the regeneration of the original tissue, the scaffold must be resorbable. The material degradation rate is generally determined based on the new tissue formation rate.
- The mechanical properties are important to provide temporary support to the loads applied *in vivo* during the regeneration process and to resist the contractile forces exerted by the cells.

Various resorbable materials, both synthetic and natural, possess good porosity for the invasion of blood vessels, in addition to providing mechanical stability during the healing process. Examples of such materials include the calcium phosphates, hydroxyapatite, bioglasses, and ceramics.

The latest studies [64, 65] describe the development of new techniques and devices to control the chemical, mechanical, and structural media, seeking the optimization of stem cell renewal and differentiation. The design of these protocols must consider flexibility, size, biodegradability, cell capacity, and the micro-/nanostructure of the substrates used in the structural matrix or scaffold. The use of technologies such as computerized tomography or magnetic resonance to obtain images of the defect region allows customization of the size and shape of the scaffold to adapt it to the site of the defect using manufacturing methods such as 3D printing [66].

10.3 Bioreactors Employed for Tissue Engineering in Guided Bone Regeneration

The increasing demand for products and therapies in tissue engineering involves the development of technologies that allow the production of new tissue through the application of synthetic bone substitutes and biomaterials on a large scale and with precise control of shape and size, maintaining the quality of the original tissue.

This context has produced the concept of the 'bioreactor', which, used *in vitro*, can be defined as a device, in which different biological or biochemical processes or both, can be reproduced under controlled conditions (pH, temperature, and pressure) [67] through the continuous supply of nutrients (e.g. glucose, amino acids), biochemical factors and oxygen, the

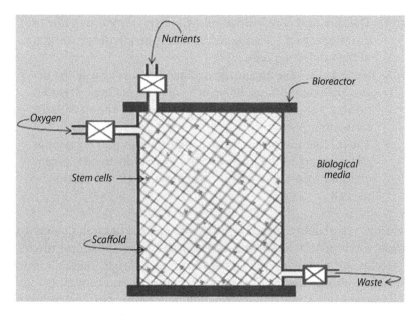

Figure 10.2 Bioreactor scheme.

diffusion of chemical species to the interior of the bioreactor, the continuous removal of the products of cell metabolism (e.g. lactic acid), and the appropriate physical stimulation of the cells contained [68] (Figure 10.2). Bioreactors can facilitate the aseptic growth of functional tissues and simultaneously offer several additional advantages over culture plates and flasks [69, 70]; they can be designed with complex three-dimensional geometries that can contain multiple cell types.

Most bioreactor research [71, 72] has focused on the development of new technologies that combine stem cells and biomaterials to achieve functional tissue-engineering systems that satisfy biocompatibility, mechanical stabilization, and safety needs for the patient.

These bioreactor systems have been widely investigated for bone engineering purposes, seeking improvements over *in vitro* culture. The research has been focused on three types of bioreactors: spinner flask, rotating wall, and perfusion.

10.3.1 Spinner Flask Bioreactors

These systems comprise a glass reservoir with side arms that can be opened to remove the scaffolds and/or the media. They have a stirring bar or other mechanism to remove the fluid media, which simulates the biological microenvironment; they also have a porous cover that permits gas exchange. Generally, the scaffolds are suspended from the container top by needles or wire. They are often used in cell culture experiments for bone tissue engineering as they have been demonstrated to increase the expression of alkaline phosphatase (early marker of osteoblastic activity) and, osteocalcin (later marker of osteoblastic activity) and the deposition of calcium minerals, in comparison with static cultures and rotating wall bioreactors [72]. It is believed that this effect is the result of nutrient transport by convection to the surface of the scaffold (which increases oxygen concentration in the whole volume of the scaffold), in comparison with transport by diffusion in static cultures, where the cells placed at the center of the scaffold receive an inefficient supply of nutrients [72].

The efficacy of this type of bioreactor is subject to the agitation rate, the type of scaffold and the pore shape and size, determining the dimensions of cell nutrition and of oxygen diffusion, as well as the distribution of forces during the agitation that can contribute to the osteogenic differentiation [73].

10.3.2 Rotating Wall Bioreactors

This model comprises two concentric cylinders: an inner cylinder that is stationary and provides for gas exchange and an outer cylinder that rotates. The space between the two cylinders is completely filled with culture media, and the scaffolds containing the cell seeding are placed in this space and allowed to move freely. The free movement of the scaffolds leads to a microgravity environment where the flow of fluid caused by the centrifugal forces of the cylinder balance the force of gravity. Although an increase in the levels of alkaline phosphatase and osteocalcin has been demonstrated [74], no increase in cell proliferation has been observed, making this method ineffective for the culture of osteoblastic cells. Furthermore, the random movement of the scaffolds, which are denser than the fluid medium, causes them to crash into the bioreactor walls, intensifying the effect of low shear stresses on cells and, consequently, the low rate of osteoblastic proliferation [72].

These two methods do not effectively perfuse media into a scaffold.

10.3.3 Perfusion Bioreactors

This system has been demonstrated to exhibit greater positive effects on osteoblastic differentiation than the two other methods. The improvement lies in the use of a pump system to perfuse media directly through the scaffold. The basic design consists of a media reservoir, a pump, a tubing circuit, and a perfusion cartridge. The perfusion cartridge houses the scaffold which is sealed so that the media cannot flow around it, but it must directly penetrate the scaffold pores. For good performance, the design of the cartridge must be customized and the pores highly interconnected, which makes the large-scale use of such bioreactors difficult. In addition to these obstacles, many perfusion bioreactor systems have been developed and proved for bone tissue engineering purposes, demonstrating increases in the calcium matrix deposition and in the rate of osteoblastic differentiation.

GFs present in the culture media play a fundamental role in osteoblastic differentiation; thus, their activity must be studied further to fully elucidate the effect of perfusion systems. The dynamic culture could not only enhance the production of GFs through cell stimulation but also potentially reduce the local concentration of soluble factors by increased mass transport [72].

10.4 Bioreactor Concept in Guided Bone Regeneration and Tissue Engineering: *In Vivo* Application

Recent studies have referred to bioreactor systems tested *in vitro* by experimental investigations, which have only been evaluated *in vivo* in large animal models. The use of the bioreactor concept grouping the membrane/scaffold/stem cell has been preliminarily studied in small animal models, specifically using noncritical-sized bone defects in rabbits [29, 30] (Figure 10.3), obtaining promising results in the field of bone regeneration and bone augmentation. In addition, some preliminary studies in humans have been developed with this concept in maxillary bones that required bone augmentation in the vertical dimension for the subsequent placement of dental implants.

In tissue engineering and regenerative medicine, bioengineering plays an important role as the design of the bioreactor components not only refers to the application of existing concepts that, used in combination, lead to the achievement of a prototype of an integral model but also includes the creative adaptation of each of these three components to achieve the objective of integrating the biological media of a defect or an injury *in vivo* with the media of the implanted device, to achieve the success in situations of bone regeneration and/or bone augmentation due to an injury or other external or internal factor.

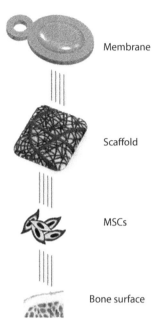

Figure 10.3 Scheme of *in vivo* bioreactor.

In these studies, the bioreactor, adapted using the components listed above, involves the application of MSCs on scaffolds, acting as a three-dimensional template for the nucleation of these progenitor cells to form new bone; with membranes used as a protective cover for the cell/scaffold complex, creating protection from the external environment and encouraging a favorable internal microenvironment for cell proliferation and blood vessels in the shortest time possible, seeking to maximize cell production and cell growth, with the aim of achieving adequate bone volume and quality for the patient's rehabilitation.

The use of microfixed nonresorbable membranes is intended to isolate and protect the augmentation site from harmful factors, and to avoid the displacement of the bioreactor from the site of placement. The application of bioengineering techniques to the design of these membranes as the outer limit of this complex has led to the use and assessment of biomaterials such as titanium and cobalt–chrome (Co–Cr) alloy.

It has been demonstrated that different surface treatments can improve biocompatibility properties and the membrane osseointegration with minerals and bone tissue. At the biomaterial level, modifications can be made to the surface oxide layer, allowing it to behave as an inert material, thus avoiding ionic exchange at the bone-implant interface, preventing a corrosion process [75] and thereby favoring the osteoconductivity [76], and improving physical, chemical and topographical surface properties that are decisively involved in biological processes [69, 77, 78]. The results highlight the importance of surface treatment to promote the speed and quality of bone tissue response [79].

The composition and surface characteristics of the material have a deeper influence in the success of the implant, as they determine the adsorption of proteins, lipids, sugars and ions present in bodily fluids. This surface characterization can be achieved by a series of processes, including sand blasting and chemical and thermal treatments, among others.

10.4.1 Sand Blasting

Sand blasting is a simple mechanical process used to modify material surface properties, cleaning, and making them rough. The method consists of 'bombing' the material surface with hard and abrasive particles at high speed. Particles can be used dry or embedded in liquid. When impacting the surface, part of the material is detached causing localized plastic deformation, which is used to create the surface roughness; to remove defects such as burrs, pores, and microcracks from the manufacturing process (machining); and to induce a surface layer of compressive residual stresses [80, 81].

One of the main variables to define prior to a sand blasting process is the nature of the projection particle. Due to the high energy of the impact, some of the particles may remain in the body of the implant material without the possibility of being removed by ultrasound, passivated or sterilized, so 'contamination' is inherent to the method.

The materials most used for bombing the biomaterials employed as containers in the bioreactor model include the aluminum oxide (Al_2O_3), the most commonly used material; silicon carbide (SiC), a biocompatible ceramic material; the titanium oxide (TiO_2), which avoids possible contamination problems, although the particle granulometry necessary for the treatment is not easy access; hydroxyapatite, which is highly bioactive and is the major component of bone, although it has been demonstrated that its mechanical properties are not sufficient to generate roughness of the desired order in these implant materials; and zirconium oxide (ZrO_2), which can be obtained in a spherical shape and allows a topography of rounded craters instead of valleys and rugged peaks, although it is an expensive material.

The nature of the selected particles must be sufficiently chemically stable, to be inert, nontoxic, and biocompatible with biological media.

While the size of the projection particles does not influence the level of surface contamination, it represents the most important parameter in controlling the final roughness of the sand blasted surface. As the particle size increases (between 3 and 5600 µm), the roughness also increases. On titanium, different particle sizes have been studied, from a few micrometers with roughness lower than 0.1 µm to average particle sizes between 25 and 75 µm, with roughness of 0.5–1.5 µm; or higher, with average particle sizes between 200 and 600 µm with a roughness of 2–6 µm, resulting in an optimal roughness value of 1.5 to 4–5 µm, suggesting that the results will not improve with higher values [82].

Because the energy transfer upon particle impact depends on their ease of breakage, as well as mass, the hardness, tenacity, and density of the projection particles may influence the roughness. The smaller the degree of particle breakage, the lower the energy loss at the moment of impact. Furthermore, the larger the impact energy, the larger the plastic deformation on the metal surface.

The adhesion of osteoblasts to sand blasted surfaces is higher than to surfaces without sand blasting [83], which highlights the importance of chemical heterogeneities to cell adhesion, proliferation and differentiation.

For the cell to respond to the microroughness, it is necessary for the cell to perceive this detail; therefore, the dimensions must be of the same order as the cells [84]. Osteoblasts identify a surface as rough when the height of the peaks is greater than 2 µm, and the distance between them does not exceed the length of the cell, approximately 10 µm, ensuring a palisade arrangement of osteoblasts.

10.4.2 Chemical Treatment

There is a wide range of surface treatment methods, including chemical, electrochemical (anodic oxidation), sol–gel, chemical vapor deposition treatments, and biochemical modification. However, the preferred method for the surface chemical treatment of biomaterials commonly employed as the outer layer of the bioreactor is chemical treatment because of its functionality, potentiality, and costs. It is based principally on the chemical reactions occurring at the interface between a metal (or alloy) and a solution. The most common chemical treatments are acids, alkalines, H_2O_2, heat, and passivation; the most usual is acid treatment.

Acid treatment is frequently used to remove oxide and contamination from previous material treatment processes, with the aim of obtaining a clean and uniform final surface. Combinations of acids are used to treat these biomaterials. A solution composed of 10–30 vol% of hydrofluoric acid (HF) in distilled water is recommended as standard acidic pretreatments. Hydrochloric acid (HCl) has been shown to act as an excellent decontamination agent because it can easily dissolve the salts from metal surfaces such as titanium, for example, without weakening the metal surface [85]. The bi-acidic treatment, using HF as a stripper agent and (HCl/H_2SO_4) to produce additional roughness, achieved uniformity and decontamination. The submicrometric topography achieved through acidic etching may influence clot retention and osseous matrix formation [79, 86]. This chemical etching generally results in the formation of a thin layer of surface oxide, less than 10 nm in thickness.

10.4.3 Heat Treatment

The passive surface oxide layer formed on the metal after the acidic treatment is highly stable. However, by reacting the TiO_2 with a basic NaOH solution, a sodium titanate gel ($Na_2Ti_5O_{11}$) is formed and can be thermally stabilized in the form of partially crystallized sodium titanate. After the formation of this layer, the substrate is covered with Ti–OH groups through the release of Na^+ ions from the metal surface into the body fluid via exchange with H_3O^+ ions in the fluid [87]. After a series of specific reactions within the body, the contact of the surface with the body fluids leads to the formation of apatite. This apatite layer shows compositional gradients from the outer apatite layer, through the hydrated titania and titanium oxide, to pure titanium. These gradients imply that the layer is integrated into the titanium substrate and leads not only to a strong bond between the apatite film and the titanium substrate but also to the production of a uniform stress gradient from the bone to the implanted material. The apatite formation in the material surface is considered a prerequisite for the direct bond to the bone [88].

The chemical reactions and subsequent heat treatment can be considered modify the surface topography in a nanometric order, achieving a triple topography. The combined effects of this geometry collaborate to increase the charge density at the surface, favoring apatite nucleation. The greater surface area offers more negative charges and Ti–OH groups to the ions of the medium. Surfaces subjected to this treatment have demonstrated adequate behavior with respect to cell adhesion, proliferation and differentiation, without compromising the material cytocompatibility. The improvement in cell differentiation is principally related to the significant increase in surface hydrophilicity after the formation of sodium titanate gel, with its high density of hydroxyl groups. These effects exhibit a synergistic response with the substrate roughness [89].

The assessment of the membrane influence in bone augmentation procedures demonstrated the existence of positive effects in bone defect regeneration [90], although in some cases it is doubted whether a barrier membrane must be used to cover the augmented site [91]. Thus, the engineering design, topography, and surface treatments play an important role in the performance of these membranes during the regeneration process, as mentioned above because these characteristics create a favorable cell microenvironment for bone augmentation. A membrane's surface topography alters OPG expression through adhesive MG63 cells [92] causing signal transduction, transcription, regulation, proliferation, apoptosis and cytoskeleton formation; it can also affect osteoclast number and activity, showing that the progenitor cell population is influenced by the membrane surface [93]. The surface oxide layer may exert a favorable effect on the matrix architecture and on cell activity close to the membrane. The placement of occlusive osteoconductive non-resolvable membranes facilitates a significant bone formation [94], favoring the induction of bone regeneration and augmentation, and improving the outcomes of techniques that utilize bone substitutes [95].

Microperforations of the bone surface provide a blood supply to the area and contribute to the migration of mesenchymal cells from the base of this *in vivo* 'bioreactor', further ensuring the presence of GFs in the microenvironment created into the implanted system. These bone defects, represented by microperforations (Figure 10.4), are noncritical because they are made with 6 mm drills or less, which constitutes the limit between a critical-sized and a noncritical-size bone defects. In small animals such as rabbits and rats, drilling the cortical bone layer and accessing the bone

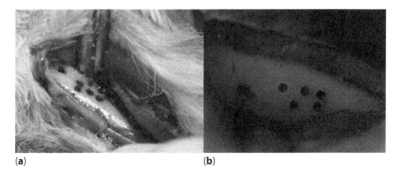

(a)   (b)

Figure 10.4 Microperforations with 2.25 mm diameter drill.

marrow, causing the invasion of blood from the interior of the bone, which facilitates the migration of MSCs, blood, and GFs toward the bone surface [29, 30, 96], has demonstrated improvements in new bone formation. The blood, upon contact with the membrane, triggers the coagulation cascade, resulting in thrombin generation [97], and initiates the healing process. Thrombin stimulates platelet secretion and aggregation and the induction of platelet degranulation produces the consequent release of GFs contained in their granules, producing vasoconstriction and benefitting osteoblast activation.

During the assessment of the *in vivo* performance of Ti and Co–Cr membranes subjected to these series of treatments, different degrees of external bone formation were obtained over the bioreactor in most cases under study (Figure 10.5), possibly due to the nature of the membrane and surface treatments performed, although such behavior might also originate in the periosteum, which is in contact with the exterior side of the membrane, once soft tissue is placed in direct contact at the end of the surgical procedure. However, an important consideration is whether the membrane architecture can influence osteoclast inhibition and osteoblast proliferation on its surface or in its surroundings. This question remains open, and should be researched in future short-term studies. It is expected that in the near future, the adhesion and proliferation of osteoblasts can be improved, further controlling the presence and differentiation of osteoclasts in compromised bone.

In reference to the bone regeneration matrix, represented in this bioreactor complex by the scaffolds, recent research concerning the comparison between different materials assessed the use of two natural scaffolds (whole blood and PRP) and of natural (whole blood) and artificial (tricalcium phosphate, TCP) scaffolds for tissue engineering (Figure 10.6), by performing prospective studies using four different assessment times: 30, 45, 60, and 110 days, in which bone augmentation, quality, architecture,

Figure 10.5 Bioreactor external bone augmentation (Co–Cr membrane).

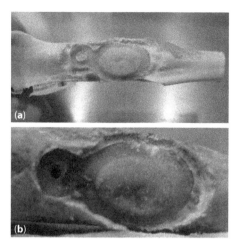

Figure 10.6 Bone augmentation from *in vivo* application of bioreactor in rabbit tibia (Co–Cr membranes): (a) natural scaffold and (b) TCP phosphate.

and cellular activity were compared over the time course of the study. Good results were obtained for the autologous whole blood scaffold, with a good response observed for GFs derived from the administration of PRP, although there was not statistically significant difference in its time evolution, bone height and mineralization beyond 90 days with respect to the intact whole blood. There was a statistically significant difference in the average number of osteoblasts, suggesting a higher osteogenic activity in the blood potentiated with PRP. However, it has been demonstrated that both osteoblast activity and blood vessel formation are improved by the addition of this platelet concentrate, possibly leading to better bone quality, attributing these characteristics to the early release of GFs from the platelet granules during platelet interactions with agents such as collagen or thrombin within the first days after supply. The advantage is that the autologous nature of PRP causes no risk of allergies or tissue versus host reactions [98] in contrast with other osteoinductive agents. Additionally, it has been demonstrated that each of the GFs contained in the platelet granules may produce different positive effects on new bone formation [99]. However, the positive effects of PRP are highly conditioned by the heterogeneity of delivery methods, the biomaterials used, and the obtention methods.

The bone quality of newly formed bone has not shown statistically significant differences between the values of the new bone formed in the different scaffolds, although the density profile analysis over the augmentation in the vertical direction has shown higher values in the extremes of the augmentation, i.e. toward the scaffold periphery and near the membrane borders, decreasing toward the center. The same situation was found in the horizontal direction, where the density values decrease with the distance from the bone surface, corresponding to the newly formed bone toward the center of the bioreactor, characterizing a process of bone centripetal formation due to the migration and placement of the cells on the matrix.

In addition, the whole blood scaffold acts as an extracellular matrix, promoting cell growth and accelerating the regenerative process through the influence of GFs present in the blood clot itself [100]. Thus, the blood clot formed spontaneously in the defect is sufficient to achieve bone healing and regeneration.

In the case of PRP, although several improvements in the bone repair process have been observed, more studies are needed to determine whether their use has any statistically significant benefit in clinical applications [101].

Artificial scaffolds such as TCP have been widely studied and used as scaffolding matrixes for bone tissue engineering techniques. Under normal conditions, the TCP placed in the defect undergoes degradation by the osteoclasts [102]. TCP resorption depends on the ability of these cells to generate an acidic extracellular compartment, with a consequent decrease in pH, essential for bone mineral dissolution. The primary cellular mechanism responsible for this acidification is the active release of protons into the extracellular space [103]. The osteoclast activity is regulated and stimulated by hormones and cytokines [103]. This process, initiated by the osteoclasts, could decelerate bone regeneration in the early stages of restoration in comparison with autologous bone grafts [104]. The high rate of TCP resorption *in vivo* is due to its porosity and small granule size [103], which are determinants of nutrient and oxygen diffusion, the insertion, migration and differentiation of osteogenic cells and the modulation of osteoclastic activity.

In the bioreactor model presented, with the TCP as scaffold, although autologous bone shows a higher total bone volume in the early stages of bone healing, bone substitutes approach this result over time. After 9 months, no statistically significant differences are found between the two types of grafts [105].

In most of the cases studied, the differentiation of MSCs into osteoblasts is induced *in vitro* from bone marrow aspirates, and the bone progenitor cells are subsequently transplanted into living organisms. However, in the cases presented above, the differentiation and proliferation of MSCs into osteoblasts is based both on the natural migration of undifferentiated MSCs from bone marrow to the site of interest through the microperforations, based on the application of undifferentiated MSCs from bone marrow aspirated by appealing to a later differentiation of bone cells at the site of interest. Good results were observed in both cases, with no significant differences in bone repair [106].

Once transplanted into the bioreactor and under optimal conditions, cells differentiate and form a mineralized matrix on the scaffold. The MSCs can differentiate into various lineages depending on local factors, oxygen tension and the mechanical stimuli to which they are exposed [107]. If there is good vascularization, close contact of bone edges and stability, bone regeneration occurs by intramembranous ossification. Defect instability and low oxygen tension favor differentiation toward the chondrogenic lineage. This tissue is then replaced with bone by endochondral ossification.

This subsection can be concluded by indicating that bone augmentation has been accomplished simply by providing the tissue with adequate space through the use of a barrier membrane. The added blood supply of the site delivering GFs by clot formation or the use of artificial scaffolds and release of GFs from platelet granules may be sufficient elements to achieve bone augmentation during a period of 3 months in small animal models.

The recent adaptation of the 'bioreactor complex' for application in techniques of *in vivo* bone regeneration and bone augmentation in preliminary studies on small animal models has yielded suggestive results. The next studies in this field will focus on seeking new applications for this system to expand the range of services offered for the treatment of other pathologies.

Furthermore, the feasibility of the bioreactor's use or its components as substance carriers should be investigated as it represents a good functionality.

10.5 New Multidisciplinary Approaches Intended to Improve and Accelerate the Treatment of Injured and/or Diseased Bone

The results achieved in preliminary studies represent greater progress in the area of bone tissue engineering, which has motivated researchers to expand their investigations and to direct their studies to the potentialities of the bioreactor system in humans.

Each particular application requires the adaptation of the bioreactor complex to permit optimal results depending on the area in which bone augmentation will be needed.

10.5.1 Application of Bioreactor in Dentistry: Therapies for the Treatment of Maxillary Bone Defects

The contributions and progress achieved in this field have encouraged researchers to perform a pilot study on patients with maxillary bone defects.

When a tooth fracture exists, the chronic inflammation that occurs causes the bone in the buccal wall of the maxillary to reabsorb (Figure 10.7a and b). When the tooth is extracted (Figure 10.7c), a cavity was observed in previous location as well as on its buccal wall (Figure 10.7d). If this diagnosis is not treated, bone resorption occurs, in both the vertical and horizontal directions, where no bone exists. This clinical situation is accompanied by soft tissue collapse, which follows the bone behavior.

This condition is avoided when, immediately upon extraction, the bioreactor complex is implanted (Figure 10.8).

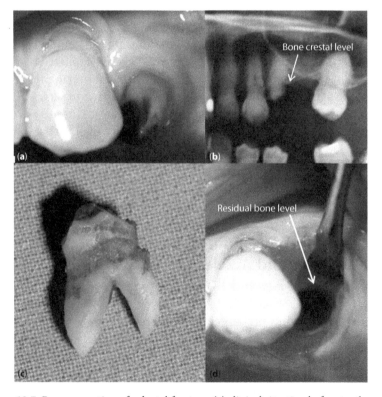

Figure 10.7 Bone resorption of a dental fracture: (a) clinical situation before tooth extraction, (b) radiographic representation before extraction, and (c) crestal bone resorption and bone cavity.

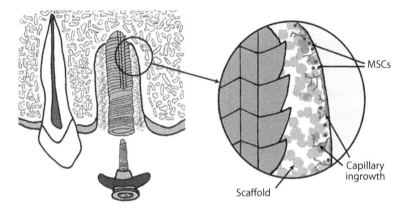

Figure 10.8 Bioreactor *in vivo*.

It is expected that the results from this study will provide valuable information for the design and development of new, less traumatic techniques and methods for bone augmentation prior to implant placement, compared to the gold standard procedures, as autologous bone grafts increase the risk of major morbidity, and the recovery period is lengthy in aesthetic terms. Furthermore, the recovery of soft tissue resorption is very difficult.

The application of this new concept when performing the extraction and during the subsequent placement of the bioreactor in the defect site allows regeneration of the missing bone and maintenance of the soft tissue, achieving a better aesthetic outcome and bone regeneration without the need for autologous bone grafts.

In the case of oral bone regeneration, the bioreactor consists of the MSCs provided by the blood clot formed after the dental extraction, which also acts as a scaffold and can be used in combination with artificial scaffolds such as TCP crystals (Figure 10.9a) to maintain the implant in position (Figure 10.9b). The implant acts as a skeleton for the complex of stem cells/scaffold.

Furthermore, taking advantage of the microfixation implant-abutment, a membrane is screwed into position (Figure 10.10a). The membrane acts as a protective cover for the bioreactor (Figure 10.10b) and accordingly consist of titanium. This membrane consists of a thin layer of this metal and is hand-moldable and easily adapts to the defect site.

This system allows maintenance of the soft tissue at its initial level (after extraction, at gum level), avoiding its resorption and improving the results in both functional and aesthetic terms (Figure 10.11a and b).

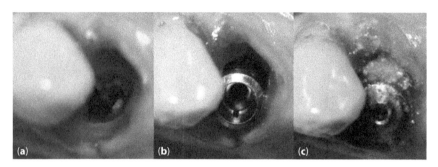

Figure 10.9 (a) Placement of the TCP scaffold. (b and c) Implant placement and fixation.

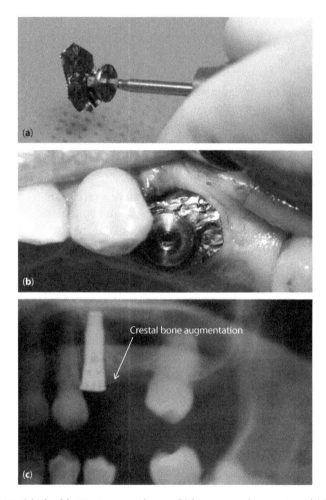

Figure 10.10 (a) Flexible titanium membrane, (b) bioreactor placement, and (c) post-implant radiography.

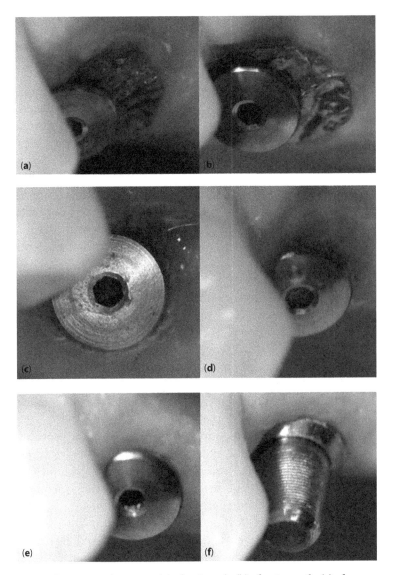

Figure 10.11 Bioreactor placement: (a) after 1 week, (b) after 1 month, (c) after membrane extraction (1 month), (d) after 2 months, (e) after 3 months, and (f) implant activation.

10.5.2 Application of Bioreactor in Cases of Osteoporosis

Osteoporosis is an illness that affects approximately 40% of the world's adult population, principally postmenopausal Caucasian women. It is characterized by the systemic impairment of bone mass, density and microarchitecture, resulting in fragility fractures [108].

Because the bone loss occurs insidiously and is initially asymptomatic, osteoporosis is often only diagnosed after the first clinical fracture has occurred; thus, the aim of the actual therapy is generally the prevention of further fractures.

Early assessment of an individual's risk of developing osteoporosis is therefore important to develop preventive actions to minimize the patient's morbidity. The clinician must consider various risk factors, such as age, low body mass index, previous fragility fractures, family history, the use of glucocorticoids, and active cigarette smoking.

Most of the actions around this disease are preventive, before its development. However, once osteoporosis is detected, the available treatments are generally experimental and based on drugs, and the purpose is to inhibit bone resorption by osteoclasts and/or promote osteoblastic activity. These approaches lack properly tested benefits, and in many occasions, have demonstrated disadvantages.

The quality of life of patients with osteoporosis and their families is reduced due to the limitations encountered daily. Because the disease has no treatment that allows a successful rehabilitation, patients' activities are restricted because of the potential risk of fracture to which they are daily exposed. Fracture and the subsequent loss of mobility and autonomy in their daily life often represent a major drop in quality of life, strongly affecting the patient's psychological state, and can lead to different psychological disorders. In addition to the higher morbidity, osteoporosis is associated with high economic costs in health systems.

Osteoporotic fractures of the hip and spine represent a probability of up to 20% mortality during the 12-month period after the fracture, especially in older patients [109] because they require hospitalization and have subsequently enhanced risk of other complications, such as pneumonia or thromboembolic disease due to chronic immobilization.

The use of Co–Cr alloy in different biomedical implants, alone or in combination with other elements, has been widely diffused. Its main application is mostly in hip prostheses. Many studies have been developed based on its use [110], principally focused on the mechanical performance offered during the replacement of the function or the structure.

Many researchers have dedicated their efforts to studying the influence of the Co–Cr alloy particles released from implanted prostheses, finding some encouraging results [110, 111]. They have demonstrated that the wear particles released from the articulating surface of total joint prostheses into the environment cause inflammatory and toxicity reactions both locally and systematically. The additional corrosion of the particles released by the wear may alter the biochemical environment of the tissue surrounding an implant to favor bone resorption. Under these circumstances, the reactions of the surrounding tissue can play an important role in the loss of the prosthesis. However, when Co–Cr membranes are implanted, no ion release is observed because there is no friction between surfaces; no frictional forces are generated. As the membranes can be removed once the desired bone formation is achieved, this placement is temporary.

Based on previous studies (mentioned in Section 10.4) and on the demonstrated potential of the Co–Cr alloy, both in bone augmentation and the repair of noncritical-sized bone defects in rabbit tibia, it has been possible to conclude that these membranes and the bioreactor complex itself provide an adequate space to allow bone growth and that the combination of the membranes with several growth-promoting factors produces good results in terms of bone height and volume as well as bone quality, as the presence of osteons with good circumferential organization and laminar disposition compatible with cortical bone was observed in all cases. Additionally, Haversian formation was found, with the expression of Volkmann and Havers canals with defined edges and concentric osteocytes.

The possibility of rehabilitation in patients with advanced osteoporosis through the use of such bioreactors for the regeneration of bone at sites affected by the disease progression can significantly improve the quality of life for patients with an increased risk of fractures in the short term. Previous studies [29, 30] have concluded that Co–Cr membranes could have '*per se*' an inductive effect in bone formation based on the observation of bone covering on the membranes and on its upper side, demonstrating that the Co–Cr alloy has the mechanical properties to support the membrane's own efforts.

For all of the above reasons, short-term investigations have focused on the use of Co–Cr membranes and plates to promote new bone formation, with similar or identical qualities to healthy bone on the preexisting osteoporotic bone, on which these implants will be placed. This approach has not yet been studied in osteoporotic animals or humans. Thus, in the first stage, the research will focus on whether the new bone formation has the appearance and characteristics of healthy bone, as well as on assessing its temporal evolution (i.e. whether the newly formed bone has good quality and architecture to avoid the risk of fractures in the short and long terms) in rabbits. In the second stage, a study may be proposed in humans.

10.6 Computational Modeling: An Effective Tool to Predict Bone Ingrowth

Numerical studies based on theoretical reaction–diffusion models for the analysis of transport phenomena in the growth of bone tissue between bone and implant metal material coincide in their results with experimental observations [112–115]. The numerical results show that surfaces with greater capacity to capture GFs accelerate the process of bone growth around the implant. This qualitative equivalence between the experimental results and theoretical predictions of other authors encourage the production of their own theoretical perspectives to explain, based on recent studies, the origin of the factors influencing bone growth.

Complementarily, initial guidelines for a theoretical model based on previous numerical studies will be proposed [112–115]. Thus, it is expected that the phenomena governing the dynamics of the experimental observations can be explained by this model.

There exist no theoretical biological models that describe the dynamics of bone growth induced only by the membranes. More recent studies on bone growth in the space between a dental implant and the bone surface [112–115] have modeled the dynamics of such growth considering the space to be completely occupied by blood, with the presence of GFs in different concentrations. These works, from a continuous perspective, qualitatively and quantitatively predict the cascade of events in the development of bone tissue inside these spaces and have concluded that higher growth is produced by high concentrations of blood factors (as fibrin) and higher contact with the implant surface (high surface microtopography), factors potentiated by the presence of external stimuli. These stimuli can have different origins, with mechanical being the most traditional. However, the general proposal of the model [114] admits the possibility that the stimuli can derived from other sources, e.g. electromagnetically or from factors related to the surface characteristics of the material (e.g. surface free energy, surface charge).

The model characteristics used by some authors [114, 115] indicate that adequate values of the parameters can be applied to predict the observations reported in other works [29]. However, with the bone–implant space occupied by blood, the domain where the equations are resolved is predefined, which is not the case in this work. It is therefore proposed to modify the existing model [114, 115] to describe the dynamics of bone growth under the planning conditions in the bioreactor system, considering the absence of blood in the space between the membrane and the bone surface.

References

1. Sfeir, C., Ho, L., Doll, B.A., Azari, K., Hollinger, J.O., et al., Fracture Repair. *Bone Regeneration and Repair: Biology and Clinical Applications*, J.R. Lieberman, G.E. Friedlaender, (eds.), pp. 21–44, Humana Press, New Jersey, 2005.
2. Ishidou, Y., Kitajima, I., Obama, H., Maruyama, I., Murata, F., Imamura, T., Yamada, N., Duke, P.T., Miyazono, K., Sakou, T, Enhanced expression of type I receptors for bone morphogenetic proteins during bone formation. *J. Bone Miner. Res.*, 10, 11, 1995.
3. Simonet, W.S., Lacey, D.L., Dunstan, C.R., Kelley, M., Chang, M.S., Luthy, R., Nguyen, H.Q., Wooden, S., Bennett, L., Boone, T., Shimamoto, G., DeRose, M., Elliott, R., Colombero, A., Tan, H.L., Trail, G., Sullivan, J., Davy, E., Bucay, N., Renshaw-Gegg, L., Hughes, T.M., Hill, D., Pattison, W., Campbell, P., Sander, S., Van, G., Tarpley, J., Derby, P., Lee, R., Boyle, W.J. Osteoprotegerin: a novel secreted protein involved in the regulation of bone density. *Cell*, 89, 309–319, 1997.
4. Tsuda, E., Goto, M., Michizuki, S., Yano, K., Kobayashi, F., Morinaga, T., Higashio, K. Isolation of a novel cytokine from human fibroblasts that specifically inhibits osteoclastogenesis. *Biochem. Biophys. Res. Commun.*, 234, 137–142, 1997.
5. Lacey, D.L., Timms, E., Tan, H.L., Kelley, M.J., Dunstan, C.R., Burgess, T., Elliott, R., Colombero, A., Elliott, G., Scully, S., Hsu, H., Sullivan, J., Hawkins, N., Davy, E., Capparelli, C., Eli, A., Qian, Y.-X., Kaufman, S., Sarosi, I., Shalhoub, V., Senaldi, G., Guo, J., Delaney, J., Boyle, W.J. Osteoprotegerin ligand is a cytokine that regulates osteoclast differentiation and activation. *Cell*, 93, 165–176, 1998.
6. Lam, J., Takeshita, S., Barker, J. E., Kanagawa, O., Ross, F. P., Teitelbaum, S. L. TNF-α induces osteoclastogenesis by direct stimulation of macrophages exposed to permissive levels of RANK ligand. *J. Clin. Invest.*, 106, 12, 2000.
7. Lehmann, W., Edgar, C.M., Wang, K., Cho, T.J., Barnes, G.L., Kakar, S., Graves, D.T., Ruege, J.M., Gerstenfeld, L.C., Einhorn, T.A. Tumor necrosis factor alpha (TNF-α) coordinately regulates the expression of specific matrix metalloproteinases (MMPS) and angiogenic factors during fracture healing. *Bone*, 36, 2, 2005.
8. Gerstenfeld, L.C., Cullinane, D.M., Barnes, G.L., Graves, D.T., Einhorn, T.A., Fracture healing as a post-natal developmental process: molecular, spatial, and temporal aspects of its regulation. *J. Cell. Biochem.*, 88, 2003.
9. Thirunavukkarasu, K., Miles, R. R., Halladay, D. L., Yang, X., Galvin, R. J., Chandrasekhar, S., Martin, T.J., Onyia, J. E, Stimulation of osteoprotegerin (OPG) gene expression by transforming growth factor-β (TGF-β). Mapping of the OPG promoter region that mediates TGF-β effects. *J. Biol. Chem.*, 276, 39, 2001.

10. Nilsson, O.S., Urist, M.R., Dawson, E.G., Schmalzried, T.P., and Finerman, G.A., Bone repair induced by bone morphogenetic protein in ulnar defects in dogs. *J. Bone Joint. Surg.*, 68, 4, 1986.
11. AI-Aql, Z.S., Alag, A.S., Graves, D.T., Gerstenfeld, L.C., Einhorn, T.A., Molecular Mechanisms Controlling Bone Formation during Fracture Healing and Distraction Osteogenesis. *J. Dent. Res.*, 87, 2, 2008.
12. Suri, C., Jones, P.F., Patan, S., Bartunkova, S., Maisonpierre, P.C., Davis, S., Sato, T.N., Yancopoulos, G.D., Requisite role of angiopoietin-1, a ligand for the TIE2 receptor, during embryonic angiogenesis. *Cell*, 87, 1996.
13. Ferrara, N., Davis-Smyth, T., The biology of vascular endothelial growth factor. *Endocr. Rev.*, 18, 1997.
14. Yeh, L.C., Lee, J.C., Osteogenic protein-1 increases gene expression of vascular endothelial growth factor in primary cultures of fetal rat calvaria cells. *Mol. Cell. Endocrinol.*, 153, 1999.
15. Deckers, M.M., Van Bezooijen, R.L., Van Der Horst, G., Hoogendam, J., Van Der Bent, C., Papapoulos, S.E., Löwik, C. W., Bone morphogenetic proteins stimulate angiogenesis through osteoblast-derived vascular endothelial growth factor A. *Endocrinology*, 143, 1545–1553, 2002.
16. Lehmann, W., Edgar, C.M., Wang, K., Cho, T.J., Barnes, G.L., Kakar, S., Graves, D.T., Rueger, J.M., Gerstenfeld, L.C., Einhorn, T.A., Tumor necrosis factor alpha (TNF-alpha) coordinately regulates the expression of specific matrix metalloproteinases (MMPS) and angiogenic factors during fracture healing. *Bone*, 36, 300–310, 2005.
17. Fagiani, E., Christofori, G., Angiopoietins in angiogenesis. *Cancer Lett.*, 328, 18–26, 2013.
18. Hung, S.P., Ho, J.H., Shih, Y.R.V., Lo, T., Lee, O.K., Hypoxia promotes proliferation and osteogenic differentiation potentials of human mesenchymal stem cells. *J. Orthop. Res.*, 30, 2, 2012.
19. Beltrán, V., Engelke, W., Prieto, R., Valdivia-Gandur, I., Navarro, P., Manzanares, M.C., Borie, E., Fuentes, R., Augmentation of intramembranous bone in rabbit calvaria using an occlusive barrier in combination with demineralized bone matrix (DBM): A pilot study. *Int. J. Surg.*, 12, 378–383, 2014.
20. Weissman, I.L., Stem cells: units of development, units of regeneration and units in evolution. *Cell*, 100, 1, 2000.
21. Blau, H.M., Brazelton, T.R., Weinmann, J.M., The evolving concept review of a stem cell: entity or function? *Cell*, 105, 829–841, 2001.
22. Panek M., Marijanović I., Ivković A., Stem cells in bone regeneration. *Period. Biol.*, 117, 1, 2015.
23. Pittenger, M.F., Mackay, A.M., Beck, S.C., Jaiswal, R.K., Douglas, R., Mosca, J.D., Moorman, M.A., Simonetti, D.W., Craig, S., Marshak, D.R., Multilineage potential of adult human mesenchymal stem cells. *Science*, 284, 5411, 1999.
24. Gnecchi, M., Danieli, P., Cervio, E., Mesenchymal stem cell therapy for heart disease. *Vasc. Pharmacol.*, 57, 1, 2012.

25. Venkataramana, N.K., Kumar, S.K., Balaraju, S., Radhakrishnan, R.C., Bansal, A., Dixit, A., Rao, D.K., Das, M., Jan, M., Kumar Gupta, P., Totey, SM., Open-labeled study of unilateral autologous bone-marrow-derived mesenchymal stem cell transplantation in Parkinson's disease. *Transl. Res.*, 155, 2, 2010.
26. Yu, J., Yin, S., Zhang, W., Gao, F., Liu, Y., Chen, Z., Zhang, M., He, J., Zheng, S. Hypoxia preconditioned bone marrow mesenchymal stem cells promote liver regeneration in a rat massive hepatectomy model. *Stem. Cell Res. Ter.*, 4, 4, 2013.
27. Yeatts, A.B., Choquette, D.T., Fisher, J.P., Bioreactors to influence stem cell fate: augmentation of mesenchymal stem cell signaling pathways via dynamic culture systems. *Biochim. Biophys. Acta*, 1830, 2, 2470–2480, 2013.
28. Sun, Y., Chen, C.S., Fu, J. Forcing stem cells to behave: a biophysical perspective of the cellular microenvironment. *Annu. Rev. Biophys.*, 41, 2012.
29. Ripamonti, U. Soluble and insoluble signals sculpt osteogenesis in angiogenesis. *World J. Biol. Chem.*, 1, 5, 2010.
30. Decco, O., Beltrán, V., Zuchuat, J., Cura, A., Lezcano, M.F., Engelke, W., Bone augmentation in rabbit tibia using microfixed Cobalt-Chromium membranes with whole blood and platelet-rich plasma. *Materials*, 8, 8, 2015.
31. Decco, O., Cura, C., Beltrán, V., Lezcano, F., Engelke, W., Bone augmentation in rabbit tibia using microfixed cobalt-chromium membranes with whole blood, tricalcium phosphate and bone marrow cells. *Int. J. Clin. Exp. Med.*, 8, 1, 2015.
32. Kima, S., Parkc, M.S., Jeonc, O., Choib, C.Y., Kim, B., Poly(lactide-co-glycolide)/hydroxyapatite composite scaffolds for bone tissue engineering. *Biomaterials*, 27, 8, 2006.
33. Phippsa, M.C., Clemb, W.C., Grundac, J.M., Clinesc, G.A., Bellis, S.L. Increasing the pore sizes of bone-mimetic electrospun scaffolds comprised of polycaprolactone, collagen I and hydroxyapatite to enhance cell infiltration. *Biomaterials*, 33, 2, 2012.
34. Petite, H., Viateau, V., Bensaïd, W., Meunier, A., Pollak, C., Bourguignon, M., Oudina, K., Sedel, L., Guillemin, G., Tissue-engineered bone regeneration. *Nat. Biotechnol.*, 18, 959–963, 2000.
35. Botchwey, E.A., Pollack, S.R., Levine, E.M., Laurencin, C.T., Bone tissue engineering in a rotating bioreactor using microcarrier matrix system. *J. Biomed. Mater. Res.*, 55, 2001.
36. Michelson A.D. (ed). Platelets. 3rd. Edition. White J.G. Chapter 7: Platelet Structure, pp. 117–144, 2013.
37. Zhang, N., Wu, Y.P., Qian, S.J., Teng, C., Chen, S., Li, H., Research progress in the mechanism of effect of PRP in bone deficiency healing. *Sci. World J.*, 2013.
38. Albanese, A., Licata, M.E., Polizzi, B., Campisi, G., Platelet-rich plasma (PRP) in dental and oral surgery: from the wound healing to bone regeneration. *Immun. Ageing*, 1, 10, 2013.
39. Malhotra, A., Pelletier, M.H., Yu, Y., Walsh, W.R., Can platelet-rich plasma (PRP) improve bone healing? A comparison between the theory and experimental outcomes. *Arch. Orthop. Trauma Surg.*, 133, 2, 2013.

40. Yun, J.H., Yoo, J.H., Choi, S.H., Lee, M.H., Lee, S.J., Song S.U., Oh, N.-S., Synergistic effect of bone marrow-derived mesenchymal stem cells and platelet-rich plasma on bone regeneration of calvarial defects in rabbits. *J. Tissue Eng. Regen. Med.*, 9, 1, pp. 17–23, 2012.
41. Makhdom, A.M., Hamdy, R.C., The role of growth factors on acceleration of bone regeneration during distraction osteogenesis. *Tissue Eng. Part B Rev.*, 19, 5, 2013.
42. Ishihara, A., Zekas, L.J., Litsky, A.S., Weisbrode, S.E., Bertone, A.L., Dermal fibroblast-mediated BMP2 therapy to accelerate bone healing in an equine osteotomy model. *J. Orthop. Res.*, 28, 3, 2010.
43. Battinelli, E.M., Markens, B.A., Italiano J.E. Release of angiogenesis regulatory proteins from platelet alpha granules: modulation of physiologic and pathologic angiogenesis. *Blood*, 118, 5, 2011.
44. Marx, R.E., Garg, A.K. et al., Bone graft physiology with use of platelet-rich plasma and hyperbaric oxygen, in: *The Sinus Bone Graft*, O.T. Jensen (ed), p. 183, Quintessence, Chicago, 1999.
45. Lynch, S.E., Marx, R.E., Nevins, M., Wisner-Lynch, L.A. (eds.), *Tissue Engineering: Applications in Oral and Maxillofacial Surgery and Periodontics*. Quintessence, China, 2008.
46. Anitua, E., Sánchez, M., Orive, G., Potential of endogenous regenerative technology for in situ regenerative medicine. *Adv. Drug Deliv. Rev.*, 62, 7, 741–752, 2010.
47. Raggatt, L.J., Partridge, N.C., Cellular and molecular mechanisms of bone remodeling. *J. Biol. Chem.*, 285, 2010.
48. Daar, I.O. (ed), *Advances in Cancer Research: Guidance Molecules in Cancer and Tumor Angiogenesis*, Academic Press, USA, Vol. 114, 2012.
49. Sagalovsky, S., Physiological role of growth factors and bone morphogenetic proteins in osteogenesis and bone fracture healing: a review. *Calendar Clin. Med.*, 38, 2015.
50. Fei, Y., Gronowicz, G., Hurley, M.M. Fibroblast growth factor-2, bone homeostasis and fracture repair. *Curr. Pharm. Des.*, 19, 19, 3354–3363, 2013.
51. Marsell, R., Einhorn, T. A., The role of endogenous bone morphogenetic proteins in normal skeletal repair. *Injury*, 40, p. S4–S7 2009.
52. Carreira, A.C., Lojudice, F.H., Halcsik, E., Navarro, R.D., Sogayar, M.C., Granjeiro, J.M., Bone morphogenetic proteins facts, challenges, and future perspectives. *J. Dent. Res.*, 2014.
53. Gazzerro, E., Canalis, E., Bone morphogenetic proteins and their antagonists. *Rev. Endocr. Metab. Disord.*, 7, 1–2, 2006.
54. Lissenberg-Thunnissen, S.N., De Gorter, D.J., Sier, C.F., Schipper, I.B., Use and efficacy of bone morphogenetic proteins in fracture healing. *Int. Orthop.*, 35, 9, 2011.
55. Neffe, A.T., Julich-Gruner, K.K., Lendlein, A., Combinations of biopolymers and synthetic polymers for bone regeneration, in: *Biomaterials for Bone Regeneration: Novel Techniques and Applications*, Dubruel, P., Van Vlierberghe, S. (eds.), WoodHead Publishing, United Kingdom, 2014.

56. Rakhmatia, Y. D., Ayukawa, Y., Furuhashi, A., Koyano, K., Current barrier membranes: titanium mesh and other membranes for guided bone regeneration in dental applications. *J. Prosthodont. Res.*, 57, 1, 2013.
57. Dimitriou, R., Mataliotakis, G.I., Calori, G.M., Giannoudis, P.V., The role of barrier membranes for guided bone regeneration and restoration of large bone defects: current experimental and clinical evidence. *BMC Med.*, 10, p. 1–24, 2012.
58. Rocchietta, I., Simion, M., Hoffmann, M., Trisciuoglio, D., Benigni, M., Dahlin, C., Vertical bone augmentation with an autogenous block or particles in combination with guided bone regeneration: A clinical and histological preliminary study in humans. *Clin. Implant Dent. Relat. Res.*, 2015.
59. Khan, R., Khan, M.H., Use of collagen as a biomaterial: an update. *J. Indian Soc. Periodontol.*, 17, 4, 539, 2013.
60. Bernabé, P.F.E., Melo, L.G.N., Cintra, L.T.A., Gomes-Filho, J.E., Dezan, E., Nagata, M.J.H. Bone healing in critical-size defects treated with either bone graft, membrane, or a combination of both materials: a histological and histometric study in rat tibiae. *Clin. Oral Impl. Res.*, 23, 2012.
61. Trombelli, L., Farina, R., Marzola, A., Itro, A., Calura, G., GBR and autogenous cortical bone particulate by bone scraper for alveolar ridge augmentation: a 2-case report. *Int. J. Oral Maxillofac. Implants*, 23, 2008.
62. Holowka, E.P., Bhatia, S.K., *Drug Delivery: Materials Design and Clinical Perspective*. Chapter 1: Introduction. Springer, New York, 1–6, 2014.
63. Bottino, M.C., Thomas, V., Membranes for periodontal regeneration—a materials perspective. *Front. Oral Biol.*, 17, 90–100, 2015.
64. Schroeder, J.E., Mosheiff, R., Tissue engineering approaches for bone repair: concepts and evidence. *Injury*, 42, 2011.
65. Liu, Y., Lim, J., Teoh, S.H. Review: development of clinically relevant scaffolds for vascularised bone tissue engineering. *Biotechnol. Adv.*, 31, 5, 688–705, 2013.
66. Billström, G.H., Blom, A.W., Larsson, S., Beswick, A.D., Application of scaffolds for bone regeneration strategies: current trends and future directions. *Injury*, 44, S28–S33, 2013.
67. Cox, S.C., Thornby, J.A., Gibbons, G.J., Williams, M.A., Mallick, K.K., 3D printing of porous hydroxyapatite scaffolds intended for use in bone tissue engineering applications. *Mater. Sci. Eng.*, 47, 2015.
68. Gardel, L.S., Serra, L.A., Reis, R.L., and Gomes, M.E., Use of perfusion bioreactors and large animal models for long bone tissue engineering. *Tissue Eng. Part B*, 20, 2, 2014.
69. Korossis, S.A., Bolland, F., Kearney, J.N., Fisher, J., Ingham, E., Bioreactors in tissue engineering. In: *Topics in Tissue Engineering*, N. Ashammakhi, R.L. Reis (eds), Expertissues, Leeds, 2005.
70. Ratcliffe, A., Niklason, L.E., Bioreactors and bioprocessing for tissue engineering. *Ann. NY Acad. Sci.*, 961, 1, 2002.
71. Pei, M., Solchaga, L.A., Seidel, J., Zeng, L., Vunjak-Novakovic, G., Caplan, A.I., and Freed, L.E. Bioreactors mediate the effectiveness of tissue engineering scaffolds. *FASEB J.*, 16, 12, 1691–1694, 2002.

72. Goncalves, A., Costa, P., Rodrigues, M.T., Dias, I.R., Reis, R.L., Gomes, M.E., Effect of flow perfusion conditions in the chondrogenic differentiation of bone marrow stromal cells cultured onto starch based biodegradable scaffolds. *Acta Biomater.*, 7, 2011.
73. Yeatts, A.B., Fisher, J.P., Bone tissue engineering bioreactors: dynamic culture and the influence of shear stress. *Bone*, 48, 2, 171–181, 2011.
74. Wang, T.W., Wu, H.C., Wang, H.Y., Lin, F.H., Sun, J.S., Regulation of adult human mesenchymal stem cells into osteogenic and chondrogenic lineages by different bioreactor systems. *J. Biomed. Mater. Res. A*, 88, 935–946, 2009.
75. Yu, X.J., Botchwey, E.A., Levine, E.M., Pollack, S.R., Laurencin, C.T., Bioreactor-based bone tissue engineering: the influence of dynamic flow on osteoblast phenotypic expression and matrix mineralization. *Proc. Natl. Acad. Sci.*, 101, 2004.
76. Paz, A., Martín, Y., Pazos, L.M., Parodi, M.B., Ybarra, G.O., González, J.E., Obtención de recubrimientos de hidroxiapatita sobre titanio mediante el método biomimético. *Rev. de. Met.*, 47, 2, 2011.
77. Parodi, M.B., Conterno, G., Pazos, L., Roth, M., Corengia, P., Egidi, D.A., Crosta, R., Características de Superficie de Titanio atacado con Ácido Sulfúrico y Ácido Clorhídrico. Centro de Investigación y Desarrollo en Mecánica, *INTI*, 157, p. B1650WAB, 2004.
78. Yan Guo, C., Hong Tang, A.T., Pekka, M.J., Insights into Surface Treatment Methods of Titanium Dental Implants. *J. Adhes. Sci. Technol.*, 26, 2012.
79. Raines, A.L., Olivares Navarretea, R., Wielandb, M., Cochranc, D.L., Schwartza, Z., Boyan, B.D., Regulation of angiogenesis during osseointegration by titanium surface microstructure and energy. *Biomaterials*, 31, 18, 2010.
80. Hung W.L., *Podwer Metallurgical Processing of Titanium and Its Alloys*. Depart. of Process Eng. & Applied Science, Dalhousie University, Halifax, Nova Scotia, 2011.
81. Mahagaonkar, S.B., Brahmankar, P.K., Seemikeri, C.Y., Effect on fatigue performance of shot peened components: an analysis using DOE technique. *Int. J. Fatigue*, 31, 4, 2009.
82. Pazos, L., Corengia, P., Svoboda, H., Effect of surface treatments on the fatigue life of titanium for biomedical applications. *J. Mech. Behav. Biomed. Mater.*, 3, 6, 2010.
83. Badenas, Conrado José Aparicio. Tratamientos de superficie sobre titanio comercialmente puro para la mejora de la osteointegración de los implantes dentales. Universitat Politècnica de Catalunya, 2005.
84. Logan, N., Sherif, A., Cross, A.J., Collins, S.N., Traynor, A., Bozec, L., Parkin, I.P., Brett, P., TiO2-coated CoCrMo: Improving the osteogenic differentiation and adhesion of mesenchymal stem cells *in vitro*. *J. Biomed. Mater. Res. Part A*, 103, 2015.
85. Xu, S., Xiaoyu, Y., Yuan, S., Minhua, T., Jian, L., Aidi, N., Xing, L., Morphology improvement of sandblasted and acid-etched titanium surface and osteoblast attachment promotion by hydroxyapatite coating. *Rare. Metal Mater. Eng.*, 44, 1, 2015.

86. Takeuchi, M., Abe, Y., Yoshida, Y., Nakayama, Y., Okazaki, M., Akagawa, Y., Acid pretreatment of titanium implants. *Biomaterials*, 24, 10, 2003.
87. Chen, J., *Osteoblasts Response to Anodized Commercially Pure Titanium In Vitro*. Department of Dentistry, Oral and Maxillofacial Surgery Tübingen, Alemania, (Doctoral dissertation) p 1–86, 2005.
88. Decco, O., Beltrán, V., Zuchuat, J., Gudiño, R., Comparative *in vitro* study of surface treatment of grade II titanium biomedical implant. *IIFMBE Proceedings*, 2015.
89. Komasa, S., Yingmin, S., Taguchi, Y., Yamawaki, I., Tsutsumi, Y., Kusumoto, T., Nishizaki, H., Miyake, T., Umeda, M., Tanaka, M., Okazaki, J., Bioactivity of titanium surface nanostructures following chemical processing and heat treatment. *J. Hard. Tissue Biol.*, 24, 3, 257–266, 2015.
90. Xiong, L., Yang, L., Xingdong, Z., Jinrui, X., Ling, Q., Chun-wai, C., Comparative study of osteoconduction on micromachined and alkali-treated titanium alloy surfaces *in vitro* and *in vivo*. *Biomaterials*, 26, 1793–1801, 2005.
91. McAllister, B.S., Haghighat, K.J., Bone augmentation techniques. *J. Periodontol.*, 78, 1793–1801, 377–396, 2007.
92. Chiapasco, M., Abati, S., Romeo, E., Vogel, G., Clinical outcomes of autogenous bone blocks or guided bone regeneration with e-PTFE membranes for the reconstruction of narrow edentulous ridges. *Clin. Oral Implant Res.*, 4, 1999.
93. Pittenger, M.F., Mackay, A.M., Beck, S.C., Jaiswal, R.K., Douglas, R., Mosca, J.D., Moorman, M.A., Simonetti, D.W., Craig, S., Marshak, D.R., Multilineage potential of adult human mesenchymal stem cells. *Science*, 284, 1999.
94. Davies, J.E., Schupbach, P., Cooper, L., *et al.*, The implant surface and biological response, in: *Osseointegration and Dental Implants*, A. Jokstad (Ed.), pp. 213–223, Wiley-Blackwell: Oxford, UK, 2009.
95. Van Steenberghe, D., Johansson, C., Quirynen, M., Molly, L., Albrektsson, T., Naert, I., Bone augmentation by means of a stiff occlusive titanium barrier. *Clin. Oral Implant Res.*, 14, 63–71, 2003.
96. Rakhmatia, Y.D., Ayukawa, Y., Furuhashi, A., Koyano, K., Current barrier membranes: Titanium mesh and other membranes for guided bone regeneration in dental applications. *J. Prosthodont. Res.*, 57, 3–14, 2013.
97. Rompen, E.H., Biewer, R., Vanheusden, A., Zahedi, S., Nusgens, B., The influence of cortical perforations and of space filling with peripheral blood on the kinetics of guided bone generation. A comparative histometric study in the rat. *Clin. Oral Implants Res.*, 10, 85–94, 1999.
98. Ogawa, M., Tohma, Y., Ohgushi, H., Takakura, Y., Tanaka, Y., Early fixation of cobalt-chromium based alloy surgical implants to bone using a tissue-engineering approach. *Int. J. Mol. Sci.*, 13, 5528–5541, 2012.
99. Anitua, E., Sanchez, M., Nurden, A.T., Zalduendo, M., Reciprocal actions of platelet secreted TGF-beta1 on the production of VEGF and HGF by human tendoncells. *Plast. Reconstr. Surg.*, 119, 950–959, 2007.

100. Gandhi, A., Bibbo, C., Pinzur, M., Lin, S.S., The role of platelet-rich plasma in foot and ankle surgery. *Foot Ankle Clin.*, 10, 2005.
101. Ishidou Y., Kitajima I., Obama H., Maruyama I., Murata F., Imamura T., Yamada N., ten Dijke P., Miyazono K. and Sakou T. Enhanced expression of type I receptors for bone morphogenetic proteins during bone formation. *J. Bone Miner. Res.*, 1995, 10, 1651–1659.
102. Miloro, M.; Haralson, D.J.; Desa, V. Bone healing in a rabbit mandibular defect using platelet-rich plasma. *J. Oral Maxillofac. Surg.*, 2010, 6, 1225–1230.
103. Okazaki K, Shimizu Y, Xu H and Ooya K. Blood-filled spaces with and without deproteinized bone grafts in guided bone regeneration. A histomorphometric study of the rabbit skull using non-resorbable membrane. *Clin. Oral Impl. Res.*, 2005, 16, 236–243.
104. Detsch, R.; Mayr, H.; Ziegler, G. (2008). Formation of osteoclast-like cells on HA and TCP ceramics. *Acta Biomater.*, 2008, 4(1), 139–148.
105. Tamimi F.M., Torres J., Tresguerres I., Clemente C., López-Cabarcos E. and Blanco L.J. Bone augmentation in rabbit calvariae: comparative study between Bio-Oss and a novel b-TCP/DCPD granulate. *J. Clin. Periodontol.* 2006, 33, 922–928.
106. Jensen, S.S., Broggini, N., Hjørting-Hansen, E., Schenk, R., Buser, D., Bone healing and graft resorption of autograft, anorganic bovine bone and beta-tricalcium phosphate. A histologic and histomorphometric study in the mandibles of minipigs. *Clin. Oral Implants Res.*, 17, 237–243, 2006.
107. Betoni, W., Queiroz, T.P., Luvizuto, E.R., Valentini-Neto, R., Garcia-Júnior, I.R., Bernabé, P.F., Evaluation of centrifuged bone marrow on bone regeneration around implants in rabbit tibia. *Implant Dent.*, 21, 481–485, 2012.
108. Handschel, J., Simonowska, M., Naujoks, C., Depprich, R.A., Ommerborn, M.A., Meyer, U., Kübler, N.R., A histomorphometric meta-analysis of sinus elevation with various grafting materials. *Head Face Med.*, 11, 2009.
109. Rachner, T.D., Khosla, S., Hofbauer, L.C., Osteoporosis: now and the future. *Lancet*, 377, 9773, 2011.
110. Teloken, M.A., Bissett, G., Hozack, W.J., Sharkey, P.F., Rothman R.H., Ten to fifteen-year follow-up after total hip arthroplasty with a tapered cobalt-chromium femoral component (tri-lock) inserted without cement. *J. Bone Joint. Surg. Am.*, 84, 12, 2002.
111. Shahgaldi, B.F., Heatley, F.W., Dewar, A., Corrin, B., *In vivo* corrosion of cobalt-chromium and titanium wear particles. *J. Bone Joint. Surg.*, 77, 6, 1995.
112. Allen, M.J., Myer, B.J., The effects of particulate cobalt, chromium and cobalt-chromium alloy on human osteoblast-like cells *in vitro*. *J. Bone Joint. Surg.*, 79, 3, 1997.
113. Prokharau, P.A., Vermolen, F.J., García-Aznar, J.M., Numerical method for the bone regeneration model, defined within the evolving 2D axisymmetric physical domain. *Comput. Methods Appl. Mech. Eng.*, 253, 117–145, 2013.

114. Prokharau, P.A., Vermolen, F.J., García-Aznar, J.M., Model for direct bone apposition on preexisting surfaces, during peri-implant osseointegration. *J. Theor. Biol.*, 304, 131–14, 2012.
115. Moreno, P., García-Aznar, J.M., Doblaré, M., Bone ingrowth on the surface of endosseous implants. Part 1: Mathematical model. *J. Theor. Biol.*, 260, 1–12, 2009.
116. Moreno, P., García-Aznar, J.M., Doblaré, M., Bone ingrowth on the surface of endosseous implants. Part 2: Theoretical and numerical analysis. *J. Theor. Biol.*, 260, 2009.

11
Stem Cell-based Medicinal Products: Regulatory Perspectives

Deniz Ozdil[1,2] and Halil Murat Aydin[3]*

[1]Institute of Science, Bioengineering Division, Hacettepe University, Ankara, Turkey
[2]BMT Calsis Health Technologies Co., Ankara, Turkey
[3]Environmental Engineering Department & Bioengineering Division and Center for Bioengineering, Hacettepe University, Ankara, Turkey

Abstract

Regulatory control and strictures are fundamentally intended to ensure a balance between the risks and benefits of medical products. It is equally important, however, that regulatory systems allow for innovative advancements in medical technologies whilst ensuring public safety. The legal infrastructure governing the field of cell-based medicinal products has been developed in order to prevent, correct and account for all non-conforming actions, potential risks and ethical concerns related to the clinical use of such products. Collectively, the rules and regulations set by regulatory authorities impact all quality, clinical and non-clinical issues related to the final product. It is crucial for all relevant professionals to familiarize themselves with the standards regulating the stem cell therapy field. A comprehensive and stable regulatory framework is critical for enabling and monitoring advancements. This chapter aims to provide an overview of some global and regional regulatory perspectives regarding stem cell-based medicinal products.

Keywords: Stem cells, advanced therapy medicinal products, regulatory perspective, directives

11.1 Introduction

Discovering the inherent ability of stem cells to develop into different types of mature cells that comprise the body's tissue and organs has lead to the

---
Corresponding author: hmaydin@hacettepe.edu.tr

Ashutosh Tiwari, Bora Garipcan and Lokman Uzun (Eds.) Advanced Surfaces for Stem Cell Research, (323–342) © 2017 Scrivener Publishing LLC

creation of one of the most exciting areas of research of our century. These unique cells may be found in a number of different sources within the body including bone marrow, umbilical cord blood and peripheral blood. Although embryonic stem cells in particular have greater appeal in regenerative medicine due to their pluripotency (ability to differentiate into any cell type), products involving adult stem cell therapies have become more widely developed, making up for the majority of current stem cell-based treatments [1]. However, the wide-scale industrial uptake of stem cell-based therapies has been largely hindered by the ethical restrictions on their research and use. The USA removed the ban on federal funding for embryonic stem cell research only in March 2009. In that same year, Geron Corporation (USA) was granted Food and Drug Administration (FDA) clearance to commence Phase I safety trials in complete subacute spinal cord injury patients using human embryonic stem cells (hESCs). This was the first study in the world involving the use of hESC-based therapy in humans.

Nonetheless, recent years have witnessed the rapidly increasing momentum of the stem cell therapy industry. While the initial focus was on the utilization of autogenic stem cells sourced from the patient, off-the-shelf (allogeneic) products are now also in development. In contrast to the time-consuming and costly procedures, such as bone marrow aspiration, required for obtaining the patient's own stem cells, allergenic stem cell products would be readily available and reduce the variability in the amount of viable cells present per sample of taken from the patient. Due to the growing interest and support for stem cell-based therapies, the regulatory issues surrounding medicinal products that incorporate stem cells have become paramount. As it is with the pharmaceutical industry, this relatively new field of medical technology is governed by numerous directives, standards, regulations, and guidelines. The legal infrastructure governing the field of cell-based medicinal products has been developed in order to prevent, correct, and account for all nonconforming actions, potential risks, and ethical concerns related to the manufacture, distribution, marketing and clinical use of such products. As such, the legal framework is designed to guide manufacturers through all stages of the product development process such that product specifications meet the expectations of regulators. Collectively, the rules and regulations set by competent and regulatory authorities impact all quality, clinical, and nonclinical issues related to the final product. All stakeholders of a particular stem cell-based product are thus affected by relevant regulatory affairs. An engineer seeking to design and introduce a technology to the market must be aware of the regulatory requirements for a product type and design the product accordingly, improving the efficiency of the regulatory process to be observed and strengthening the translation of research and development outcomes to the clinic [2]. The surgeon aims to obtain clinical success, relying on the approved performance and safety declaration of the product. The profitability of manufacturers and insurers is also directly impacted by the duration, requirements, and outcomes of regulatory assessment. Therefore, it is crucial for all relevant professionals to familiarize themselves with both the organizations and standards regulating the stem cell therapy field, so as to both maximize operational efficiency and ensure the regulatory conformity of their products.

It is also necessary to understand that regulatory control and strictures are fundamentally intended to ensure a balance between the risks and benefits of medical products. It is equally important, however, that regulatory systems can allow for innovative advancements in medical technologies whilst ensuring public safety. Alternatively, the potential for functionally more effective and/or economically more feasible products may be prematurely barred. The regulatory uncertainties or gray areas that may exist for stem cell technologies have, and largely continue to, impact the appeal for financial investment into stem cell-based therapeutic products. A comprehensive and stable regulatory framework is critical to supporting advancements in this field of medicine. This chapter aims to provide an overview of some global and regional regulatory perspectives regarding stem cell-based medicinal products, focusing on the widely used and accepted systems of Europe and the USA.

11.2 Defining Stem Cell-based Medicinal Products

There are two emerging categories of products in the field of biomedicine and biotechnology: cell therapy products and gene therapy products. Although these terms may show some degree of variation or be used in combination in the literature, there are key regulatory guidelines which provide us with acceptable definitions for these terms. For the purposes of this chapter, products such as decellularized tissues, enhanced reproductive technologies (*in vitro* fertilization), nonmammalian-based products, vaccines such as live attenuated virus, and products manufactured by cells or recombinant DNA (rDNA) technology (noncell, nongene products) will be excluded from our discussion. The general consensus as to what constitutes as a cell-based therapy product is any product containing living mammalian cells as one of their active ingredients. In tissue regeneration applications, cells are usually combined with either natural or synthetic materials that act as delivery vehicles and/or scaffolds for the implanted cells. The primary aim of the cell therapy product is to replace, augment, or modify the function of the patient's compromised, damaged, missing, or dysfunctional cell and/or cellular structures.

The US Pharmacopeia has categorized cell-based products, including stem cell-based products, according to their known applications [3] (Table 11.1).

Table 11.1 Types of cell-based therapy products according to the US Pharmacopeia.

Indication	Product
Bone marrow transplantation	Devices and reagents to propagate stem and progenitor cells, to select stem and progenitor cells, or to remove diseased (cancerous) cells
Cancer	T cells, dendritic cells, or macrophages exposed to cancer-specific peptides to elicit an immune response Autologous or allergenic cancer cells injected with a cytokine and irradiated to elicit an immune response
Pain	Cells secreting endorphins or catecholamines (encapsulated in a hollow fiber)
Diabetes	Encapsulated β-islet cells secreting insulin in response to glucose levels
Wound healing	Sheet of autologous keratinocytes or allergenic fibroblasts in a biocompatible matrix
	Sheet of allergenic keratinocytes layered on a sheet of dermal fibroblasts
Tissue repair	
Focal defects in knee cartilage	Autologous chondrocytes
Cartilage-derived structures	Autologous or allergenic chondrocytes in a biocompatible matrix
Bone repair	Mesenchymal stem cells in a biocompatible matrix
Neurodegenerative diseases	Allergenic or xenogeneic neuronal cells
Liver assist (temporary; for bridging until liver transplant recovery)	Allergenic or xenogeneic hepatocytes in an extracorporeal hollow fiber system
Infectious disease	Activated T cells

The European Medicine Agency (EMA) regards cell-based products as advanced therapy medicine products (ATMPs). ATMPs are described as 'innovative, regenerative therapies which combine aspects of medicine, cell biology, science and engineering for the purpose of regenerating, repairing or replacing damaged tissues or cells'. ATMPs may fall into one of the following categories:

- *Gene therapy medicinal product*: involves the transfer of a prophylactic, diagnostic, or therapeutic gene to human or animal cells and its subsequent expression *in vivo*.
- *Somatic cell therapy medicinal product*: *(CTMPs)* contains autologous, allergenic, or xenogeneic somatic living cells. These cells are manipulated (includes expansion or activation of cell populations *ex vivo*), such that their biological characteristics are substantially altered in order to obtain therapeutic, diagnostic, or preventive effects via metabolic, pharmacological, and immunological actions.
- *Tissue engineered medicinal product*: includes products containing engineered cells or tissues (viable/nonviable, human, or animal origin or both) used for the regeneration, repair, or replacement of human tissue. May also contain other biological (cellular products, biomolecules) or nonbiological (biomaterials, chemicals) substances. Excludes products containing or consisting exclusively of nonviable human or animal cells and/or tissues that do not contain any viable cells or tissues and that do not act principally by pharmacological, immunological, or metabolic action.

According to EU legislation, for cells-based technologies to be considered and regulated as medicinal products, they must fulfill one of the following criteria:

- Cells have undergone 'substantial' manipulation. Substantial manipulation is defined as processes which alter the biological characteristics, functions, or properties relevant to the therapeutic effect

OR

- Cells are ultimately used in the recipient for a purpose that is different to which it fulfilled in the donor

AND

- The product has qualities capable of treating, preventing, or diagnosing a disease through the action of the cells or tissues it incorporates

With the aim of providing further clarity, Annex I of Regulation EC No. 1394/2007 states that the following processes do not qualify as substantial manipulation of cellular tissues: cutting, grinding, shaping, centrifugation, soaking in antibiotic or antimicrobial solutions, sterilization, irradiation, cell separation, concentration or purification, filtering, freezing, cryopreservation, and vitrification.

11.3 Regional Regulatory Issues for Stem Cell Products

Global trends in the development of stem cell products have been shaped by the various regulatory approaches of different jurisdictions. The variation of regulations, in both principle and practice, presents a significant issue within itself for those technologies with global uptake potential or intention. There are several geopolitical factors, as well as regional cultural attitudes and health economics, which affect the regulatory standards established and implemented in each country. The practice or concept of stem cell research itself may even cause serious bioethical dilemmas in some regions. There is a growing awareness for the need to recognize the current and future social impacts of such emerging technologies [4]. Even with such concerns regarding stem cell-based medicinal products, international research efforts in stem cell related technologies are charging at a rapid pace and in parallel regulatory control issues are becoming evermore significant.

The regulatory pathway chosen for cell-based technologies depends primarily on the product itself. The type, components, methods or action, potential risks, intended use and indicated use of a product are the main determinants of the classification and regulatory process applicable. Products which incorporate both biological substances and nonbiological carriers are subjected to a more complex regulation process. Products designed to merely collect or separate cell-based materials, such as bone marrow concentration systems, may follow much simpler regulatory protocols. In general, the approach to cell-based therapy regulation is a risk-based one. Product classifications, testing requirements (nonclinical/clinical), and even post-market surveillance activities all dependent on the associated level of risk a medical device poses to the end users. It is important to note that key regulatory elements in the assessment of cell-based products is not only on the intended use, but also the indications and known/foreseen contraindications.

Although the need for regulatory authorities and policies is indisputable, those relevant to stem cell-based products have occasionally warranted further clarity and improvement. These systems are evolving parallel to the advancements in stem cell technologies. It is particularly important, however, for the regulatory approval applicant to ensure that they understand the views of the regulatory body regarding a particular type product, even prior to product development. For example, although Nu Vasive's Osteocel Plus and Orthofix's Trinity Evolution products include stem cells and a matrix (normally classed as either biologics or demineralized bone matrix), the FDA classifies it as an allograft for which certain approval requirements are not deemed necessary. Similarly, the DeNovo NT graft (Zimmer/ISTO Technologies), although containing juvenile cartilage cells is considered by the FDA as a minimally manipulated allograft. The Therapeutic Goods Administration (TGA) of Australia has been the first regulatory body in the world to approve the manufacture and supply of a stem cell-based product, in July 2010 – Mesenchymal Precursor Cell Products (Mesoblast, Australia). In Australia, the term 'Therapeutic Goods' include medicines, devices, and biologics. Human cellular and tissue products are regulated, with some exemptions and exceptions, as biologics. Recently, the TGA also revised its regulatory framework to accommodate the emergence of various biotherapies [5].

11.4 Regulatory Systems for Stem Cell-based Technologies

Risk management is generally the central theme to regulatory control. It forms the basis and protocols for defining, identifying, analyzing, and managing risks by regulators. This is a greater challenge for technologies which incorporate living cells as the potential risks of such devices are harder to predict. The challenge is even greater for products based on stem cells. There are several limitations and deficiencies associated with the field of cellular therapies which contribute to this predicament. Preclinical testing protocols, which are always required prior to the approval of clinical trials, may not always successfully reflect the true efficacy or potential hazards of such medical devices. The complexity of the assessing stem cell-based products is also compounded by the lack of relevant established clinical data, as well as probable variations to be expected in clinical outcomes due to variability of age, gender, or health status in patient populations. In addition to the poor availability of preclinical *in vitro* testing data, there are also issues associated with the testing of human cell-based products in xenogenic milieu, i.e. *in vivo* tests. Such problems include the immuno-compromisation requirement for experimental animals and the fact that there may not be an appropriate animal model for the disease that is targeted by a particular product. Therefore, *in vivo* testing may also not always be successful in terms of generating data apt for use in planning and defining, theoretically, certain endpoints to be achieved in clinical trials.

Some guidelines, such as the EMEA/CHMP/SWP/28367/07 of the European Medicines Agency, provide methods of mitigating risks related to first-in-human clinical trials for investigational medical products. A system for approaching, assessing, verifying, validating, and monitoring stem cell-based medical products is thus as complex as it is necessary, in order to confirm any potential risks and benefits to patients. There are countless cases of unmet medical needs which traditional treatment options do not successfully address. Regenerative medicine, particularly stem cell-based technologies, holds great promise for effectively solving such medical problems. However, the translation of such technical innovations into clinical benefits depends, first and foremost, on regulatory structures that recognize and actualize solutions to current regulatory dilemmas.

11.4.1 The US Regulatory System

The US regulatory framework for medical devices, founded almost entirely on the FDA, covers the entire life cycle of a product, from research and development to premarket approval, and from market presence and product obsolescence. In the USA, federal law (Federal Food, Drug, and Cosmetic Act, section 513) provides a system for classifying medical products according the degree of risk it may pose to a patient. Essentially, the US regulatory system for medical products consists of three main elements:

1. Statues: Statutes are laws that are in effect. Cell-based medicinal products are legally bound by the Food, Drug and Cosmetic Act (FDCA) and the Public Health Service Act (PHSA).
2. Regulations: Regulations, e.g. Code of Regulations (21 CFR), provide the specificities of the general laws and are drafted by and approved by the FDA.

3. Guidance Documents: Guidance documents (published by the FDA) provide nonbinding advice with regard to the comprehension of regulations.

All medical products must comply with relevant laws and regulations governing all activities of the product development cycle, including clinical investigations and market authorization processes. US policies and guidelines for medical products are provided to the public via the Federal Register and or the FDA website under FDA Guidance Documents.

As previously mentioned, considerations for a product are not only based on the intended use of a particular device but also the indications for use. There are three classes for medical devices which are subjected to different tiers of regulatory control, namely, Class I (General Controls), Class II (General Controls and Special Controls), and Class III (General Controls and Pre-Market Approval). Medical devices of low risk, medium risk, and high risk to patients are assigned to, Class I, Class II, and Class III, respectively. There are two methods for discerning the classification and corresponding regulations of a particular device.

A search for the device name on the FDA classification database or, if the 'device panel' or medical specialty of the device is known, then a search on the listing for a particular panel may be performed. By definition, Class III products are those which have insufficient information established regarding their use, safety, and efficacy. These products intend to support or sustain human life, prevent impairment of human health, and may pose unreasonable risks of illness or injury. As such, stem cell-based technologies intended to impact on the body for the purposes of repairing, restoring, or replacing lost functionality of cells/tissues or for the treatment of certain diseases would be considered belonging to this high-risk category.

A further consideration is whether the product involves laboratory processing or is used directly at the point of care in the operation room. Products belonging to the former application must adhere to the Good Manufacturing Practices (GMPs) and objective quality assurance and are exempt from the 510 (k) premarket notification (PMN).

There are two main regulatory routes for any medical device to be launched on the US market. The first, which the majority of medical devices are assessed with, is known as PMN or 510 (k) (with reference to the Section 510 (k) of the Food, Drug and Cosmetic Act). A product, of which a 'substantial equivalent' already exists, may be given market approval provided that it meets the relevant general and special controls of its class. The second route, known as premarket approval, applies to new products for which no such data exist and it often involves the conduction of clinical trials before approval for market release is granted.

The Center for Biologics Evaluation and Research (CBER) within the FDA is responsible for regulating cellular therapy products, human gene therapy products, and certain devices related to cell and gene therapy. In fact, it is the Office of Cellular, Tissue and Gene Therapies (OCTGT) within the CBER who oversees applications and issues related to stem cell-based products. Other centers under the Office of the Commissioner at the FDA relevant to cell-based medicinal products include the Center for Drug Evaluation and Research (CDER) and Center for Devices and Radiological Health (CDHR). The OCTGT is responsible for a range of different product types:

- Somatic cell therapies [includes stem cells (hematopoietic, mesenchymal, embryonic), chondrocytes, myoblasts, keratinocytes, pancreatic islets, hepatocytes],
- Gene therapies (includes gene-modified cells, iPS cells, plasmids, viral vectors, bacterial vectors),
- Cancer vaccines and immunotherapies,
- Cell–device combination products [tissue-engineered and regenerative medicine (TERM) products],
- Devices (such as tissue/cell processing/separation, cell selection, cell delivery, companion diagnostics),
- Tissues and tissue-based products,
- Xenotransplantation products.

Cell-based products approved so far by the CBER include Apligraf® (skin graft), Carticel® (autologous cultured chondrocytes) and Epicel® (cultured epidermal autografts). Products such as these must obtain a Biologics License Approval (BLA) prior to market availability by first establishing safety, purity, and potency according to the products declared intended use. For clinical application in humans, these products must have an 'investigational new drug' application in effect, approved by the CBER. The 'investigational new drug' scheme allows the use of products for which the pharmacological activity and acute toxicity potential in animals has been established, to be tested in humans, so that they may eventually obtain premarket approval. The FDA has not yet, however, approved of any stem cell-based products for use, other than the use of cord blood-derived hematopoietic progenitor cells (blood forming stem cells) for particular indications. Some examples of products currently available under the approval of CBER include Ducord (HPC Cord Blood, Duke University School of Medicine), Provenge (sipuleucel-T) (autologous cellular therapy, Dendreon Corp.), and IMLYGIC (oncolytic virus therapy, BioVex, Inc.).

In the USA, IND regulations (21 CFR 312), biologics regulations (21 CFR 600), and cGMP (21 CFR 211) of the Code of Federal Regulations (CFR) are used to codify cell therapy products. The PHSA divides cell-based products under two categories: '361 products' and '351 products'. Effective since 2005, Part 1271 of the CFR [6] provides all US regulations regarding cells, tissues, and cellular and tissue-based products (HCT/Ps). The CFR defines HCT/Ps as 'Articles containing or consisting of human cells or tissues that are intended for implantation, transplantation, infusion, or transfer to a human recipient'. In the USA, stem cell products are classified as HCT/P, but more specifically, somatic cell therapy products (autogenic or allogenic). These products may contain stem cells that have been more than minimally manipulated, including mesenchymal stem cells, adipose stem cells, and allergenic pancreatic islet cells and certain bone marrow cells. The HCT/P category generally describes single-entity products. There are also 'combined products' (device–biologic, biologic–drug, drug–device, biologic–drug–device, etc.) in which all components of the combined product must be present to achieve the intended therapeutic effect of the product. Cell-scaffold products or cells and gene-delivery devices are some examples of combined products. Following the initial preclinical development of these products, investigators must lodge an application for 'investigational new drug' exemption from the FDA prior to proceeding with Phase 1, 2, or 3 studies [7]. As such, these products are not regulated solely under section 361 of the PHSA but are also subjected to other premarket review and licensure processes by classifying as biological products.

As mentioned, the level of regulation for a cell-based product is based on risk assessment. 'Low-risk' cell-based products, such as tissues, are subjected to Section 371 of the PHSA, do not require premarket review, and are regulated solely under the 21 CRF 1271. 351 HCT/P-based 'therapeutic' products (can be either biologic or device), are considered to carry higher risks and are subjected to Sections 351 & 361 of the PHSA, the FDCA and are regulated under the 21 CFR, Parts 1271, 600, 200, 312, and 812 (if device). If the product is a therapeutic biologic, it is obtain a BLA, and if it is device based, it will require PMA, 501 (k) or Humanitarian Use Device (HUD) exemption. HUDs are those devices that are intended to

Table 11.2 Regulatory requirements for HCT/Ps.

	Law	Regulation	Marketing approval
361 HCT/P – Tissue	361 PHS Act	21 CFR 1271	Premarket review not required
351 HCT/P – Biologic therapeutic	361 PHSA 351 PHSA FDCA	21 CFR 1271 21 CFR 600 21 CFR 200 21 CFR 300	21 CFR 800
351 HCT/P – Device	FDCA	21 CR 800	PMA 510 (k) HDE

treat diseases or conditions which manifest in less than 4 000 patients a year. This information has been summarized in Table 11.2.

Currently, there are only two scenarios permitted by the FDA in which stem cell products may be administered by physicians. The first scenario is based on investigational drugs and biological products used in treatment of medical conditions, for which clinical trials are currently performed elsewhere. This is part of the FDA's expanded access program. The alternative is off-label prescription of FDA-approved stem cell products where the FDA does not have regulatory authority over the medical practice and physicians are allowed to exercise their professional judgment with regard to the use of such products.

Cell-based products are subjected to the assessment and approval of several different governing bodies in the area of cell therapy standards. In the USA, each of these bodies (or their related standards) is involved in a particular aspect of product assessment. The International Standards Organization (ISO) and American Society for Testing and Materials (ASTM), for example, provides the manufacturing standards which the manufacturing facility and process are subjected to. The Pharmocopeial Standards (USP) outline the approved standards for the use of raw and ancillary materials in cellular and tissue products. Facilities working with cell-based products may also require accreditation from organizations such as the Foundation for the Accreditation of Cellular Therapy. Manufactured products will also need to conform to the standards of a number of other authorized organizations with regard to their development, characterization, labeling, and preparation for translational research.

Some of the current standards applicable to cell-based products in the USA and their short descriptions are listed below:

ISO – International Organization for Standardization: Develops and provides international standards related to medical devices (including cell therapy and tissue-engineered products), laboratory testing. There are several ISO standards relevant to stem cell technologies, including: ISO 13022:2012 Medical products containing viable human cells – Application of risk management and requirements for processing practices, ISO 22442-1:2007 Medical devices utilizing animal tissues and their derivatives – Part 1: Application of risk management, ISO 22442-2:2015 Medical devices utilizing animal tissues and their derivatives – Part 2: Controls on sourcing, collection and handling, ISO 22442-3:2007 Medical devices utilizing animal tissues and their derivatives – Part 3: Validation of the elimination and/or inactivation of viruses and transmissible spongiform encephalopathy (TSE) agents and SO/TR 22442-4:2010 Medical devices utilizing animal tissues and their derivatives – Part 4: Principles for elimination and/or inactivation of TSE agents and validation assays for those processes.

ASTM International – American Society for Testing and Materials: leading international standards organization. The standards for cell therapy and tissue engineering are developed by the ASTM Subcommittee F04.43 (these include the ASTM F2210 – Standard Guide for Processing Cells, Tissues, and Organs for Use in Tissue Engineered Medical Products, ASTM F2739 – Standard Guide for Quantitating Cell Viability Within Biomaterial Scaffolds, ASTM F2315 – Standard Guide for Immobilization or Encapsulation of Living Cells or Tissue in Alginate Gels and ASTM F2944 – Standard Test Method for Automated Colony Forming Unit (CFU) Assays).

USP – US Pharmacopeial Convention: Provides standards for the use of ancillary and raw materials for cellular and tissue products.

GBSI – Global Biological Standard Institute: Develops standards for various areas of the life sciences.

ATCC – American Type Culture Collection: Manufactures and provides materials (including cells), develops biological standards for basic and translational research.

BSI – British Standards Institution: Develops regenerative medicine definitions and guidelines for the characterization of clinical cell-based products (European).

ISCT – International Society for Cellular Therapy: Provides guidelines and recommendations regarding the development, characterization, and quality of cell-based products.

AABB – American Association of Blood Banks – Center for Cellular Therapies: Similar to FACT.

FACT – Foundation for the Accreditation of Cellular Therapy: Provides standards for the collection and processing cell-based products. Authorized to accredit laboratories working with stem cells, cord blood banks, and facilities where 'minimal manipulation" cell therapies are used.

ICCBBA – International Council for Commonality in Blood Bank Automation: Responsible for the management of the ISBT-128 Standard (the standard covers the terminology, identification, coding, and labeling of medical products of human origin).

In the presence of such a large number of guidelines, recommendations, and regulations for the standardization of cell-based products, there have been efforts to coordinate and harmonize the reference materials of relevant regulatory agencies. While some standards are concurrent with others, as the ASTM is with ISO, others are not. There are presently some organizations with resources and activities designed to provide a more coordinated approach to the conformity assessment of cell-based products. These include the Alliance for Harmonization of Cellular Therapy Accreditation (AHCTA) and the USP which has a number of collaborative efforts (e.g. USP-ISCT workshop). There are also others such as the International Society for Stem Cell Research (ISSCR), which publishes nonbinding codes of practice and guidelines regarding stem cell research, and the Hinxton Group, who has published 'An international consortium on stem cells, ethics and law' [8].

The Committee for Advanced Therapies (CAT) has published (2009) an article entitled 'Reflection paper on stem cell-based medicinal products' which highlights for professionals the key considerations when designing, manufacturing and applying for approval of stem cell-based products. These may be summarized as follows:

1. Quality Considerations
    1.1 Starting Materials (cell line derivation and cell banking)
    1.2 Manufacturing Process
    1.3 Process Validation
    1.4 Characterization and Quality Control
        1.4.1 Identity of Stem Cells
        1.4.2 Purity of the cell-based product
        1.4.3 Potency of a stem cell-based product
        1.4.4 Tumorigenicity and genomic stability

2. Nonclinical Considerations
   2.1 Animal models reflecting the therapeutic indication
   2.2 Nonclinical biodistribution studies of stem cells
   2.3 Tumorigenicity
   2.4 Differentiation *in vivo*
   2.5 Immune rejection and persistence
3. Clinical Considerations
   3.1 Pharmacodynamics
   3.2 Pharmacokinetics
   3.3 Dose finding studies
   3.4 Clinical trials to study efficacy
   3.5 Clinical safety requirements
   3.6 Pharmacovigilance (specific safety issues, including lack of efficacy)

The current Good Manufacturing Practices (cGMPs) are minimal standards applicable manufacturing, controls, testing, and documentation or all cell-based therapeutics. They are made available the FDA with the intention of ensuring that all quality considerations, ranging from facility design to personal training and environmental monitoring, are covered by the manufacturer, and that the final product is ready to assume nonclinical and clinical testing. At this point, the Good Clinical Practices (GCPs) come into effect, stating the minimal regulatory standards which are required while conducting a clinical study.

The interactions with and processes involving the FDA remain throughout the life cycle of the product. In terms of FDA processes, there are four main phases relating to the product life cycle. These phases and the relevant processes are summarized in Figure 11.1.

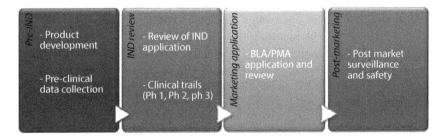

Figure 11.1 Key steps in the FDA regulatory process.

11.5 Stem Cell Technologies: The European Regulatory System

The European system to regulating medical devices differs in many ways to the US FDA-approval regime. Both national and centralized authorities work to introduce, implement, and regulate EU legislature regarding cell-based products and therapies. The European Union (EU) and European Free Trade Association (EFTA) are responsible for the approval and control of all medical devices introduced and circulating within the EU. The European Medicines Agency (EMA) is responsible for the protection of public and animal health in EU Member States through the scientific evaluation, supervision, and safety monitoring of medicines used in the EU. In fact, it operates within a regulatory network of scientific experts and national regulatory authorities, including the European Commission, known as the European Medicines Regulatory Network to coordinate the establishment and implementation of medical policies. Unlike the FDA, the EMA does not have direct mandate to regulate development-related issues for cell-based products such as preclinical development or clinical trial authorization, nor does it cover research related issues of such products. The EMA contains seven committees and expert parties who have specific scopes. The EMA not only evaluates product marketing applications, but it also coordinates collaborations with product developers and inspection bodies (Good Laboratory Practice (GLP), GCP, GMP) to develop centralized procedures.

There are two types of legislative acts that compose the EU regulatory framework: Directives and Regulations. Directives are 'all-inclusive' general requirements which must be abided by all EU Member States. Each state is responsible for developing its own national legislative acts for the implementation of the directives. Regulations, on the other hand, must be applied directly and uniformly by all Member States, regardless of national legislative acts. While there may be a degree of variability between the implementation of Directives between Member States, the adoption and implementation of Regulations may not be compromised by the any Member State. Although the Regulations set by the EMA are binding, the regulations do not override any decisions of Member States as to the prohibition or restriction of supply, sales, or use of any medicinal products consisting or derived from particular cells such as embryonic stem cells or xenogeneic cells.

Three core directives constitute the legal framework for medical devices in the EU:

- Directive 90/385/EEC regarding active implantable medical devices,
- Directive 93/42/EEC regarding medical devices,
- Directive 98/79/EC regarding *in vitro* diagnostic medical devices.

The Competent Authority of each member state is responsible for ensuring that the regulatory standards outlined in the medical directives are incorporated into national law and implemented accordingly. For medical devices in general there four classes: Class I (Is (sterile) and Im (nonsterile)), Class IIa, Class IIb, and Class III, as defined by Article IX of the Council Directive 93/42/EEC. Classes represent low to high-risk products, respectively. In the EU, although Class I devices (those that are non sterile and have no measuring function) can be marketed directly with self-certification of the manufacturer, other medical class devices must obtain for declared products a Certificate of Conformity issued by a Notified Body and be identifiable with the CE Mark. A stem cell therapy design dossier, submitted to a notified body for approval, typically consists of five main modules, namely, Module 1 (specific administrative data), Module 2 (quality, nonclinical, and clinical summaries), Module 3 (chemical, pharmaceutical, and biological information), Module 4 (nonclinical reports), and Module 5 (clinical study reports).

There are, however, more specific directives of the EU which govern regulatory aspects of cell-based products:

- Directive 2003/63/EC amending Directive 2001/83/EC of the European Parliament and of the Council on the Community code relating to medicinal products for human use. It defines and specifies requirements for clinical cell therapy products, introduced 2003.
- Directive 2001/20/EC on the approximation of the laws, regulations, and administrative provisions of the Member States relating to the implementation of GCP in the conduct of clinical trials on medicinal products for human use. The directive establishes the basis and requirement for approval of clinical trials for clinical cell therapy products, introduced 2001.
- Directive 2004/23/EC on setting standards of quality and safety for the donation, procurement, testing, processing, preservation, storage, and distribution of human tissues and

cells. The directive covers regulations for all aspects of handling human tissues and cells, introduced 2004.
- Directive 2002/98/EC on setting standards of quality and safety for the collection, testing, processing, storage, and distribution of human blood and blood components and amending Directive 2001/83/EC. The directive is intended for ATMPs which are manufactured from blood cells or blood components, introduced 2003.

In December 2008 the Regulation (EC) No. 1394/2007 on advanced therapy medicinal products and amending Directive 2001/83/EC and Regulation (EC) No. 726/2004 came into effect, binding to all member states of the union. This regulation provides a centralized authorization procedure for all Advanced Therapy Medicinal Products (ATMPs). The Directive involves a single scientific evaluation of the quality, safety, and efficacy of ATMPs carried out to the highest possible standard.

The European Commission defines ATMPs as those products based on genes (gene therapy), somatic cells (cell therapy) (as defined in Directive 2001/83/EC), or tissues (tissue engineering).

Previously, ATMPs lingered between the EU categories of medicinal products and medical devices, poorly represented by either. Those ATMPs existing prior to Regulation (EC) No. 1394/2007 received transitional protection until December 2011 (or December 2012 for tissue-engineered products), after which they needed to comply with the new regulations.

There is an interesting exemption in Regulation (EC) No. 1394/2007 [Article 3 (7)] which exempts some ATMPs from centralized marketing authorization. Such ATMPs are those which are prepared on a non-routine basis (according to traceability, quality, and pharmacovigilance standards equivalent to those for ATMPs with a centralized marketing authorization) and implemented in a hospital within the same Member State under the exclusive responsibility of a medical practitioner and in accordance to the medical prescription for the specific. This 'Hospital Exemption' (HE) is intended to allow for the use of customized innovative products in rare situations where effective treatment of a patient is impaired due to lengthy product validation times that are not currently feasible. Implementation of the HE must be authorized by the EU Member State who must ensure that specific requirements of the regulation are met at the national level.

With regard to European regulation, the stem cell-based product must first qualify as a medical product. It must then either have been subjected to substantial manipulation such that its biological characteristics, physiological functions, or structural properties are relevant to the intended function, or it must not be intended to be used for the same essential function(s) in the recipient as the donor. Minimally manipulated cells or tissues are not considered medicinal products by the EU and are regulated by Directive 2004/23/EC which sets standards of quality and safety for the donation, procurement, testing, processing, preservation, storage, and distribution of human tissues and cells. Classification is still, however, based on both existing scientific knowledge and the case-by-case assessment of claims and supporting scientific rationale. The quality, safety, and efficacy assessment of ATMP applications submitted to the EMA is performed by the CAT within EMA. The draft opinion prepared by CAT is then provided to the Committee for Medicinal Products for Human Use (CHMP), who then presents a final opinion on the granting, variation, suspension, or revocation of a marketing authorization for the product. All ATMPs commercially produced and sold within the EU must complete a Marketing Authorization Application (MAA). The EMA has expressed [9] that all stem cell-based products should be produced in accordance to the risk-based approach outlined in Annex I, part IV of Directive 2001/83/EC and under quality control conditions sufficient in enabling consistent and reproducible outputs. The Commission Regulation (EC) No. 668/2009 on implementing Regulation (EC) No. 1394/2007 of the European Parliament and of the Council with regard to the evaluation and certification of quality and nonclinical data relating to advanced therapy medicinal products developed by micro-, small-, and medium-sized enterprises, is a procedure related only to products of advanced therapy classification and relates to the certification of quality and nonclinical data by CAT. Although it is not legally binding, certification will promote clinical trials and MAAs based on the same data. It is important to mention that although ATMP regulations govern regulatory issues related to stem cell-based products, each product is still liable for meeting relevant laws and regulations of each EU Member State. For example, the UK has published the UK Stem Cell Tool Kit intending to guide applicants through the approval processes relevant to the UK. The roles and responsibilities of national competent authorities and the EMA/CAT in the ATMP field can be summarized in Table 11.3.

ChondroCelet (Tigenix, Belgium) was the first ATMP to receive marketing approval on 5 October 2009. It is a cell-based medicinal product for use in autologous chondrocyte implantation in cartilage defects. It was only in 2015 that the European Commission gave conditional marketing authorization to Holoclar® (ChiesiFarmaceutici S.p.A, Italy), an ATMP

Table 11.3 The responsibilities of NAC and EMA/CAT.

National competent authority	EMA/CAT
Inspection and authorization of cell and tissue donation, procurement, processing	Classification, evaluation, and certification of ATMPs.
Inspection of GLP in preclinical development	Evaluation of ATMPs in transition periods
Inspection of GCP and GMP and authorization of clinical development	
Issuing production license to HE ATMPs	

designed to treat moderate to severe limbal stem cell deficiency (LSCD), a rare eye condition resulting in blindness. It is the first tissue-engineering product containing stem cells that is authorized for marketing in Europe.

References

1. Ahrlund-Richter, L., et al., Isolation and production of cells suitable for human therapy: challenges ahead. *Cell Stem Cell*, 4(1), pp. 20–6, 2009.
2. Johnson, P.C., et al., Awareness of the role of science in the FDA regulatory submission process: a survey of the TERMIS-Americas membership. *Tissue Eng Part A*, 20(11–12), pp. 1565–82, 2014.
3. U.S Pharmacopeia.
4. van Osch, G.J., J.A. Burdick, and W. Liu, Emerging issues in translating laboratory experiments to applications for society. *Tissue Eng Part A*, 20(19–20), pp. 2547–8, 2014.
5. Sturm, M., Regulation policy for cell and tissue therapies in Australia. *Tissue Eng Part A*, 21(23–24): pp. 2797–801, 2015.
6. Administration, F.a.D., *21 CFR Part 1271 – Human Cells, Tissues, and Cellular and Tissue-Based Products*. Food and Drug Administration.
7. Gee, A., Mesenchymal stem-cell therapy in a regulated environment. *Cytotherapy*, 3(5), pp. 397–8, 2001.
8. Hinxton, G., An international consortium on stem cells, ethics and law. *Rev Derecho Genoma Hum*, 2006(24), pp. 251–5.
9. (CAT), C.f.A.T., *Reflection Paper on Stem Cell-based Medicinal Products*. C.f.A.T. (CAT), Editor. 2011, European Medicines Agency.

12
Substrates and Surfaces for Control of Pluripotent Stem Cell Fate and Function

Akshaya Srinivasan[1], Yi-Chin Toh[1]*, Xian Jun Loh[2,3,4]* and Wei Seong Toh[5,6]*

[1]Department of Biomedical Engineering, National University of Singapore, Singapore, Singapore
[2]Institute of Materials Research and Engineering (IMRE), Singapore, Singapore
[3]Department of Materials Science and Engineering, National University of Singapore, Singapore, Singapore
[4]Singapore Eye Research Institute, Singapore, Singapore
[5]Faculty of Dentistry, National University of Singapore, Singapore, Singapore
[6]Tissue Engineering Program, Life Sciences Institute, National University of Singapore, Singapore, Singapore

Abstract

Pluripotent stem cells (PSCs) provide unique opportunities for understanding basic biology, for developing tissue models for drug testing, and for clinical applications in regenerative medicine. These require the development of robust platforms and protocols for maintenance of their self-renewal as well as for differentiation to a specific cell fate. Cell fate and functions are influenced by components of the surrounding microenvironment, including soluble factors, extracellular matrix (ECM), cell–cell and cell–matrix interactions, and mechanical forces. Various culture substrates and surfaces based on feeder cells, decellularized matrices, ECM proteins, and cell adhesion molecules have been reported, with profound effects on proliferation and differentiation of stem cells. Synthetic substrates, based on peptides and polymers, have also emerged as alternative platforms to provide a chemically defined and xeno-free culture system for expansion and long-term maintenance of stem cells. These advances in surface engineering will continue to grow and guide the development of processes for stem cell application.

Keywords: Substrates, surfaces, stem cells, extracellular matrix, adhesion

---

Corresponding authors: dentohws@nus.edu.sg; lohxj@imre.a-star.edu.sg; biety@nus.edu.sg

12.1 Introduction

Human pluripotent stem cells (hPSCs) have the unique capacity for unlimited self-renewal and differentiation into specialized cell types [1, 2]. These fundamental characteristics make hPSCs a potential cell source for applications in regenerative medicine, drug discovery, disease modeling, and for studies aimed to better understand human development and biology [1, 2]. Human embryonic stem cells (hESCs) are isolated from the inner cell mass of human blastocyst, while human induced pluripotent stem cells (hiPSCs) are generated from somatic cells by overexpression of key transcription factors. Both hESCs and hiPSCs share similarities in their morphology, expression of key transcription factors including Oct4, Sox2 and Nanog, cell surface markers such as the stage-specific embryonic antigens (SSEA)-3 and -4, the keratin sulfate-related antigens (TRA-1-60 and -81), high telomerase and alkaline phosphate activity, as well as capacity for indefinite propagation and lineage-specific differentiation under specific conditions *in vitro* [1, 2]. Notably, several diverse cell types have been derived from hPSCs, including endothelial cells [3, 4], cardiomyocytes [5, 6], osteoblasts [7, 8], chondrocytes [9–12], neurons [13, 14], keratinocytes [15, 16], and hepatocytes [17, 18].

To maximize the potential of hPSCs in various applications in regenerative medicine and disease modeling, it is critical that the culture conditions are consistent, chemically defined and/or non-xenogeneic for purposes of scale-up, reproducibility, safety, and clinical compliance. Conventional cultures of hPSCs made use of xenogenic mouse feeder cells and animal medium components that pose difficulties in defining the culture system and in clinical translation [19]. Over the years, significant advances have been made in optimization of surfaces and substrates for the maintenance of pluripotency of hPSCs [19]. Here, we describe the various culture substrates and surfaces based on feeder cells, decellularized matrices, ECM proteins, and cell adhesion molecules (CAMs) that have been developed and used to support the proliferation and differentiation of hPSCs. Synthetic substrates, based on peptides and polymers, that have emerged as alternative platforms to provide a chemically defined and xeno-free culture system will also be described.

12.2 Pluripotent Stem Cells

The fundamental characteristics of PSCs include self-renewal and multi-lineage differentiation *in vitro* and *in vivo* [1, 2]. In animals such as mouse, *in vivo* differentiation capacity can contribute to all somatic lineages and produce germ-line chimera. Human PSCs including hESCs and hiPSCs can generate cell lineages of all three germ layers (ectoderm, mesoderm, and endoderm) both *in vitro* and *in vivo*. In vitro differentiation of PSC usually occurs through nonadherent culture of PSCs to induce formation of embryoid bodies. Alternatively, *in vitro* differentiation of PSCs can also be directed toward certain cell fates under specific culture conditions [3–18]. *In vivo* differentiation potential of PSCs is usually determined by implantation into an immunodeficient mouse [10]. The formation of teratoma, a benign tumor compromising of cell types representative of the three germ layers, remains one of the gold standard characterization assays to validate the differentiation potential of the PSCs. Other standard characterization assays include daily visual assessment of colony morphology (tightly packed colonies of cuboidal-shaped cells with high nuclei-to-cytoplasm ratio) and evaluation of molecular markers including Oct4, Nanog, Sox2, and Rex1, and surface markers such as SSEA-3, SSEA-4, TRA-1-60, and TRA-1-81 [20, 21]. However, some of these markers such as SSEA-4 may also be expressed in other cell types [22], making the assay nonspecific for PSCs. Therefore, a panel of standard *in vitro* assays and *in vivo* teratoma assay is usually required to fully validate the pluripotency of hPSCs. In addition, genomic integrity of long-term cultures would require G-banding and/or fluorescence *in situ* hybridization (FISH) analysis that detect chromosomal changes [23].

It is well established that *in vitro* maintenance of self-renewal and pluripotency of PSCs requires culture in supportive medium within a favorable microenvironment. The supportive media consist of soluble growth factors such as the fibroblast growth factor (FGF)-2 are involved in the regulation of critical signaling pathways for maintenance of pluripotency [24]. Another critical element of the microenvironment is the culture substrate for adhesion of hPSCs. Traditionally, hPSCs were cultured on inactivated mouse embryonic fibroblast (MEF) feeder cells or on animal-derived complex ECMs such as the Matrigel™ with conditioned medium (MEF-CM) from MEF feeder cells [1, 2]. The presence of poorly defined and animal-derived or xenogeneic components renders the hPSC lines unsuitable for clinical applications. Over the years, significant efforts have been made in the development of xeno-free and chemically defined substrates that will advance the development of clinically compliant hPSCs for therapeutic applications.

12.3 Substrates for Maintenance of Self-renewal and Pluripotency of PSCs

12.3.1 Cellular Substrates

Co-culture of isolated stem cells with MEF feeder cells was the first technique used for successful establishment of hESC lines [1]. The same culture system would also be applicable for culture of hiPSCs. Besides growth factors and cytokines being secreted, the MEF feeder cells also provide the critical ECM support and CAMs necessary for maintenance of hPSCs in their undifferentiated state [19]. Although long-term undifferentiated growth of hPSCs on MEF feeder cells has been demonstrated in several studies [1, 24, 25], there is still an inherent risk of xenogeneic contamination that hampers potential clinical applications [19, 26].

To improve the culture system, human and autologous/autogenic feeders have been explored as cellular substrates to support culture of hPSCs [25, 27–32]. Notably, in a study by Richards and co-workers, a panel of 11 different human adult, fetal, and neonatal feeders for hESC culture were evaluated [28]. In that study [28], differences in feeder support exist among different human cell types and sources, with adult skin fibroblast feeders demonstrating superior ability to support prolonged undifferentiated hESC culture for over 30 weekly passages. Several studies have shown comparable supportive ability of human feeders as the MEF feeder for undifferentiated hPSC culture [25, 27, 28]. However, there still exist risks of disease transmission from different human donor sources. On this note, autologous/autogenic feeders (fibroblasts derived from hPSCs used as feeder cells) might be advantageous [30–32]. Autologous fibroblast cells are usually derived through spontaneous differentiation with or without embryoid body (EB) formation [31]. The conditions for derivation of autologous feeders have a significant effect on support of undifferentiated hESC growth [31]. Notably, Fu *et al.* showed that autologous fibroblast feeder cells derived through EB formation secreted more FGF-2 than those derived directly from hESCs and were able to better support the undifferentiated state of hESCs and differentiation potential [31].

FGF-2 remains a necessary growth factor as it regulates the expression of TGF-β1, Activin A, and BMP-4 antagonist which are important for the undifferentiated growth of hPSCs [24]. It has been shown that different levels of FGF-2 production by different feeder cells would affect the long-term culture of hPSCs. Notably, human mesenchymal stem cells (MSCs) demonstrated higher production of FGF-2 than human foreskin fibroblasts and enabled further culture of hPSCs on Matrigel™ without the need of additional exogenous FGF-2 [33].

To obtain a consistent and reproducible source of feeders for hPSC expansion and research, immortalization is an attractive approach and has been demonstrated feasibility in several studies [34–36]. Immortalization enables the use of the same feeder cell line for prolonged periods of time under standardized culture conditions, a critical issue in the maintenance and stability of hPSCs. Notably, immortalization has been performed on primary MEF [34], human foreskin fibroblasts [35], and human MSCs [36] and has been shown to maintain the FGF-2 production by these feeder cells that is critical for supporting self-renewal and undifferentiated growth of hPSCs [35].

12.3.2 Acellular Substrates

While the culture of PSCs was first established on cellular substrates like MEFs, which were described in the previous section, the use of cellular substrates is both time and labor intensive. Moreover, the secretion and expression of factors from feeder cells are highly variable. Gamma irradiation of the feeder layers induces apoptosis in the cells as well as inhibits proliferation, which could affect the profile of factors secreted by the cells [26]. Therefore, they are limited in consistency and scalability. In view of these limitations, acellular substrates provide a good alternative for the maintenance of PSCs. There are a variety of acellular substrates currently in use and they are detailed below.

While the focus of this study is on the different substrates used for the maintenance of PSCs, culture medium and soluble factors are also crucial in maintaining the self-renewal and pluripotency of PSCs. It is the combination of the culture medium and substrate which collectively maintain PSC pluripotency. Thus, this section will also consider the matrix–medium combinations used for PSC maintenance (Table 12.1).

12.3.2.1 Biological Matrices

Biological matrices are ECMs that are derived directly from animals or animal cells. The ECM is a complex mixture of proteins, which provides structural and biochemical support for cells. The animal ECM consists of two components: the interstitial matrix and the basement membrane.

12.3.2.1.1 Matrigel™
Matrigel is a basement membrane derived from Engelbreth–Holm–Swarm (EHS) mouse sarcomas. It consists of various ECM components, such as

Table 12.1 Substrate–medium combinations for maintenance of PSCs.

Substrate	Medium	Remarks	Refs.
Matrigel/ Geltrex	MEF-CM	Matrigel	[39, 40, 50]
	mTeSR1	Matrigel	[41–43, 50]
	E8	Matrigel	[58]
	StemPro	Matrigel [64], Geltrex [45]	[64, 45]
Vitronectin	E8	Truncated recombinant vitronectin	[58]
	mTeSR1, MEF-CM	Full length, purified from human plasma	[50, 56]
	StemPro	Recombinant fragment (SMB domain)	[57]
Fibronectin	MEF-CM	Full length, from human plasma	[56]
	Chemically defined medium with KO-SR, TGF-β1 and FGF-2	Full length, from human plasma	[60]
Laminin	MEF-CM	Laminin 111 [49], Laminin purified from EHS [39]	[39, 49, 56]
	StemFit	Laminin 511 E8 fragments	[53]
	mTeSR1, TeSR2, StemPro	Laminin 511 and 322 E8 fragments	[54]
	mTeSR1	Full length Laminin 511	[52]
Collagen IV	MEF-CM	Full length, from human placenta	[56]
Collagen I	Defined medium with FGF-2 and Heparin	Full length, from porcine tendon	[63]
HA	MEF-CM	Methacrylated HA	[62]

laminin, fibronectin, collagen IV, entactin, and heparin sulfate [37] and also a multitude of growth factors that are bound to these matrix components. Matrigel has been used since 1982 for the culture of many different cell types [38]. Its primary component is Laminin 511. It is commercially marketed by Becton and Dickinson.

Basement membranes are the first ECM produced in the developing embryo with laminin being expressed as early as the two cell stage [37], providing impetus for the use of Matrigel as a substrate for PSC maintenance. While early experiments first established the capability of Matrigel to maintain hESC self-renewal and proliferation for long periods of continuous culture in the presence of MEF conditioned medium (CM) [39, 40], it was later found to sustain growth even in chemically defined TeSR1 medium [41]. Some modifications were later made to this medium to produce mTeSR1 [42] and also a xeno-free medium TeSR2, wherein bovine serum albumin (BSA) was replaced with the human serum albumin (HSA), both of which formed a compatible system with Matrigel. The Matrigel-mTeSR 1 combination is the most widely used culture system for hPSCs today. A study that compared various ECM components, biomaterials, and human and animal sera as culture substrates along with Matrigel concluded that Matrigel was the best substrate in terms of sustaining hPSC attachment, morphology and pluripotency [43]. A variant of Matrigel designed for hPSC cultures named hESC qualified Matrigel is currently the 'gold standard' substrate for feeder-free culture. Every batch of hESC qualified Matrigel is tested and quality is controlled is ensure consistency and reproducibility [44]. Geltrex® is another commercially available substrate similar to Matrigel that is derived from the mouse EHS tumor. It has also been demonstrated to support hPSC culture in conjunction with StemPro medium [45].

While there have been efforts to define the components that make up Matrigel [46], it remains highly undefined. Apart from its ECM constituents, it is believed to contain a variety of growth factors, which further confound efforts to decipher its biological active components. This jeopardizes the use of hPSCs cultured on Matrigel in clinical applications. It is also postulated the Matrigel may expose cultured hPSCs to nonhuman immunogenic sialic acid.

12.3.2.1.2 Human Serum

Stojkovic *et al.* demonstrated the use of a human serum (HS) matrix, which is derived from male clotted blood, along with a fibroblast CM for hPSC culture [47]. They concluded that hPSCs cultured on the substrate could maintain their characteristic morphology, undifferentiated state and proliferative abilities for a prolonged period. This study could not be validated by Hakala *et al.* (2009), who found that the HS matrix did not allow cell attachment while using a chemically defined medium (with knockout serum). Although the hESCs did attach on the HS substrate in the presence of CM, excessive differentiation was seen and the culture could not be extended beyond three passages [43]. However, differences in cell lines used and in the method of collecting CM could account for these discrepancies.

12.3.2.2 ECM Components

While the cellular substrates and animal-derived biological matrices described in the previous sections have been widely used for hPSC culture, there are several drawbacks associated with their use. The cell-secreted factors that are bound to these biological matrices show inconsistent expression and are undefined. Also, feeder cells have been shown to express Neu5Gc—immunogenic nonhuman sialic acid [48]. This has led to concerns that cells derived from hPSCs grown on feeder cells and subsequently transplanted would elicit an immune response. Unlike cellular substrates and biological matrices, purified ECM components provide a more defined substrate for hPSC growth and maintenance. They also have improved batch-to-batch consistency as compared to biological matrices. Recombinant human ECM proteins could provide a xeno-free alternative for hPSC culture.

Integrins are cell surface receptors, which couple cell–ECM adhesion to intracellular signaling pathways. This cell adhesion-mediated intracellular signaling is crucial for the maintenance of hPSC pluripotency. Various publications have profiled the integrins that are abundantly present on the surface of undifferentiated hPSCs [39, 49–51]. Thus, the selection of ECM components for hPSC growth is in large part based on their affinity to the integrins expressed by hPSCs [39].

12.3.2.2.1 Laminin

Laminin is a glycoprotein found in basement membranes consisting of three chains (known as α, β, and γ), which are covalently linked. They can form 15 known isoforms with different combination of these chains (α 1–5, β 1–3, and γ 1–3). A study compared Matrigel, laminin, fibronectin, and Collagen IV to test undifferentiated growth and self-renewal of hESCs cultured in MEF-CM [39]. They showed that only hESCs grown on laminin (derived from mouse EHS sarcomas) and Matrigel retained their pluripotency and self-renewal capacity. Rodin in 2010 demonstrated that Laminin 511 actually showed better adhesive properties to hPSCs than Matrigel, and could maintain three hESC lines and two hiPSC lines [52]. Miyazaki et al. (2008) investigated various human recombinant laminin isoforms and their effectiveness in maintaining hPSCs [49]. They concluded that Laminin 511, 332, and 111 could support hPSC culture mainly due to their high affinity to integrin $\alpha_6\beta_1$ expressed in hPSCs. Among the three isoforms, Laminin 111 showed significantly weaker cell adhesive ability. The same study found that Laminin 211 and 411 did not promote hPSC adhesion.

A recombinant truncated version of Laminin 511 known as Laminin E8 was also shown to support hPSC culture [53, 54]. These E8 fragments were proved to be the minimum fragments that conferred integrin binding ability and were sufficient to support long-term self-renewal of hESCs and hiPSCs in three different chemically defined media. In fact, E8 fragments of Laminin 511 and Laminin 322 were shown to support hESC cell adhesion better than Matrigel™ [54].

12.3.2.2.2 Vitronectin
Vitronectin is a glycoprotein present in blood and ECM, which binds glycosaminoglycans, collagen, plasminogen, and the urokinase receptor, and also stabilizes the inhibitory conformation of plasminogen activation inhibitor-1. It has been known to support cell adhesion, spreading, and migration in a variety of cell types [55]. Braam et al. (2008) established that a human recombinant vitronectin (full-length) substrate [56] could support the self-renewal of hESCs. Rowland (2010) extended the study to show that vitronectin could also support the growth of hiPSCs [50]. Both studies showed that the attachment of hPSCs to the substrate was modulated by integrin $\alpha_v\beta_5$. Another study produced recombinant vitronectin consisting of its SMB domain along with an RGD sequence (Arg–Gly–Asp) which is known to bind to integrins [57]. They showed that hPSCs could be maintained undifferentiated for long periods on this recombinant vitronectin substrate using a chemically defined commercial medium (StemPro).

Chen et al. (2011) demonstrated the use of two truncated recombinant vitronectin constructs named VTN-NC and VTN-N and showed that they could both support hESC growth, hiPSC derivation, and growth [58]. They developed a new chemically defined medium for use with these substrates, named the E8 medium as it consists only of eight essential components. This medium drastically cut down on the number of components used in other chemically defined media, such as mTeSR1 (18 components), therefore translating to lower costs.

12.3.2.2.3 Fibronectin
Fibronectin is an abundant protein found in human plasma (in its soluble form) as well as in the ECM (in its insoluble form). It has been demonstrated to mediate a wide variety of cell interactions with the ECM [59]. Amit et al. (2006) used human fibronectin from plasma for the culture of hESCs in the presence of knockout serum medium supplemented with growth factors (TGF-β1 and FGF-2). They found that fibronectin could support hESC proliferation and maintain pluripotency for up to 1 year [60]. Other studies could also maintain hESCs on fibronectin substrates

via integrin $\alpha_5\beta_1$ mediated adhesion with the help of MEF-CM [39, 56]. However, they seemed to be less adept at maintaining pluripotency than other ECM components and Matrigel [39]. They were also unable to support hESC growth in the defined mTeSR1 medium [56].

12.3.2.2.4 Hyaluronic Acid

Hyaluronic acid (HA) is a large polysaccharide found in the ECM, which has been shown to play a role in regulating cell behavior during embryonic development [61]. Due to its role in early development, Gerecht *et al.* evaluated the use of 3D HA hydrogels for hESC culture in the presence of MEF-CM [62]. The cells that were encapsulated in the HA hydrogel maintained their undifferentiated state, normal karyotype, and differentiation abilities.

12.3.2.2.5 Collagen I

Furue *et al.* reported the use of a Collagen I substrate in a defined serum-free medium that was supplemented with heparin [63]. This system could support the growth of hPSCs for up to 25 passages. Similar to HA substrates, the use of collagen I for PSC culture has not been extensively studied.

12.3.2.2.6 ECM Combinations

A recombinant human matrix consisting of a combination of different purified ECM components (collagen IV, fibronectin, laminin, and vitronectin) has been described [41]. Although the hPSCs could maintain pluripotency and were able to differentiate into all three germ layers, they showed karyotypical abnormalities on prolonged culture. There was an absence of Neu5Gc in cells derived on Matrigel and on then cultured on this human matrix.

12.3.2.3 Decellularized Matrices

Biological matrices such as the Matrigel™ have been commonly used for support of undifferentiated growth of hPSCs, including hESCs and iPSCs [64]. However, these biological matrices are often poorly defined and subjected to batch variability. Furthermore, inherent xenogeneic components present in these matrices pose risks of immune rejections and disease transmission that largely hamper the prospects for clinical applications [19, 26]. As previously described, several feeder systems including the mouse and human feeders have been used to support the undifferentiated growth of hPSCs [27–32]. However, the major limiting factor of such systems is the risk of cross-contamination. To overcome this issue, decellularization of the feeders to preserve only the ECMs seems to be a technically easy, fast, safe, and cheap method for maintaining a refined feeder-free system for culture of hPSCs [65].

Several technical advances have been made to improve ECM deposition as well as in the decellularization techniques [65, 66]. It has been well established that ECM contains proteins and factors vital for the adhesion, proliferation, and differentiation of PSCs. Of these ECM proteins, the basement membrane proteins such as laminins are necessary to support self-renewal and undifferentiated growth of hPSCs [67]. Chemicals such as ascorbic acid [68], dextran sulfate [65, 69], and Ficoll [65] are commonly added to induce ECM deposition before decellularization. Peng et al. showed that addition of dextran sulfate accelerated the deposition of ECM by human lung fibroblasts, and the generated decellularized matrix was able to support undifferentiated growth of hESCs under chemical-defined medium (mTeSR1) conditions [69]. Several methods have also been employed in decellularization, which include chemical methods using sodium dodecyl sulfate (SDS) [70] and ammonium hydroxide [71], as well as physical methods such as freeze-thawing [68]. Fu et al. showed that autogenic feeder cells from hESCs could be easily decellularized through a freeze-thaw procedure, and the generated decellularized matrix, together with animal component-free mTeSR2 medium were able to support long-term growth of H1 and H9 hESC lines for up to 20 passages [68]. In summary, decellularized matrices provide a cost-effective and easy-to-prepare alternative for hPSC culture and differentiation. Furthermore, these decellularized matrices could be stored for off-the-shelf use, thus facilitating large-scale hPSC culture for clinical and translational applications.

12.3.2.4 Cell Adhesion Molecules

CAMs are cell surface proteins that mediate interactions with nearby cells and ECM proteins through extracellular ligands [72]. CAMs provide two-way signaling from cells to their surrounding environment and from one cell to another. The main CAM family members including integrins and cadherins have been identified on the surface of PSCs and found to modulate their self-renewal and pluripotency in culture [19, 72]. Integrins are a family of transmembrane heterodimeric glycoproteins that are composed of α and β chains. Eighteen α and 8 β chains have been identified, combining to form 24 known types of integrins [72]. In mouse ESCs, simultaneous stimulation of integrins—$\alpha_5\beta_1$, $\alpha_6\beta_5$, $\alpha_6\beta_1$, and $\alpha_9\beta_1$ within 3D poly(ethylene glycol) (PEG) scaffold greatly increased Akt1 and Smad1/5/8 activation, which resulted in prolonged self-renewal of the mESCs in the absence of leukemia inhibitory factor (LIF) [73]. Similarly, integrins $\alpha_5\beta_1$, $\alpha_6\beta_1$, $\alpha_v\beta_3$, and $\alpha_v\beta_5$ have been reported to mediate hPSC self-renewal and pluripotency on defined culture surfaces [19, 49–51].

The other class of CAMs is the cadherins, transmembrane glycoproteins that form calcium-dependent cell–cell and cell–matrix homophilic binding junctions [19, 72]. Among the cadherins (>100 members), E-cadherin is the primary cadherin expressed by PSCs, with important roles in intercellular adhesion and colony formation of mESCs [72]. Accordingly, E-cadherin expression decreases during differentiation of hPSCs [74] and has been considered as a molecular marker for pluripotency [75] and target CAM for development of defined culture surfaces to support undifferentiated growth of hPSCs. Notably, disruption of E-cadherin signaling has been demonstrated to result hESC blebbing and apoptosis [76–78]. Furthermore, establishment of E-cadherin-mediated cellular adhesion has been shown in studies to play a pivotal role in the control of hPSC survival or death following dissociation [76–78]. As such, a recombinant fusion protein (hE-cad-Fc) composed of human E-cadherin and the Fc region of IgG1 antibodies has been developed as surface coating to mediate cellular adhesion of hPSCs. Using this system, hPSCs maintained similar morphology and growth rate to those grown on Matrigel™ under the same mTeSR1 medium condition, and retained all pluripotency features, including *in vitro* EB differentiation and *in vivo* teratoma formation [79] (Figure 12.1).

12.3.2.5 Synthetic Substrates

Synthetic substrates are ideally designed for stem cell attachment, proliferation, and differentiation for tissue regeneration [19]. The consideration of the material and the biological requirements are important for the design of synthetic substrates for stem cell culture. There are several aspects to note when considering the type of materials [80–82]. They should be materials which are biocompatible and nontoxic especially since they are in such close proximity to the living cells. They should possess a desired degradation rate, surface properties, processability, porosity, and mechanical properties. Biological considerations point to nonimmunogenicity of the synthetic substrates and prevention of foreign body reactions. These materials should also possess bioactivity, nutrient availability, and access to growth factors.

There have been significant research efforts on new polymeric substrates for the maintenance of undifferentiated growth of PSCs. In order to design cost-effective and commercially ready culture surfaces, fully synthetic plastic surfaces without peptide fragments were studied [83–87]. Among the reported synthetic surfaces, poly[2-(methacryloyloxy)ethyl

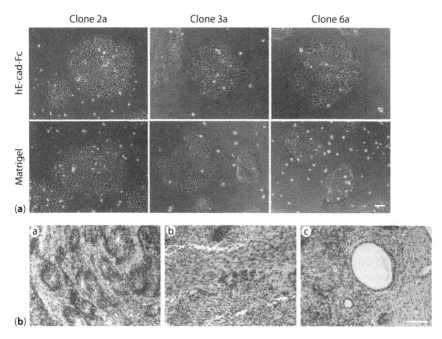

Figure 12.1 Maintenance of human induced PSCs on an hE-cad-Fc surface. (a) Morphology of human iPS cell clones (2a, 3a, and 6a) on surfaces coated with Matrigel or hE-cad-Fc. Scale bar indicates 100 µm. (b) Characterization of teratomas from iPS cells (clone 2a) cultured on an hE-cad-Fc-coated surface. Hematoxylin and eosin staining of teratomas showed the differentiation into various tissues, including immature neuroblastic tissue with neuronal rosettes (a), striated muscle (b), and columnar epithelium (c). Bar indicates 100 µm. Reproduced with permission [79].

dimethyl-(3-sulfopropyl)ammonium hydroxide] (PMEDSAH) [86] and aminopropylmethacrylamide (APMAAm) [83] showed the most promising results. These polymers support the long-term expansion of multiple hiPSC lines without compromising pluripotency or inducing abnormal karyotypes. The underlying mechanism and contributing factors are not well understood, but it was found that mechanical/surface properties such as stiffness, wettability, and rigidity and chemical properties of synthetic surfaces influenced cell attachment and proliferation by mimicking the biological microenvironment as other ECM substrates.

The design of functional synthetic surfaces brings us a step closer toward large-scale robotic adhesion cultures as they eliminate the need for the coating procedure and allow storage and handling at room temperature. It is also critical that the materials are compatible with sterilization techniques used in manufacturing processes, such as electron beam- and γ-radiation. As these sterilization techniques are usually harsh, it is not suitable for substrates or surfaces containing protein or peptide components due to the potential denaturation and degradation of these components. Additionally, these sterilization techniques possibly affect the physical and chemical properties of the synthetic or the plastic surface. Therefore, the surfaces that have been proven to be capable for bulk sterilization, such as PMEDSAH, facilitate using cGMP in the production of commercial disposable vessels for large-scale expansion of clinical-grade hPSCs. Polymerization from surfaces has been used to prepare six acrylate-based surfaces through ozone-activation of tissue culture polystyrene (TCPS) and subsequent surface-initiated polymerization with a range of acrylate monomers [87]. One of these materials (PMEDSAH) could sustain the long-term culture of hPSCs in serum-free defined mTeSR1 medium. H9 cells that were supported through 10 passages using StemPro medium showed normal karyotype, and expressed levels of pluripotency markers that were similar to cells cultured on Matrigel™. However, no large-scale demonstration was reported for this material. An APMAAm-based coating grafted to TCPS dishes was fabricated using a photoinitiated addition polymerization [83]. H1 and H9-hOCT-pGZ hESC lines were cultured in mTeSR1 medium for 10 passages, maintained typical cell morphology, and grew in colonies similar to Matrigel™-cultured cells. BSA was proposed to play a crucial role in attachment of hPSCs achieved in the culture medium, and quartz crystal microbalance with dissipation experiments were used to identify the adsorption of BSA to the growth substrate from the mTeSR1 medium.

Recently, a high-throughput combinatorial approach was used to identify and develop a defined, synthetic polymeric substrate that supports pluripotency and expansion of hPSCs through serial passages in commercially defined media without the need for protein pre-adsorption [88]. This was achieved for both hESCs and hiPSCs. Additionally, directed differentiation was achieved on the hit polymer, poly(HPhMA-co-HEMA) (HPhMA: (N-(4-hydroxyphenyl)methacrylamide); HEMA: 2-hydroxyethyl methacrylate), to representatives of each of the three germ layers, including spontaneous beating clusters of cardiomyocytes (mesoderm), hepatocyte-like cells (endoderm), and neuro-ectoderm (ectoderm). It appears that the compatibility of this substrate with pluripotent cell expansion is consistent with the ready differentiation of these cells under the influence of soluble factors. Thus, poly(HPhMA-co-HEMA) fulfills all the current culture requirements for the clinical use of stem cells within regenerative medicine and can be scaled up by coating onto tissue culture surfaces to be used off-the-shelf, providing a cost-effective alternative to commercially available substrates. The expansion of hPSCs and production of terminally differentiated cell types without the influence of undefined and xenogenic matrix protein coatings provides a robust platform for the industrial scale production of hPSCs for regenerative medicine applications and therapies.

To examine the effects of substrate properties on hiPSC behavior, PCL, PET, PEKK, and PCU were electrospun to produce various fibrous substrates with distinctive physiochemical characteristics [89]. These polymer precursors were selected due to their biocompatibility as well as significant differences in bulk elastic modulus to render a large range of stiffness among the substrates after the electrospinning process. Additionally, the ester linkage present in these polymers allowed for the modification of their surface chemistry through collagen conjugation *via* NHS-EDAC chemistry. Generally, as the stiffness of the substrate increased, the proliferation also increased, reaching a similar cell number to those cultured on Geltrex®-coated TCPS (a typical iPSC cell culture substrate). The results showed that the mechanical properties of substrates determine the stem cell colony morphology that subsequently influences cellular behaviors.

12.4 Substrates for Promoting Differentiation of PSCs

12.4.1 Cellular Substrates

To induce lineage-specific differentiation of PSCs, co-culture with cellular substrates either through direct contact or culture in CM is commonly explored [90–96]. The cellular substrate to direct differentiation of PSCs is usually the somatic cell of interest. For instance, co-culture with OP9 stromal cells has been employed to induce differentiation of hESCs to hematopoietic [90] and MSC lineages [91–93]. Other studies have also explored co-culture with MS5 and PA6 cells to induce neural crest and neuronal cell lineages [95]. Notably, Domev *et al.* showed that co-culture of hESCs with OP9 stromal cells followed by CD73 sorting generated highly purified CD73+ MSCs with functional capacity for vascularized bone formation *in vivo* [92]. Additionally, Vazin *et al.* demonstrated co-culture with PA6 cells induced dopaminergic neuronal differentiation of hESCs. In that study, factors including stromal cell-derived factor 1 (SDF-1), pleiotrophin, insulin-like growth factor 2(IGF-2), and ephrin B1 secreted by PA6 cells were identified, mediating the stromal-derived inducing activity necessary for DA neuronal induction [96]. The use of cellular substrates offer high efficiency of differentiation, but the roles of feeder cell contact and morphogenetic factors involved are often poorly understood [93]. Furthermore, many of these cellular substrates are derived from animal sources that may pose risk of xenogeneic contamination and hamper prospects for clinical applications.

12.4.2 Acellular Substrates

The differentiation of PSCs can either be accomplished in 3D (in an EB) or by 2D monolayer culture of the cells on a substrate. An EB is a 3D structure that consists of cells from all three germ layers and allows for multicellular interactions. This mimics the *in vivo* environment in early embryogenesis and has thus emerged to be a major tool in the differentiation of PSCs *in vitro* [97]. However, cells differentiate spontaneously into all three germ layers due to the inherent environmental gradient present in EBs. This inherent heterogeneity presents a challenge in generating a pure cell population of application interest, e.g. motor neurons or cardiomyocytes. Therefore, directed differentiation of a monolayer of hPSCs cultured on a substrate is increasingly being employed.

In directed differentiation, the goal is to mimic not only the biochemical soluble factors but also the ECM components present in embryonic developmental processes. Studies have shown that the ECM displays a wide range of macromolecular cues, which guide stem cell proliferation, migration, and differentiation [98]. A recent review by Reilly and Engler (2010) demonstrated how mechanical, chemical, and structural properties of the ECM can affect stem cell differentiation [99]. This holds true for the differentiation of PSCs *in vitro* as well. Thus, the selection of the appropriate combination of soluble and adhesive ECM induction factors for differentiation is of critical importance.

12.4.2.1 Biological Matrices

Various cellular matrices and ECM components are utilized for hPSC maintenance, and therefore their use is naturally extended for their differentiation *in vitro*. In the following section, we will provide a succinct summary on the use of various biological matrices to differentiate hPSCs into different lineages (Table 12.2).

12.4.2.1.1 Matrigel™

Matrigel™ has been extensively used for the differentiation of PSCs into a variety of cell types. Having an assortment of ECM components, it has been shown to aid differentiation into all three germ layers [64, 100–105]. As a native basement membrane secreted by mammalian cells, it might

Table 12.2 Substrates used for the differentiation of PSCs.

Substrate	Differentiation	Cell type	References
Matrigel	Cardiomyocyte	hiPSC	[101]
	Endothelial	Rat ESC [102], hESC [103]	[102, 103]
	Hepatocyte	hESC	[104]
	Neuronal	mESC	[105]
Vitronectin	Oligodendrocyte	hESC	[106]
	Cardiac	mESC	[107]
Collagen I	Mesoderm	hESC	[109]
	Hepatocyte	hESC	[112, 113, 104]
Collagen IV	Mesoderm	miPSC [115], hESC [116], mESC [117]	[115–117]
	Trophectoderm	mESC	[117]
	Corneal	hESC	[118]
Laminin	Cardiac	mESC	[121]
	Neuronal	hESC	[122]
	Pancreatic	mESC	[124]
Fibronectin	Endothelial	mESC	[121, 125]

also have the various ECM constituent present in optimal concentrations for differentiation [100], unlike purified singular ECM components. Hence, Matrigel has been successfully used to drive hPSCs into a variety of lineages as follows.

Cardiac
Feaster (2015) showed that the use of thick Matrigel substrates ranging from 0.4 to 0.8 mm could enable rapid generation of hiPSC-derived contractile cardiac myocytes, which displayed robust contractile responses and enhanced maturation [101].

Endothelial
Rat ESCs differentiated for 14 days on a 5-mm-thick Matrigel substrate, which was previously shown to promote tube formation of HUVEC cells *in vitro*, showed the formation of cord-like structures [102]. The cells showed 3D remodeling and formed a network of vascular-like structures, which were not seen in rat ESCs differentiated in conditions without Matrigel. The cells differentiated on Matrigel also showed high expression

of endothelial markers unlike the control cells. The formation of cord-like structures in Matrigel was also observed by Levenberg *et al.* while differentiating hESCs [103]. The cells underwent a spontaneous reorganization, and the cord structures were shown to have lumen, proving them to be tubes.

Hepatic
Ishii *et al.* cultured hESCs on collagen I, laminin, and Matrigel in the presence and absence of growth factors to evaluate which conditions were best suited for hepatocyte differentiation. They found that both Collagen I and Matrigel promoted hepatocyte differentiation even in the absence of growth factors [104]. This ability was enhanced, however, with the addition of Activin A. The cells showed endodermal differentiation and expressed early hepatocyte markers.

Neuronal
A study compared collagen, ornithine/laminin, and Matrigel (with reduced growth factors) for their ability to support the growth of mESC-derived neuronal precursor cells and their subsequent differentiation into functional neurons. They concluded that Matrigel was the most effective in supporting cell survival and promoting neurite outgrowth [105]. When these NPCs grown on Matrigel were grafted into mice, they showed the ability to form dopaminergic neurons.

12.4.2.2 ECM Components

12.4.2.2.1 Vitronectin
Vitronectin has been found to localize in late-stage glioblastomas, wound tissues, and elastin fibers of the skin. Much of the vitronectin found in tissues may have their source in serum as vitronectin shows great binding affinity to proteoglycans in the ECM. Murine cells undergoing neuronal differentiation have been shown to upregulate their expression of integrin $\alpha_v\beta_1$ and $\alpha_v\beta_3$, both integrin receptors known to bind to vitronectin [55]. Hence, vitronectin has been evaluated for their ability to support neuronal and mesenchymal differentiation.

Neuronal
Gil *et al.* (2009) showed that the use of vitronectin as a substrate could enhance the potential of hESCs to form oligodendrocytes during neurogenesis, with a four-fold increase in yield of oligodendrocytes as compared to controls. They hypothesized that the effect was due to the fact that vitronectin is expressed in the ventral part of the developing human spinal cord [106].

Cardiac
mESCs cultured on a 3D vitronectin scaffold were induced to differentiate into cardiac progenitor cells (Flk-1+). The number of Flk-1+ cells was higher on 3D scaffolds coated with vitronectin and laminin than Collagen IV. These progenitor cells could be further differentiated into functional cardiomyocytes, smooth muscle cells, and endothelial cells [107].

12.4.2.2.2 Collagen I

Collagen I is the most abundant collagen in the human body and is found mainly in connective tissues, such as the skin dermis, bone, tendon, and ligaments. Hence, it is one of the most common ECM components used in *in vitro* cell culture. In particular, *in vitro* culture of hepatocytes has been established almost exclusively with collagen I since it is abundant in the hepatocytic basal membrane [108]. Consequently, collagen I is explored in the directed differentiation of hPSCs into these lineages.

Mesoderm
Liu et al. (2012) used a thin fibrillary Collagen I substrate to direct hESCs toward MSCs without the use of any soluble factors [109]. The cells produced were positive for CD markers that are characteristic of MSCs and were also capable of tri-lineage differentiation (i.e. osteocyte, adipocyte, and chondrocyte differentiation). They postulated that the effect was due to the fact that collagen is known to stimulate epithelial–mesenchymal transition (EMT) through integrin-mediated signaling [110] and also that EMT has been used to differentiate ESCs to MSCs [111].

Hepatic
A study showed that hESC-derived hepatocytes when cultured in a Collagen I gel showed a greater tendency to form mature hepatocytes as compared to hESCs grown on Matrigel. This might be due to the fact that there is an abundance of Collagen I in the hepatocyte microenvironment *in vivo* (space of Disse). Matrigel, which contains a lot of Collagen IV but no Collagen I, seemed to preferentially direct cells toward a cholangiocyte lineage instead [112]. This was validated in another study wherein the authors cultured hESC-derived EBs on 2D collagen coatings and 3D collagen I scaffolds and induced to differentiate into hepatocytes [113]. While both systems could produce functional hepatocytes, gene expression of hepatocyte markers, and urea production was higher in the 3D system, indicating it to be a better model for hepatocyte differentiation.

12.4.2.2.3 Collagen IV

Collagen IV is the type of collagen that is prevalent in basement membrane and abundant in biological matrices [114], such as Matrigel, which has been shown to support hPSC differentiation into multiple lineages. Thus, collagen IV is explored as a single ECM component that can possibly replace biological matrices in the directed differentiation of hPSCs.

Mesoderm
Collagen IV has been widely reported to drive PSCs toward mesodermal lineages. Murine undifferentiated iPSCs directly seeded on Collagen IV coated plates have been shown to express mesodermal progenitor markers after 4 days of differentiation [115]. These cells also expressed early markers of cardiac, hematopoietic, endothelial, and smooth muscle differentiation. Upon further culture in respective differentiation media, these Collagen IV differentiated cells were induced to form functional cardiomyocytes, smooth muscle cells, endothelial cells, and hematopoietic cells. The ability of Collagen IV to promote mesoderm differentiation was also shown to be true for hESCs by Gerecht *et al.* (2003). They seeded undifferentiated hESCs onto Collagen IV substrates without any growth factors and observed an increase in mesodermal and endothelial progenitor markers [116]. These mesodermal cells were then induced to form endothelial cells and smooth muscle cells on collagen IV in differentiation medium with growth factors (VEGF and PDGF-BB, respectively)

Trophectoderm
The trophectoderm of the blastocyst in mouse embryos ultimately gives rise to extraembryonic tissues, including the trophoblast layer of the placenta. When mESCs were cultured on a Collagen IV substrate, they expectedly showed the expression of mesodermal genes (hematopoietic, endothelial, and smooth muscle lineages), but also surprisingly expressed trophoectodermal specific genes [117]. This was not seen on other substrates, such as Collagen I, Laminin, and Fibronectin. These trophoectodermal cells were successfully differentiated into a variety of trophoblast lineages.

Corneal
hESCs cultured on a Collagen IV substrate with a limbal fibroblast (a cell type found in the stroma of the human cornea) CM were shown to differentiate into corneal epithelial cells [118]. The authors credited this differentiation to the fact that Collagen IV is a major component of the basement membrane of the corneal epithelium, and along with the limbal fibroblast CM, they could create an environment that closely mimicked the human cornea.

12.4.2.2.4 Laminin
Similarly, laminin, being the other major component of basement membrane-like biological matrices [119], is evaluated as a more defined ECM

substitute in the directed differentiation of hPSCs. It appears early during embryonic development at the four-cell embryo stage and plays a major role in cell differentiation [120].

Cardiac
A study which cultured mouse ESC-derived EBs in a collagen–laminin 3D gel with constant collagen concentration found that cardiac differentiation increased with increasing laminin concentration. This was indicated by a larger number of beating cells in the EB [121]. The laminin was extracted and purified from the EHS tumor.

Neuronal
A study compared different ECM components plated along poly-D-lysine (PDL—a positively charged synthetic molecule which enhances cell attachment) to test neuronal differentiation of EBs [122]. They found that laminin and Matrigel promoted neuronal generation and neurite outgrowth more than the other substrates (fibronectin, collagen I, and only PDL) and also showed that the signaling was mediated by integrin $\alpha_6\beta_1$. Laminin in conjunction with PDL or poly-L-ornithine (PLO) is widely used for neuronal differentiation of hESC-derived neuronal precursors [123].

Pancreatic
Insulin-producing pancreatic cells have been derived from mouse ESCs on a PLO–Laminin substrate in a medium containing soluble laminin, insulin, and nicotinamide. Laminin was found to be a critical factor for pancreatic differentiation [124]. Following EB formation by the mouse ESCs and spontaneous differentiation into endoderm progenitors, the cells were directed into pancreatic cells on the laminin substrate.

Fibronectin
Fibronectin is expressed abundantly in the human body in circulating blood plasma and is also essential in vertebrate development [59]. Its ubiquitous presence in the plasma may be the reason it promotes endothelial differentiation *in vitro*.

Endothelial
3D embryoid bodies (from mouse ESCs) cultured in a collagen–fibronectin matrix showed increased endothelial differentiation in a dose-dependent response to fibronectin [121]. Also, collagen gels without fibronectin exhibited attenuated levels of endothelial differentiation. This was validated by another study, which showed the important role fibronectin played in endothelial cell development by demonstrating that mouse ESCs lacking fibronectin or integrin $\alpha 5$ were unable to form any endothelial cells [125].

Endothelial cells express a number of integrins, which are fibronectin receptors: $\alpha_5\beta_1$, $\alpha_V\beta_1$, and $\alpha_V\beta_5$.

12.4.2.3 Decellularized Matrices

To promote differentiation of PSCs, decellularized cell and tissue/organ matrices have been frequently explored [126-133]. Recently, Duan et al. investigated the role of decellularized porcine heart ECM in directing cardiac differentiation of hESCs [126]. In that study, a series of hydrogels was prepared from decellularized ECM from porcine hearts by mixing ECM and type I collagen at varying ratios. It was shown that hydrogel with high ECM content was able to increase cardiac differentiation and maturation without the need of supplemental soluble factors, making this system applicable for studies of cardiac development and potentially stem cell delivery to the heart [126]. Other studies have also explored decellularized heart [127], lung [128], and kidney [129] as scaffolds to direct organ-specific differentiation of PSCs. Notably, Zhou et al. showed that differentiated iPSCs reformed an alveolar structure and expressed surfactant protein or T1α protein, indicative of respective alveolar epithelial type II and type I cells, when cultured in decellularized mouse lung scaffold. When instilled intratracheally into a bleomycin-induced mouse acute lung injury model, the differentiated cells integrated into the lung alveolar structure and significantly reduced lung inflammation and decreased collagen deposition [128].

Decellularized matrices could also be derived from committed cells that were coaxed to deposit specific ECMs, prior to decellularization [130-133]. For instance, human osteoblasts were seeded on poly(lactic-co-glycolic acid) (PLGA) scaffolds for 14 days, followed by decellularization, leaving the mineralized bone ECM. PLGA scaffolds modified with bone ECM significantly enhanced osteogenic differentiation of hESCs, compared to the control PLGA scaffolds [131]. In a separate study [132], cell lysate of hESC-derived osteoblasts was used to coat the surface of tissue culture plates. It was found that the autogenic cell lysate containing the osteogenic matrix proteins was able to enhance osteogenic differentiation of hESCs. Similarly, decellularized matrix from hESC-derived fibroblasts has been used to create an autogenic dermo-epidermal junction-like matrix necessary for induction of hESCs toward keratinocytes [133]. Notably, the composition of the decellularized matrix exerts significant effects not only on differentiation of hESCs, but also the subsequently replicative lifespan and cellular senescence of the derived keratinocytes [133]. These findings are reminiscent of other studies that indicate critical roles of ECMs in mediating cell–matrix interactions, growth factor signaling, and cellular response to its microenvironment [134, 135]. Importantly, these findings suggest new opportunities and perspectives for applications of decellularized matrices in stem cell-based regenerative medicine.

12.4.2.4 Cell Adhesion Molecules

In recent years, there has been a surge of interest in exploring CAMs for development of defined substrates and surfaces for lineage-specific differentiation of PSCs [19, 136–140]. In a recent study, it was found that the differentiation dynamics of hESCs cultured on E-cadherin coated substrates were distinctive from those on Matrigel. When hESCs were induced to undergo mesoendoderm differentiation on substrates coated with alternating strips of E-cadherin fragment (Ecad-Fc) and Matrigel, it was observed that resultant differentiation patterns corresponded to the underlying substrates [136]. Regions of the hESC colony on Matrigel expressed the mesoendoderm marker, Brachyury, robustly whereas the regions on Ecad-Fc did not [136].

The effects of CAMs on PSC differentiation depend largely on the culture conditions [137, 139, 140]. Notably, highly homogenous population of mESCs and miPSCs could be maintained on E-cadherin-based substrate under serum-free culture conditions, and when transferred onto E-cadherin and N-cadherin hybrid substrate under differentiation culture condition, highly homogenous population of primitive ectoderm and neural progenitor cells with high expression of neural markers (βIII-tubulin, Pax6, and tyrosine hydroxylase) was achieved [137]. This finding is consistent with another study performed on hPSCs that co-culture of hESCs with E-cadherin-transfected L929 fibroblasts promoted neuronal differentiation with rapid expression of nestin and βIII-tubulin within 1 week of induction [138]. However, it was also found that E-cadherin substrate synergistically promoted stepwise differentiation of mESCs to cells with characteristics of definitive endoderm, hepatic progenitor cells and hepatocytes under endoderm and hepatocyte-differentiation medium conditions [139, 140].

12.4.2.5 Synthetic Substrates

It has been proposed that the use of hiPSCs could provide an unlimited supply of cells which are specially tailored for a particular patient. The differentiation of stem cells into specific cell types means that, in theory, replacement cells for most parts of the human body can be readily obtained and organ repair can be carried out with the patient's own cells in future. However, the clinical application of these cells will require (i) defined, xeno-free conditions for their expansion and differentiation, and (ii) scalable culture systems that enable their expansion and differentiation in numbers sufficient for regenerative medicine and drug screening purposes. Current ECM protein-based substrates for the culture of stem cells are expensive, difficult to isolate, subject to batch-to-batch variations, and, therefore, unsuitable for clinical application of PSCs. It is therefore increasingly critical to develop synthetic stem cell culture substrates for lineage-specific differentiation of stem cells. In addition, synthetic substrates could be modified easily in their biochemical and biophysical properties to tailor specifically to their intended applications.

Myocardial infarction results in extensive cardiomyocyte death which can lead to fatal arrhythmias or congestive heart failure. Cardiomyocytes derived from hiPSCs (hiPSC-CMs) are promising for human myocardial repair, disease modeling, and drug screening [6]. Current approaches of PSC differentiation to cardiomyocytes often generate immature embryonic myocytes and failed to adequately reproduce native adult cardiomyocyte phenotypes. Recently, Gupta *et al.* optimized the chemical and mechanical properties of a combinatorial polymer scaffold for differentiation of PSCs toward cardiomyogenic lineage [141]. In that study, a combinatorial polymer library was prepared by copolymerizing three distinct subunits at varying molar ratios to tune the physicochemical properties of the resulting polymer: hydrophilic PEG, hydrophobic poly(ε-caprolactone) (PCL), and negatively charged, carboxylated PCL (cPCL). It was found that mESCs on the most compliant substrate, 4% PEG–86% PCL–10% cPCL, exhibited the highest α-MHC expression as well as the most mature calcium signaling dynamics.

The similar approach was applied to optimize combinatorial polymers as synthetic substrates for culturing hiPSC-CMs [142]. It was found that cells cultured on 4 mol% PEG–96 mol% PCL showed the greatest contractility and mitochondrial function. These functional enhancements are associated with increased expression of cardiac myosin light chain-2v, cardiac troponin I, and integrin $\alpha 7$. The cells demonstrated troponin I (TnI) isoform switch from the fetal slow skeletal TnI (ssTnI) to the postnatal cardiac TnI (cTnI). The substrate also increased the expression of genes encoding intermediate filaments known to transduce integrin-mediated mechanical signals to the myofilaments. This study shows that synthetic culture matrices engineered from combinatorial polymers can be utilized to promote *in vitro* maturation of hiPSC-CMs through the engagement of critical matrix–integrin interactions.

Stem cell therapies have also emerged as a viable option for treating many incurable neurological disorders [143]. In recent years, there have been escalating efforts on development of substrates for PSC differentiation to neural lineages as well as for propagation of derived neural progenitors under defined culture conditions [144–147]. A synthetic hybrid scaffold comprised of PCL–poly(β-hydroxybutyrate) (PHB) heparinized and grafted with neuron growth factor (NGF) was found to improve the differentiation of mouse iPSCs to βIII tubulin-positive neurons and inhibited other lineage differentiation [145]. With high efficiency in neuronal differentiation, these NGF-grafted PCL–PHB scaffolds might be useful for nerve repair. hPSC provides a viable source of neural progenitor cells (hNPCs) that could be expanded indefinitely and subsequently differentiate into the various cell types of the central nervous system (CNS). This could provide an unlimited source of cells for such cell-based therapies. Recently, a synthetic polymer, poly(4-vinyl phenol) (P4VP) was identified and shown to support the long-term proliferation and self-renewal of human neural progenitor cells (hNPCs) derived from hPSCs [147] (Figure 12.2). The hNPCs cultured on P4VP maintained their morphology, expressed high levels of markers of multipotency, and retained their differentiation ability into neurons [147]. These polymeric substrates could eliminate critical obstacles for the utilization of hNPCs for human neural regenerative repair, disease modeling, and drug discovery.

Apart from lineage-specific differentiation of PSCs, there have been recent efforts on development of synthetic substrates for cellular reprogramming [148]. Notably, a group of four polymers, polylactic acid (PLA), PCL, thermoplastic polyurethane (TPU), and polypropylene carbonate (PPC), was electrospun into nanofiber sheets [148]. Protein absorption on the substrate was performed to allow the attachment of fibroblasts to all substrates. Fibroblasts on aligned substrates had elongated nuclei, but after reprogramming factor expression, nuclei became more circular. Reprogramming factors could override the nuclear shape constraints imposed by nanofibrous substrates, and the majority of substrates supported full reprogramming. Early culture on PCL and TPU substrates promoted reprogramming, and TGF-β repressed substrate effects. TGF-β and the identity of the polymer were important cues governing cellular reprogramming responses. This approach of using a nanofibrous materials library can be used to dissect molecular mechanisms of reprogramming and generate novel substrates that enhance epigenetic reprogramming.

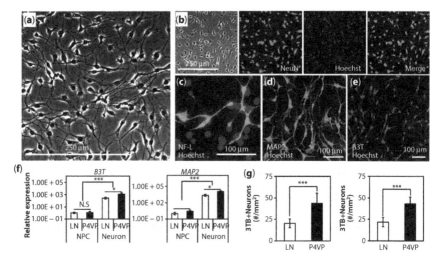

Figure 12.2 Neuronal differentiation of hNPCs on P4VP. (a) Phase contrast images of neurons differentiated on P4VP. (b) Phase contrast and immunofluorescence of mature neuronal markers NeuN, (c) NF-L, (d) MAP2, and (e) B3T in neuronal cultures differentiated on P4VP. (f) Gene expression analysis of MAP2 and B3T of neuronal cultures differentiated on P4VP and LN substrates (mean ± SEM, n = 3). Populations were compared using Student's t-test. * $p < 0.05$, **$p < 0.01$. (g) Quantification of the number of MAP2- and B3T-positive neurons generated on P4VP and LN substrates. Quantification of images was performed by counting a minimum of 9 fields at 20 × magnification. Image quantification of the data is presented as the average of these fields ± standard deviation (SD). Populations were compared using Student's t-test. ***$p < 0.001$. Abbreviations: hNPCs, human neural progenitor cells; NeuN, neuronal nuclear antigen; NF-L, neurofilament-68; MAP2, microtubule-associated protein 2; B3T, βIII tubulin; LN, laminin. Reproduced with permission [147].

12.5 Conclusions

To utilize hPSCs in various applications in regenerative medicine and disease modeling, it is critical that the culture conditions are consistent, chemically defined and/or nonxenogeneic for purposes of scale-up, reproducibility, safety, and clinical compliance. The surfaces and substrates for culture of hPSCs have evolved significantly over the years from xenogenic feeders to purified ECM proteins and synthetic substrates. Moving ahead, a deeper understanding of how different functional moieties support or direct cell behavior will lead to development of new synthetic substrates and will improve our understanding of PSC biology.

Acknowledgments

We are grateful to the Ministry of Education, National University of Singapore, National University Healthcare System, National Medical Research Council Singapore, and Agency for Science, Technology and Research (A*STAR) for funding our research programs.

References

1. Thomson J.A., Itskovitz-Eldor J., Shapiro S.S., Waknitz M.A., Swiergiel J.J., Marshall V.S., Jones J.M: Embryonic Stem Cell Lines Derived from Human Blastocysts. *Science* 282, 1145–1147, 1998.
2. Takahashi K., Tanabe K., Ohnuki M., Narita M., Ichisaka T., Tomoda K., Yamanaka S: Induction of Pluripotent Stem Cells from Adult Human Fibroblasts by Defined Factors. *Cell* 131, 861–872, 2007.
3. Rufaihah A.J., Haider H.K., Heng B.C., Ye L., Tan R.S., Toh W.S., Tian X.F., Sim E.K.-W, Cao T: Therapeutic angiogenesis by transplantation of human embryonic stem cell-derived CD133+ endothelial progenitor cells for cardiac repair. *Regenerative Medicine* 5, 231–244, 2010.
4. Rufaihah A.J., Huang N.F., Jamé S., Lee J.C., Nguyen H.N., Byers B., De A., Okogbaa J., Rollins M., Reijo-Pera R., et al: Endothelial Cells Derived From Human iPSCS Increase Capillary Density and Improve Perfusion in a Mouse Model of Peripheral Arterial Disease. *Arteriosclerosis, Thrombosis, and Vascular Biology* 31, e72–e79, 2011.
5. He J.Q., Ma Y., Lee Y., Thomson J.A., Kamp T.J: Human Embryonic Stem Cells Develop Into Multiple Types of Cardiac Myocytes: Action Potential Characterization. *Circulation Research* 93, 32–39, 2003.
6. Zhang J., Wilson G.F., Soerens A.G., Koonce C.H., Yu J., Palecek S.P., Thomson J.A., Kamp T.J: Functional Cardiomyocytes Derived From Human Induced Pluripotent Stem Cells. *Circulation Research* 104, e30–e41, 2009.
7. Cao T., Heng B.C., Ye C.P., Liu H., Toh W.S., Robson P., Li P., Hong Y.H., Stanton L.W: Osteogenic differentiation within intact human embryoid bodies result in a marked increase in osteocalcin secretion after 12 days of *in vitro* culture, and formation of morphologically distinct nodule-like structures. *Tissue & Cell* 37, 325–334, 2005.
8. Lou X: Induced Pluripotent Stem Cells as a new Strategy for Osteogenesis and Bone Regeneration. *Stem Cell Reviews and Reports* 11, 645–651, 2015.
9. Toh W.S., Yang Z., Liu H., Heng B.C., Lee E.H., Cao T: Effects of Culture Conditions and Bone Morphogenetic Protein 2 on Extent of Chondrogenesis from Human Embryonic Stem Cells. *Stem Cells* 25, 950–960, 2007.
10. Toh W.S., Guo X.M., Choo A.B., Lu K., Lee E.H., Cao T: Differentiation and enrichment of expandable chondrogenic cells from human embryonic stem cells *in vitro*. *Journal of Cellular and Molecular Medicine* 13, 3570–3590, 2009.

11. Toh W.S., Lee E.H., Guo X.M., Chan J.K.Y., Yeow C.H., Choo A.B., Cao T: Cartilage repair using hyaluronan hydrogel-encapsulated human embryonic stem cell-derived chondrogenic cells. *Biomaterials* 31, 6968–6980, 2010.
12. Ko J.Y., Kim K.I., Park S., Im G.I: *In vitro* chondrogenesis and *in vivo* repair of osteochondral defect with human induced pluripotent stem cells. *Biomaterials* 35, 3571–3581, 2014.
13. Lim M.S., Shin M.S., Lee S.Y., Minn Y.K., Hoh J.K., Cho Y.H., Kim D.W., Lee S.H., Kim C.H., Park C.H: Noggin over-expressing mouse embryonic fibroblasts and MS5 stromal cells enhance directed differentiation of dopaminergic neurons from human embryonic stem cells. *PLoS One* 10, e0138460, 2015.
14. Haque A., Adnan N., Motazedian A., Akter F., Hossain S., Kutsuzawa K., Nag K., Kobatake E., Akaike T: An engineered N-cadherin substrate for differentiation, survival, and selection of pluripotent stem cell-derived neural progenitors. *PLoS One* 10, e0135170, 2015.
15. Kidwai F.K., Liu H., Toh W.S., Fu X., Jokhun D.S., Movahednia M.M., Li M., Zou Y., Squier C.A., Phan T.T., Cao T: Differentiation of human embryonic stem cells into clinically amenable keratinocytes in an autogenic environment. *J Invest Dermatol* 133, 618–628, 2013.
16. Bilousova G., Chen J., Roop D.R: Differentiation of mouse induced pluripotent stem cells into a multipotent keratinocyte lineage. *Journal of Investigative Dermatology* 131, 857–864, 2011.
17. Asgari S., Moslem M., Bagheri-Lankarani K., Pournasr B., Miryounesi M., Baharvand H: Differentiation and Transplantation of Human Induced Pluripotent Stem Cell-derived Hepatocyte-like Cells. *Stem Cell Reviews and Reports* 9, 493–504, 2013.
18. Watt A., Forrester L: Deriving and identifying hepatocytes from embryonic stem cells. *Stem Cell Reviews* 2, 19–22, 2006.
19. Lambshead J., Meagher L., O'Brien C., Laslett A: Defining synthetic surfaces for human pluripotent stem cell culture. *Cell Regeneration* 2, 7, 2013.
20. Amit M., Itskovitz-Eldor J: Embryonic Stem Cells: Isolation, Characterization and Culture. *Engineering of Stem Cells. Volume* 114. Edited by Martin U: Springer Berlin Heidelberg; 2009: 173–184: *Advances in Biochemical Engineering/Biotechnology*].
21. Boulting G.L., Kiskinis E., Croft G.F., Amoroso M.W., Oakley D.H., Wainger B.J., Williams D.J., Kahler D.J., Yamaki M., Davidow L., *et al*: A functionally characterized test set of human induced pluripotent stem cells. *Nature Biotechnology* 29, 279–286, 2011.
22. Li J., Campbell D.D., Bal G.K., Pei M: Can Arthroscopically harvested synovial stem cells be preferentially sorted using stage-specific embryonic antigen 4 antibody for cartilage, bone, and adipose regeneration? *Arthroscopy: The Journal of Arthroscopic & Related Surgery* 30, 352–361, 2014.
23. Dekel-Naftali M., Aviram-Goldring A., Litmanovitch T., Shamash J., Reznik-Wolf H., Laevsky I., Amit M., Itskovitz-Eldor J., Yung Y., Hourvitz A., *et al*: Screening of human pluripotent stem cells using CGH and FISH reveals

low-grade mosaic aneuploidy and a recurrent amplification of chromosome 1q. *Eur J Hum Genet* 20, 1248–1255, 2012.
24. Greber B., Lehrach H., Adjaye J: Fibroblast growth factor 2 modulates transforming growth factor β signaling in mouse embryonic fibroblasts and human ESCs (hESCs) to support hESC self-renewal. *Stem Cells* 25, 455–464, 2007.
25. Ghasemi-Dehkordi P., Allahbakhshian-Farsani M., Abdian N., Mirzaeian A., Saffari-Chaleshtori J., Heybati F., Mardani G., Karimi-Taghanaki A., Doosti A., Jami M.S., *et al*: Comparison between the cultures of human induced pluripotent stem cells (hiPSCs) on feeder-and serum-free system (Matrigel matrix), MEF and HDF feeder cell lines. *Journal of Cell Communication and Signaling* 9, 233–246, 2015.
26. Villa-Diaz L., Ross A., Lahann J., Krebsbach P: Concise review: the evolution of human pluripotent stem cell culture: from feeder cells to synthetic coatings. *Stem Cells* 31, 1–7, 2013.
27. Richards M., Fong C.Y., Chan W.K., Wong P.C., Bongso A: Human feeders support prolonged undifferentiated growth of human inner cell masses and embryonic stem cells. *Nature Biotechnology* 20, 933–936, 2002.
28. Richards M., Tan S., Fong C.Y., Biswas A., Chan W.K., Bongso A: Comparative evaluation of various human feeders for prolonged undifferentiated growth of human embryonic stem cells. *Stem Cells* 21, 546–556, 2003.
29. Cheng L., Hammond H., Ye Z., Zhan X., Dravid G: Human adult marrow cells support prolonged expansion of human embryonic stem cells in culture. *Stem Cells* 21, 131–142, 2003.
30. Choo A., Ngo A.S., Ding V., Oh S., Kiang L.S: Autogeneic feeders for the culture of undifferentiated human embryonic stem cells in feeder and feeder-free conditions. In *Methods in Cell Biology*. Volume 86: Academic Press; 15–28, 2008.
31. Fu X., Toh W.S., Liu H., Lu K., Li M., Hande M.P., Cao T: Autologous feeder cells from embryoid body outgrowth support the long-term growth of human embryonic stem cells more effectively than those from direct differentiation. *Tissue Engineering Part C: Methods* 16, 719–733, 2009.
32. Du S.H., Tay J.C.K, Chen C., Tay F.C., Tan W.K., Li Z.D., Wang S: Human iPS cell-derived fibroblast-like cells as feeder layers for iPS cell derivation and expansion. *Journal of Bioscience and Bioengineering* 120, 210–217, 2015.
33. Sánchez L., Gutierrez-Aranda I., Ligero G., Martín M., Ayllón V., Real P.J., Ramos-Mejía V., Bueno C., Menendez P: Maintenance of human embryonic stem cells in media conditioned by human mesenchymal stem cells obviates the requirement of exogenous basic fibroblast growth factor supplementation. *Tissue Engineering Part C: Methods* 18, 387–396, 2011.
34. Choo A., Padmanabhan J., Chin A., Fong W.J., Oh S.K.W: Immortalized feeders for the scale-up of human embryonic stem cells in feeder and feeder-free conditions. *Journal of Biotechnology* 122, 130–141, 2006.
35. Unger C., Gao S., Cohen M., Jaconi M., Bergstrom R., Holm F., Galan A., Sanchez E., Irion O., Dubuisson J.B., *et al*: Immortalized human skin fibroblast

feeder cells support growth and maintenance of both human embryonic and induced pluripotent stem cells. *Human Reproduction* 24, 2567–2581, 2009.
36. Zou C., Chou B.K., Dowey S.N., Tsang K., Huang X., Liu C.F., Smith C., Yen J., Mali P., Zhang Y.A., et al: Efficient derivation and genetic modifications of human pluripotent stem cells on engineered human feeder cell lines. *Stem Cells and Development* 21, 2298–2311, 2012.
37. Kleinman H.K., Martin G.R: Matrigel: basement membrane matrix with biological activity. *Seminars in Cancer Biology* 15, 378–386, 2005.
38. Kleinman H.K., McGarvey M.L., Liotta L.A., Robey P.G., Tryggvason K., Martin G.R: Isolation and characterization of type IV procollagen, laminin, and heparan sulfate proteoglycan from the EHS sarcoma. *Biochemistry* 21, 6188–6193, 1982.
39. Xu C., Inokuma M.S., Denham J., Golds K., Kundu P., Gold J.D., Carpenter M.K: Feeder-free growth of undifferentiated human embryonic stem cells. *Nature Biotechnology* 19, 971–974, 2001.
40. Rosler E.S., Fisk G.J., Ares X., Irving J., Miura T., Rao M.S., Carpenter M.K: Long-term culture of human embryonic stem cells in feeder-free conditions. *Developmental Dynamics* 229, 259–274, 2004.
41. Ludwig T.E., Levenstein M.E., Jones J.M., Berggren W.T., Mitchen E.R., Frane J.L., Crandall L.J., Daigh C.A., Conard K.R., Piekarczyk M.S., et al: Derivation of human embryonic stem cells in defined conditions. *Nature Biotechnology* 24, 185–187, 2006.
42. Ludwig T.E., Bergendahl V., Levenstein M.E., Yu J., Probasco M.D., Thomson J.A: Feeder-independent culture of human embryonic stem cells. *Nature Methods* 3, 637–646, 2006.
43. Hakala H., Rajala K., Ojala M., Panula S., Areva S., Kellomaki M., Suuronen R., Skottman H: Comparison of biomaterials and extracellular matrices as a culture platform for multiple, independently derived human embryonic stem cell lines. *Tissue Engineering Part A* 15, 1775–1785, 2009.
44. Bosnakovski D.I., S.; Nandez R.; Zaidman N.; Struck M.; Dandapat A.; Kyba M: Maintenance of Human iPSC Cells in a Feeder-free Culture System. *Corning Application Notes*, Note 490, 2012.
45. Chatterjee P., Cheung Y., Liew C: Transfecting and nucleofecting human induced pluripotent stem cells. *Journal of Visualized Experiments: JoVE* 3110, 2011.
46. Hughes C.S., Postovit L.M., Lajoie G.A: Matrigel: A complex protein mixture required for optimal growth of cell culture. *Proteomics* 10, 1886–1890, 2010.
47. Stojkovic P., Lako M., Przyborski S., Stewart R., Armstrong L., Evans J., Zhang X., Stojkovic M: Human-serum matrix supports undifferentiated growth of human embryonic stem cells. *Stem Cells* 23, 895–902, 2005.
48. Martin M.J., Muotri A., Gage F., Varki A: Human embryonic stem cells express an immunogenic nonhuman sialic acid. *Nature Medicine* 11, 228–232, 2005.
49. Miyazaki T., Futaki S., Hasegawa K., Kawasaki M., Sanzen N., Hayashi M., Kawase E., Sekiguchi K., Nakatsuji N., Suemori H: Recombinant human

laminin isoforms can support the undifferentiated growth of human embryonic stem cells. *Biochemical and Biophysical Research Communications* 375, 27–32, 2008.
50. Rowland T.J., Miller L.M., Blaschke A.J., Doss E.L., Bonham A.J., Hikita S.T., Johnson L.V., Clegg D.O: Roles of integrins in human induced pluripotent stem cell growth on Matrigel and vitronectin. *Stem Cells and Development* 19, 1231–1240, 2010.
51. Assou S., Le Carrour T., Tondeur S., Ström S., Gabelle A., Marty S., Nadal L., Pantesco V., Réme T., Hugnot J.-P, *et al*: A meta-analysis of human embryonic stem cells transcriptome integrated into a web-based expression atlas. *Stem Cells* 25, 961–973, 2007.
52. Rodin S., Domogatskaya A., Strom S., Hansson E.M., Chien K.R., Inzunza J., Hovatta O., Tryggvason K: Long-term self-renewal of human pluripotent stem cells on human recombinant laminin-511. *Nature Biotechnology* 28, 611–615, 2010.
53. Nakagawa M., Taniguchi Y., Senda S., Takizawa N., Ichisaka T., Asano K., Morizane A., Doi D., Takahashi J., Nishizawa M., *et al*: A novel efficient feeder-free culture system for the derivation of human induced pluripotent stem cells. *Scientific Reports* 4, 3594, 2014.
54. Miyazaki T., Futaki S., Suemori H., Taniguchi Y., Yamada M., Kawasaki M., Hayashi M., Kumagai H., Nakatsuji N., Sekiguchi K., Kawase E: Laminin E8 fragments support efficient adhesion and expansion of dissociated human pluripotent stem cells. *Nature Communications* 3, 1236, 2012.
55. Felding-Habermann B., Cheresh D.A: Vitronectin and its receptors. *Current Opinion in Cell Biology* 5, 864–868, 1993.
56. Braam S.R., Zeinstra L., Litjens S., Ward-van Oostwaard D., van den Brink S., van Laake L., Lebrin F., Kats P., Hochstenbach R., Passier R., *et al*: Recombinant vitronectin is a functionally defined substrate that supports human embryonic stem cell self-renewal via $\alpha V\beta 5$ integrin. *Stem Cells* 26, 2257–2265, 2008.
57. Prowse A.B.J., Doran M.R., Cooper-White J.J., Chong F., Munro T.P., Fitzpatrick J., Chung T.L., Haylock D.N., Gray P.P., Wolvetang E.J: Long term culture of human embryonic stem cells on recombinant vitronectin in ascorbate free media. *Biomaterials* 31, 8281–8288, 2010.
58. Chen G., Gulbranson D.R., Hou Z., Bolin J.M., Ruotti V., Probasco M.D., Smuga-Otto K., Howden S.E., Diol N.R., Propson N.E., *et al*: Chemically defined conditions for human iPSC derivation and culture. *Nature Methods* 8, 424–429, 2011.
59. Pankov R., Yamada K.M: Fibronectin at a glance. *Journal of Cell Science* 115, 3861–3863, 2002.
60. Amit M., Itskovitz-Eldor J: Feeder-free culture of human embryonic stem cells. In *Methods in Enzymology. Volume* 420: Academic Press, 37–49, 2006.
61. Toole B.P: Hyaluronan: from extracellular glue to pericellular cue. *Nature Reviews Cancer* 4, 528–539, 2004.

62. Gerecht S., Burdick J.A., Ferreira L.S., Townsend S.A., Langer R., Vunjak-Novakovic G: Hyaluronic acid hydrogel for controlled self-renewal and differentiation of human embryonic stem cells. *Proceedings of the National Academy of Sciences of the United States of America* 104, 11298–11303, 2007.
63. Furue M.K., Na J., Jackson J.P., Okamoto T., Jones M., Baker D., Hata R.-I, Moore H.D., Sato J.D., Andrews P.W: Heparin promotes the growth of human embryonic stem cells in a defined serum-free medium. *Proceedings of the National Academy of Sciences* 105, 13409–13414, 2008.
64. Wang L., Schulz T.C., Sherrer E.S., Dauphin D.S., Shin S., Nelson A.M., Ware C.B., Zhan M., Song C.Z., Chen X., et al: Self-renewal of human embryonic stem cells requires insulin-like growth factor-1 receptor and ERBB2 receptor signaling. *Blood* 110, 4111–4119, 2007.
65. Chen C., Loe F., Blocki A., Peng Y., Raghunath M: Applying macromolecular crowding to enhance extracellular matrix deposition and its remodeling *in vitro* for tissue engineering and cell-based therapies. *Advanced Drug Delivery Reviews* 63, 277–290, 2011.
66. Toh W.S., Foldager C.B., Pei M., Hui J.H: Advances in Mesenchymal Stem Cell-based Strategies for Cartilage Repair and Regeneration. *Stem Cell Reviews and Reports* 10, 686–696, 2014.
67. Kruegel J., Miosge N: Basement membrane components are key players in specialized extracellular matrices. *Cellular and Molecular Life Sciences* 67, 2879–2895, 2010.
68. Fu X., Toh W.S., Liu H., Lu K., Li M., Cao T: Establishment of clinically compliant human embryonic stem cells in an autologous feeder-free system. *Tissue Engineering Part C: Methods* 17, 927–937, 2011.
69. Peng Y., Bocker M.T., Holm J., Toh W.S., Hughes C.S., Kidwai F., Lajoie G.A., Cao T., Lyko F., Raghunath M: Human fibroblast matrices bioassembled under macromolecular crowding support stable propagation of human embryonic stem cells. *Journal of Tissue Engineering and Regenerative Medicine* 6, e74–e86, 2012.
70. Lim M.L., Jungebluth P., Sjöqvist S., Nikdin H., Kjartansdóttir K.R., Unger C., Vassliev I., Macchiarini P: Decellularized feeders: an optimized method for culturing pluripotent cells. *Stem Cells Translational Medicine* 2, 975–982, 2013.
71. Vuoristo S., Toivonen S., Weltner J., Mikkola M., Ustinov J., Trokovic R., Palgi J., Lund R., Tuuri T., Otonkoski T: A novel feeder-free culture system for human pluripotent stem cell culture and induced pluripotent stem cell derivation. *PLoS One* 8, e76205, 2013.
72. Li L., Bennett S.A.L., Wang L: Role of E-cadherin and other cell adhesion molecules in survival and differentiation of human pluripotent stem cells. *Cell Adhesion & Migration* 6, 59–73, 2012.
73. Lee S.T., Yun J.I., van der Vlies A.J., Kontos S., Jang M., Gong S.P., Kim D.Y., Lim J.M., Hubbell J.A: Long-term maintenance of mouse embryonic stem cell pluripotency by manipulating integrin signaling within 3D scaffolds without active Stat3. *Biomaterials* 33, 8934–8942, 2012.

74. D'Amour K.A., Agulnick A.D., Eliazer S., Kelly O.G., Kroon E., Baetge E.E: Efficient differentiation of human embryonic stem cells to definitive endoderm. *Nature Biotechnology* 23, 1534–1541, 2005.
75. Park I.H., Zhao R., West J.A., Yabuuchi A., Huo H., Ince T.A., Lerou P.H., Lensch M.W., Daley G.Q: Reprogramming of human somatic cells to pluripotency with defined factors. *Nature* 451, 141–146, 2008.
76. Li D., Zhou J., Wang L., Shin M.E., Su P., Lei X., Kuang H., Guo W., Yang H., Cheng L., et al: Integrated biochemical and mechanical signals regulate multifaceted human embryonic stem cell functions. *The Journal of Cell Biology* 191, 631–644, 2010.
77. Chen G., Hou Z., Gulbranson D.R., Thomson J.A: Actin-myosin contractility is responsible for the reduced viability of dissociated human embryonic stem cells. *Cell Stem Cell* 7, 240–248, 2010.
78. Ohgushi M., Matsumura M., Eiraku M., Murakami K., Aramaki T., Nishiyama A., Muguruma K., Nakano T., Suga H., Ueno M., et al: Molecular pathway and cell state responsible for dissociation-induced apoptosis in human pluripotent stem cells. *Cell Stem Cell*, 7, 225–239.
79. Nagaoka M., Si-Tayeb K., Akaike T., Duncan S: Culture of human pluripotent stem cells using completely defined conditions on a recombinant E-cadherin substratum. *BMC Developmental Biology* 10, 60, 2010.
80. Moroni L., de Wijn J.R., van Blitterswijk C.A: Integrating novel technologies to fabricate smart scaffolds. *Journal of Biomaterials Science, Polymer Edition* 19, 543–572, 2008.
81. Badylak S.F., Freytes D.O., Gilbert T.W: Extracellular matrix as a biological scaffold material: structure and function. *Acta Biomaterialia* 5, 1–13, 2009.
82. Mano J.F., Silva G.A., Azevedo H.S., Malafaya P.B., Sousa R.A., Silva S.S., Boesel L.F., Oliveira J.M., Santos T.C., Marques A.P., et al: Natural origin biodegradable systems in tissue engineering and regenerative medicine: present status and some moving trends. *Journal of the Royal Society Interface* 4, 999–1030, 2007.
83. Irwin E.F., Gupta R., Dashti D.C., Healy K.E: Engineered polymer-media interfaces for the long-term self-renewal of human embryonic stem cells. *Biomaterials* 32, 6912–6919, 2011.
84. Mei Y., Saha K., Bogatyrev S.R., Yang J., Hook A.L., Kalcioglu Z.I., Cho S.-W., Mitalipova M., Pyzocha N., Rojas F., et al: Combinatorial development of biomaterials for clonal growth of human pluripotent stem cells. *Nature Materials* 9, 768–778, 2010.
85. Nandivada H., Villa-Diaz L.G., O'Shea K.S., Smith G.D., Krebsbach P.H., Lahann J: Fabrication of synthetic polymer coatings and their use in feeder-free culture of human embryonic stem cells. *Nature Protocols* 6, 1037–1043, 2011.
86. Villa-Diaz L.G., Brown S.E., Liu Y., Ross A.M., Lahann J., Parent J.M., Krebsbach P.H: Derivation of mesenchymal stem cells from human induced pluripotent stem cells cultured on synthetic substrates. *Stem Cells* 30, 1174–1181, 2012.

87. Villa-Diaz L.G., Nandivada H., Ding J., Nogueira-de-Souza N.C., Krebsbach P.H., O'Shea K.S., Lahann J., Smith G.D: Synthetic polymer coatings for long-term growth of human embryonic stem cells. *Nature Biotechnology* 28, 581–583, 2010.
88. Celiz A.D., Smith J.G.W., Patel A.K., Hook A.L., Rajamohan D., George V.T., Flatt L., Patel M.J., Epa V.C., Singh T., et al: Discovery of a novel polymer for human pluripotent stem cell expansion and multilineage differentiation. *Advanced Materials* 27, 4006–4012, 2015.
89. Maldonado M., Wong L.Y., Echeverria C., Ico G., Low K., Fujimoto T., Johnson J.K., Nam J: The effects of electrospun substrate-mediated cell colony morphology on the self-renewal of human induced pluripotent stem cells. *Biomaterials* 50, 10–19, 2015.
90. Choi K.-D., Yu J., Smuga-Otto K., Salvagiotto G., Rehrauer W., Vodyanik M., Thomson J., Slukvin I: Hematopoietic and endothelial differentiation of human induced pluripotent stem cells. *Stem Cells* 27, 559–567, 2009.
91. Barberi T., Willis L.M., Socci N.D., Studer L: Derivation of multipotent mesenchymal precursors from human embryonic stem cells. *PLoS Medicine* 2, e161, 2005.
92. Domev H., Amit M., Laevsky I., Dar A., Itskovitz-Eldor J: Efficient engineering of vascularized ectopic bone from human embryonic stem cell–derived mesenchymal stem cells. *Tissue Engineering Part A* 18, 2290–2302, 2012.
93. Hwang N.S., Varghese S., Lee H.J., Zhang Z., Ye Z., Bae J., Cheng L., Elisseeff J: In vivo commitment and functional tissue regeneration using human embryonic stem cell-derived mesenchymal cells. *Proceedings of the National Academy of Sciences* 105, 20641–20646, 2008.
94. Lee G., Chambers S.M., Tomishima M.J., Studer L: Derivation of neural crest cells from human pluripotent stem cells. *Nature Protocols* 5, 688–701, 2010.
95. Lim M.S., Shin M.S., Lee S.Y., Minn Y.K., Hoh J.K., Cho Y.H., Kim D.W., Lee S.H., Kim C.H., Park C.H: Noggin over-expressing mouse embryonic fibroblasts and ms5 stromal cells enhance directed differentiation of dopaminergic neurons from human embryonic stem cells. *PLoS One* 10, e0138460, 2015.
96. Vazin T., Becker K.G., Chen J., Spivak C.E., Lupica C.R., Zhang Y., Worden L., Freed W.J: A novel combination of factors, termed spie, which promotes dopaminergic neuron differentiation from human embryonic stem cells. *PLoS One* 4, e6606, 2009.
97. Joddar B., Ito Y: Artificial niche substrates for embryonic and induced pluripotent stem cell cultures. *Journal of Biotechnology* 168, 218–228, 2013.
98. Holly S.P., Larson M.K., Parise L.V: Multiple roles of integrins in cell motility. *Experimental Cell Research* 261, 69–74, 2000.
99. Reilly G.C., Engler A.J: Intrinsic extracellular matrix properties regulate stem cell differentiation. *Journal of Biomechanics* 43, 55–62, 2010.
100. Dickinson L.E., Kusuma S., Gerecht S: Reconstructing the differentiation niche of embryonic stem cells using biomaterials. *Macromolecular Bioscience* 11, 36–49, 2011.

101. Feaster T.K., Cadar A.G., Wang L., Williams C.H., Chun Y.W., Hempel J., Bloodworth N., Merryman W.D., Lim C.C., Wu J.C., et al: Matrigel mattress: a method for the generation of single contracting human-induced pluripotent stem cell-derived cardiomyocytes. Circulation Research, 117(12):995–1000, 2015.
102. Ruhnke M., Ungefroren H., Zehle G., Bader M., Kremer B., Fändrich F: Long-term culture and differentiation of rat embryonic stem cell-like cells into neuronal, glial, endothelial, and hepatic lineages. Stem Cells 21, 428–436, 2003.
103. Levenberg S., Golub J.S., Amit M., Itskovitz-Eldor J., Langer R: Endothelial cells derived from human embryonic stem cells. Proceedings of the National Academy of Sciences 99, 4391–4396, 2002.
104. Ishii T., Fukumitsu K., Yasuchika K., Adachi K., Kawase E., Suemori H., Nakatsuji N., Ikai I., Uemoto S: Effects of extracellular matrixes and growth factors on the hepatic differentiation of human embryonic stem cells. American Journal of Physiology - Gastrointestinal and Liver Physiology 295, G313–G321, 2008.
105. Uemura M., Refaat M.M., Shinoyama M., Hayashi H., Hashimoto N., Takahashi J: Matrigel supports survival and neuronal differentiation of grafted embryonic stem cell-derived neural precursor cells. Journal of Neuroscience Research 88, 542–551, 2010.
106. Gil J.E., Woo D.H., Shim J.H., Kim S.E., You H.J., Park S.H., Paek S.H., Kim S.K., Kim J.H: Vitronectin promotes oligodendrocyte differentiation during neurogenesis of human embryonic stem cells. FEBS Letters 583, 561–567, 2009.
107. Heydarkhan-Hagvall S., Gluck J.M., Delman C., Jung M., Ehsani N., Full S., Shemin R.J: The effect of vitronectin on the differentiation of embryonic stem cells in a 3D culture system. Biomaterials 33, 2032–2040, 2012.
108. Wang Y.J., Liu H.L., Guo H.T., Wen H.W., Liu J: Primary hepatocyte culture in collagen gel mixture and collagen sandwich. World Journal of Gastroenterology 10, 699–702, 2004.
109. Liu Y., Goldberg A.J., Dennis J.E., Gronowicz G.A., Kuhn L.T: One-Step Derivation of Mesenchymal Stem Cell (MSC)-Like Cells from Human Pluripotent Stem Cells on a Fibrillar Collagen Coating. PLoS One 7, e33225, 2012.
110. Medici D., Nawshad A: Type I collagen promotes epithelial–mesenchymal transition through ILK-dependent activation of NF-κB and LEF-1. Matrix Biology 29, 161–165, 2010.
111. Boyd N.L., Robbins K.R., Dhara S.K., West F.D., Stice S.L: Human embryonic stem cell-derived mesoderm-like epithelium transitions to mesenchymal progenitor cells. Tissue Engineering Part A 15, 1897–1907, 2009.
112. Nagamoto Y., Tashiro K., Takayama K., Ohashi K., Kawabata K., Sakurai F., Tachibana M., Hayakawa T., Furue M.K., Mizuguchi H: The promotion of hepatic maturation of human pluripotent stem cells in 3D co-culture using type I collagen and Swiss 3T3 cell sheets. Biomaterials 33, 4526–4534, 2012.
113. Baharvand H., Hashemi S.M., Kazemi Ashtiani S., Farrokhi A: Differentiation of human embryonic stem cells into hepatocytes in 2D and 3D culture systems in vitro. The International Journal of Developmental Biology 50, 645–652, 2006.

114. Kühn K: Basement membrane (type IV) collagen. *Matrix Biology* 14, 439–445, 1995.
115. Schenke-Layland K., Rhodes K.E., Angelis E., Butylkova Y., Heydarkhan-Hagvall S., Gekas C., Zhang R., Goldhaber J.I., Mikkola H.K., Plath K., MacLellan W.R: Reprogrammed mouse fibroblasts differentiate into cells of the cardiovascular and hematopoietic lineages. *Stem Cells* 26, 1537–1546, 2008.
116. Gerecht-Nir S., Ziskind A., Cohen S., Itskovitz-Eldor J: Human embryonic stem cells as an *in vitro* model for human vascular development and the induction of vascular differentiation. *Laboratory Investigation* 83, 1811–1820, 0000.
117. Schenke-Layland K., Angelis E., Rhodes K.E., Heydarkhan-Hagvall S., Mikkola H.K., MacLellan W.R: Collagen IV induces trophoectoderm differentiation of mouse embryonic stem cells. *Stem Cells* 25, 1529–1538, 2007.
118. Ahmad S., Stewart R., Yung S., Kolli S., Armstrong L., Stojkovic M., Figueiredo F., Lako M: Differentiation of human embryonic stem cells into corneal epithelial-like cells by *in vitro* replication of the corneal epithelial stem cell niche. *Stem Cells* 25, 1145–1155, 2007.
119. Timpl R., Rohde H., Robey P.G., Rennard S.I., Foidart J.M., Martin G.R: Laminin—a glycoprotein from basement membranes. *J Biol Chem* 254, 9933–9937, 1979.
120. Kleinman H.K., Cannon F.B., Laurie G.W., Hassell J.R., Aumailley M., Terranova V.P., Martin G.R., DuBois-Dalcq M: Biological activities of laminin. *Journal of Cellular Biochemistry* 27, 317–325, 1985.
121. Battista S., Guarnieri D., Borselli C., Zeppetelli S., Borzacchiello A., Mayol L., Gerbasio D., Keene D.R., Ambrosio L., Netti P.A: The effect of matrix composition of 3D constructs on embryonic stem cell differentiation. *Biomaterials* 26, 6194–6207, 2005.
122. Ma W., Tavakoli T., Derby E., Serebryakova Y., Rao M.S., Mattson M.P: Cell-extracellular matrix interactions regulate neural differentiation of human embryonic stem cells. *BMC Developmental Biology* 8, 90–90, 2008.
123. Hazeltine L.B., Selekman J.A., Palecek S.P: Engineering the human pluripotent stem cell microenvironment to direct cell fate. *Biotechnology Advances* 31, 1002–1019, 2013.
124. Schroeder I.S., Rolletschek A., Blyszczuk P., Kania G., Wobus A.M: Differentiation of mouse embryonic stem cells to insulin-producing cells. *Nature Protocols* 1, 495–507, 2006.
125. Francis S.E: Central roles of alpha5beta1 integrin and fibronectin in vascular development in mouse embryos and embryoid bodies. *Arteriosclerosis, Thrombosis, and Vascular Biology* 22, 927–933, 2002.
126. Duan Y., Liu Z., O'Neill J., Wan L., Freytes D., Vunjak-Novakovic G: Hybrid gel composed of native heart matrix and collagen induces cardiac differentiation of human embryonic stem cells without supplemental growth factors. *Journal of Cardiovascular Translational Research* 4, 605–615, 2011.
127. Ng S.L.J., Narayanan K., Gao S., Wan A.C.A: Lineage restricted progenitors for the repopulation of decellularized heart. *Biomaterials* 32, 7571–7580, 2011.

128. Zhou Q., Ye X., Sun R., Matsumoto Y., Moriyama M., Asano Y., Ajioka Y., Saijo Y: Differentiation of mouse induced pluripotent stem cells into alveolar epithelial cells *in vitro* for use *in vivo*. Stem Cells Translational Medicine 3, 675–685, 2014.
129. Ross E.A., Williams M.J., Hamazaki T., Terada N., Clapp W.L., Adin C., Ellison G.W., Jorgensen M., Batich C.D: Embryonic stem cells proliferate and differentiate when seeded into kidney scaffolds. Journal of the American Society of Nephrology 20, 2338–2347, 2009.
130. Li J., Hansen K.C., Zhang Y., Dong C., Dinu C.Z., Dzieciatkowska M., Pei M: Rejuvenation of chondrogenic potential in a young stem cell microenvironment. Biomaterials 35, 642–653, 2014.
131. Rutledge K., Cheng Q., Pryzhkova M., Harris G.M., Jabbarzadeh E: Enhanced differentiation of human embryonic stem cells on extracellular matrix-containing osteomimetic scaffolds for bone tissue engineering. Tissue Engineering Part C: Methods 20, 865–874, 2014.
132. Heng B.C., Toh W.S., Pereira B.P., Tan B.L., Fu X., Liu H., Lu K., Yeo J.F., Cao T: An autologous cell lysate extract from human embryonic stem cell (hESC) derived osteoblasts can enhance osteogenesis of hESC. Tissue and Cell 40, 219–228, 2008.
133. Movahednia M.M., Kidwai F.K., Zou Y., Tong H.J., Liu X., Islam I., Toh W.S., Raghunath M., Cao T: Differential effects of the extracellular microenvironment on human embryonic stem cell differentiation into keratinocytes and their subsequent replicative life span. Tissue Engineering Part A 21, 1432–1443, 2015.
134. Lutolf M.P., Hubbell J.A: Synthetic biomaterials as instructive extracellular microenvironments for morphogenesis in tissue engineering. Nature Biotechnology 23, 47–55, 2005.
135. Wang H., Luo X., Leighton J: Extracellular matrix and integrins in embryonic stem cell differentiation. Biochemistry Insights, 8(Suppl 2): 15–21, 2015.
136. Toh Y.C., Xing J., Yu H: Modulation of integrin and E-cadherin-mediated adhesions to spatially control heterogeneity in human pluripotent stem cell differentiation. Biomaterials 50, 87–97, 2015.
137. Haque A., Yue X.-S., Motazedian A., Tagawa Y.-i., Akaike T: Characterization and neural differentiation of mouse embryonic and induced pluripotent stem cells on cadherin-based substrata. Biomaterials 33, 5094–5106, 2012.
138. Moore R.N., Cherry J.F., Mathur V., Cohen R., Grumet M., Moghe P.V: E-Cadherin-expressing feeder cells promote neural lineage restriction of human embryonic stem cells. Stem Cells and Development 21, 30–41, 2011.
139. Haque A., Hexig B., Meng Q., Hossain S., Nagaoka M., Akaike T: The effect of recombinant E-cadherin substratum on the differentiation of endoderm-derived hepatocyte-like cells from embryonic stem cells. Biomaterials 32, 2032–2042, 2011.
140. Dasgupta A., Hughey R., Lancin P., Larue L., Moghe P.V: E-cadherin synergistically induces hepatospecific phenotype and maturation of embryonic

stem cells in conjunction with hepatotrophic factors. *Biotechnology and Bioengineering* 92, 257–266, 2005.
141. Gupta M.K., Walthall J.M., Venkataraman R., Crowder S.W., Jung D.K., Yu S.S., Feaster T.K., Wang X., Giorgio T.D., Hong C.C., *et al*: Combinatorial polymer electrospun matrices promote physiologically-relevant cardiomyogenic stem cell differentiation. *PLoS One* 6, e28935, 2011.
142. Chun Y.W., Balikov D.A., Feaster T.K., Williams C.H., Sheng C.C., Lee J.-B., Boire T.C., Neely M.D., Bellan L.M., Ess K.C., *et al*: Combinatorial polymer matrices enhance *in vitro* maturation of human induced pluripotent stem cell-derived cardiomyocytes. *Biomaterials* 67, 52–64, 2015.
143. Ross C.A., Akimov S.S: Human-induced pluripotent stem cells: potential for neurodegenerative diseases. *Human Molecular Genetics* 23, R17–R26, 2014.
144. Lei Y., Schaffer D.V: A fully defined and scalable 3D culture system for human pluripotent stem cell expansion and differentiation. *Proceedings of the National Academy of Sciences* 110, E5039–E5048, 2013.
145. Kuo Y.C., Huang M.J: Material-driven differentiation of induced pluripotent stem cells in neuron growth factor-grafted poly(ε-caprolactone)-poly(β-hydroxybutyrate) scaffolds. *Biomaterials* 33, 5672–5682, 2012.
146. Kuo Y.C., Wang C.T: Neuronal differentiation of induced pluripotent stem cells in hybrid polyester scaffolds with heparinized surface. *Colloids and Surfaces B: Biointerfaces* 100, 9–15, 2012.
147. Tsai Y., Cutts J., Kimura A., Varun D., Brafman D.A: A chemically defined substrate for the expansion and neuronal differentiation of human pluripotent stem cell-derived neural progenitor cells. *Stem Cell Research* 15, 75–87, 2015.
148. Cordie T., Harkness T., Jing X., Carlson-Stevermer J., Mi H.Y., Turng L.S., Saha K: Nanofibrous electrospun polymers for reprogramming human cells. *Cellular and Molecular Bioengineering* 7, 379–393, 2014.

13
Silk as a Natural Biopolymer for Tissue Engineering

Ayşe Ak Can[1,2]* and Gamze Bölükbaşi Ateş[2]

[1]*Department of Biomedical Engineering, Engineering Faculty, Erzincan University, Erzincan, Turkey*
[2]*Institute of Biomedical Engineering, Boğaziçi University, Istanbul, Turkey*

Abstract

The general principles of tissue engineering involve the use of various cells, biomaterials, and bioactive factors to facilitate the regeneration of damaged tissue by providing controlled environment. Many natural and synthetic biomaterials have been used in tissue engineering including ceramics, metals, synthetic polymers, and natural polymers. The ideal biomaterial should mimic native tissue as much as possible. Therefore, it should enhance cell attachment, proliferation, and differentiation. It should also have biocompatible and biodegradable properties in a controlled environment. The success of tissue development highly depends on these properties. Natural materials have advantages due to their biological recognition through receptor–ligand interactions, proteolysis, and low toxicity. Therefore, the host can successfully metabolize the implanted substance through either enzymatic or hydrolytic degradation. Natural materials can also be modified and cross-linked to have better properties for cell environment. In recent years, several natural materials, e.g. chitosan, hyaluronic acid, and silk, have become more attractive for tissue engineering. Silk is produced by spider and silkworm and has received significant attention as a versatile natural polymer due to its strength and slow biodegradation. Silk, a protein polymer, is naturally synthesized, composed of a filament core protein, silk fibroin (SF), and wax-like sericin protein. It has been used as sutures and artificial ligaments in clinic for decades. The potential of native and regenerated SF-based biomaterials has been studied in various forms such as films, fibers, hydrogels, 3D scaffolds, and silk combined with other biomaterials in tissue engineering. Recent studies have shown that SF has unique structural, mechanical, and biochemical properties *in vitro* and *in vivo* tissue engineering.

Corresponding author: ayseak@erzincan.edu.tr; ayseak980@gmail.com

Studies with silk have demonstrated that silk is biocompatible and suitable matrix for the cultivation, osteogenesis, and chondrogenesis of mesenchymal stem cells. Finally, in research for potential alternatives to synthetic-based polymers, SF with unique structure seems to be a promising candidate for the wide range of tissue-engineering needs. A better understanding of silk fibers effects on stem cells will provide greater insight into the appropriate silk biomaterials design for future medical applications.

Keywords: Silk fibroin, biopolymer, stem cell, tissue engineering

13.1 Introduction

Silk proteins are produced by an enormous variety of insect and spider including ants, fleas, and crickets. Cultivation of domestic *Bombyx mori* is easier; therefore, *B. mori* is the most popular source for silk [1, 2]. Silk from silkworms is classified into mulberry and nonmulberry silks. Silk includes two primary proteins: fibroin and sericin. These proteins contain glycine, alanine, and serine amino acids [3].

Silk fibroin (SF) produced by silk gland cells is basically composed of two chains of amino acids: a highly repetitive heavy (H) chain (325 kDa) and a nonrepetitive light (L) chain (25 kDa). These two chains are linked by a disulfide bond which is thought to take part in efficient secretion of SF. A thin cocoon and 0.3% of less secretion are observed in the silkworm carrying the Nd-s or Nd-s^D mutations [4].

Another glycoprotein, P25, is noncovalently linked to these chains via hydrophobic interactions and plays an important role in maintaining the integrity [5].

The amino acids composing the highly repetitive heavy chain from *B. mori* are glycine (Gly) (44%), alanine (Ala) (30%), and serine (Ser) (12%). The presence of high glycine content provides its rigid structure. The Gly-X repeats are the building blocks of crystalline regions of the whole fiber. The repeated sequences are organized into 12 domains, and there are 11 amorphous regions between them. The amorphous regions contain negatively charged, polar, and hydrophobic residues [6].

The central region of the protein is mostly hydrophobic, while the nonrepetitive amino- and carboxy-terminal domains are more hydrophilic. The heavy chain contains numerous small ß-sheet crystals. It is proposed that the details of the structure of the protein are the base for its mechanical and physical properties. SF can arrange itself in three different structures: silk I is the crystalline form before spinning, and it is less stable and easily transformed to silk II (ß-sheet form) structural form. The silk II form is the structure after spinning which is more stable. The Raman spectrum has shown that 50% of the *B. mori* silk is composed of anti-parallel β-sheet conformation. The remaining 50% are noncrystalline regions or amorphous regions. This hydrophobic structure dominated by ß-sheets makes SF stronger compared to other commonly used biomaterials. The ultimate tensile strength of *B. mori* silk is 740 MPa, which is very high compared to collagen tensile strength (0.9–7.4 MPa) [7].

The structural protein fibroin is covered by another protein sericin which is more hydrophilic and with glue-like properties. It is characterized by the presence of 32% of serine, and it occupies 25–30% of the weight of *B. mori* silk fiber. By the degumming process, this sericin coating is removed.

The SF-based materials can be prepared in different forms such as films, hydrogels, membranes, scaffolds, or porous sponges. For each format, first sericin must be removed from raw silk, and SF must be extracted from cocoons. At the end of extraction procedure, an aqueous solution of pure SF is obtained which can be lyophilized or used for further processes [8].

13.1.1 Mechanical Properties

The *B. mori* cocoons have a nonwoven surface structure with multiple layers composed of random fiber arrangement which are maintained together with sericin. The layers are connected to each other by fewer fibers. This multiple layer structure is thought to be related to its mechanical properties [9]. The individual cocoon layers can be easily separated because the inter-layer bounding is weaker than the intra-layer bonding. The sericin covers the surface of the fibroins. Fourier transform infrared spectroscopy (FTIR) results reveals that the inner layers of the cocoon have significantly less sericin than the outer layers. The porosity also decreases from the outside to the inside layers. As a result, the fiber network has less bonding. The fibers have less modulus, lower strength, and higher breaking strain so that they unravel rather than breaking [10]. Chen *et al.* have observed 25 types of cocoon and concluded that the inter-fiber variability of properties is similar to that of an individual fiber type, and the connectivity of inter-fiber bonding plays a dominant role in stress–strain properties [10].

A typical silkworm silk of *B. mori* has a tensile strength of about 0.5 GPa, a breaking elongation of 15%, and a toughness of 6×10^4 J/kg [11]. According to the literature, the exceptional mechanical properties of silk mainly depend on its hierarchical structures with highly conserved poly-(Gly–Ala) and poly-Ala domains. The crystal structure of SF may contain polar-antiparallel, polar-parallel, and antipolar-antiparallel β-sheets [1]. In an antiparallel arrangement, the N-terminus of one β-strand is adjacent to the C-terminus of the next. In a parallel arrangement, all of the N-termini of successive strands are oriented in the same direction. Polar β-sheets have a Gly face and Ala face; on the other hand, antipolar β-sheets have two identical faces. These β-sheet crystallites which contain hydrogen bonds along with van der Waals' and hydrophobic interactions make SF strong and stiff. The presence of more intramolecular β-sheet crystallites increases mechanical strength. Simulation and experimental data agree that SF with antiparallel β-sheet crystallites shows more strength and stiffness due to stronger hydrogen bonds which are shorter and linear in this arrangement [12].

Modeling and simulation studies demonstrate that β-sheet nanocrystals confined to a few nanometers achieve higher stiffness, strength, and mechanical toughness than larger nanocrystals [13]. Moreover, breaking stress of silkworm is related to the orientation of β-sheet crystallites which are oriented along the fiber axis [14].

13.1.2 Biodegradation

Slow degradation is another desired characteristic for good biomaterials. The rate of formation of new tissue should match the rate of degradation for an ideal biomaterial. A biomaterial is defined as absorbable if it loses most of its tensile strength within 60 days. It takes over 60 days for silk to lose most of its tensile strength *in vivo*, so silk is classified as nondegradable. SF degrades over longer time periods (typically within 1 year) both proteolytically and by gradual absorption. Proteolytic enzymes such as α-chymotrypsin, collagenase IA, and protease XIV can degrade SF, but each attacks at different regions and at a different rate; α-chymotrypsin is the least effective [15].

Silk samples incubated in protease (Protease XIV) *in vitro*, significantly decrease in mass and UTS after a week incubation, and decreased mechanical integrity is correlated by increased fragmentation of fibroin filaments. Horan *et al.* have demonstrated that the substantial loss in mechanical integrity can be correlated to mass loss, the changes in fibroin diameter, failure strength, and cycles *in vitro* studies [16].

This predictable degradation allows the engineers to design scaffolds that meet particular requirements [17].

Other *in vivo* studies have shown that the rate of absorption depends on the implantation site, mechanical environment, and the diameter of the silk fiber. Processing methods and surface features are other factors that influence *in vivo* degradation of SF. Kim *et al.* have showed that three-dimensional (3D), aqueous-derived scaffolds fully are degraded if scaffolds are being exposed to protease during 21 days, unlike the scaffolds are prepared from organic solvent processing [18].

Gamma irradiation is another investigated technique for SF biodegradation. Irradiation of SF degrades glycine and alanine in amorphous region selectively, and mechanical properties and morphology are influenced by irradiation [19].

In another study, when silk from the native Thai silkworm, *B. mori* var. *nangnoi Sisaket-1*, is irradiated with various doses before *in vitro* biodegradation testing, their results show that the biodegradation of SF increases with increasing irradiation intensity [20].

13.1.3 Biocompatibility

Although silk has been used in sutures for many years, it can cause foreign body response, hypersensitivity, and variable degradation rates at different implantation sites and immunological problems, since it is not from mammalian origin. Their clinical applications have been diminished during the past years. However, some recent studies suggest that the undesirable immunogenic reactions are associated with the sericin but not by the SF which can be easily separated from it. Purified SF studies have shown that silk can be a promising biocompatible material [21, 22].

13.2 SF as a Biomaterial

Three-dimensional scaffolds are one of the key steps in tissue engineering. Prior to implantation *in vivo*, the tissue must be constructed in a microenvironment that will support cell attachment and proliferation and will guide them into functional tissues. Besides being biocompatible, the scaffold biomaterial must maintain mechanical and structural properties in all stages, and the degradation time must match tissue regeneration period without activating host immune response. Both synthetic and natural polymers are used for different types of tissue; hence, each type has its own requirements.

For decades, silk, which has good physical and chemical properties, has been extensively used as suture material in eye and lip surgery, intraoral surgery, and skin wound healing. Recently, it gains more importance for biomedical applications. Silk-based composites, as the most popular candidates, have been researched in various application areas from tissue engineering to drug delivery systems to implanted devices. Various researches have showed that SF-based scaffolds have better mechanical properties

Table 13.1 Biocompatible forms and properties of SF.

Biomaterial	Properties
Hydrogels, sponges	Injectable for *in vivo* High water concentration Controllable gelation Controllable porosity
Films, membranes	Permeability of water and oxygen Available for surface modification Structural stability Controlled morphological a structural features
Nanofibers, microfibers	Controlled fiber size Porous structure Available for surface modification High strength and structural stability Mimic the *in vivo* physiological micro environment

than other biodegradable polymeric scaffolds [e.g. collagen, chitosan, and hyaluronan (HA)] [2, 21, 23, 24].

Fibroin is the main component of nonbioabsorbable biomedical silk material without inducing adverse immunological effects and shows good biocompatibility in cell types by inducing adhesion, proliferation, migration, and differentiation [1, 21].

Different forms of SF-based biomaterial including hydrogel, sponge, film, membrane, nonwoven, and woven net/mat/fiber have been developed and searched *in vitro* and *in vivo*. These forms and their properties are summarized in Table 13.1 [21, 22].

13.2.1 Fibroin Hydrogels and Sponges

Fibroin hydrogels are important forms and polymeric networks for biomedical applications. Hydrogels are capable of absorbing large amounts of water and provide delivery of cells and cytokines in tissue engineering [1, 22, 25]. Hydrogels can be prepared by a sol–gel transition with acid, ions, and other additives [26].

Fibroin that contains glycine, serine, and alanine amino acids gains hydrophobic character, and this situation makes fibroin to be easier gel without adding any agent. Gelation rate of fibroin depends on various factors and gelation process such as changes of pH level, temperature, and ionic strength, vortexing, sonication, and freeze gelation or electrogelation [22, 25, 27, 28].

Mechanical properties of silk hydrogels are suitable for preparing scaffolds for tissue engineering such as cartilage regeneration. Different polymers have been combined with SF to enhance the mechanical properties of regenerated fibroin. SF hydrogels combined with synthetic polymers such as polyacrylamide, polyethylene oxide, and polyethylene glycol have advantages of the resulting improved mechanical strength, enhanced gelation rate, and injectable form, respectively. Hybrid hydrogels with natural polymers (e.g. collagen, chitosan, and HA) provide thermal stability, pH, ion-sensitive forms, and controllable degradation [1, 3, 25].

Hydrogels can induce osteogenic differentiation of human mesenchymal stem cells (MSCs) without osteogenic stimulants. Furthermore, injectable hydrogels have been demonstrated to accelerate the remodeling processes in rabbit distal femurs. *In vivo* studies demonstrate that vascular epidermal growth factor (VEGF) and bone morphogenetic protein 2 (BMP-2) with SF hydrogels promote vascularization and bone formation in rabbits [22]. It has also been shown that acidic and sonicated hydrogels have biocompatibility and biodegradation properties when implant to knee of rabbit and rat, respectively [29].

In addition to regenerated SF hydrogels with other polymers, porous SF sponges can be prepared from hydrogels and evaluated for cell culture and tissue engineering. Sponges have highly porous structure and large surface area, providing a scaffold for cell attachment, proliferation, and migration [1, 28].

They can be prepared by using porogens (NaCl), gas forming, and freeze-drying, freeze drying foaming and electrospun fibers [26].

Porous sponge-like scaffolds provide better cell attachment, proliferation, and migration and allow nutrient and waste transport. Processing parameters can be varied by using different solvents for regeneration of SF including organic solvents such as hexafluoroisopropanol (HFIP) or aqueous solution. It has been demonstrated that they have a better bone tissue formation with increased levels of osteopontin, collagen type I, bone sialoprotein, increased calcium deposition, and mineralized ECM volume, if HFIP-derived SF scaffolds are seeded with adipose-derived stem cells (ASCs) [22].

Advances in process methods, control of mechanical and chemical properties, and possibility of combining cells provide silk hydrogels and sponges more suitable constructs for tissue engineering. Nevertheless, there are limitations that need to be solved such as poor compatibility. It results with inhomogeneous mixtures, phase separation, and adverse tissue reactions [25, 27].

13.2.2 Fibroin Films and Membranes

Fibroin films can be prepared by different techniques namely dry-casting, spin-coating, or spin-assisted layer-by-layer assembly [1, 27].

Fibroin films can be used as wound dressings and skin or corneal replacement grafts because of their high water content, oxygen permeability, lower inflammatory response, and good optical transparency. When compared to collagen films and porcine wound dressing, fibroin films induce higher cell proliferation and lower inflammatory response. Fibroin films possess extremely good optical transparency, essential for cornea tissue engineering. Blending with chitin, SF films enhance adhesion of keratinocytes and fibroblasts [1, 27].

Fibroin films may also prepared with nanoparticles to provide more biocompatible material *in vitro*. When fibroin films are combined with silver or titanium dioxide, they have anti-bacterial effects; therefore, they can be used for bone implants surgical coatings as well as wound healing and cosmetics. When seed with osteoblast cells, fibroin films also enhance bone tissue growth *in vitro* [3, 7].

13.2.3 Nonwoven and Woven Silk Scaffolds

Nonwoven fibrous SF nets/mats/membranes can be prepared by spinning SF solution into micro- or nanofibers in an electric field. Solvent system (e.g. methanol, formic acid), fibroin concentrations, electric field, and spinning distance influence fiber properties. These scaffolds provide large surface area and rough topography for cell seeding. There dimensional scaffolds of nanofibers can be used in vessel grafts and nerve guides. Mats also support bone mesenchymal stem cells (BMSCs) proliferation and increase adhesion and proliferation of pre-osteoblasts and alkaline phosphatase (ALP) activity of osteoblasts after 14 days of incubation [1, 22, 26–28].

To form woven matrices such as braided cords and knitted mats, silk fibers are degummed followed by braiding, or first knitted followed by degumming. These scaffolds provide a surface for human bone marrow stromal cell (BMSC) to adhere, proliferate, and differentiate [1].

13.2.4 Silk Fibroin as a Bioactive Molecule Delivery

To use a biomaterial for drug delivery applications, it is important to show its clearance pathway from the body. SF can be considered an ideal vehicle for transport and delivery of bioactive molecules because of having unique properties such as slow biodegradation rate without any side effects, controllable composition [24, 30]. Molecular weight, pore size, hydrophobicity, and form of silk-based scaffold affect controlled release rate of drug molecules [25, 27]. SF scaffolds as a platform are exciting candidates for controlled release [26].

SF has been investigated to determine its feasibility for drug release system. BMP-2 is important to promote osteogenesis. The delivery of BMP-2 using SF has induced the formation of bone, increased bone density and promoted neo-osteogenesis after 8 weeks of implantation [22, 24, 30]. It has also been shown that release of BMP-2 from fibroin scaffold induces higher bone formation if human mesenchymal cells are seeded with scaffold [22]. To design osteoconductive matrices, growth factor adsorption has also been prepared by coating silk scaffolds with BMP-2 or BMP-7 adenoviruses [3, 29, 65].

Insulin growth factor-I (IGF-I) is also another important protein for cartilage repair. IGF-I releasing SF matrices have potential to induce chondrogenic stimuli for hMSCs [3, 26].

In order to achieve bone regeneration, besides BMP-2, VEGF plays a vital actor by promoting neovascularization and osteoblastic differentiation. It has also been found that silk scaffold-loaded VEGF induce cell adhesion, proliferation *in vitro*, blood vessel formation, and mineralization *in vivo* [2, 22, 24].

13.3 Biomedical Applications of Silk-based Biomaterials

SF proteins possess alterable features to design a scaffold and may be formed in different geometries for various tissue-engineering applications. Different forms of SF-derived biomaterials have been investigated for multiple biomedical applications (Table 13.2). In the following section, we discuss the application of SF in tissue engineering.

13.3.1 Bone Tissue Engineering

Bone tissue, as a specialized form of connective tissue, includes calcified extracellular matrix that has collagen type I and hydroxyapatite (HAp) and bone cells including osteoprogenitor, osteoblast, osteocytes, and osteoclasts. A suitable scaffold should be biocompatible to host, be biodegrade at a controlled rate, support cell attachment, migration, proliferation, and differentiation in the bone tissue engineering [32].

Table 13.2 Current applications of SF as biomaterials.

Target tissues	SF forms
Bone	Films, sponges, electrospun fibers [21], hydrogel [22], nonwoven mats [27]
Cartilage	Sponges, films, electrospun fibers [21], hydrogels [25], fibroin mats [1]
Anterior cruciate ligament (ACL)	Silk fiber wire-rope, knitted silk mesh [21] Braided silk fiber tube filled with collagen–HA porous scaffold [2], Braided cord, knitted mats with sponges [1]
Cardiac tissue	Microparticle patches, sponges [21], electrospun silk sheet treated with MeOH treated [2]
Skin	Films, electrospun fibers [21], lyophilized porous silk scaffolds [2], fibroin–alginate-blended scaffolds, water vapor-treated silk nanofibrous mats [27], microfibers, chitin/SF blends [26], nonwoven meshes [31]
Soft tissue	Gels, nanofibrous electrospun scaffold [21]
Nerve tissue	Fiber, film [21, 26], composite of SF with chitosan, nanofibrous SF [27]
Liver	Films [27], micro-/nanofibrous nonwoven scaffold [21]
Vascular tissue	Electrospun fibers [27], tubes [21, 26]
Adipose-like tissue	Sponges [21]
Tendon	Braided fibers [21]

Good biocompatibility, controllable biodegradability, modifiable mechanical properties, and low inflammatory response of SF provide unique material for bone tissue engineering [24, 26, 27].

Plain SF-based scaffold which is engineered within a bioreactor is able to provide an appropriate environment and induces development of bone tissue, showing a lack of differentiation into mature osteoblast [33]. In another bioreactor study, it has been shown that SF/gelatin/hydroxyapatite scaffold enhances rat MSCs to differentiate into osteoblastic cells [32].

To evaluate bone regeneration ability *in vivo*, SF has been proved as porous scaffold, electrospun material, and hydrogel in different models comprising mainly mice, rats, rabbits, and sheep with calvarial, mandibular, or femoral defects [22, 27].

It has been shown that SF scaffolds have potential for bone regeneration in nude mice [34–36]. Bone regeneration with implanted SF membrane,

SF nanofibrous meshes, and SF hydrogel has also been shown in calvarial defects of rabbit model without inflammatory effects [24, 26, 37–39].

Silk proteins can be modified to provide more bioactive material for bone implants and to induce bone formation. SF fibers with polyethylene oxide scaffold with BMP-2 give rise to formation of bone-like tissue in human MSCs [26, 27]. Incorporation of hydroxyapatite nanoparticles to the SF solution in fibers improves bone formation *in vivo* [26, 27]. SF can be also blended with poly(L-aspartate) as a template for the growth of apatite crystals [26].

It has been shown in a study that silk-coated poly(ε-caprolactone) (PCL) nanofibers have effects to enhance the mechanical and biological behaviors of biphasic calcium phosphate (BCP) scaffolds. Moreover, application of BCP/PCL silk scaffold gives rise to proliferation and differentiation of osteoblasts when compared to BCP/PCL and BCP scaffolds [3, 24].

It has also been shown that pure SF and SF with nHAp composite scaffolds can support attachment, proliferation, and osteogenic differentiation of BMSCs *in vitro*. Therefore, SF with HA scaffolds has been demonstrated to be excellent biocompatible and osteoinductive [27, 40, 41, 42].

Studies with human MSCs indicate that a combination of SF and HA enhances regenerative stimuli and provides cell attachment and proliferation [24, 43].

Although pure SF has potential usage for bone tissue engineering, it seems that combining SF with synthetic or natural polymers may improve the characteristics of SF for bone tissue regeneration.

Silk can be considered as attractive biomaterial because of having potential to stimulate proliferation, differentiation, and maintenance of stem cells [22, 24]. Some outcomes of silk matrices, their properties, and compatibility with different stem cell lines are shown in Table 13.3.

13.3.2 Cartilage Tissue Engineering

Cartilage, a nonvascular and noninnervated tissue, is mainly composed of type II collagen and chondrocyte cells. Cartilage tissue can be supported by chondrocyte-based cartilage tissue engineering [25, 27, 47]. Due to physical and chemical properties and mechanical stability of SF, it can maintain the structure of cartilage [48].

Hydrogels, sponges, electrospun mats and fibers, SF scaffolds with HA, and SF blended with various polymers have ability for mimicking the extracellular matrix components and promoting chondrocyte-based cartilage regeneration *in vitro* and *in vivo* [1, 25, 28, 27, 48, 49]. SF scaffolds incorporated with stem cells and encapsulated growth factor (e.g. TGF-B, IGF, FGF,

Table 13.3 Studies on the properties and compatibility of stem cells with silk-based matrices.

Stem cell type	Outcomes
bone marow stromal cells	Bone formation after surgery in a month, repairment after 12 months, improved cell attachment, differentiation of BMSCs into the osteogenic and adipogenic cell line [24]
human bone marrow stromal osteoprogenitor stem cells	Increased osteoconductive results such as apatite and BMP-2 content *in vitro,* hMSC differentiation, formation of templates for bone-like tissue *in vitro,* formation of mineralized bone components, rapid degradation, enhanced osteogenic differentiation of BMSCs, high cancellous bone formation, upregulation of osteogenic markers, forming templates for trabecular-like bone matrix *in vitro,* induction of bone formation within 5 weeks in critical size calvarial defects, induced cell attachment and proliferation, advanced calcium deposition and formation of bone-like trabeculae with cuboid cells, induction of osteogenic differentiation in hBMSC expressing BMP7 [24, 44, 65]
induced pluripotent stem cells	Effective bone regeneration [24]
human mesencyhmal stem cells	Enhanced osteogenic differentiation of hMSCs [24], vascularization *in vivo* model, calvarial defect in nude mice [45], enhanced cell adhesion and spreading with fibronectin [46]
placenta-derived mesenchymal stem cells	Osteogenic differentiation of PMSCs *in vitro,* improved fracture healing [44]
human adipose-derived stem cells	Good bone tissue formation [24]

hBMSCs (human bone marrow stromal osteoprogenitor cells), BMSCs (bone marrow stromal cells), iPSCs (induced pluripotent stem cells), PMSCs (placenta-derived mesenchymal stem cells), hASCs (human adipose-derived stem cells).

EGF, and PDGF) can also be applied at the site of defect [25]. Although SF scaffolds have good mechanical properties, poor biocompatibility and slow biodegradability and adverse response of host cause negative results in cartilage tissue [50].

Bioreactors can also be used to provide nutrients and growth factors to cultured cells, support chondrogenesis, and promote in cartilage regeneration in the rabbits by using SF hydrogel [51]. Nevertheless, all these scaffolds need further testing to prove their performance for the cartilage tissue engineering.

13.3.3 Ligament and Tendon Tissue Engineering

Tissue-engineering strategies can also be used for treatment of anterior cruciate ligament (ACL), rotator cuff, and Achilles tendon tears [2].

SF fibers, a promising candidate material for ligament and tendon scaffolds, can enhance progenitor cell growth and differentiation of tendon cell types in silk-based engineering of ACL [2, 27, 52].

Pure silk scaffolds have limited potential to induce ligament, tendon healing, and rotator cuff repair; therefore, silk scaffolds are combined with collagen matrix, poly-lactic-co-glycolic acid (PLGA), or encapsulated with tendon cell types and loaded growth factors to increase tissue regeneration [3, 27].

It is important to choose cell source in silk-based tissue regeneration. For ligament tissue engineering, anterior cruciate ligament fibroblasts (ACLFs) and BMSCs, which are seeded on the combined silk scaffolds to investigate the biological response, have been compared to choose the most appropriate cell source for ligament tissue engineering. It has been found that BMSCs have more advantages in cell proliferation, gene and protein expression for ligament-related ECM markers [26, 53].

Applying time of mechanical stimulation to induce mesenchymal stromal cell (MSC) differentiation on silk matrix is also important for ligament tissue engineering [54].

Silk-based materials have also been investigated for tendon tissue engineering. Silk–RGD films have potential to induce attachment and proliferation of human tenocyte as compared to unmodified silk films [49]. To improve cell attachment, growth, and differentiation of cells, silk is combined with collagen in order to have properties for tendon engineering. Therefore, silk collagen increases tendon reconstruction and revascularization. Although there are many researches on silk-based tendon engineering, it is difficult to say that silk scaffolds can be used in clinical application [2].

13.3.4 Cardiovascular Tissue Engineering

Cardiovascular disorders are still common leading cause of death in worldwide. Loss of cardiomyocytes leads to reduced heart function [27, 55]. There are several approaches for treatment, and one of them is tissue engineering to repair infarcted cardiac tissue. In order to treat damaged myocardium, various stem cell therapies have been investigated *in vivo* and *in vitro*. Due to low survival rates of cells, these therapies should have been developed. For this purpose, new biomaterials have been investigated to encapsulate cells [56, 57]. Among the natural biomaterials, SF possesses good mechanical properties and biocompatibility and tunable geometry [55].

SF/polysaccharide hybrid films, scaffolds, or bio-absorbable cardiac patches have been investigated to improve proliferation of MSCs and their cardiomyogenic differentiations. Yang *et al.* have shown that SF/chitosan (SF/CS) or SF/CS–hyaluronic acid (SF/CS–HA) patches significantly enhance the growth rate of MSCs and improved the cardiac muscle gene expressions compared to SF patches *in vitro* [58].

It has been reported that CD44 surface marker of rat MSCs, which bind to extracellular HA, may modulate cellular behaviors such as cell adhesion, matrix assembly, endocytosis, and cell signaling but especially cardiomyogenic differentiation. This effect tested by Yang *et al.* is complemented with SF/HA patches. They have determined that CD44 surface marker of rMSCs highly influences the proliferations, fibronectin expressions, and cardiomyogenic differentiation of rMSCs cultivated on cardiac SF/HA patches [59].

Moreover, bone marrow MSCs have been added to the SF/hyaluronic acid (BMSC/SH) hybrid patches, and they are used to enhance left (LV) remodeling and cardiac repair after myocardial infarction (MI) in rat hearts. The results have indicated that wall thickness of MI rat hearts has been increased, and neo-vascularization has been seen in SH patches groups compared to the untreated groups. The effect has been enhanced using BMSC/SH patches with only a minor inflammation compared with those of the SH patches [56].

Silk-based vascular tissue engineering gains importance in tissue engineering. Results from human aortic endothelial (HAECs) and human coronary artery smooth muscle cells (HCASMC) on silk fiber scaffolds show that SF can be used as a biomaterial for tissue-engineered vascular grafts [60, 61].

In another study, cylindrically shaped tubes using electrospun silk have been prepared and mechanical tests have been performed. Although electrospinning parameters influence the elastic and strength properties of the SF fibers, the scaffolds have the ability to withstand arterial pressures [62].

It has been demonstrated that SF scaffolds have good anticoagulant activity and platelet response and support endothelial cell spreading and proliferation. To improve blood compatibility of SF, it may be combined with water-soluble polymers or coated with fibronectin or collagen to support endothelial cell attachment, maintenance of phenotype and the formation of microvessel-like structures [49].

Therefore, SF films have suitable mechanical properties and biocompatibility to use as artificial blood vessels.

13.3.5 Skin Tissue Engineering

Skin, as a biggest organ in human, is a barrier against the environment [27]. Many dressing and grafts have been developed to speed and improve the wound-healing process. Recently, various forms of silk have been tested *in vivo* for their efficacy improving wound healing. When porous silk scaffolds are implanted, it has been shown that silk scaffolds induce vascular growth with decreased inflammatory response. Silk scaffold seeded with adipose-derived MSCs also increase wound healing in diabetic mice [2].

Several studies demonstrate that SF porous matrices can accelerate wound healing, improve adhesion, and spreading of normal human keratinocytes and fibroblasts without secretion of pro-inflammatory interleukins when coated with chitin, collagen, and alginate [26, 27, 49]. Eventually, SF porous materials have potential applications for skin tissue repairing.

13.3.6 Other Applications of Silk Fibroin

SF porous materials have been also studied in some other fields. SF porous materials can be regarded as cornea material if films are chemically attached RGD peptide, giving rise to highly transparent and mechanically stable cornea tissue construct [26, 27, 63, 64].

If nonwoven fibroin mats are coated with various polymers (e.g. polyaniline, polythiophene, and polypyrrole), they also have good electrical conductivity for neutrite growth and extension [1]. It has also been demonstrated that silk may be a supportive material for growth and proliferation of Schwann cells and dorsal root ganglia [26, 27, 49]. Silk scaffold can be evaluated in soft tissue regeneration, especially for breast implants [2].

13.4 Conclusion and Future Directions

Up to now, we have discussed this natural biomaterial as promising candidate in tissue engineering. All researches of SF demonstrate its potential ability in clinical applications among other biomaterials. Tunable structure, physical and chemical features, its biocompatibility and biodegradability, availability of different morphologies make it amazing biomaterial. Silk biomaterials, which are appropriate for desired applications, can be produced. We are sure that silk-based scaffolds can be developed with advances in processing techniques and stem cell-based strategies to meet different needs. We believe that new SF-based materials will be applied clinically to rehabilitate suffering patients.

References

1. Koh, L.D., Cheng, Y., Teng, C.-P., Khin, Y.W., Loh, X.J., Tee, S.Y., Low M., Ye, E., Yu, H.D., Zhang, Y.W., Han, M.Y., Structures, mechanical properties and applications of SF materials. *Progress in Polymer Science*, 46, 86–110, 2015.
2. Thurber, A.E., Omenetto F.G., Kaplan, D.L., *In vivo* bioresponses to silk proteins. *Biomaterials*, 71, 145–157, 2015.
3. Kundu, B., Kurland N.E., Bano, S., Patra, C., Engel, F.B., Yadavalli, V.K., Kundu, S.C., Silk proteins for biomedical applications: Bioengineering perspectives. *Progress in Polymer Science*, 39, 251–267, 2014.
4. Mori, K., Tanaka, K., Kikuchi, Y., Waga, M., Waga, S., Mizuno, S., Production of a chimeric fibroin light-chain polypeptide in a fibroin secretion-deficient naked pupa mutant of the silkworm *Bombyx mori*. *Journal of Molecular Biology*, Aug 11;251(2), 217–28, 1995.
5. Inoue, S., Tanaka, K., Arisaka, F., Kimura, S., Ohtomo, K., Mizuno, S., Silk Fibroin of *Bombyx mori* is secreted, assembling a high molecular mass elementary unit consisting of H-chain, L-chain, and P25, with a 6:6:1 molar ratio. *Journal of Biological Chemistry*, 275, 40517–40528, 2000.
6. Asakura T., Okushita, K., Williamson, M.P., Analysis of the structure of *Bombyx mori* silk fibroin by NMR. *Macromolecules*, 48, 2345–2357, 2015.
7. Altman, G. H., Diaz, F., Jakuba, C., Calabro, T., Horan, R. L., Chen, J., Lu, H., Richmond, J., Kaplan, D. L., Silk-based biomaterials. *Biomaterials*, 24, 401–416, 2003.
8. Rockwood, D.N., Preda, R.C., Yucel, T., Wang, X., Lovett M.L., Kaplan, D.L., Materials fabrication from *Bombyx mori* silk fibroin. *Nature Protocols*, 6 (10), 1612–1631, 2011.
9. Chen, F., Porter, D. & Vollrath, F. Silk cocoon (*Bombyx Mori*): Multi-layer structure and mechanical properties. *Acta Biomaterialia*, 8, 2620–2627, 2012.
10. Chen, F., Porter, D., Vollrath, F. Structure and physical properties of silkworm cocoons. *Journal of The Royal Society Interface*, 9, 2299–308, 2012.
11. The, C., Shao, Z., Vollrath, F. Surprising strength of silkworm silk. *Nature*, 418, 741–741, 2002.
12. Xiao, S., Stacklies, W., Cetinkaya, M., Markert, B., Gräter, F. Mechanical response of silk crystalline units from force-distribution analysis. *Biophysical Journal*, 96, 3997–4005, 2009.

13. Keten, S., Xu, Z., Ihle, B., Buehler, M. J. Nanoconfinement controls stiffness, strength and mechanical toughness of [beta]-sheet crystals in silk. *Nature Materials*, 9, 359–367, 2010.
14. Lefèvre, T., Rousseau, M.E., Pézolet, M. Protein secondary structure and orientation in silk as revealed by Raman spectromicroscopy. *Biophysical Journal*, 92, 2885–2895, 2007.
15. Li, M., Ogiso, M., Minoura, N. Enzymatic degradation behavior of porous silk fibroin sheets. *Biomaterials*, 24, 357–365, 2003.
16. Horan, R. L., Antle, K., Collette, A. L., Wang, Y., Huang, J., Moreau, J. E., Volloch, V., Kaplan, D. L., Altman, G. H., In vitro degradation of silk fibroin. *Biomaterials*, 26, 3385–3393, 2005.
17. Wang, Y., Rudym, D. D., Walsh, A., Abrahamsen, L., Kim, H. J., Kim, H. S., Kirker-Head, C., Kaplan, D. L. In vivo degradation of three-dimensional silk fibroin scaffolds. *Biomaterials*, 29, 24, 3415–3428, 2008.
18. Kim, U. J., Park, J., Joo Kim, H., Wada, M. & Kaplan, D. L. Three-dimensional aqueous-derived biomaterial scaffolds from silk fibroin. *Biomaterials*, 26, 2775–2785, 2005.
19. Ishida, K., Takeshita, H., Kamiishi, Y. & Yoshi, F. Kume, T., Radiation degradation of silk. *Jaeri-Conf.*, 005, 130–138, 2001.
20. Kojthung, A., Meesilpa, P., Sudatis, B., Treeratanapiboon, L., Udomsangpetch, R., & Oonkhanond, B., Effects of gamma radiation on biodegradation of *Bombyx mori* silk fibroin. *International Biodeterioration & Biodegradation*, 62, 487–490, 2008.
21. Maghdouri-White, Y., Bowlin, G.L., Lemmon, C.A., Dréau, D., Bioengineered silk scaffolds in 3D tissue modelling with focus on mammary tissues. *Materials Science and Engineering C*, 59, 1168–1180, 2016.
22. Melke, J., Midha, S., Ghosh, S., Ito, K., Hofmann, S., Silk fibroin as biomaterial for bone tissue engineering. *Acta Biomaterialia*, 31, 1–16, 2016.
23. Li, Z.H., Ji, S.C., Wang, Y.Z., Shen, X.C., Liang, H., Silk fibroin-based scaffolds for tissue engineering. *Frontiers of Materials Science*, 7(3), 237–247, 2013.
24. Mottaghitalab, F., Hosseinkhani, H., Shokrgozar, M.A., Mao, C., Yang, M., Farokhi, M., Silk as a potential candidate for bone tissue engineering. *Journal of Controlled Release*, 215, 112–128, 2015.
25. Kapoor, S., Kundu, S.C., Silk protein-based hydrogels: Promising advanced materials for biomedical applications. *Acta Biomaterialia*, 31, 17–32, 2016.
26. Zhang, Q., Yan, S., Li, M., Silk fibroin based porous materials. *Materials*, 2, 2276–2295, 2009.
27. Kundu, B., Rajkhowa, R., Kundu, S.C., Wang X., Silk fibroin biomaterials for tissue regenerations. *Advanced Drug Delivery Reviews*, 65, 457–470, 2013.
28. Wang, Y., Kim, H.J., Vunjak-Novakovic, G., Kaplan, D.L., Stem cell-based tissue engineering with silk biomaterials. *Biomaterials*, 27, 6064–6082, 2006.
29. Yucel, T., Lovett, M., Kaplan, D.L., Silk-based biomaterials for sustained drug delivery. *Journal of Controlled Release*, 190, 381–397, 2014.

30. Mottaghitalab, F., Farokhi, M., Shokrgozar, M.A., Atyabi, F., Hosseinkhani, H., Silk fibroin nanoparticle as a novel drug delivery system. *Journal of Controlled Release*, 206, 161–176, 2015.
31. Vepari C., Kaplan, D.L., Silk as a Biomaterial. *Progress in Polymer Science*, 32 8–9, 991–1007, 2007.
32. Sinlapabodin, S., Amornsudthiwat, P., Damrongsakkul, S., Kanokpanont, S., An axial distribution of seeding, proliferation, and osteogenic differentiation of MC3T3-E1 cells and rat bone marrow-derived mesenchymal stem cells across a 3D Thai silk fibroin/gelatin/ hydroxyapatite scaffold in perfusion bioreactor. *Materials Science and Engineering C*, 58, 960–970, 2016.
33. Meinel, L., Fajardo, R., Hoffman, S.,Langer, R., Chen, J., Snyder, B., Vunjak-Novakovic, G., Kaplan, D.L., Silk implants for the healing of critical size bone defects. *Bone*, 37, 688–698, 2005.
34. Meinel, L., Betz, O., Fajardo, R., Hofmann, S., Nazarian, A., Cory, E., Hilbe, M., McCool, J., Langer, R., Vunjak-Novakovic, G., Merkle, H.P., Rechenberg, B., Kaplan, D.L., Kirker-Head, C., Silk based biomaterials to heal critical sized femur defects. *Bone*, 39, 922–931, 2006.
35. Etienne, O., Schneider, A., Kluge, J.A., Bellemin-Laponnaz, C., Polidori, C., Leisk, G.G., Kaplan D.L., Garlick, J.A., Egles, C., Soft tissue augmentation using silk gels: an *in vitro* and *in vivo* study. *Journal of Periodontology*, 80, 1852–1858, 2009.
36. Lai, G.J., Shalumon, K.T., Chen, J.P., Response of human mesenchymal stem cells to intrafibrillar nanohydroxyapatite content and extrafibrillar nanohydroxyapatite in biomimetic chitosan/SF/nanohydroxyapatite nanofibrous membrane scaffolds. *International Journal of Nanomedicine*, 12, 567–584, 2015.
37. Song, J.Y., Kim, S.G., Lee, J.W., Chae, W.S., Kweon, H., Jo, Y.Y., Lee, K.G., Lee, Y.C., Choi, J.Y., Kim, J.Y., Accelerated healing with the use of a silk fibroin membrane for the guided bone regeneration technique. *Oral Surgery, Oral Medicine, Oral Pathology, Oral Radiology, and Endodontology*, 112, 26–33, 2011.
38. Kim, K.H., Jeong, L., Park, H.N., Shin, S.Y., Park, W.H., Lee, S.C., Kim, T.I., Park, Y.J., Seol, Y.J., Lee, Y.M., Biological efficacy of SF nanofiber membranes for guided bone regeneration. *Journal of Biotechnology*, 120, 327–339, 2005.
39. Fini, M., Motta, A., Torricelli, P., Giavaresi, G., Nicoli Aldini N., Tschon M., *et al.*, The healing of confined critical size cancellous defects in the presence of silk fibroin hydrogel. *Biomaterials*, 26 (17), 3527–3536, 2005.
40. Liu, H., Xu, G.W. Wang, Y.F., Zhao, H.S., Xiong, S., Wu, Y., Heng, B.C., An C.R., Zhu, G.H., Xie, D.H., Composite scaffolds of nano-hydroxyapatite and silk fibroin enhance mesenchymal stem cell-based bone regeneration via the interleukin 1 alpha autocrine/paracrine signaling loop. *Biomaterials*, 49, 103–112, 2015.
41. Wei, K., Li, Y., Kim, K.O., Nakagawa, Y., Kim, B.S., Abe, K., Chen, G.-Q., Kim, I.S., Fabrication of nano-hydroxyapatite on electrospun silk fibroin

nanofiber and their effects in osteoblastic behavior. *Journal of Biomedical Material Research Part A*, 97, 272–280, 2011.
42. Qian, J., Suo, A., Jin, X., Xu, W., Xu, M., Preparation and *in vitro* characterization of biomorphic silk fibroin scaffolds for bone tissue engineering. *Journal of Biomedical Materials Research Part A*, 102, 2961–2971, 2014.
43. Garcia-Funtes, M., Meinel, A.J., Hilbe, M., Meinel, L., Merkle, H.P., Silk fibroin/hyaluronan scaffolds for human mesenchymal stem cell culture in tissue engineering. *Biomaterials*, 30, 5068–5076, 2009.
44. Jin, J., Wang, J., Huang, J., Huang, F., Fu, J., Yang, X., Miao, Z., Transplantation of human placenta-derived mesenchymal stem cells in a silk fibroin/hydroxyapatite scaffold improves bone repair in rabbits. *Journal of Bioscience and Bioengineering*, 118 (5), 593–598, 2014.
45. Hoffman, S., Hilbe, M., Fajardo, R.J., Hagenmuller, H., Nuss, K., Arras, M., Muller, R., von Rechenberg, B., Kaplan, D.L., Merkle, H.P., Meinel, L., Remodelling of tissue-engineered bone structures *in vivo*. *European Journal of Pharmaceutics and Biopharmaceutics*, 85 (1), 119–129, 2013.
46. Meinel, A.J., Kubow, K.E., Klotzsch, E., Garcia-Fuentes, M., Smith, M.L., Vogel, V., Merkle, H.P., Meinel, L., Optimization strategies for electrospun silk fibroin tissue engineering scaffolds. *Biomaterials*, 30, 3058–3067, 2009.
47. Sundelacruz, S., Kaplan, D.L., Stem cell- and scaffold-based tissue engineering approaches to osteochondral regenerative medicine. *Seminars in Cell & Developmental Biology*, 20(6), 646–655, 2009.
48. Jaipaew J., Wangkulangkul, P., Meesane, J., Raungrut, P., Puttawibul, P., Mimicked cartilage scaffolds of silk fibroin/hyaluronic acid with stem cells for osteoarthritis surgery: Morphological, mechanical, and physical clues. *Materials Science and Engineering C*, 64, 173–182, 2016.
49. Kearns V., MacIntosh, A.C., Crawford, A., Hatton, P.V., Silk-based biomaterials for tissue engineering, in *Topics in Tissue Engineering*, Vol. 4. Eds. N Ashammakhi, R Reis, F Chiellini, 1–19, 2008. http://www.oulu.fi/spareparts/ebook_topics_in_t_e_vol4/
50. Chao, P.H., Yodmuang, S., Wang, X., Sun, L., Kaplan, D.L., Vunjak-Novakovic, G., Silk hydrogel for cartilage tissue engineering. *Journal of Biomedical Materials Research Part B: Applied Biomaterials*, 95, 84–90, 2010.
51. Shangkai, C., Naohide, T., Koji, Y., Yasuji, H., Masaaki, N., Tomohiro, T., Yasushi, T., Transplantation of allogeneic chondrocytes cultured in fibroin sponge and stirring chamber to promote cartilage regeneration. *Tissue Engineering*, 13, 483–492, 2007.
52. Altman, G., Horan, R., Lu, H., Moreau, J., Martin, I., Richmond, J., Kaplan, D.L., Silk matrix for tissue engineered anterior cruciate ligaments. *Biomaterials*, 23, 4131–4141, 2002.
53. Liu, H., Fan, H., Toh, S.L., Goh, J.C.H., A comparison of rabbit mesenchymal stem cells and anterior cruciate ligament fibroblast responses on combined silk scaffolds. *Biomaterials*, 29, 1443–1453, 2008.
54. Chen, J., Horan R.L, Bramono, D., Moreau, J.E., Wang Y., Geuss, L.R., Collette, A.L., Volloch, V., Altman, G.H., Monitoring mesenchymal stromal

cell developmental stage to apply on-time mechanical stimulation for ligament tissue engineering. *Tissue Engineering*, 12 (11), 3085–3095, 2006.
55. Patra, C., Talukdar, S., Novoyatleva, T., Velagala, S.R., Mühlfeld, C., Kundu, B., Kundu, S.C., Engel, F.B., Silk protein fibroin from *Antheraea mylitta* for cardiac tissue engineering. *Biomaterials*, 33, 2673–2680, 2012.
56. Chi, N.H., Yang, M.C., Chung, T.W., Chen, J.Y., Chou N.K., Wang, S.S., Cardiac repair achieved by bone marrow mesenchymal stem cells/silk fibroin/hyaluronic acid patches in a rat of myocardial infarction model. *Biomaterials*, 33, 5541–5551, 2012.
57. Rahimi, M., Mohseni-Kouchesfehani, H., Zarnani, A.H., Mobini, S., Nikooand, S., Kazemnejad, S., Evaluation of menstrual blood stem cells seeded in biocompatible *Bombyx mori* silk fibroin scaffold for cardiac tissue engineering. *Journal of Biomaterials Applications*, 29(2) 199–208, 2014.
58. Yang, M.C., Wang, S.S., Chou, N.K., Chi, N.H., Huang, Y.Y., Chang, Y.L., Shieh, M.J., Chung, T.W., The cardiomyogenic differentiation of rat mesenchymal stem cells on silk fibroin–polysaccharide cardiac patches *in vitro*. *Biomaterials*, 30, 3757–3765, 2009.
59. Yang, M.C., Chi, N.H., Chou, N.K., Huang Y.Y., Chung, T.W., Chang, Y.L., Liu, H.C., Shieh, M.J., Wang S.S., The influence of rat mesenchymal stem cell CD44 surface markers on cell growth, fibronectin expression, and cardiomyogenic differentiation on silk fibroin – Hyaluronic acid cardiac patches. *Biomaterials*, 31, 854–862, 2010.
60. Lee, S.J., Yoo, J.J., Lim, G.J., Atala, A., Stitzel, J., *In vitro* evaluation of electrospun nanofiber scaffolds for vascular graft application. *Journal of Biomedical Materials Research Part A*, 83A, 999–1008, 2007.
61. Jin, H.J., Fridrikh, S.V., Rutledge, G.C., Kaplan, D.L., Electrospinning *Bombyx mori* silk with poly(ethylene oxide). *Biomacromolecules*, 3, 1233–1239, 2002.
62. Soffer, L., Wang, X., Zhang, X., Kluge, J., Dorfmann, L., Kaplan, D.L., Leisk, G., Silk-based electrospun tubular scaffolds for tissue-engineered vascular grafts. *Journal of Biomaterials Science Polymer Edition*, 19(5), 653–664, 2008.
63. Lawrence, B.D., Marchant, J., Pindrus, M.A., Omenetto, F., Kaplan, D.L., Silk film biomaterials for cornea tissue engineering. *Biomaterials*, 30, 1299–1308, 2009.
64. Gil, E.S., Park, S.H., Marchnat, J., Omenetto, F., Kaplan, D.L., Response of human corneal fibroblasts on silk film surface patterns. *Macromolecular Bioscience*, 10, 664–673, 2010.
65. Zhang,Y., Fan, W., Ma, Z., Wu, C., Fang, W., Liu, G., Xiao, Y., The effects of pore architecture in silk fibroin scaffolds on the growth and differentiation of mesenchymal stem cells expressing BMP7. *Acta Biomaterialia*, 6, 3021–3028, 2010.

14
Applications of Biopolymer-based, Surface-modified Devices in Transplant Medicine and Tissue Engineering

Ashim Malhotra[1]*, Gulnaz Javan[2] and Shivani Soni[3]

[1]*School of Pharmacy, College of Health Professions, Pacific University Oregon, Hillsboro, OR, USA*
[2]*Forensic Science Program, Physical Science Department, Alabama State University, Montgomery, AL, USA*
[3]*Department of Biological Sciences, Alabama State University, Montgomery, AL, USA*

Abstract

In this chapter, we discuss biopolymer-based surface modification of implements employed in reconstructive, surgical, and research-based therapeutic approaches to disease. In particular, we will explore the indications, ideal characteristics, manufacture process, and targeted delivery for, and cutting-edge discovery and innovation research in novel transplant therapeutics of surface-modified devices, implants, and stem cell-based novel artificial grafts using biopolymers with particular reference to the field of cardiovascular and osteopathic science. Discussion includes both natural biopolymers and artificial polymers employed for enhancing cellular contact, vascular regrowth, and ease of implantation. Finally, modified biopolymer-based nanodelivery constructs employed for targeted delivery of therapeutic cargo and in the manufacture of artificial scaffolds for stem cell-based reconstructive cardiovascular implantation are discussed.

Keywords: Biopolymer, surface-modified devices, nanodelivery, cardiovascular implantation

---

Corresponding author: ashim.malhotra@pacificu.edu

14.1 Introduction to Cardiovascular Disease

Cardiovascular disease is defined as pathology involving any structural or functional component of the heart or the blood vessels. Cardiovascular disease may be congenital or may develop during an individual's lifetime. Multiple causes have been attributed to the development of cardiovascular disorders and range from genetic or familial disposition to the contribution of behavioral factors such as routine exercise, chronic cigarette, alcohol and/or drug consumption, or the long-time consumption of lipid-rich diets.

Diseases of the heart encompass conditions such as heart failure, acute coronary syndrome, angina, myocardial infarction, dilated and hypertrophic cardiomyopathies, valvular dysfunction, arrhythmias, infections, myocarditis, and rare tumors of the heart. Diseases affecting the vascular system include anatomical disorders such as coarctation of the aorta and aneurysms, while other vascular diseases include atherosclerosis, thrombosis, and peripheral artery disease. These diseases may ultimately affect overall systemic processes and the general health status and quality of life of an individual by indirectly influencing circulation, blood pressure, organ and tissue perfusion, oxygen availability, exercise and stress tolerance, and the possibility of complications as a consequence of the involvement of other organs. Taken together, these factors make cardiovascular disease a formidable medical challenge – a fact that is supported by the high mortality associated with cardiovascular disease in developed countries [1]. For instance, in the United States alone, around 600,000 people die due to heart disease every year as per the Centers for Disease Control [2]. Among the types of cardiovascular diseases mentioned above, acute coronary syndrome alone accounted for 380,000 deaths in 2010 [1]. The collective economic burden encompassing costs incurred due to cardiovascular mortality, morbidity, loss of productivity, hospitalizations, and insurance is a whopping $108.9 billion a year [2]!

14.2 Need Assessment for Biopolymer-based Devices in Cardiovascular Therapeutics

The composite of cardiovascular diseases often requires a higher burden of invasive procedures, whether for diagnosis, imaging, or treatment than most other organ-system pathologies. The use of implantable devices and implements is on the rise and coupled with the enormous variety of uses and time schedules for these devices, complications following the procedures also show a corresponding increase.

A major factor influencing the success of an implanted cardiovascular device is the extent to which tissue re-colonization occurs around the device. For instance, one issue common to most implantable cardiovascular devices is that of tissue compatibility, which involves adequate regrowth and re-colonization by cells in the graft region, in turn allowing neovascular growth in the area adjacent to the grafted device. Adsorbable biopolymers provide a ready solution to this problem as they may be used as coating materials on such devices. They allow for enhanced cell contact *in situ* enabling cellular regrowth and tissue integration and make an otherwise incompatible structure biocompatible. However, interestingly, the reverse may be equally true as occurs following stenting, for instance, to improve the structural quality of a compromised coronary artery. Base metal stents (BMS), such as those composed of stainless steel, were traditionally used to enhance the structural integrity of coronary vessel walls, however, restenosis, which is the regrowth of cells in the stented area was an undesirable effect. The biological progression of this pathology was explained by Virmani and Farb in 1999 as the regrowth of the intima, which is the inner most layer of the vessel wall, due to vascular smooth muscle cell hyperplasia [3]. The emergence of this need in the cardiovascular field was also addressed by the development of specific advanced materials, in particular pyrrole–imidazole polyamide-based systems. Pyrrole–imidazole polyamides prevent cell growth by silencing genes such as tumor growth factor β (TGFβ) essential for intimal revascularization [4, 5]. Thus, coating stents and other implantable cardiovascular devices with pyrrole–imidazole polyamides was suggested to ameliorate stent thrombosis and restenosis.

14.3 Emergence of Surface Modification Applications in Cardiovascular Sciences: A Historical Perspective

Practice innovations and advancements in interventional cardiothoracic technology, coupled with progress in drug formulation and delivery modalities heralded a new epoch in cardiovascular therapeutics. That the emergence of modern technology and the development of innovative therapeutic and pharmaceutical approaches was a pressing need in the cardiovascular field is supported by the high incidence of cardiovascular disease in the United States as mentioned earlier. The huge financial burden associated with cardiovascular morbidity and mortality provided further impetus to the search for biocompatible, nontoxic, stable, and flexible materials that would lend themselves to the design and manufacture of devices, implements, and tools that could be used in diagnostic, therapeutic, and ultimately interventional cardiovascular medicine. Historically, a number of discoveries and inventions over the past three decades in disparate fields of science, ranging from life sciences to engineering, enabled the birth of modern approaches for treating cardiovascular diseases. Instances of the latter abound and range from the invention of drug-eluting stents (DES) to nanoparticle-based scaffolding platforms used for reconstructive tissue engineering for the replacement of injured cardiac or vascular tissues [6].

Of the various forces that were working in tandem during this time period, perhaps the most significant that aided in ushering in the modern era of cardiovascular therapeutics, with its sophisticated use of surface modified polymers, could be ascribed to a finite few discoveries. Prime among these were the advent of (1) molecular biology since the early 1990s that offered a firm foundation for understanding the fundamental basis of disease processes, (2) of biomolecular imaging techniques that could provide a level of diagnostic ability previously unheard of, and (3) nanotechnology that enabled the fashioning of diagnostic, therapeutic, and recently, of tissue-engineering capabilities at the same dimensional level as the pathological processes.

Since the 1970s, research into advanced materials that could be used to fabricate devices and implements used in interventional cardiology started gaining impetus as the focus shifted to coronary artery bypass surgery (CABG) and ballooning angioplasty. The 1980s saw the use of BMS to bolster coronary artery walls to overcome damage due to conditions such as atherosclerosis and acute coronary syndrome, but upon the discovery of extensive restenosis following intracoronary delivery of BMS, the field shifted to the use of DES, which were essentially controlled release devices containing immunosuppressive and antiproliferative agents such as sirolimus and paclitaxel, released gradually over time, to prevent the regrowth of cells in the stent area [5–8]. However, inventions such as DES required an extensive refocusing in the engineering of the stents, which in turn was heavily dependent upon the discovery and novel use of advanced materials. For instance, many DES in use today are self-assembling nanoparticle-based systems that simultaneously achieve targeted delivery to a particular site-specific region of the compromised vessel wall. Such an achievement was only possible due to simultaneous advancements across multiple science disciplines.

Gupta et al. [9] and Prasad et al. [10], among others, demonstrated that most nonbiological materials like metals commonly used for device construction are not readily amenable to biological and tissue integration following implantation procedures. This is especially true in the context of the absence of biological contact points, surface attachments required for anchorage, contact inhibition, and surface adsorption and attachment, all of which may occur with the employment of steel and other nonbiological material-based implants. Additionally, the tissue epithelium, along with the extracellular matrix (ECM), actively sculpts the nature and physiology of the organ by not only serving as an anchoring material, but also by influencing the secretion of multiple proteins. This presents problems in the event of nonbiological material-based implements and devices, which due to their noncellular composition result in an absence of the requisite secretory proteins and generation of the appropriate microenvironment milieu. This further results in a lack of development of adequate cell-to-cell contact, preventing recolonization of implant. The consequent effect becomes particularly visible in long term usage, as persistent lack of contact and alterations in tissue microenvironment evince as lack of revascularization and integration of implants over time.

14.4 Nitric Oxide Producing Biosurface Modification

Recently, an interesting and technically significant advancement has occurred in the field of vascular and epithelial integration following implantation. This has been in the context of continuous secretion of molecular factors essential for the maintenance of normal physiology. Nitric oxide, NO, also known as endothelium-dependent relaxing factor (EDRF), is released from epithelial cells lining the intima of blood vessels, particularly arteries, and plays a central role in influencing vascular epithelial cell relaxation and vasodilatation. Normal vascular epithelium has been demonstrated to release 0.1 nmol cm^{-1} min^{-1} of NO [11]. The absence of NO release, which occurred following the employment of nonbiological material for implantation, presented a major hurdle for tissue integration. Biopolymer-based surface modification of implant material aided in overcoming this deficit. About a decade ago, Oh and Meyerhoff created an ingenious system of coating implant material surface with hydrophobic polymer with specific surface modifications. In particular, catalyst pockets were etched on the surface of the coating biopolymer, with the special property of contact-associated activation and continuous production of NO. This was made possible due to the use of endogenous biological sources of NO in the circulating bloodstream as precursors, such as S-nitrosothiols, using Cu(I) mediated reduction of endogenous nitrite ions. Using the lipophilic chelating agent known as dibenzo[e,k]-2,3,8,9-tetraphenyl-1,4,7,10-tetraaza-cyclododeca-1,3,7,9-tetraene (DTTCT) to anchor Cu(II) to the polymeric surface, they created a modified surface coating for implants used in cardiovascular surgery. The anchored Cu (II) was shown to catalytically convert to Cu (I) ions upon contact with endogenous ascorbate or thiolate ions, and finally the bioproduction of NO [11].

It is important to note that NO is not only a significant molecular player orchestrating vasodilatation but also regulates platelet activation and aggregation. Since most invasive cardiovascular procedures, especially those involving nonbiological material-based devices, have a high incidence of thrombotic and obstructive complications, the ability to engineer materials for the continuous release of NO is particularly significant.

Several modifications of the above procedure have been successfully employed for surface modification of implant materials to augment biocompatibility. Lee *et al.* discovered that dip coating polymers and various materials in a solution of dopamine allowed adequate first layer surface adsorption and coating of material surfaces. They also demonstrated the flexibility of this simple technique in its applicability to a host of disparate materials ranging from noble metals, oxides, and polymers, to semiconductors and ceramics. Following adsorption of an initial layer by dip-coating to create surface-adherent polydopamine films, self-assembling, high molecular weight, long-chain polymers were used to create monolayers to enhance surface coating [12]. Weng *et al.* reported a simplification of this procedure by the immersion of materials into dilute, aqueous solutions of dopamine, with pH similar to marine solutions, making it easier and more cost effective to modify and enhance surface chemistry of materials used in cardiovascular implantation [13]. In some instances, instead of using biopolymers, the use of cystamine and selenocystamine, especially on titanium dioxide devices and implantable materials was shown to result in sustained and measured release of NO [14].

14.5 Surface Modification by Extracellular Matrix Protein Adherence

ECM is the ground substance of any tissue that contains the cellular and protein array that anchors the cells of the tissue. But its function is not limited to mere anchorage, as has emerged over the past few decades, particularly from studies of cancer metastases. Cells secrete special proteins into the ECM that function as molecular cues that regulate a variety of physiological processes in both health and disease.

Mimicking biological constructs was considered a valid approach in the design and fabrication of surface enhanced materials engineered for biocompatibility. This was achieved by coating materials, especially metals, with usually one major adhesion protein. The rationale for this undertaking was the observation that the epithelial lining of an organ is anchored via a basement membrane that is a thick milieu of proteins such as laminin, type IV collagen, heparin sulfate proteoglycan, and the glycoproteins fibronectin and vitronectin. Usually, these proteins are secreted by the epithelial cells themselves and help to maintain not only anchorage but cellular polarity. Thus, mimicking such protein coating on nonbiological implant material was designed to improve tissue integration and revascularization.

14.6 The Role of Surface Modification in the Construction of Cardiac Prostheses

Cardiac surgery and cardiovascular sciences have profited in a big way from the advent and the use of carefully designed, experimentally tested and validated cardiac prostheses that serve to replace worn or damaged anatomical cardiac components. One common example of a much used cardiac prosthetic device is the prosthetic heart valve. The human heart has two atrioventricular (AV) and two semi-lunar valves. The AV valves separate the atria from the ventricles, while the semi-lunar valves guard the orifices of the major vessels (the aorta and the pulmonary artery). Both serve similar functions of preventing regurgitation of blood. Although cardiac muscle is fairly resilient to wear, the cartilaginous valves are somewhat more prone to damage either following infection, or injury or lipoid disease or other factors, such as age-related wear and tear.

Advances in technology have made it possible to construct and surgically implant prosthetic heart valves in human patients. These valves fall into three categories: (1) bioprosthetic, (2) mechanical, or (3) polymeric valves. Although polymeric valves are more biocompatible, less immunogenic, flexible, and stable, they also suffer from the major disadvantages of thrombolysis, pathological calcification and ameliorated durability. Although a number of synthetic polymers, polyurethane prime among them, have been used more commonly for the manufacture of prosthetic heart valves, recently biopolymers have been used either for full construction of heart valves or for surface modification to lessen some of the disadvantages mentioned above.

14.7 Biopolymer-based Surface Modification of Materials Used in Bone Reconstruction

Due to its highly specific matrix, osseous tissue needs its own category for discussion of biopolymer materials that can be used for surface modification. This is partly due to the process of bone repair and mineral deposition that is a continuous process and one that is regulated at various levels of control by the immediate biological microenvironment, and in part due to the spatial distribution of living cells in the otherwise mineral (dead) deposit of bone. In order to fully understand the significance of biopolymer-based surfaces in the repair of bones, particularly in the context of metal-based implants such as pins, needles, and other implements, it is imperative to learn the processes of bone growth, repair and regeneration.

Perhaps the most interesting aspect of bone growth is the spatial orchestration of a number of cells in the growth of bone tissue, especially considering that release of molecular signals from bone cells may only have an influence once these can be adequately moved from one micro-location to the other. This is due to the fact that compact bone is composed of cylinder-like structures called osteon or Haversian canals, conspicuous under the microscope. The Haversian canal is hollow in the center, with this space being occupied by nerves and vessels. Surrounding the central lumen are multiple small "hole" like structures called lacunae which house bone cells called osteocytes. Lacunae are connected by minute spider-web-like channels called canaliculi. Cells of the bones communicate with one another and regulate bone growth by secreting molecules into the canaliculi.

This rigid structure of the bone makes the surgical and interventional process perhaps more difficult when compounded with the knowledge of the complex process of bone growth, regeneration, and repair. For instance, bone growth occurs by two distinct mechanisms across two axes. Longitudinal growth and growth in girth or "appositional growth", as the latter is more commonly known as occur via two different mechanisms. One of these involves yet another cell type, this time of a different tissue called cartilage. Cartilage producing cells, called chondrocytes, upon becoming activated secrete fibrillar proteins such as collagen and elastin. These proteins constitute the ground substance or matrix of cartilage tissue. Appositional growth of bone occurs by gradual replacement of chondrocyte secreted collagen and elastin with calcium-rich material known as hydroxyapatite. A number of extraneous factors such as inflammation, and the recruitment of inflammatory cells, along with the enrichment of the local milieu with inflammatory cytokines and chemokines influence bone growth following injury or damage.

Implantation of pins, needles, and plates, generally made of metals such as titanium or stainless steel are often used in bone setting following injury to provide not only mechanical and tensile strength, but also a spatial three-dimensional support system. A variety of biopolymers may be used as coating materials, which aid in the integration of the artificial and supportive mechanical matrices. There are three main materials.

1. **Chondroitin sulfate**: Chondroitin sulfate is an interesting biopolymer composed of a repeating disaccharide dimer composed of N-acetylgalactosamine and D-glucuronic acid which is attached to proteins, creating its overall complex glycosaminoglycan structure. The galactosamine residues of chondroitin are sulfated in positions 4 or 6 and the negative charge that sulfate imparts to this molecule is considered to be important for its interaction with other molecules, such as growth factor [15] and inflammatory chemokines and cytokines [16]. Coating materials intended for bone-tissue implants with chondroitin sulfate lends a number of characteristics that enhance integration into the re-growing matrix. Briefly, in addition to enhancing interaction with signal and hormone molecules in the ECM, chondroitin sulfate increases the adsorption of bone cells such as osteocytes and osteoblasts to the implants [17]. Interestingly, these studies were extended to animal models, including small animals such as rats, and other experiments in larger animals such as sheep, proving the validity and clinical relevance, at least in a veterinary system. In rats, Rammelt *et al.* showed that implanting a device coated with chondroitin sulfate and collagen led to improved adherence of a number of cell types including fibroblasts to the regrowing bone. This enhanced cellular recruitment resulted in a positive environment with a milieu rich in secreted chemokines that further aided cell adherence to the implement. For instance, in the sheep experimental model, following implantation with a chondroitin sulfate-coated implant, the number of osteoblasts recruited to the interface between the new implant and the older bone mass over time [17, 18].
2. **Collagen Type I**: Collagen is a fibrillar protein, found in great abundance in the bone matrix, where it plays more than a mere structural role. A number of studies in small animal models of bone disease have provided evidence that coating

metal-based implants such as titanium and steel with collagen results in enhanced recruitment of cells in the inflammatory reaction [19–21]. Importantly, studies have compared the difference between collagen-coated and collagen-uncoated implants in animal models following implantation. Rammlet *et al.* demonstrated that 4 days after implantation, the collagen-coated constructs resulted in a 70% recruitment of "granulation"-rich matrix, composed of cells of the inflammatory reaction as well as osteocytes, to the implant, compared with the uncoated control implants which did not show this granulation reaction [17]. This experiment showed the importance of biopolymer-based surface modification of implants, particularly in the augmentation of tissue integration. Repetition of this experimental premise in various animal models, proved the universality of use of collagen as a surface modifying biopolymer in mammals. In experimental models of sheep [17] and goats [21], collagen coating caused similar improvements in tissue regeneration in bone.

3. **RGD peptide sequence**: In a tissue, a variety of proteins aid in cell anchoring. Divided into a number of protein families based on their structures, these family of proteins serve to provide tissue-wide cohesiveness and prevent cell migration. In fact, dysregulation of some of these proteins, such as altered protein expression, usually acts as a mechanism in diseases processes. However, since these proteins are an important component of the extra cellular matrix as they serve to enhance mechanical tissue strength, and by the 1970s modern technology had provided a process for the construction of protein-coated platforms, it seemed plausible that coating surgical implant surfaces with biopolymers of such proteins would enhance tissue integration of the implant. Of note to the arena of bone implants was the observation that common to many ECM proteins such as collagen and fibronectin was a three amino acid peptide sequence that was reported to result in enhanced cell adhesion. This was the RGD or the arginine–glycine–aspartate amino acid-based peptide sequence that was ubiquitous in expression [22], and was empirically demonstrated to be important for the interaction and binding of cell surface receptors to adhesion proteins such as integrin [23]. Unsurprisingly, coating titanium-based implants for use in

bone-resetting and regrowth experiments in a rats resulted in an increased recruitment of bone cells to the implant [24] and overall enhanced tissue integration. In similar experiments, Ferris *et al.* demonstrated augmentation of the bone mass of the femur bone in rats, following the use of RGD-coated titanium-based implants [25], which was an event subsequent to the enhanced recruitment of bone cells to the biomaterial-coated implant surface. Further support for this model was provided by the work of Elmengaard *et al.*, who demonstrated enhanced bone regrowth in the well-established bone gap model in dogs [26].

Interestingly, in this context, the nature of the implant material seems to be at least as important as the peptide sequence used for coating and enhancement of tissue integration. For instance, adoption of the same strategy of coating implant material for use in bone-growth experimental studies where hydroyapetite was employed as the implant base material instead of titanium resulted in a failure of promotion of bone growth [27]. Thus, biopolymer-based surface modification of implants may be used as a viable method to increase tissue integration and overall implant stability, but it seems to also depend upon the substance the implant is composed of.

14.8 The Use of Biopolymers in Nanotechnology

The term nanotechnology encompasses the process for the manufacture, application, and validation of particles in the 10^{-9} m ranged. Originating in the 1980s, this technology has shown remarkable adaptability as it can be used across multiple scientific disciplines, in fields as diverse as pharmaceutical formulation, medicine, and mechanical engineering. Coupled with rapid advances in the field of molecular biology, nanotechnology has afforded a particularly powerful tool to change the face of modern medicine by providing a means to innovate and improve delivery, targeting and absorption of drugs, including for the first time in history, that of genes for the much anticipated gene therapy.

One reason of the success of nanoparticle-based platform is in medicine stems from the dimension of these particles. The scale of these particles makes intervention possible at the levels of proteins, making is easier to reach physiological targets as different as the intracellular, or the interstitial (between cells in a tissue) micro-environments, improving the ultra-local delivery and thus targeting of therapeutic modalities. In addition to the above, apart from scale, nanoparticles offer various other advantages. They are easy and cost effective to manufacture, their surface can be readily modified by chemical processes or by physical adsorption, and combined with other technologies, they can be assessed and the process validated with relative ease in the human patient. Surface modification of nanoparticles affords malleability to their use and improves site-specific delivery, while simultaneously reducing toxicity. Such surface augmentation can be achieved by either adsorption of chemical moieties such as proteins, lipids, or polysaccharides to the nanoparticles, while the body of the nanoparticle encloses a drug to be delivered to a specific tissue. One example of the latter is the delivery of anticancer drugs to various cancerous tissues such as the breast, or pancreas, employing antibody-coated nanoparticle vehicles to ensure targeting of specific proteins in the cancerous tissue. Advances in materials science and process engineering technologies have provided productive means for *en masse* surface modification of nanoparticles, generally by newer coating or adsorption technologies.

Alternate methodologies for surface modification of nanoparticles are based in chemical modifications of either the nanoparticle surface itself, or follow attachment of molecules, for instance synthetic or natural polymers, that lend themselves to easy chemical modification. For example, natural polysaccharides such as chitosan, alginate, and heparin, natural polypeptides such as collagen, and protein-based polymers such as gelatin can be either adsorbed by physical or chemical means on to nanoparticles or can be chemically modified to achieve flexibility in nanoparticle design.

14.8.1 Protein Nanoparticles

Nanoparticles formed from proteins and used for therapeutic purposes offer a number of advantages over synthetic nanoparticles. These include biocompatibility, biodegradation, reduced toxicity, and easily obtained bioavailability. Another major advantage of protein-based nanoparticles is that they are more amenable to further surface modification by chemical or physical adsorption allowing the achievement of site-specific delivery and drug targeting. This is made possible by the attachment of specific antibodies and other biologically active molecules as needed. Protein nanoparticles are usually made from one of the following materials: albumin, gelatin, collagen, or silk proteins. Many of these afford surface modification possibilities and are discussed in detail below.

14.8.1.1 Albumin-based Nanoparticles and Surface Modification

Albumin is a naturally occurring protein, commonly found in human plasma, where it helps to transport various drugs in the bound state. The use of albumin for the construction of nanoparticles imparts the properties of biocompatibility due to low immunogenic potential, biodegradability, and bioavailability. Interestingly, albumin can be used both as base material for the construction of nanoparticles, and also for the surface modification of nanoparticles formed from other polymers. Albumin used for surface modification helps to increase hydrophilic behavior and thus solubility characteristics of a number of hydrophobic nanoparticle systems. This in turn improves their bioavailability and uptake. Manoocheri *et al.* coated the surface of nanoparticles formed of poly lactic-co-glycolic acid (PLGA) with human serum albumin. Although PLGA particles are emerging as important nanovectors carrying drug cargo for therapeutic intervention, the system suffers from hydrophobicity that restricts bioavailability and systemic transport. Human serum albumin-based surface modification of docetaxel-loaded PLGA nanoparticles resulted in improvement in three key areas: (1) provision of a hydrophilic surface due to surface adsorption of albumin increasing bioavailability and (2) extravasation and accumulation of the nanoparticles in the leaky blood vessels of the tumorous tissues. The authors concluded that surface modification of PLGA nanoparticles through HSA conjugation resulted in more cytotoxicity against tumor cell lines compared with free docetaxel and unconjugated PLGA nanoparticles [28]. In a similar fashion, Mao *et al.* demonstrated an increase in site-specific delivery of nanoparticles to the liver by conjugating the biological molecule glycyrrhizin to the surface of albumin nanoparticles [29].

Conversely, nanoparticles can also be designed using albumin, and subsequently the surface decorated with biologically active molecules to achieve targeted delivery. Bae *et al.* constructed the anticancer drug doxorubicin enclosing albumin nanoparticles and conjugated tumor necrosis factor α (TNF-α)-related apoptosis inducing ligand (TRAIL), and the protein transferrin, onto the surface of this nanoparticle. This "double-tagging" resulted in an enhanced uptake of the nanoparticles by cancerous tissue in their rat model [30]. Another application of surface-modified albumin nanoparticles in medicine is their use for the delivery of drug or diagnostic dye cargo to regions of the body rationally considered difficult to reach in the context of drug distribution. A well-known example of this is delivery across the extremely selective blood–brain barrier (BBB) into the brain. Michaelis *et al.* provided an effective solution to this problem by designing surface-modified albumin nanoparticles carrying an embargo of loperamide. These nanoparticles contained the lipoprotein molecule apolipoprotein E conjugated by biotinylation onto the surface of the nanoparticle. This was made possible by one of two means: chemical modification of purified nanoparticles using the cross-linker NHS-PEG3400-Mal, which opened up some chemical groups on the nanoparticle surface, making it easy to tag apolipoprotein E. Alternatively, biotin–avidin molecules were directly conjugated onto the surface of the nanoparticles. The authors provided evidence that regardless of the method used for surface modification, the conjugation of apolipoprotein E to the surface of albumin nanoparticles enhanced the delivery of loperamide to the brain in an experimental animal model [31].

The different approaches described above serve to demonstrate the ease of surface manipulation and the flexibility of use of albumin nanoparticles.

14.8.1.2 Collagen-based Nanoparticles and Surface Modification

Collagen is a connective tissue protein in mammals and is found in the greatest abundance among all proteins in humans, being an important component of bones and cartilage. Due to its presence in the human body, using this protein for the construction of nanoparticles offers the distinct advantage of low immunogenicity and reduced rejection response. Additionally, the chemistry and physical nature of the protein, namely that it forms a triple helix fiber, makes it easy to modify the surface of nanoparticles made from collagen. Among the other proteins that have been conjugated to collagen nanoparticles, fibronectin, glycosaminoglycans, and elastin are commonly employed. The addition of these proteins not only aids to augment physical and chemical properties of collagen nanoparticles but also improved their pharmacokinetic behavior including biodegradability [32].

As discussed for the protein albumin, the converse use of collagen biopolymers to coat the surface of nanoparticles composed of other materials is another use of collagen. As mentioned, the triple-helix nature of collagen makes it amenable for further modification, a fact that can be advantageously manipulated for the construction of site-specific or targeted-release nanoparticles. Aime *et al.* demonstrated that collagen triple helices can be confined onto the surface of sulfonate-modified silica nanoparticles for the construction of bio-hybrid networks. The authors noted that the self-assembly property of collagen further aided in the manufacturing process [33]. Additionally, Kim *et al.* successfully demonstrated the flexibility of use of collagen in the surface modification of polycaprolactone-based matrix for ultimate use in bone grafting in skulls using an experimental rat model [34].

14.8.1.3 Gelatin-based Nanoparticle Systems

Gelatin is a macromolecule derivative of collagen, obtained by partial heating and hydrolysis. It is a very popular material for use in pharmaceutical applications due to almost absent irritability, excellent biocompatibility, and lack of cytotoxicity when used in formulation design. Importantly, gelatin naturally contains chemical groups that can be activated or modified allowing the addition of multiple functional groups. Thus, surface modification of gelatin nanoparticles is a commonly used technique that results in the synthesis of nanoparticles for use in various medical applications. It is important to note that generation from collagen results in the formation of two types of gelatin, known as gelatin A and B and both may be employed for surface modification, depending on the final need.

14.8.2 Polysaccharide-based Nanoparticle Systems

Polysaccharides are polymers of carbohydrates, in which multiple monosaccharide units are chemically bonded to create a long molecule. Polysaccharides can be classified as cationic, anionic, or nonionic, based on their ability to form charged particles.

14.8.2.1 The Use of Alginate for Surface Modifications

Alginate is a linear, unbranched polysaccharide that adheres to mucous membranes and is biocompatible, with excellent immunogenic tolerance. Derived from seaweed, it has a number of chemical groups, such as hydroxyl and carboxyl functional groups, that may be "activated" and which in turn enhance interaction with proteins and other chemical moieties on cell surfaces in the body. It is a polymer of uronic acid but is composed of three distinct combinations of this monomeric unit. Alginate-based nanoparticles were first designed for the delivery of oligonucleotides (gene sequences) to cells in the much anticipated genetherapy [35]. Similar to the proteins discussed above, alginate can be used both for the manufacture of nanoparticles, as well to modify the surface of nanoparticles and other pharmaceutical constructs. Kodiyan *et al.* took advantage of the multiple functional chemical groups on alginate to create a gold–nanoparticle construct with enhanced interaction with cell surfaces. Their invention was to employ the cysteine activated alginate polysaccharide to stabilize gold nanoparticles instead of the more commonly employed poly(ethylene) glycol, with added advantage of cell interaction. The authors reported augmented stability and cellular interaction of the alginate-coated instead of the PEGylated gold nanoparticles [36]. Jain and Amiji constructed a sodium alginate nanoparticle system, where induction of gelation was the method of choice for the encapsulation of DNA. They conjugated the peptide-sequence tuftsin onto the surface of the alginate nanoparticles and demonstrated greater delivery to macrophage cells due to this surface enhancement. They reported a successful delivery of gene sequences to the designated population of cells using their surface-enhanced alginate-based nanoparticle system [37].

An interesting study employing alginate will be briefly mentioned here to emphasize the flexibility of application of coating techniques to various pharmaceutical platforms, not only nanoparticles. Mahlicli and Altinkaya demonstrated the versatility of polysaccharide coatings, in particular alginate, for surface adsorption of bioactive molecules, heparin in this specific instance. Interestingly, instead of spherical nanoparticles, they deposited alginate–heparin along with synthetic polymers onto polysulfone membranes and employed these in animal models to prevent thrombosis by the creation of a blood-compatible, heparinized, uncoagulating surface [38].

14.8.2.2 The Use of Chitosan-based Nanoparticles and Chitosan-based Surface Modification

Natural repair and regeneration processes for the restoration of dermal and epidermal tissues include dynamic and complex processes like hemostasis, inflammation, and proliferation for restoration of normal conditions, while simultaneously combating possible complications that might occur due to compounding external factors. Examples of possible complications include infections that might extend to adjacent tissues or internal organs, disrupted skin formation, and so on. As an important and primary protection against infection, both human skin and man-made biomaterials mimicking its function must possess characteristics such as adequate gas exchange capability, adequate moist environment, protection against infection, permeability to necessary nutritional products and necessary regulatory agents like cytokines and growth factors, biocompatibility, non-toxicity, and elasticity. It is in this context of dermal-use biomaterials in regenerative medicine, implantable materials, and scaffolds for tissue engineering that chitosan products are in the forefront of scientific studies and development [39].

Multiple material characteristics of the chitosan polymer make it desirable for use as an anchorage-based surface modification polymer. Some of these advantages include biocompatibility, biodegradability, and non-toxicity. This natural macromolecule has a wide range of medical and pharmaceutical uses because of its bactericidal, fungicidal, and antitumor properties [40]. N-deacetylation of chitin, the most naturally abundant polymer after cellulose, is the widely used method of producing chitosan. Studies have shown that free-standing membranes can be used in proliferation of skeletal myoblasts for osteoinductivity [41]. Cross-linked polysaccharides alginate (ALG) and chitosan (CHI) using 1-ethyl-3-(3-dimethylaminopropyl (EDC) carbodiimide hydrochloride) was employed to make free standing membrane utilizing layer-by-layer (LBL) technique. This resultant membrane was loaded with osteoinductive growth factor BMP-2 (bone morphogenetic protein 2) to witness growth factor release over a month by diffusion.

Chitosan can be used as a DNA carrier where gene delivery is needed. It protects DNA degradation by nuclease by condensing nucleic acid to stable complexes with diameter between 100 and 250 nm. Chitosan's adhesive and transport properties in GI tract make it an ideal oral gene carrier.

Studies have shown [42] that chitosan can induce cell apoptosis by directly acting on tumor cells and inhabiting cell growth and metabolism. For example, chitosan has been used in treatment of bladder tumor. It also shown to have antitumor activity because of its ability to improve body's immune function by inducing cytokines and chemokines. Chitosan induces both cell-mediated and humoral immune responses when is used in subcutaneous vaccination. Therefore, chitosan or its derivatives are suitable adjuvants for immunotherapy.

Chitosan cross-linked with nanoparticles can be used as drug carriers. *In vivo* this complex of chitosan nanoparticle is capable of crossing biological barriers like BBB that makes it an ideal candidate to deliver drugs to the site of intended treatment [43]. Porous morphological and microspheres structure of chitosan enable it to adsorb more protein as compared to ones without a poly porous structure. Studies shown that structures with high porosity have higher adsorption ability. For instance chitosan scaffold along with load of rhBMP-2 initially resulted in a minimal burst release which followed by moderate release which makes it is desirable scaffold and excellent carrier of rhBMP-2. The silica and chitosan combination has shown to be an osteoconductive and in control of growth factors release behavior over an extended time frame. This hybrid showed a sustained release of the growth factors, which lasted weeks without any initial burst. These new designs, concepts, and various techniques used in devolving scaffold for regeneration of bone tissue promises a better outcomes for patients, as scientists are examining to include proteins and growth factors within scaffolds in order to make them osteoinductive.

14.8.2.3 The Use of Chitin-based Nanoparticles and Chitin-based Surface Modification

During the past two decades, fabrication of chitin nanoparticles has garnered many studies of their potential biomedical applications. Chitin is a long-chain, amide-containing N-acetylglucosamine derivative of glucose and is second to cellulose as the most abundant biopolymer found in nature. It is a widely distributed, natural component of the exoskeletons of arthropods such as crabs, prawns, and insects as well as the cell walls of fungi. Chitin has very high availability and low cost because it is mined during numerous industrial processes at a rate of 10–100 billion tons per year [44, 45]. Chitinous biopolymers have a high utility as nanofibers and nanowhiskers due to their high surface area, high porosity, and highly crystalline structures. Pure chitin is extracted from shells by several physical and chemical assembly techniques which include mechanical disintegration in a top-down, electrospinning process, molecular self-assembly, phase separation techniques, microcontact printing by dissolving in 1,1,1,3,3,3-hexafluoro-2-propano (HFIP), mechanical treatment in a grinder following acidic or basic conditions, ultrasonication of β-chitan under acidic conditions, and 2,2,6,6-tetramethylpiperidinooxy (TEMPO)-mediated oxidation of β-chitin [46].

Chitinous biopolymers have several properties that make them advantageous as bio-based polymers unlike metals and other nonbiological-based composite material [47]. For example, chitin is renewable; that is, they are made of natural materials. They also have excellent functionality due to their biodegradability, nontoxic, and odorless. Chitin has poor solubility in aqueous solutions and various organic solvents; therefore, surface modifications must be done to increase its bioapplicablity. They are also highly biocompatible with exceptional ability to bind to cells for 3D tissue-engineering scaffolds designed for skin, bone, and cardiac tissue and for 2D wound dressing applications [48, 49]. Chitinous tissue material is also biodegradable; thus, after the wound therapeutic process, native tissue can replace chitin biomaterial [50]. Chitin is also well documented for its antimicrobial properties that inhibit the Gram-positive and negative bacteria, yeast, and fungi [51].

In the past decade, there have been numerous studies investigating the introduction of permanent chemical modifications to the surface of chitin to formulate nanoparticles with biological functions. Notably, Mendes *et al.* [52] recently demonstrated that combining chitin with plasmid DNA resulted in the aggregation of the chitin to the DNA surface. When the aggregate is placed in agar containing *Escherichia coli*, the polymer is absorbed through sliding friction, known as the Yoshida effect, which results in a transformation of the *E. coli* and a novel way to introduce genes into bacteria. Also, Muzzarelli *et al.* showed that oxidizing the surface of chitin with sodium hypochlorite in the presence of TEMPO produces beads that precipitate proteins, particularly lysozymes and hydrolases [53].

Extensive scientific research into modified chitin has shown the remarkable properties of chitin in tissue engineering. Early studies demonstrated that chitin-based hydrogel tubes can be fabricated as biodegradable nerve grafts for tissue engineering for the nervous system [54]. Also, nanofibers fabricated from silk fibroin, chitin, and osteogenic and morphogenetic protein-2 (BMP-2) produced a core-sheath that showed excellent potential in tissue engineering and drug release [55].

14.8.2.4 *The Use of Cellulose-based Nanoparticles and Cellulose-based Surface Modification*

Cellulose is a polysaccharide comprised of a linear chain extending up to ten thousand D-glucose monomers linked in a (1→4) ester linkage. It is isolated from plant cell walls. It is categorically the most abundant organic biopolymer in Nature, with approximately 1.5 trillion tons produced each year [56]. Therefore, it presents a low-cost, inexhaustible raw material for nanoparticle bio-applications. However, cellulose nanoparticles have several noted disadvantages such as high water absorbing ability, poor wettability, incompatibility with other polymeric materials, and limited heat capacity [57]. Moreover, although cellulose has been used as a skin substitute in treating extensive burns, it does not have antimicrobial properties as does other biopolymers like chitin and chitosan [58].

There are three different types of nanocellulose investigated for their potential biomedical applications, which are nanocrystals, nanofibrils, and bacterial cellulose [59]. Cellulose is commonly modified on its surface by acetate anhydride to form cellulose acetate.

Several noteworthy investigations show the biocompatible and bioactivity properties of functionalized cellulose in tissue engineering. Lee *et al.* formed regenerated cellulose by modifying carboxymethyl cellulose into poly(ethylene oxide) (CMC-PEO) that produced porous films that performed well in *in vitro* adhesion for tissue engineering [60]. Cellulose-modified cardiac tissue engineering was recently demonstrated using bacterial cellulose grafts produced nonthrombogenic and noninflammatory-inducing scaffold for a three-layered vascular wall with the potential to be employed in cardiovascular surgery [61]. Also, Novotna *et al.* study fabricated 6-carboxycellulose further functionalized with arginine and chitosan and were seeded with vascular smooth muscle cells with applications in vascular prostheses [62]. To evaluate the controlled delivery of ampicillin and gentamycin, Kaplan *et al.* formulated bacterial cellulose membranes that demonstrated prolonged drug release in *E. coli, Enterococcus feacalis, Streptococcus aureus, Staphylococcus epidermidis,* and *Pseudomonas aeruginosa* up to days after injection. Also, a novel study recently demonstrated that NO_2-mediated oxidized cellulose powder produced a biomaterial that generated a bactericidal effect against *S. aureus* and demonstrated rapid blood coagulation for hemorrhage uses [63].

Other noteworthy applications for cellulose are those with potential as drug excipient and/or as a drug delivery particle. Nanocrystalline cellulose crystallites were prepared by acid hydrolysis and successfully bound to tetracycline and doxorubicin that were absorbed into KU-7 cells and were released during a period of 1 day [64].

References

1. Murphy S.L., Xu J.Q., Kochanek K.D. Deaths: Final data for 2010. *Natl. Vital Stat. Rep.* 61(4), 2013. http://www.cdc.gov/nchs/data/nvsr/nvsr61/nvsr61_04.pdf.
2. Heidenreich P.A., Trogdon J.G., Khavjou O.A., *et al.* Forecasting the future of cardiovascular disease in the United States: a policy statement from the American Heart Association. *Circulation* 123: 933–44, 2011.
3. Virmani R., Farb A. Pathology of in-stent restenosis. *Curr. Opin. Lipidol.* 10, 499–506, 1999.
4. Yao E.H., Fukuda N., Ueno T., Matsuda H., Nagase H., Matsumoto Y., *et al.* A pyrrole-imidazole polyamide targeting transforming growth factor-beta1 inhibits restenosis and preserves endothelialization in the injured artery. *Cardiovasc. Res.* 81(4), 797–804, 2009.
5. Cohen D.J., Bakhai A., Shi C., Githiora L., Lavelle T., Berezin R.H. *et al.* Cost effectiveness of sirolimus-eluting stents for treatment of complex coronary stenoses: results from the Sirolimus-Eluting Balloon Expandable Stent in the Treatment of Patients with De Novo Native Coronary Artery Lesions (SIRIUS) trial. *Circulation* 110, 508–14, 2004.
6. Regar E., Serruys P.W., Bode C., Holubarsch C., Guermonprez J.L., Wijns W. *et al.* Angiographic findings of the multicenter Randomized Study with the Sirolimus-Eluting Bx Velocity Balloon-Expandable Stent (RAVEL): sirolimus-eluting stents inhibit restenosis irrespective of the vessel size. *Circulation.* 106, 1949–56, 2002.

7. Colombo A., Drzewiecki J., Banning A., Grube E., Hauptmann K., Silber S. et al. Randomized study to assess the effectiveness of slow- and moderate-release polymer-based paclitaxel-eluting stents for coronary artery lesions. Circulation. 108, 788, 2003.
8. Park S.J., Shim W.H., Ho D.S., Raizner A.E., Park S.W., Hong M.K. et al. A paclitaxel-eluting stent for the prevention of coronary restenosis. N. Engl. J. Med. 348, 1537–45, 2003.
9. Gupta A., Majumdar P., Amit J., Rajesh A., Singh S.B., Chakraborty M. Cell viability and growth on metallic surfaces: in vitro studies. Trends Biomater. Artif. Organs. 20, 84–9, 2006.
10. Prasad C.K., Krishnan L.K. Regulation of endothelial cell phenotype by biomimetic matrix coated on biomaterials for cardiovascular tissue engineering. Acta Biomater. 4, 182–91, 2008.
11. Oh B.K., Meyerhoff M.E. Catalytic generation of nitric oxide from nitrite at the interface of polymeric films doped with lipophilic Cu(II)-complex: a potential route to the preparation of thromboresistant coatings. Biomaterials 25, 283–93, 2004.
12. Lee H., Dellatore S.M., Miller W.M., Messersmith P.B. Mussel-inspired surface chemistry for multifunctional coatings. Science. 318, 426–30, 2007.
13. Weng Y.J., Song Q., Zhou Y.J., Zhang L.P., Wang J., Chen J.Y., et al. Immobilization of selenocystamine on TiO_2 surfaces for in situ catalytic generation of nitric oxide and potential application in intravascular stents. Biomaterials. 32, 1253–63, 2011.
14. Zhou Y.J., Weng Y.J., Zhang L.P., Jing F.J., Huang N. Cystamine immobilization on TiO_2 film surfaces and the influence on inhibition of collagen-induced platelet activation. Appl. Surf. Sci. 258, 1776–83, 2011.
15. Andres J.L., DeFalcis D., Noda M., Massagué J. Binding of two growth factor families to separate domains of the proteoglycan betaglycan. J. Biol. Chem. 267(9), 5927–30, 1992.
16. Rhee S.H., Tanaka J. Self-assembly phenomenon of hydroxyapatite nanocrystals on chondroitin sulfate. J. Mater. Sci. Mater. Med., 13(6), 597–600, 2002.
17. Rammelt S., Heck C., Bernhardt R., Bierbaum S., Scharnweber D., Goebbels J., et al. In vivo effects of coating loaded and unloaded Ti implants with collagen, chondroitin sulfate, and hydroxyapatite in the sheep tibia. J. Orthop. Res. 25(8), 1052–61, 2007.
18. Bierbaum S., Douglas T., Hanke T., Scharnweber D., Tippelt S., Monsees T.K., et al. Collageneous matrix coatings on titanium implants modified with decorin and chondroitin sulfate: characterization and influence on osteoblastic cells. J. Biomed. Mater. Res. A. 77(3), 551–62, 2006.
19. Geissler U., Hempel U., Wolf C., Scharnweber D., Worch H., Wenzel K. Collagen type I-coating of Ti_6Al_4V promotes adhesion of osteoblasts. J. Biomed. Mater. Res. 51(4), 752–60, 2000.
20. Roehlcke C., Witt M., Kasper M., Schulze E., Wolf C., Hofer A., et al. Synergistic effect of titanium alloy and collagen type I on cell adhesion, proliferation and differentiation of osteoblast-like cells. Cells Tissues Organs. 168, 178–87, 2001.

21. Bernhardt R., van den Dolder J., Bierbaum S., Beutner R., Scharnweber D., Jansen J., Beckmann F., Worch H. Osteoconductive modifications of Ti-implants in a goat defect model: characterization of bone growth with SR muCT and histology. *Biomaterials* 26(16), 3009–19, Jun 2005.
22. Pierschbacher M.D., Ruoslahti E. Cell attachment activity of fibronectin can be duplicated by small synthetic fragments of the molecule. *Nature* 309(5963), 30–3, 1984.
23. Horton M.A. Interactions of connective tissue cells with the extracellular matrix. *Bone* 17(Suppl), 51S–3S, 1995.
24. Rammelt S., Illert T., Bierbaum S., Scharnweber D., Zwipp H., Schneiders W. Coating of titanium implants with collagen, RGD peptide and chondroitin sulfate. *Biomaterials* 27(32), 5561–71, 2006.
25. Ferris D.M., Moodie G.D., Dimond P.M., Gioranni C.W., Ehrlich M.G., Valentini R.F. RGD-coated titanium implants stimulate increased bone formation *in vivo*. *Biomaterials* 20, 2323–31, 1999.
26. Elmengaard B., Bechtold J.E., Søballe K. *In vivo* effects of RGD-coated titanium implants inserted in two bone-gap models. *J. Biomed. Mater. Res.* A. 75, 249–55, 2005.
27. Hennessy K.M., Clem W.C., Phipps M.C., Sawyer A.A., Shaikh F.M., Bellis S.L. The effect of RGD peptides on osseointegration of hydroxyapatite biomaterials. *Biomaterials* 29(21), 3075–83, 2008.
28. Manoochehri S., Darvishi B., Kamalinia G., Amini M., Fallah M., Ostad S.N., *et al.* Surface modification of PLGA nanoparticles via human serum albumin conjugation for controlled delivery of docetaxel. *Daru.* 21(1), 58, 2013.
29. Mao S.J., Hou S.J., He R., Zhang L.K., Wei D.P., Bi Y.Q. *et al.* Uptake of albumin nanoparticle surface modified with glycyrrhizin by primary cultured rat hepatocytes. *World J. Gastroenterol.* 11(20), 3075–79, 2005.
30. Bae S., Ma K., Kim T.H., Lee E.S., Oh K.T., Park E.S., *et al.* Doxorubicin-loaded human serum albumin nanoparticles surface-modified with TNF-related apoptosis-inducing ligand and transferrin for targeting multiple tumor types. *Biomaterials* 33(5), 1536–46, 2012.
31. Michaelis K., Hoffmann M.M., Dreis S., Herbert E., Alyautdin R.N., Michaelis M., *et al.* Covalent linkage of Apolippoprotein E to albumin nanoparticles strongly enhances drug transport into the brain. *J. Pharmacol. Exp. Ther.* 317(3), 1246–1253, 2006.
32. Barbani N., Giusti P., Lazzeri L., Polacco G., Pizzirani G. Bioartifical materials based on collagen: 1. Collagen cross-linking with gaseous glutaraldehyde. *J. Biomater. Sci. Polym. Ed.* 7(6), 461-9, 1995.
33. Aimé C., Mosser G., Pembouong G., Bouteiller L., Coradin T. Controlling the nano–bio interface to build collagen–silica self-assembled networks. *Nanoscale* 4, 7127–34, 2012.
34. Kim J.H., Linh N.T., Min Y.K., Lee B.T. Surface modification of porous polycaprolactone/biphasic calcium phosphate scaffolds for bone regeneration in rat calvaria defect. *J. Biomater. Appl.* 29(4), 624-35, 2014.

35. Aynie I.C., Vauthier C., Fattal E., Foulquier M., Couvreur P. Alginate nanoparticles as a novel carrier for antisense oligonucleotide, in: J.E. Diederichs, R. Muler (Eds.), *Future Strategies of Drug Delivery With Particulate Systems*, Med-pharm Scientific Publisher, Stuttgart, pp. 5–10, 1998.
36. Kodiyan A., Silva E.A., Kim J., Aizenberg M., Mooney D.J. Surface modification with alginate-derived polymers for stable, protein-repellent, long-circulating gold nanoparticles. *ACS Nano.* 6(6), 4796–805, 2012.
37. Jain S., Amiji M. Tuftsin-modified alginate nanoparticles as a noncondensing macrophage-targeted DNA delivery system. *Biomacromolecules* 13(4), 1074–85, 2012.
38. Mahlicli F., Altinkaya S. Surface modification of polysulfone based hemodialysis membranes with layer by layer self-assembly of polyethyleneimine/alginate-heparin: a simple polyelectrolyte blend approach for heparin immobilization. *J. Mater. Sci. Mater. Med.* 24(2), 533, 2013.
39. Ahamed M.N., Sankar S., Kashif P.M., Basha S.H., Sastry T.P. Evaluation of biomaterial containing regenerated cellulose and chitosan incorporated with silver nanoparticles. *Int. J. Biol. Macromol.* 72, 680–6, 2015.
40. Gibot L., Chabaud S., Bouhout S., Bolduc S., Auger F.A., Moulin V.J. Anticancer properties of chitosan on human melanoma are cell line dependent. *Int. J. Biol. Macromol.* 72, 370–9, 2015.
41. Caridade G.S., Monge C., Almodovar J., Guillot R., Lavaud J., Josserand V., et al. Myoconductive and osteoinductive free standing polysaccharide membranes. *Acta Biomaterialia.* 15, 139–149, 2015.
42. Romagnoli C., D'Asta F., Brandi M.L. Drug delivery using composite scaffolds in the context of bone tissue engineering. *Clin. Cases Miner. Bone. Metab.* 10(3), 155–61, 2013.
43. Bolhassani A., Javanzad S., Saleh T., Hashemi M., Aghasadeghi M.R., Sadat S.M. Polymeric nanoparticles: Potent vectors for vaccine delivery targeting cancer and infectious diseases. *Hum. Vaccin. Immunother.* 10(2), 321–32, 2014.
44. Ifuku S. Chitin and chitosan nanofibers: Preparation and chemical modifications. *Molecules* 19(11), 18367–80, 2014.
45. Ifuku S., Saimoto H. Chitin nanofibers: preparations, modifications, and applications. *Nanoscale* 4(11), 3308–18, 2012.
46. Ding F., Deng H., Du Y., Shi X., Wang Q. Emerging chitin and chitosan nanofibrous materials for biomedical applications. *Nanoscale* 6(16), 9477–93, 2014.
47. Dash M., Chiellini F., Ottenbrite R.M., Chiellini E. Chitosan—A versatile semi-synthetic polymer in biomedical applications. *Prog. Polym. Sci.* 36(8), 981–1014, 2011.
48. Jayakumar R., Prabaharan M., Kumar P.S., Nair S.V., Tamura H. Biomaterials based on chitin and chitosan in wound dressing applications. *Biotechnol. Adv.* 29(3), 322–37, 2011.
49. Azuma K., Izumi R., Osaki T., Ifuku S., Morimoto M., Saimoto H., et al. Chitin, chitosan, and its derivatives for wound healing: old and new materials. *J. Funct. Biomater.* 6(1), 104–42, 2015.

50. Croisier F., Jérôme C. Chitosan-based biomaterials for tissue engineering. *Eur. Polym. J.* 49(4), 780–92, 2013.
51. Ignatova M., Manolova N., Rashkov I. Novel antibacterial fibers of quaternized chitosan and poly (vinyl pyrrolidone) prepared by electrospinning. *Eur. Polym. J.* 43(4), 1112–22, 2007.
52. Mendes G.P., Vieira P.S., Lanceros-Mendez S., Kluskens L.D., Mota M. Transformation of Escherichia coli JM109 using pUC19 by the Yoshida effect. *J. Microbiol. Methods.* 115, 1–5, 2015.
53. Muzzarelli R.A., Mehtedi M.E., Mattioli-Belmonte M. Emerging biomedical applications of nano-chitins and nano-chitosans obtained via advanced eco-friendly technologies from marine resources. *Marine Drugs.* 12(11), 5468–502, 2014.
54. Freier T., Montenegro R., Koh H.S., Shoichet M.S. Chitin-based tubes for tissue engineering in the nervous system. *Biomaterials.* 26(22), 4624–32, 2005.
55. Park K.E., Jung S.Y., Lee S.J., Min B.M., Park W.H. Biomimetic nanofibrous scaffolds: preparation and characterization of chitin/silk fibroin blend nanofibers. *Int. J. Biol. Macromol.* 38(3), 165–73, 2006.
56. Kim J., Yun S., Ounaies Z. Discovery of cellulose as a smart material. *Macromolecules.* 39(12), 4202–6, 2006.
57. Siqueira G., Bras J., Dufresne A. Cellulose whiskers versus microfibrils: influence of the nature of the nanoparticle and its surface functionalization on the thermal and mechanical properties of nanocomposites. *Biomacromolecules.* 10(2), 425–32, 2009.
58. Czaja W., Krystynowicz A., Bielecki S., Brown R.M. Microbial cellulose the natural power to heal wounds. *Biomaterials.* 27(2), 145–51, 2006.
59. Lin N., Dufresne A. Surface chemistry, morphological analysis and properties of cellulose nanocrystals with gradiented sulfation degrees. *Nanoscale.* 6(10), 5384–93, 2014.
60. Lee S.Y., Bang S., Kim S., Jo S.Y., Kim B.C., Hwang Y., et al. Synthesis and in vitro characterizations of porous carboxymethyl cellulose-poly (ethylene oxide) hydrogel film. *Biomater. Res.* 19(1), 12, 2015.
61. Scherner M., Reutter S., Klemm D., Sterner-Kock A., Guschlbauer M., Richter T., et al. In vivo application of tissue-engineered blood vessels of bacterial cellulose as small arterial substitutes: proof of concept? *J. Surg. Res.* 189(2), 340–7, 2014.
62. Novotna K., Havelka P., Sopuch T., Kolarova K., Vosmanska V., Lisa V., et al. Cellulose-based materials as scaffolds for tissue engineering. *Cellulose.* 20(5), 2263–78, 2013.
63. Kaplan E., Ince T., Yorulmaz E., Yener F., Harputlu E., Laçin N.T. Controlled delivery of ampicillin and gentamycin from cellulose hydrogels and their antibacterial efficiency. *J. Biomater. Tissue Eng.* 4(7), 543–9, 2014.
64. Jackson J.K., Letchford K., Wasserman B.Z., Ye L., Hamad W.Y., Burt H.M. The use of nanocrystalline cellulose for the binding and controlled release of drugs. *Int. J. Nanomed.* 6, 321, 2011.

15
Stem Cell Behavior on Microenvironment Mimicked Surfaces

M. Özgen Öztürk Öncel and Bora Garipcan*

Institute of Biomedical Engineering, Boğaziçi University, Istanbul, Turkey

Abstract

The development of novel biomaterials to control over stem cell behavior is holding a promising hope for regenerative medicine and tissue-engineering fields. As stem cells are very useful tools in tissue regeneration, drug investigation studies, and cell-based therapies, the control of their cellular behavior and selective differentiation outside of the body is a great challenge. To do that, synthetic biomaterials with different properties have been synthesized. In some studies, researchers do not show sufficient attention on the natural physicochemical parameters of the cells. However, cells regulate their own physiology such as adhesion, viability, proliferation, and differentiation, according to the signals coming from their microenvironment, which means that simple biomaterials are providing only support for the cells and do not mimic these important signals between the extracellular matrix (ECM), and cells cannot promote functional cellular physiology. The recent research on this area is therefore more focused on the preparation of cell microenvironment-like novel biomaterials with bioinspired and biomimicked approaches. With the novel biomaterials prepared by using these approaches, natural-like microenvironment of stem cells and the targeted differentiated cells are obtained in mechanical, topographical, and biochemical manners. Building novel biomaterials similar to the own microenvironment of cells mimics the original interaction mechanism between cells and their ECM. Thus, culturing and differentiating stem cells on these substrates enhance *in vivo*-like conditions, which promotes their natural cellular behavior and selective differentiation.

Keywords: Stem cells, biomaterials, cellular behavior, Biomimicked and bioinspired materials

---

Corresponding author: bora.garipcan@boun.edu.tr

15.1 Introduction

It is very fascinating that, many parts of the human body have the ability to heal themselves in certain cases [1]. Highly complex processes are happening inside the body to maintain the original internal environment and homeostasis. Cells, which are the fundamental units of life, dynamically monitor and work to keep up this balance, by secreting some molecules or replicating themselves to replace the destroyed areas. However, this natural self-regeneration process is limited and due to traumas or diseases. This balance may be damaged, causing the *functio laesa* "loss of tissues or organ functions" [2, 3]. Nowadays, successful treatments for such excessive tissue loss are done by the use of mechanical devices and artificial prostheses which do not integrate to the original tissue. Another treatment is the organ transplantation, which has the challenges of suitable organ deficiency, organ rejection, or the lifelong immunosuppressive treatment. Thus, tissue-engineering and regenerative medicine fields have attracted much interest among the scientists whereby tissues or organs can be regenerated with the aid of specific novel biomaterials and differentiated or/and undifferentiated cells. The developed tissue substitutes are not only important for repairing damaged body parts, but also for the drug design studies. Introducing novel medication to functional tissues would accelerate the research on new drugs and personalized medicine [4].

Tissue engineering is a multidisciplinary field, which is mainly the development of tissue substitutes *in vivo* or *in vitro* by controlling the biological, chemical, mechanical, and physical parameters of a specific cell culture with appropriate scaffolds and biomolecules [1]. In tissue engineering, cells are seeded into scaffolds, which behave as templates for tissue regeneration, and then these compounds are either cultured *in vitro* to be implanted later or implanted directly to the damaged sites to repair the tissues *in vivo*. The interaction of seeded cells and scaffolds ultimately determines the cellular behavior, so the choice of a cell type with an appropriate scaffold is crucial for the development of viable tissue constructs [5].

Stem cells are very useful tools in tissue regeneration with their differentiation and proliferation capacities. However, it is a great challenge for the researchers to utilize these benefits of stem cells and control their behavior in *in vitro* culture conditions [6]. In order to overcome this challenge and to increase the success of the regeneration, scaffold properties are very important, as they provide support to the stem cells and promote them to proliferate, differentiate, and form an extracellular matrix (ECM) [7–10].

All cell types in the body are located in their natural scaffolds, known as microenvironment [11, 12]. Different biochemical and physical signals coming from their microenvironment regulate cellular behavior directly or indirectly [13]. Specific microenvironment of stem cells dynamically provides the proliferation, self-renewal, and differentiation abilities [6, 14]. Surface-bound signaling factors, cell–cell interactions, biochemical, mechanical, and topological cues create specific stem cell microenvironments which can be sensed by stem cells and regulate their fate [15–17]. In order to provide these natural interactions, scaffolds designed to support stem cells temporary must mimic the biochemical, mechanical, and topological signals of stem cell microenvironment.

15.2 Stem Cells

15.2.1 Definition and Types

Stem cells are undifferentiated cells, which are capable of self-renewal, proliferation, and differentiation into specialized cells. They act as a repair system in the body. Each time a stem cell divides, two new cells both have the potential to differentiate into a special cell type or remain as an undifferentiated stem cell. Due to these unique properties, stem cells have attracted much attention in regenerative medicine and tissue-engineering fields that they will take part in the treatment of incurable diseases and organ loss. New research and therapies in these fields are mainly focused on a better understanding of the natural mechanisms of stem cells and the control and regulation of their behavior *in vivo* or *in vitro* [18].

During the development and the whole lifetime of human body, many different stem cell types are assigned to produce new stem cells and at the same time differentiated and functional cells, such as bone, cardiac, nerve, and skin cells (Figure 15.1).

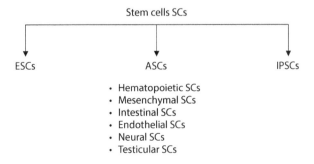

Figure 15.1 Types of stem cells.

Stem cells are classified into three general types: embryonic stem cells (ESCs), adult stem cells (ASCs), and induced pluripotent stem cells (IPSCs). The main difference between them is that ESCs can produce the three embryonic germ layers, as ectoderm, endoderm, and mesoderm where ASCs can only differentiate into specific cell types from the tissue in which they are found. IPSCs, on the other hand, are produced by reprogramming adult cells to express specific genes which provide them to differentiate into many cell types [19, 20].

15.2.1.1 Embryonic Stem Cells

In the first stage of human body's development, zygote, which is a totipotent cell, divides continuously. This specific cell has the ability to build the whole body, including placenta. After 3–5 days, it forms a hollow ball of cells called blastocyst. The blastocyst adheres to the wall of the uterus and developed into embryo. The outer cells of blastocyst generate placenta. The inner cell mass of the blastocyst is where ESCs are derived [18, 21]. ESCs are pluripotent, meaning that they are capable of producing all specialized cell types [20].

ESCs are very promising cell sources in tissue engineering to provide tissue regeneration and also for drug testing and other cellular therapies. One of the unique and superior characteristics of ESCs is that they can preserve their undifferentiated state purely while proliferation for long time. Also, because of their pluripotent nature, they have the ability to produce all types of cells in the body [23]. With these important features, ESCs attract much interest in tissue regeneration research. Their use in *in vitro* culture conditions shows a significant sufficiency same as in *in vivo* culture [14, 22, 24].

15.2.1.2 Adult Stem Cells

ASCs, also called tissue-specific stem cells or somatic stem cells, are more specialized stem cells than ESCs. ASCs can give rise in the cell types, which they reside in. They are found in the body after embryonic development and differentiate into specific cell types in their microenvironment to regenerate the areas of dying cells [25]. As other types of stem cells, ASCs also have the abilities of self-renewal and differentiation into mature functional cells. The distinguishing feature of ASCs from ESCs is that they can only produce cells that they are derived from. This property of ASCs is called multipotency [26].

Before the ASCs are differentiated completely, progenitor cells are obtained like intermediate cells. These cells are capable of replicating themselves and thereby induce differentiation [26, 27].

Generally, the differentiation of ASCs is tissue specific, except the mesenchymal stem cells (MSCs), which can be derived from bone marrow and adipose tissue [27, 28]. MSCs can differentiate into chondrocytes, osteoblasts, skeletal muscle cells, adipocytes, and myocytes. This feature of differentiating into multiple cell types is known as plasticity or transdifferentiation [29, 30]. Hematopoietic stem cells (HSCs) can also give rise to different cell types including neurons, liver cells, skeletal, and muscle cells. Plasticity of stem cells is an important advantage in tissue engineering, which enables the possibility of multiple tissue regeneration by using the same type of cell source [14].

Using ASCs in therapies and research have gained more acceptance than ESCs because of the ethical issues. In the ASC studies, cells are derived directly from donors where in ESC therapies from embryos, so there is no need to destroy embryos. Also, ASCs can be obtained from the same patient, which is called as an autograft. In these cases, the risk of tissue rejection is nonexistent.

15.2.1.3 Reprogramming and Induced Pluripotent Stem Cells

For the production of any kind of tissue *in vitro*, successful generation of pluripotent cells from healthy as wells as diseased human of any age has gained attention of many scientists in the field of regenerative medicine, toxicology, and pharmacology [31, 32]. In 2007, human IPSCs, which can differentiate into virtually any cell type of the human body without prompting ethical concerns usually associated with embryonic cells, were first described [33].

Forcing adult cells, like fibroblasts, to express pluripotent stem cell-specific genes, produces IPSC. In the earliest research on IPSC, adult mouse fibroblasts were introduced with the genes encoding transcription factors, called Oct4, Sox2, Klf4, and c-Myc by retroviruses. The transcription factors Oct3 and Sox2 are responsible from the pluripotency preservation whereas Klf4 and c-Myc enables Oct4 and Sox2 to connect the target by changing the chromatin structure (Figure 15.2) [31–33].

15.2.2 Stem Cell Niche

Stem cells are the regeneration and repair mechanism of the body. They are undifferentiated cells, which can produce differentiated functional cells and new stem cells capable of self-renewal. These important behavior and function of stem cells are not only designated by their gene profiles, but also arranged with the signals coming from their niche.

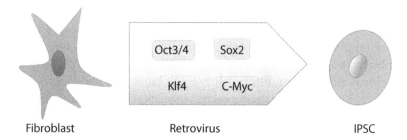

Figure 15.2 Generation of IPSCs from fibroblast cells [33].

Although stem cells have great potential to divide, their stem cell specialized niche helps them to lower their metabolism to prevent from exhaustion [11, 13]. In addition, niche interacts with stem cells for the requirements of their surrounding tissues. Recent findings show that the size of stem cell groups is not fixed at a constant number by divisions, but they usually have the ability to proliferate. The stem cell niche is responsible of this control mechanism. Cell surface molecules and different types of growth factors generated by niche share this specific information between the surrounding tissues and stem cells to regulate their functions [6, 36, 37].

The properties of a niche depend on the type of its specific type of stem cell. The stem cell niche is produced by stem cell and by supportive cells together with ECM. ECM defines the biochemical, mechanical, physical, and structural properties of niches. In addition to ECM, cell surface receptors, soluble and immobilized signaling factors and some physical factors, such as temperature, pH and oxygen species, accomplish the niches. Also, as stem cells are responsible to produce their own niche and secrete ECM, their cell to cell interactions are also important signals for their own niche [13, 19].

15.3 Stem Cells: Microenvironment Interactions

Physical, mechanical, chemical, and biochemical cues of dynamic and specialized stem cell microenvironment determine the stem cell fate by regulating stem cell behavior, survival, and the balance between their differentiation and self-renewal capacity [34–36]. Principles factors regulating stem cell–microenvironment interactions are summarized in Figure 15.3.

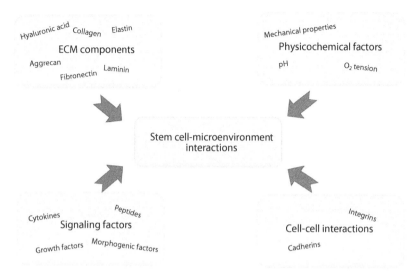

Figure 15.3 Stem cell–microenvironment interactions (adopted from Stevens and Mwenifumbo et al. [37]).

15.3.1 Extracellular Matrix

ECM is a cell-secreted material, and it provides cells structural three-dimensional support and regulates specific functions of stem cells including cellular morphology, migration, self-renewal, and differentiation. The macromolecules found in the ECM help to maintain the stem cell homeostasis and produce a specific environment for the signals between cells and their ECM [12, 15]. These interactions between stem cells and ECM are mediated by specific integrins and nonintegrin cell surface receptors. These receptors are responsible for the protein adhesion, and they build a bridge from the stem cell niche to the cytoskeleton of stem cells. Integrin receptors take part in the migration, proliferation, self-renewal, and differentiation of stem cells via focal adhesion kinase and phosphoinositide 3-kinase signaling [39].

Collagens are one of the most important macromolecules in the niche. They are the structural proteins with elastin, laminin, and fibronectin [38]. These proteins are responsible and have an important role for the adhesion, migration, shape recovery, and wound healing [19].

15.3.2 Signaling Factors

In the stem cell microenvironment, there are some small soluble and immobilized signaling factors, such as growth factors, morphogenic factors, enzymes, and cytokines. They also take part in the stem cell functions *in vivo*.

Signaling factors in ECM are proteins, which initiate cellular proliferation and differentiation. These signaling factors activate cellular response in various cell types. Some different kinds of these factors are: transforming growth factors-beta, transforming growth factors-alpha, nerve growth factor, epidermal growth factor, Wnts (Wingless-related integration site), and hedgehog proteins. Each of them has different stimulating functions. For example, bone morphogenetic protein 4 (BMP-4) mediates the self-renewal of ESCs, whereas Activin-A induces dorso-anterior mesoderm differentiation [41, 42],

Signaling soluble factors are very important for the interactions between differentiated cells and stem cells. Signals coming from the differentiated cells are capable of regulating the proliferation of stem cells to keep the balance [19, 41].

15.3.3 Physicochemical Composition

The physicochemical environment of stem cell niche depends on stem cell types. Physicochemistry of the niche includes specific number and types of proteins, in addition to the oxygen tension and pH [19].

15.3.4 Mechanical Properties

Focal adhesions connect the ECM to cytoskeleton of stem cells with adhesion receptors, known as integrins [43]. This link between these adhesion receptors and the cytoskeleton is generated by talin and vinculin, which are the adapter proteins. Integrins have two different chains: α and β subunits. These subunits are combined to form various integrin heterodimers. Different types of integrin heterodimers have shown different binding affinity on ECM proteins, and in addition to that, they also take part in the mechanical sensing and accordingly the behavior of stem cells [44]. For example, some *in vitro* studies have mentioned that stem cells on medium stiffness undergo myogenic differentiation by $\beta 3$ integrins, where on stiff substrates they differentiate into osteoblasts by $\alpha 2$ modulation [44–46].

In the mechanical sensing of stem cells, microtubules compress first according to the matrix stiffness, and they transfer forces to the ECM across the focal adhesions. Then, the integrin receptors activate the signal transduction pathways. These pathways initiate actin filament polymerization and alter the contractility of stress fibers [44]. The tension within the cytoskeleton provides the cellular shape to the stem cells and the responses to the mechanical forces direct stem cell behavior [13].

15.3.5 Cell–Cell Interactions

Cell–cell interactions are considered to be the most influential mechanism on stem cell behavior. Cell–cell adhesions are mediated by cadherins, which are calcium-dependent homophilic adhesion molecules. In addition to cadherins, Ephrin and Notch receptors are responsible for the signaling of the interactions between stem cells and their microenvironment [19, 41]

ESC differentiation and proliferation are highly affected by the cell–cell contacts. Depending on the differentiation stage of ESC, different types of cadherins can be obtained. For instance, cell–cell contacts in the generation of embryoid bodies (EBs) are observed by E-cadherin, which is downregulated when the differentiation takes place [41, 47].

15.4 Biomaterials as Stem Cell Microenvironments

Stem cells *in vivo* are situated in their specific microenvironment. The biochemical, mechanical, and physical signals coming from their microenvironment regulate their cellular behavior, such as proliferation, self-renewal, migration, and differentiation. In fields of tissue engineering and regenerative medicine, control of the stem cell fate mainly depends on properties of the chosen biomaterial. Biomaterials provide structural support to the cells in culture conditions as the ECM in the natural microenvironment of cells. In addition to that, ECM determines cellular behavior through biochemical, physical, and mechanical cues. Thus, these different properties must be considered when designing biomaterials for cell culture [9].

15.4.1 Surface Chemistry

Cells have the ability to sense and respond to the differences in the surface chemistry of their microenvironment, so the chemical composition of cell scaffolds is a very important parameter for the stem cell fate in culture conditions. Within this knowledge, materials with different chemistries, surface coatings, and incorporation of materials with different functional groups, such as $-CH_3$, $-COOH$, $-NH_2$, and $-OH$, have been designed and influences of these parameters on stem cell fate have been investigated [48].

For controlling the cell behavior in culture conditions, different materials have been designed to incorporate some signals found in the ECM. Many macromolecules in the ECM, including collagen, fibronectin, hyaluranic acid, and laminin have been used in the design of biomaterials

[49–53]. The motivation of using these molecules is that the ECM and cell interactions depend on specific ligand and receptor interactions, which is very important for stem cell fate. Collagen type I, for example, is a structural protein, which induces cell attachment, proliferation, and differentiation in natural conditions, promotes the differentiation of stem cells into osteoblasts in culture conditions [54, 55]. Collagen (type I)-coated substrates also induce cell adhesion and migration [56, 57].

Besides proteins, polysaccharides have also attracted much interest in the biomaterial design for enhanced cellular behavior. They provide structural support to the ECM like proteins. Many different types of polysaccharides have been used for stem cell culture depending on the monomer composition [55]. Chitosan is the most common polysaccharide as a scaffold material with its tunable chemical composition and mechanical properties. The biocompatibility, adjustable surface amine content, and glycosaminoglycans (GAGs)-like structure make it a favorable scaffold material for the chondrogenic and neural differentiation of MSCs and NSCs [49, 55, 58].

Introducing some integrin binding peptides into biomaterials can influence the cell–material interactions. For example, Arg–Gly–Asp (RGD) modification of poly(ethylene glycol) diacrylate (PEODA) hydrogel has shown increased osteogenic differentiation of MSCs [62].

Modifications of biomaterial surfaces with functional chemical groups (e.g. $-CH_3$, $-COOH$, $-NH_2$, and $-OH$) have investigated to effect the focal adhesions which in turn modulate cellular functions [50]. Ren *et al.* demonstrated a study to show the relation between glass surfaces modified with different chemical groups, including hydroxyl, sulfonic, amino, carboxyl, mercapto, and methyl groups, and the behavior of neural stem cells (NSCs). The results they obtained showed that NSCs on $-SO_3H$-modified glass surfaces had largest contact area and were inclined to differentiate into oligodendrocytes where $-NH_2$-modified surfaces promoted NSCs to differentiate into neurons. Cells on the $-OH$ functional group modified surfaces provided the weakest migration. These findings indicated that different functional group-modified materials have the ability to regulate stem cell migration, adhesion, and differentiation providing some chemical cues to prepare biomaterials for NSCs with required cellular effects (Table 15.1) [63].

In conclusion, the chemical properties of biomaterial and its modifications with either proteins, peptides, or functional groups influence the stem cell behavior. Thus, in the design of biomaterials for *in vivo* or *in vitro* cell culture, the chemical composition and surface properties should be taken into consideration.

STEM CELL BEHAVIOR ON MIMICKED SURFACES   435

Table 15.1 Effects of surface functional groups on cellular behavior of SCs.

Substrate material	Functional group	Target SC	Effects on Cellular Behavior	Ref.
Glass	$-CH_3$, $-OH$, $-COOH$, $-NH_2$, $-SH$, $-SO_3H$	NSCs	$-CH_3$: Round morphology, Promoted differentiation into neurons, astrocytes, and oligodendrocytes. $-OH$: Weakest ability to migrate $-COOH$: Promoted differentiation into neurons, astrocytes, and oligodendrocytes. $-NH_2$: Promoted differentiation into neurons $-SH$: Promoted differentiation into neurons, astrocytes, and oligodendrocytes. $-SO_3H$: Promoted differentiation into oligodendrocytes	58
Gold	$-COOH$, $-NH_2$, $-OH$	MSCs	$-COOH$: Decreased ALP activity and matrix mineralization, increased proliferation $-NH_2$: Promoted downregulation of osteogenesis genes, increased cell adhesion $-OH$: Increased ALP activity	59
Gold	$-CH_3$, $-COOH$, $-NH_2$, $-OH$	NSCs	$-CH_3$: Similar promoted differentiation into astrocytes, neurons, and oligodendrocytes $-COOH$: Similar promoted differentiation into astrocytes, neurons and oligodendrocytes $-NH_2$: Promoted differentiation into astrocytes	60
Hydroxyapatite	$-CH_3$, $-COOH$, $-NH_2$	ADSCs	$-CH_3$: Less effective to promote osteogenic differentiation $-COOH$: Promoted adhesion and spreading of cells $-NH_2$: Promoted osteogenic differentiation	61

15.4.2 Surface Hydrophilicity and Hydrophobicity

Wettability is an important characteristic of a biomaterial surface, as it effects the behavior of molecules, proteins, and cells. It is determined by the surface free energy and characterized by contact angle measurements. Contact angle is defined as the angle between a solid and a liquid droplet interface (Figure 15.4). In water contact angle measurements, if this angle is higher than 90°, the surface is known as hydrophobic, and if this angle is lower than 90°, the surface is known as hydrophilic. Each molecule in a liquid droplet is pulled equally in every direction which makes the net force zero. However, when a liquid droplet touches a solid surface, each molecule in the droplet cannot balance the net force. Thus, liquid contracts to the surface in order to achieve the lowest surface free energy [65].

Since the cell–biomaterial interaction occurs through the cell–surface interface, wettability properties of biomaterial surfaces have great influence in the attachment and proliferation of stem cells as well as the adsorption of ECM proteins. For instance, it is found that human adipose-derived stem cells (hADSCs) adhered better on hydrophilic polyethylene surfaces than hydrophobic ones. In addition, their proliferation was higher on hydrophilic surfaces [64]. Hydrophobic surfaces, on the other hand, are known to prevent protein adsorption and cellular attachment. Because of this nature, they are used in formation of EBs. It was found that, methyl-terminated hydrophobic surfaces were very useful for the differentiation of ESCs into EB and also provided better cellular behavior [66].

Ayala *et al.* showed the effect of biomaterial hydrophobicity by alteration of stem cell–substrate interactions through increasing or decreasing –CH_2 groups of surfaces. They observed that changes in surface hydrophobicity affected the cytoskeleton of MSC in addition to adhesion and spreading [67].

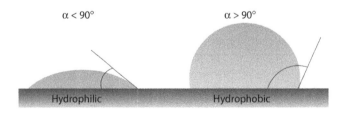

Figure 15.4 Water contact angles of hydrophilic and hydrophobic surfaces.

15.4.3 Substrate Stiffness

Cells can "feel" the mechanical alterations of their microenvironments through integrin adhesion receptors and accordingly change their cytoskeletal organization, which initiates signal-transducing pathways. Thus, mechanical properties of microenvironment modulate some specific functions of cells, such as differentiation, adhesion, and proliferation [46, 68, 69].

The effects of cell substrate stiffness on the differentiation of stem cells have investigated extensively. In these studies, stem cells have seeded on substrates with varying Young's modulus data and their influences on differentiation have been recorded as MSC cultured on materials with brain-like stiffness tended to differentiate into neurons, on bone-like stiffness into osteoblasts, and on muscle-like stiffness undergone myogenic differentiation [70].

Apart from the effect of substrate stiffness on stem cell differentiation, it also influences adhesion, spreading, morphology, and proliferation of cells. hESC culture studies on substrates with varying stiffness showed that proliferation and attachment of the cells were increased with increased stiffness [71].

The stiffness of the substrate surface determines the cellular behavior of cultured cells, such as attachment, structure, proliferation, and differentiation. However, these effects are depended on the cell type and the adhesion receptors of cells.

15.4.4 Surface Topography

Cells are capable of distinguishing and responding the micro- and macro-features of their environment. Thus, surface topography of cell substrates is also an important parameter for controlling cellular behavior in addition to the chemical and mechanical properties.

Natural ECM of cells is composed of many different and cell type-specific macromolecules. These molecules include protein fibrils and fiber networks [72]. The nano- and microscale structures of ECM matrix components provide structural support to cells in addition to guidance of their specific functions [73].

Effects of surface topography on different cell types showed that cellular behavior and functions could be controlled by shape and roughness of surface. For instance, cell attachment and orientation can be modulated with varying surface microtopographies [74]. Proliferation and osteoblastic cellular activity were found higher in the culture of rat osteoblasts on rough surfaces, than smooth ones [75]. Same as rat osteoblasts, human osteoblasts also proliferated and spread more on rough surfaces [76].

Attachment, proliferation, and differentiation of stem cells can be modulated by topographical cues. For example, Takahashi et al. investigated the effects of polyethylene terephthalate (PET) microfiber diameters on MSCs. They found that the attachment of MSCs was higher in thick fibers, their proliferation was depended on the fiber diameter and porosity and their differentiation into osteoblasts was also influenced by these fiber properties [78].

Various surface topography patterns have been designed by electrospinning, photolithography, and spin casting processes, like steps, grooves, pillars, pits, and fibers. For example, nanofibers, which can be generated by using electrospinning, provide ECM-like fibrous structures with porous and large surfaces [77]. Studies showed that nanofiber scaffolds could enhance proliferation, attachment, and spreading of cells [79].

15.5 Biomimicked and Bioinspired Approaches

Biomimetics means mimicking nature by understanding the functions and structure of naturally occurring objects to design artificial materials and devices. In bioinspiration, desired properties of natural materials are combined to produce better materials or devices. They are interdisciplinary fields composed of biology, chemistry, physics, and engineering. The idea of mimicking or inspiring nature developed many centuries ago. For example, Leonardo da Vinci investigated the flying mechanism of birds to design flying machines [81]. Nowadays, many products are produced by mimicking or inspiring nature. Super-hydrophobic, self-cleaning surfaces have developed inspired by a flower; lotus [82] or low drag swimsuits have produced mimicking sharkskin [83].

Biomimicked and bioinspired approaches have also taken part in regenerative medicine and tissue-engineering fields. In the natural microenvironment of stem cells, ECM is responsible from the structural support, cell attachment, proliferation, migration, and differentiation. In order to maintain the survival, attachment, self-renewal capacity, and differentiation of stem cells in *in vivo* or *in vitro* culture conditions, biomimetic or bioinspired materials have to be designed. Novel scaffolds with biomimetic properties can provide the required chemical, physical, and mechanical signals to stem cells. With bioinspired approaches, on the other hand, stem cell behavior can be modulated by preparation of synthetic scaffolds in which willing to solve the limitations of natural microenvironments (Table 15.2) [6, 80].

Table 15.2 Different biomimicked and bioinspired approaches in biomaterial design and their effects on stem cell behavior.

Target Sc	Biomaterial properties					Cellular behavior	Ref.
	Type	Stiffness	Topography	Chemistry	Differentiation		
MSCs	Hydrogel	80 kPa, 2.1 MPa, and 320 MPa	Nanofibers	Growth factors (TGF-β1, BMP-7, IGF-7) Hyaluronic acid (HA)	Chondrogenic	Enhanced zone-specific chondrogenic differentiation.	[84]
MSCs	P(LLA-CL)/collagen type I	-	Nanofibers	Collagen-I	Chondrogenic	Induced proliferation, orientation and adhesion of MSCs. Increased GAG content and Col-II expression.	[85]
MSCs	PLGA	-	-	Col-I, Col-II, aggrecan	Chondrogenic	Moderate amount of Col-I and aggrecan induced chondrogenesis. High amount of Col-II and aggrecan inhibited chondrogenesis.	[86]
Neural stem/progenitor cells (NSPCs)	Methacrylamide chitosan	-	-	Growth factors (N-terminal tagged fusion proteins), acrylated laminin	Neuronal	Enhanced differentiation and neuronal development.	[87]
Cardiosphere-derived cells	Hydrogel	50–60 kPa	ECM-mimicking fibers	-	Cardiac	Stimulated cardiac differentiation.	[88]

(Continued)

Table 15.2 Cont.

Target Sc	Type	Biomaterial properties				Differentiation	Cellular behavior	Ref.
		Stiffness	Topography	Chemistry				
MSCs and NMCs	PHBHV/gelatin constructs	4–5 MPa	Decellularized suine cardiac tissue-like	PHBHV/gelatin		Cardiac	Enhanced stem cell adhesion and cardiac differentiation.	[89]
MSCs	Methylcellulose-based scaffold	0.2–1 MPa	Porous	Methylcellulose-based		Osteogenic	Scaffolds with higher stiffness accelerated osteogenic differentiation.	[90]
hBMSCs	PVA/HA hydrogel	20 and 200 kPa	-	PVA-HA		Specific	20 kPa for neural, 40 kPa for myogenic, 80 kPa for chondrogenic, and 190 kPa for osteogenic differentiation.	[91]
MSCs	Polyimide	-	Nanogrooves	Polyimide		Osteogenic and adipogenic	Enhanced osteogenic and adipogenic differentiation.	[92]
hIPSCs	Hydroxyapatite and poly(L-lactic acid)/poly(lactic-co-glycolide)	-	Porous	Hydroxyapatite and poly(L-lactic acid)/poly(lactic-co-glycolide)		Osteogenic	Enhanced bone formation.	[93]

15.5.1 Bone Tissue Regeneration

Combination of tissue engineering and stem cells with osteogenic differentiation potential is one of the promising solutions for bone repair. The fundamental requirement in bone tissue engineering is the integration of designed scaffold into the host tissue. Besides, scaffold should be strong enough to meet the mechanical requirements of bone and act as a microenvironment to promote the cells to produce ECM. Thus, scaffolds for bone tissue regeneration should mimic the chemical, mechanical, and physical properties of natural bone tissue microenvironment to support and promote cellular adhesion, proliferation, and differentiation of stem cells [74, 75].

Hydrogels are widely used as a scaffold material for bone tissue engineering. They are injectable, capable of chemical modification, and have tunable mechanical properties [72]. Fang et al. produced biomimetic gelatin methacrylamide hydrogels, which can mimic the physical and chemical properties of original bone ECM. Their results showed that these mimicked scaffolds induced osteogenic differentiation of adipose-derived stem cells efficiently [73].

Bone tissue is a complex structure, composed of mineral phase, hydroxyapatite, and an organic phase (mainly formed of collagen type I, cellular phase, osteoblasts, soluble factor phase, and growth factors) [75]. In designing scaffolds for bone tissue engineering, this natural structure of bone tissue should be considered. Common scaffold materials used in bone tissue engineering are poly(lactide-co-glycolide), poly(glycolic acid), poly(lactic acid), calcium phosphates, collagen, alginate, fibrin, and silk [94–98]. Apart from these, titanium is also an inspiring scaffold material for bone tissue regeneration with its mechanical properties [99,100]. Surface modifications of bone scaffolds are generally done with the peptides and proteins in the ECM of bone tissue, such as RGD peptide and collagen type I [91, 92]. In addition, functional groups, such as $-CH_3$ and $-SH$ have also found to promote osteogenesis of MSCs [64]. Scaffold mechanical properties have been also shown to regulate osteogenic stem cell differentiation. For instance, Zhao et al. indicated that substrate stiffness similar to the bone tissue's *in vivo* microenvironment promotes osteogenic differentiation of hMSCs [103]. Besides the effects of chemical and mechanical cues on osteogenic differentiation of stem cells, surface topography also manipulates the behavior of stem cells through the nanosized integrin receptors of stem cells [104]. Fiedler et al. investigated the effects of topographical cues on the osteogenic differentiation of MSCs Nanopillars were produced with various diameters (10/30 nm), distances (50–120 nm), and heights (25–30 nm), whereas MSCs on higher nanopillars (50 nm) were observed to effect osteogenic differentiation the most [105].

15.5.2 Cartilage Tissue Regeneration

Cartilage is an avascular tissue. Because of this nature, its regenerative capacity is relatively very low. As there are no long-term clinical solutions to cartilage defects, cell-based therapies, and tissue engineering have been attracted so much interest among researchers [106]. For the regulation of stem cell behavior in culture conditions, many different biomaterials have been prepared with various parameters to achieve a better cellular behavior and a better stimulation for the differentiation [107–110]. Many natural polymers, such as hyaluronic acid, collagen, gelatin, chitosan, and chondroitin, have been used to produce scaffolds for chondrogenic differentiation. Fan *et al.* developed porous scaffolds by using natural polymers for chondrogenic MSC differentiation. They aimed to mimic cartilage ECM by using poly-(lactic-co-glycolic acid) (PLGA-GCH)-gelatin/chondroitin/hyaluronate scaffolds. The comparison of *in vitro* MSC proliferation was done between PLGA and PLGA-GCH scaffolds. *In vitro* studies presented that PLCA-GCH scaffolds enhanced proliferation and GAGs synthesis, significantly. For the *in vivo* studies, autologous differentiated MSC–scaffold complexes implanted to rabbits. Their histology results indicated that morphology, integration, and thickness of formed cartilage were significantly better in the PLGA-GCH scaffold group, showing that ECM-like materials enhanced cellular behavior in regeneration [111].

Annamalai *et al.* showed that supplementation of collagen type II, which is the dominant collagen in natural ECM of cartilage tissue, to microbeads with agarose promoted chondrogenic differentiation of MSCs [106]. Another distinctive parameter of the ECM is the architecture. Different biomaterials have prepared to show the relationship between topographic cues of substrates and their inductive role in chondrogenic differentiation of MSCs such as nanogrill, nanohole, nanopillars, and cylindrical microtextures on the substrate or producing the substrate itself as nanofibers [110, 111]. Wu *et al.* found that biomimetic forms of nanotopographies coated with chondroitin sulfate, which is a sulfated GAG found in cartilage tissue induce MSC behavior in morphological and structural manners ensuring the chondrogenic differentiation [109]. As stated before, stiffness directs the chondrogenesis of stem cells, also. Singh *et al.* showed that MSCs cultured on specific blend composition of 75:25, cellulose and silk, significantly increases chondrogenic differentiation [112].

15.5.3 Cardiac Tissue Regeneration

The idea that ECM and the signaling that it provokes through membrane-bound receptors could influence the fate determination of cardiac stem cells is has been an interesting concept in cardiac regeneration. Reports by the group of Wobus have shown that cardiac differentiation of murine ES cells is highly impaired if the cells have lost beta1-integrin as a major compound of the matrix-recognizing integrin receptors [112].

Changing the cell substrate and surface properties of the substrate, such as chemistry, roughness, stiffness, and topography can alter cell–substrate interfacial characteristics and probably effect cellular behavior and function of cardiomyocytes as well as chondrocytes and stem cells discussed above. These factors are very crucial when synthesizing and developing novel materials and surfaces for cardiac tissue-engineering applications as well as prosthetic devices [114].

Biomaterials as instructive extracellular microenvironments have been used for controlled differentiation of stem cells in the form of polymers and natural materials for cardiac tissue-engineering applications [115–117]. Biomaterial substrates can be prepared in different surface chemistry and biochemistry, surface topography, surface stiffness, and micro-nanopattern designs those which have a great influence on the differentiation of stem cells to cardiomyocytes and other cell types.

Li *et al.* suggested a study including the preparation of hydrogels with native heart tissue-like stiffness (1–140 kPa) and stimulating MSCs differentiation into cardiac cells in these gels. They produce hydrogels by using *N*-isopropylacrylamide, *N*-acryloxysuccinimide, acrylic acid, and poly(trimethylene carbonate)-hydroxyethyl methacrylate with three different stiffness values, as 16, 45, and 65 kPa. MSC culture results showed that gels with 65 kPa provided cardiac differentiation the most [118].

Bautista *et al.* investigated long-term culture of cardiomyocytes on classic polystyrene. Their results showed that after five weeks of culture, cells lost sarcomere integrity which was caused from high amounts of mechanical stress. In order to eliminate this issue, polyacrylamide hydrogels were prepared with the matrix stiffness of cardiac tissue and coated with a matrix protein, fibronectin. With these novel substrates, long-term IPSC-derived cardiomyocytes culture was provided, and contractility was preserved by using 55 kPa hydrogels [119].

Since the cardiac differentiation of stem cells is dependent on their specific matrix conditions, many studies have been designed to mimic these conditions in biomechanical, morphological, and/or biochemical ways. Effects of the fiber arrangement on cardiac differentiation were observed on adipose-derived MSCs (ADMSCs) by Safaeijavan *et al.* They aimed to mimic the ECM of cardiomyocytes by preparing random and aligned nanofibrous polycaprolactone (PCL) scaffolds. Their results confirmed the success of cardiomyocyte differentiation of ADSCs on nanofibers. In addition, aligned nanofibrous scaffolds were observed to induce proliferation and cardiomyocyte differentiation of ADSCs [120]. Another study was designed to observe the relation between cardiosphere-derived cells (CDCs, a type of a cardiac progenitor cell) and fibers with different mechanical properties. Scaffolds were prepared by mixing soft N-isopropylacrylamide-based hydrogel with relatively stiff polyurethane to achieve tensile moduli range from 48 to 461 kPa which is in the range of heart tissue (10–500 kPa). Fibers were fabricated by electrospinning and electrospraying with approximately same fiber diameter (>1 μm) and different fiber densities (22.2–46.1 (#/100 μm^2)). The fibrous scaffolds with a modulus of 48 kPa, higher fiber density, and lower alignment were found to promote cardiac differentiation of CDCs which confirms the effects of ECM properties on cardiac differentiation [121].

15.6 Conclusion

Stem cells are very promising tools for tissue-engineering and regenerative medicine applications with their self-renewal and differentiation capacities. The main drawback of using stem cells is that their control outside of the body is very challenging. As they are found in their specific microenvironments, naturally, the idea of controlling their behavior by using their own microenvironment-mimicked scaffolds has attracted much attention among researchers. It is highly desirable to mimic stem cell type-specific microenvironment as closely as their own one by using chemical, topographical and mechanical signals and also to overcome the limitations of natural microenvironment with bioinspired approaches. With the design of such scaffolds, it can be possible to regenerate cells similar to their natural environment.

References

1. Langer, R., Vacanti, J. P., Tissue engineering. *Science*, 260, pp. 920–7, 1993.
2. Vacanti, J., Tissue engineering and regenerative medicine: from first principles to state of the art. *Journal of Pediatric Surgery*, 45, pp. 291–4, 2010.
3. Baddour, J.A., Sousounis, K., Tsonis, P.A., Organ repair and regeneration: an overview. *Birth Defects Research C: Embryo Today.* 96, pp. 1–29, 2012.

4. Yeni Griffith, L.,G., Naughton, G., Tissue engineering—current challenges and expanding opportunities. *Science*, 295, pp. 1009–14, 2002.
5. O'Brien, F.J., Biomaterials & scaffolds. *Materials Today*, 14, pp. 88–95, 2011.
6. Fisher, O., Khademhosseini, A., Langer, R., Peppas, N.A., Bioinspired materials for controlling stem cell fate. *Accounts of Chemical Research*, 43, pp. 419–28, 2010.
7. Agrawal, C. M., Ray, R. B., Biodegradable polymeric scaffolds for musculoskeletal tissue engineering. *Journal of Biomedical Materials Research*, 55, pp. 141–50, 2001.
8. Dawson, E., Mapili, G., Erickson, K., Taqvi, S., Roy, K., Biomaterials for stem cell differentiation. *Advanced Drug Delivery Reviews*, 60, pp. 215–28, 2007.
9. Shin, H. Jo, S., Mikos, A.G., Biomimetic materials for tissue engineering, *Biomaterials*, 24, pp. 4353–64, 2003.
10. Drury, J.L., Mooney, D.J., Hydrogels for tissue engineering: scaffold design variables and applications, *Biomaterials*, 24, 4337–51, 2003.
11. Ozbek, S., Balasubramanian, P.G., Chiquet-Ehrismann, R., Tucker, R.P., Adams, J.C., The evolution of extracellular matrix. *Molecular Biology of the Cell*, 21, pp. 4300–5, 2010.
12. Watt, F.M., Huck, W.T.S., Role of the extracellular matrix in regulating stem cell fate. *Nature Reviews Molecular Cell Biology*, 14, pp. 467–73, 2013.
13. Gattazzo, F., Urciulo, A., Bonaldo, P., Extracellular matrix: a dynamic microenvironment for stem cell niche. *Biochimica et Biophysica Acta*, 1840, pp. 2506–19, 2014.
14. Bradman, D.A., Constructing stem cell microenvironments using bioengineering approaches. *Physiological Genomics*, 45, pp. 1123–35, 2013.
15. Sun, Y., Chen, C.S., Fu, J.F., Forcing stem cells to behave: a biophysical perspective of the cellular microenvironment. *Annual Review of Biophysics* 41, pp. 519–42, 2012.
16. Discher, D.E., Mooney, D.J., Zandstra, P.W., Growth factors, matrices, and forces combine and control stem cells. *Science* 324, pp. 1673–7, 2009.
17. Guilak, F., Cohen, D.M., Estes, B.T., Gimble, J.M., Liedtke, W., Chen, C.S., Control of stem cell fate by physical interactions with the extracellular matrix. *Cell Stem Cell* 5, pp. 17–26, 2009.
18. Smith, A.G., Embryo-derived stem cells: of mice and men. *Annual Review of Cell and Developmental Biology* 17, pp.435–62, 2001.
19. Fernandes, T.G., Diogo, M., Cabral, J., *Stem Cell Bioprocessing: For Cellular Therapy, Diagnostics and Drug Development*, Woodhead Publishing, 2013.
20. Li, S., L'Heureux, N., Elisseeff, J., *Stem Cell and Tissue Engineering*, pp. 1–11, World Scientific Publishing Company, 2011.
21. Evans M., Kaufman, M., 1981. Establishment in culture of pluripotent cells from mouse embryos. *Nature* 292, pp. 154–6, 1981.
22. Keller G., *In vitro* differentiation of embryonic stem cells. *Current Opinion in Cell Biology* 7, pp. 862–9, 1995.

23. Keller, G., Embryonic stem cell differentiation: emergence of a new era in biology and medicine. *Genes & Development*, 19, pp. 1129–55, 2005.
24. Bongso, A., Lee, E.H., *Stem Cells from Bench to Bedside*, World Scientific Publishing, pp. 1–13, 2005.
25. Czyz, J.,Wobus, A.M., Embryonic stem cell differentiation: the role of extracellular factors. *Differentiation*, 68, pp. 167–74, 2001.
26. Young, H.E., Black, A.J., Adult stem cells, *The Anatomical Record Part A: Discoveries in Molecular, Cellular, and Evolutionary Biology*, 276A, pp. 75–102, 2004.
27. Pittenger, M. F., Mackay, A. M., Beck, S. C., Jaiswal, R. K, Douglas, R., Mosca, J. D., Moorman, M. A., Simonetti, D. W., Craig, S.,Marshak, D.R., Multineage potential of adult human mesenchymal stem cells. *Science*, 284, pp. 143–7, 1999.
28. Jiang, Y., Jahagirdar, B. N., Reinhardt, R. L., Schartz, R. E., Keene, C. D., Ortiz-Gonzalez, X. R., Reyes, M., Lenvik, T., Lund, T., Blackstad, M., Pluripotency of mesenchymal stem cells derived from adult marrow. *Nature*, 418, pp. 41–9, 2002.
29. Filip, S., English, D., Mokrý, J., Issues in stem cell plasticity. *Journal of Cellular and Molecular Medicine*, 8, pp. 572–7, 2004.
30. Caplan, A.I., Bruder, S.P., Mesenchymal stem cells: building blocks for molecular medicine in the 21st century. *Trends in Molecular Medicine*, 7, 259–64, 2001.
31. Zwi-Dantsis, Z., Gepstein, L., Induced pluripotent stem cells for cardiac repair. *Cellular and Molecular Life Sciences*, 69, pp. 3285–99, 2012.
32. Sinnecker, D., Laugwitz, K.L. Moretti, A. Induced pluripotent stem cell-derived cardiomyocytes for drug development and toxicity testing. *Pharmacology & Therapeutics*, 143, pp. 246–52, 2014.
33. Takahashi K., Tanabe K., Ohnuki M., Narita M., Ichisaka T., Tomoda K., Yamanaka, S., Induction of pluripotent stem cells from adult human fibroblasts by defined factors. *Cell*, 5, pp. 861–72, 2007.
34. Fuchs, E., Tumbar, T., Guasch, G., Socializing with the neighbors: stem cells and their niche. *Cell*, 116, pp. 769–78, 2004.
35. Morrison, S.J., Spradling, A.C., Stem cells and niches: mechanisms that promote stem cell maintenance throughout life, *Cell*, 132, pp. 598–611, 2008.
36. Gattazzo, A. U., Bonaldo, P., Extracellular matrix: a dynamic microenvironment for stem cell niche, *Biochimicaet Biophysica Acta (BBA)* 1840, pp. 2506–19, 2014.
37. Yeni Stevens, M.M., Mwenifumbo, S., ECM Interactions with cells from the macro-to nanoscale, *Biomedical Nanostructures*, Gonsalves, K., Halberstadt, C., Laurencin, C., Nair, L., Wiley, pp. 225–52, 2008.
38. Orford, K.W., Scadden, D.T., Deconstructing stem cell self-renewal: genetic insights into cell-cycle regulation. *Nature Reviews Genetics*, 9, pp. 115–28, 2008.
39. Legate, K.R., Wickström, S.A., Fässler, R., Genetic and cell biological analysis of integrin outside-in signaling. *Genes & Development*, 23, pp. 397–418, 2009.

40. Votteler, M., Kluger, P.J., Walles, H., Schenke-Layland, K., Stem cell microenvironments – unveiling the secret of how stem cell fate is defined. *Macromolecular Bioscience Special Issue: Instructive Materials for Functional Tissue Engineering*, 10, pp. 1302–15, 2010.
41. Prakash, S., Shum-Tim, D., *Stem Cell Bioengineering and Tissue Engineering Microenvironment*. World Scientific, 2011.
42. Johansson, B.M., Wiles, M.V., Evidence for involvement of activin A and bone morphogenetic protein 4 in mammalian mesoderm and hematopoietic development. *Molecular and Cellular Biology*, 15, pp.141–51, 1995.
43. Hynes, R.O., Integrins: versatility, modulation, and signaling in cell adhesion. *Cell*, 69, pp. 11–25, 1992.
44. Lv, H., Li, L., Sun, M., Zhang, Y., Chen, L., Rong, Y., Li Y., Mechanism of regulation of stem cell differentiation by matrix stiffness. *Stem Cell Research & Therapy*, 6, pp. 103, 2015.
45. Yu, H.Y., Lui, Y.S., Xiong, S.J., Leong, W.S., Wen, F., Nurkahfianto, H., Insights into the role of focal adhesion modulation in myogenic differentiation of human mesenchymal stem cells. *Stem Cells and Development*, 22, pp. 136–47, 2013.
46. Shih, Y.R., Tseng, K.F., Lai, H.Y., Lin, C.H., Lee, O.K., Matrix stiffness regulation of integrin-mediated mechanotransduction during osteogenic differentiation of human mesenchymal stem cells. *Journal of Bone and Mineral Research*, 26, pp. 230–8, 2011.
47. Dang, S.M., Gerecht-Nir, S., Chen, J., Controlled, scalable embryonic stem cell differentiation culture. *Stem Cells*, 22, pp. 275–82, 2004.
48. Keselowskya,B.G., Collard, D.M., Garcia, A.J., Surface chemistry modulates focal adhesion composition and signaling through changes in integrin binding, *Biomaterials*, 25,pp. 5947–54, 2004.
49. Ragetly, G. R., Griffon, D. J., Lee, H.-B., Fredericks, L. P., Gordon-Evans, W., Chung, Y.S., Effect of chitosan scaffold microstructure on mesenchymal stem cell chondrogenesis, *Acta Biomaterialia*, 6, pp. 1430–6, 2010.
50. Kim, H.J., Kim, U.J., Kim, H.S., Bone tissue engineering with premineralized silk scaffolds. *Bone*, 42, pp. 1226–34, 2008.
51. Chiu, L.L.Y., Janic, K., Radisic, M., Engineering of oriented myocardium on threedimensional micropatterned collagen-chitosan hydrogel, *International Journal Of Artificial Organs*, 35, pp. 237–50, 2012.
52. Sawatjui, N., Damrongrungruang, T., Leeanansaksiri, W., Jearanaikoon, P., Hongeng, S., and T. Limpaiboon, Silk fibroin/gelatin–chondroitin sulfate–hyaluronic acid effectively enhances *in vitro* chondrogenesis of bone marrow mesenchymal stem cells, *Materials Science and Engineering: C*, 52, pp. 90–6, 2015.
53. Bray, L.J., George, K.A., Ainscough, S.L., Hutmacher, D.W., Chirila, T.V. Harkin, D.G., Human corneal epithelial equivalents constructed on Bombyx mori silk fibroin membranes, *Biomaterials*, 32, pp. 5086–91, 2011.
54. Wei, H., Tan, G., Manasi, Qiu, S., Kong, G., One-step derivation of cardiomyocytes and mesenchymal stem cells from human pluripotent stem cells. *Stem Cell Research*, 9, pp. 87–100, 2012.

55. Singh A, Elisseeff J. Biomaterials for stem cell differentiation. *Journal of Materials Chemistry*, 20, 8832–47, 2012.
56. Evans, M. D. M., McFarland, G. A., Taylor, S., Johnson, G., McLean, K. M., The architecture of a collagen coating on a synthetic polymer influences epithelial adhesion. *Journal of Biomedical Materials Research Part A*, 56, pp. 461–8, 2001,
57. Tija, J.S., Aneskievich, B.J., Moghe, P.V., Substrate-adsorbed collagen and cell secreted fibronectin concertedly induce cell migration on poly (lactide-glycolide) substrates. *Biomaterials*, 20, pp. 2223–33, 1999.
58. Li, H., Wijekoon, A., Leipzig, N.D., Encapsulated neural stem cell neuronal differentiation in fluorinated methacrylamide chitosan hydrogels. *Annals of Biomedical Engineering*, 42, pp. 1456–69, 2013.
59. Li, J.J., Kawazoe, N., Chen, G., Gold nanoparticles with different charge and moiety induce differential cell response on mesenchymal stem cell osteogenesis. *Biomaterials*, 54, pp. 226–36, 2015.
60. Yao, S., Liu, X., He, J., Wang, X., Wang, Y., Cui, FZ, Ordered self-assembled monolayers terminated with different chemical functional groups direct neural stem cell linage behaviours. *Biomedical Materials*, 11, 14107, 2015.
61. Liu, X., Feng, Q., Bachhuka, A., Vasilev, K., Surface chemical functionalities affect the behavior of human adipose derived stem cells *in vitro*. *Applied Surface Science*, 270, pp. 473–9, 2013.
62. Yang, F., Williams, C.G., Wang, D., Lee, H., Manson, P.N., Elisseeff, J., The effect of incorporating RGD adhesive peptide in polyethylene glycol diacrylate hydrogel on osteogenesis of bone marrow stromal cells. *Biomaterials*, 26, pp. 5991–8, 2005.
63. Ren, Y.J., Zhang, H., Huang, H., Wang, X.M., Zhou, Z.Y., Cui, F.Z., An, Y.H., In vitro behavior of neural stem cells in response to different chemical functional groups. *Biomaterials*, 30, pp. 1036–44, 2009.
64. Ahn, H.H., Lee, W., Lee, H.B., Kim, M.S., Cellular behavior of human adipose-derived stem cells on wettable gradient polyethylene surfaces. *International Journal of Molecular Sciences*, 15, pp. 2075–86, 2014.
65. Yuan, Y., Lee, R., Contact angle and wetting properties, in: *Surface Science Techniques*, Bracco, G., Holst, B. (eds.), pp. 3–34, Berlin Heidelberg, 2013.
66. Valamehr, B., Jonas, S.J., Polleux, J., Qiao, R., Guo, S., Gschweng, E.H., Stiles, B., Kam, K., Luo, T.M., Witte, O.N., Liu, X., Dunn, B., Wu, H., Hydrophobic surfaces for enhanced differentiation of embryonic stem cell-derived embryoid bodies. *PNAS*, 105, pp. 14459–5564, 2008.
67. Ayala, R., Zhang, C., Yang, D., Hwang, Y., Aung, A., Shroff, S.S., Arce, F.T., Lal, R., Arya, G., Varghese, S., Engineering the cell–material interface for controlling stem cell adhesion, migration, and differentiation. *Biomaterials*, 32, pp. 3700–11, 2011.
68. Discher, D.E., Janmey, P., Wang, Y.L. Tissue cells feel and respond to the stiffness of their substrate. *Science*, 310, pp.1139–43, 2005.

69. Wells, R.G., The role of matrix stiffness in regulating cell behavior. *Hepatology* 47, pp. 1394–400, 2008.
70. Engler, A.J., Sen, S., Sweeney, H.L. and D.E. Discher, Matrix elasticity directs stem cell lineage specification. *Cell*, 126, pp. 677–89, 2006.
71. Eroshenko, N., Ramachandran, R., Yadavalli, Y.K. and R.R. Rao, Effect of substrate stiffness on early human embryonic stem cell differentiation. *Journal of Biological Engineering*, 7(1), 7, 2013.
72. Ko, E. and S.W. Cho, Biomimetic polymer scaffolds to promote stem cell-mediated osteogenesis. *International Journal of Stem Cells*, 6(2), pp. 87–91, 2013.
73. Fang, X., Xie, J., Xhong, L., Li, J. Rong, D., Li, X. and J. Ouyang, Biomimetic gelatin methacrylamide hydrogel scaffolds for bone tissue engineering. *Journal of Materials Chemistry B*, 4, pp. 1070–80, 2016.
74. Anselme, K., Osteoblast adhesion on biomaterials biomaterials: review. *Biomaterials*, 21(7), pp. 667–81, 2000.
75. Bououdina, M., *Emerging Research on Bioinspired Materials Engineering*, pp. 104–32, IGI Global, 2016.
76. Amini, A.R., Laurencin, C.T. and S.P. Nukavarapu, Bone tissue engineering: recent advances and challenges. *Critical Reviews in Biomedical Engineering*, 40(5), pp. 363–408, 2012.
77. Wang, P., Zhao, L., Liu, J., et al., Bone tissue engineering via nanostructured calcium phosphate biomaterials and stem cells. *Bone Research*, 2, article no: 14017, 2014.
78. Takahashi, Y., Tabata, Y., Effect of the fiber diameter and porosity of non-woven PET fabrics on the osteogenic differentiation of mesenchymal stem cells. *Journal of Biomaterials Science Polymer Edition*, 15, pp. 41–57, 2004.
79. Wang, P., Liu, X., Zhao, L., et al., Bone tissue engineering via human induced pluripotent, umbilical cord and bone marrow mesenchymal stem cells in rat cranium. *Acta Biomaterials*, 18, pp. 236–48, 2015.
80. Liao, S, Chan, C.K., Ramakrishna, S., Stem cells and biomimetic materials strategies for tissue engineering, *Materials Science and Engineering: C*, 28, pp. 1189–202, 2008.
81. Bhushan, B., Biomimmetics: lessons from nature-an overview. *Philosophical Transactions of the Royal Society A*, 367, pp. 1445–86, 2009.
82. Bhushan, B., Jung, Y.C., Koch, K., Micro-,nano-, and hierarchical structures for superhydrophobicity, self-cleaning and low adhesion. *Philosophical Transactions of the Royal Society A*, 367, pp. 1631–72, 2009.
83. Mark, R.C., Kim,S., Design and fabrication of multi-material structures for bioinspired robots. *Philosophical Transactions of the Royal Society A*, 367, pp. 1799–813, 2009.
84. Moeinzadeh, S., Pajoum, S.R., Jabbari, E., Comparative effect of physicomechanical and biomolecular cues on zone-specific chondrogenic differentiation of mesenchymal stem cells. *Biomaterials*, 92, pp. 57–70, 2016.

85. Zheng, X., Wang, W., Liu, J., Li, F., Cao, L., Liu, X., Mo, X., Fan, C., Enhancement of chondrogenic differentiation of rabbit mesenchymal stem cells by oriented nanofiber yarn-collagen type I/hyaluronate hybrid. *Materials Science and Engineering: C*, 58, pp. 1071–6, 2016.
86. Cai, R., Nakamoto, T., Kawazoe, N., Chen, G., Influence of stepwise chondrogenesis-mimicking 3D extracellular matrix on chondrogenic differentiation of mesenchymal stem cells. *Biomaterials*, 52, pp. 199–207, 2015.
87. Li, H., Koenig, A.M., Sloan, P., Leipzig, N.D., In vivo assessment of guided neural stem cell differentiation in growth factor immobilized chitosan-based hydrogel scaffolds. *Biomaterials*, 35, pp. 9049–57, 2014.
88. Xu, Y., Patnaik, S., Guo, X., Li, Z., Lo, W., Butler, R., Claude, A., Liu, Z., Zhang, G., Liao, J., Anderson, P.M., Guan, J., Cardiac differentiation of cardiosphere-derived cells in scaffolds mimicking morphology of the cardiac extracellular matrix. *Acta Biomaterialia*, 10, pp. 3449–62, 2014.
89. Cristallini, C., Rocchietti, C.C., Accomasso, L., Folino, A., Gallina, C., Muratori, L., Pagliaro, P., Rastaldo, R., Raimondo, S., Saviozzi, S., Sprio, A.E., Gagliardi, M., Barbani, N., Giachino, C., The effect of bioartificial constructs that mimic myocardial structure and biomechanical properties on stem cell commitment towards cardiac lineage. *Biomaterials*, 35, pp. 92–104, 2014.
90. Shen, H., Ma, Y., Luo, Y., Liu, X., Zhang, Z., Dai, J., Directed osteogenic differentiation of mesenchymal stem cell in three-dimensional biodegradable methylcellulose-based scaffolds. *Colloids and Surfaces B: Biointerfaces*, 135, pp. 332–8, 2015.
91. Oh, S.H., An, D.B., Kim, T.H., Lee, J.H., Wide-range stiffness gradient PVA/HA hydrogel to investigate stem cell differentiation behavior. *Acta Biomaterialia*, 35, pp. 23–31, 2016.
92. Abagnale, G., Steger, M., Nguyen, V.H., Hersch, N., Sechi, A., Joussen, S., Denecke, B., Merkel, R., Hoffmann, B., Dreser, A., Schnakenberg, U., Gillner, A., Wagner, W., Surface topography enhances differentiation of mesenchymal stem cells towards osteogenic and adipogenic lineages. *Biomaterials*, 61, pp. 316–26, 2015.
93. Lee, E.S., Park, J., Lee, H., Hwang, N.S., Osteogenic commitment of human induced pluripotent stem cell-derived mesenchymal progenitor-like cells on biomimetic scaffolds. *Journal of Industrial and Engineering Chemistry*, in press, 2016.
94. Jabbarzadeh, E., Starnes, T., Khan, Y.M., Jiang, T., Wirtel, A.J. and M. Deng, Induction of angiogenesis in tissue-engineered scaffolds designed for bone repair: a combined gene therapy–cell transplantation approach. *Proceedings of National Academy of Sciences*, 105(32), pp. 11099–104, 2008.
95. Osathanon, T., Linnes, M.L., Rajachar, R.M., Ratner, B.D., Somerman M.J. and C.M. Giachelli, Microporous nanofibrous fibrin-based scaffolds for bone tissue engineering. *Biomaterials*, 29, pp. 4091–99, 2008.
96. Guofu, X., Shenzhou, M., Zhimin, Y., Lingping, Z., Fuzhai, C. and L. Susan, Biomimetic strengthening polylactide scaffold materials for bone tissue engineering. *Frontiers of Chemistry in China*, 2(1), pp. 27–30, 2007.

97. Meinel, L., Karageorgiou, V., Fajardo, R., Snyder, B., et al., Bone tissue engineering using human mesenchymal stem cells: effects of scaffold material and medium flow. *Annals of Biomedical Engineering*, 32(1), pp. 112–22, 2004.
98. Kim, H.J., Kim, U.J., Kim, H.S., et al., Bone tissue engineering with premineralized silk scaffolds. *Bone*, 42(6), pp. 1226–34, 2008.
99. Liu, C., Xia, Z., and J.T. Czernuszka, Design and development of three-dimensional scaffolds for tissue engineering. *Chemical Engineering Research and Design*, 85A, pp. 1051–64, 2007.
100. Davies, J.E., Bone bonding at natural and biomaterial surfaces. *Biomaterials*, 89, pp. 5058–67, 2007.
101. Mano, J.F. and R. L. Reis, Osteochondral defects: present situation and tissue engineering approaches. *Journal of Tissue Engineering and Regenerative Medicine*, 1(4), pp. 261–73, 2007.
102. Pang, L., Hu, Y., Yan, Y., Xiong, Z., et al., Surface modification of PLGA/β-TCP scaffold for bone tissue engineering: hybridization with collagen and apatite. *Surface and Coatings Technology*, 201(24), pp. 9549–57, 2007.
103. Zhao, W., Li, X., Liu, X., Zhang, N. and X. Wen, Effects of substrate stiffness on adipogenic and osteogenic differentiation of human mesenchymal stem cells. *Materials Science and Engineering: C*, 40, pp. 316–23, 2014.
104. Metavarayuth, K., Sitasuwan, P., Zhao, X., Lin, Y. and Q. Wang, Influence of Surface Topographical Cues on the Differentiation of Mesenchymal Stem Cells In Vitro. *ACS Biomaterials Science & Engineering*, 2(2), pp. 142–51, 2016.
105. Fiedler, J., Özdemir, B., Brenner, R.E., et al., The effect of substrate surface nanotopography on the behavior of multipotent mesenchymal stromal cells and osteoblasts. *Biomaterials*, 34(35), pp. 8851–59, 2013.
106. Annamalai, R.T., Mertz, D.R., Daley, E.L.H. and J.P. Stegemann, Collagen Type II enhances chondrogenic differentiation in agarose-based modular microtissues. *Cytotherapy*, 18(2), 263–277, 2016.
107. Bornes, T.D., Adesida, A.B., and N.M. Jomha, Mesenchymal stem cells in the treatment of traumatic articular cartilage defects: a comprehensive review. *Arthritis Research & Therapy*, 16(5), p.432, 2014.
108. Zheng, D., Dan, Y., Yang, S. and P.K. Chu, Controlled chondrogenesis from adipose derived stem cells by recombinant transforming growth factor-β3 fusion protein in peptide scaffolds. *Acta Biomaterialia*, 11, pp. 191–203, 2015.
109. Huang, Z., Nooeaid, P., Kohl, B., and G. Schulze-Tanzil, Chondrogenesis of human bone marrow mesenchymal stromal cells in highly porous alginate-foams supplemented with chondroitin sulfate. *Materials Science and Engineering: C*, 50, pp. 160–72, 2015.
110. Wu, Y.N., He, A.Y., Hong Y.L. and H. Lee, Substrate topography determines the fate of chondrogenesis from human mesenchymal stem cells resulting in specific cartilage phenotype formation. *Nanomedicine: Nanotechnology, Biology, and Medicine*, 10, pp. 1507–16, 2014.
111. Uematsu, K., Hattori, K., Ishimoto, Y., Yamauchi, J., Habata, T., Takakura, Y., Ohgushi, H., Fukuchi, T., Sato, M., Cartilage regeneration using mesenchymal

stem cells and a three-dimensional poly-lactic-glycolic acid (PLGA) scaffold. *Biomaterials*, 26, pp. 4273–79, 2005.
112. Yin, Z., Chen, X., Song, H., Hu, J. and H.W. Ouyang, Electrospun scaffolds for multiple tissues regeneration *in vivo* through topography dependent induction of lineage specific differentiation. *Biomaterials*, 44, pp. 173–85, 2015.
113. Singh N., Rahatekar, S.S., Koziol, K.K.K., Patil, A.J., *et al.*, Directing chondrogenesis of stem cells with specific blends of cellulose and silk. *Biomacromolecules*, 14(5), pp. 1287–98, 2013.
114. Fässler, R., Rohwedel, J., Maltsev, V., Bloch, W., Lentini, S., Guan, K., Gullberg, D., Hescheler, J., Addicks, K. and A.M. Wobus, Differentiation and integrity of cardiac muscle cells are impaired in the absence of beta 1 integrin. *Journal of Cell Science*, 109, pp. 2989–99, 1996.
115. Marsano, A., Maidhof R., Wan, L.Q., Marsano, A., Maidhof, R., *et al.*, Scaffold stiffness affects the contractile function of three-dimensional engineered cardiac constructs. *Biotechnology Progress*, 26, pp. 1382–90, 2010.
116. Prabhakaran, M.P., Venugopal, J.K., Dan, P., Molamma P., Venugopal, J., Kai, D., Molamma P., Prabhakaran, J. and D.K. Venugopal, Biomimetic material strategies for cardiac tissue engineering. *Materials Science & Engineering C— Materials for Biological Applications*, 31, pp. 503–13, 2011.
117. Mooney, E., Mackle, J.N., Blond, D.J. -P. Blond, *et al.*, The electrical stimulation of carbon nanotubes to provide a cardiomimetic cue to MSCs. *Biomaterials*, 33, pp. 6132–9, 2012.
118. Li., Z., Guo, X., Palmer, A.F., Das, H., Guan, J., High-efficiency matrix modulus-induced cardiac differentiation of human mesenchymal stem cells inside a thermosensitive hydrogel. *Acta Biomaterialia*, 8, pp. 3586–95, 2012.
119. Heras-Bautista, C.O., Katsen-Globa, A., Schloerer, N.E., Dieluweit, S., Hescheler, J. and K. Pfannkuche, The influence of physiological matrix conditions on permanent culture of induced pluripotent stem cell-derived cardiomyocytes. *Biomaterials*, 35, pp. 7374–85, 2014.
120. Safaeijavan, R, Soleimani, M., Divsalar, A., Eidi, A., Ardeshirylajimi, A., Comparison of random and aligned PCL nanofibrous electrospun scaffolds on cardiomyocyte differentiation of human adipose-derived stem cells. *Iranian Journal of Basic Medical Sciences*, 17, pp. 903–11, 2014.
121. Xu, Y., Patnaik, S., Guo, X., Li, Z., Lo, W., Butler, R., Claude, A., Liu, Z., Zhang, G., Liao, J., Anderson, P.M., Guan, J., Cardiac differentiation of cardiosphere-derived cells in scaffolds mimicking morphology of the cardiac extracellular matrix. *Acta Biomaterialia*, 10, pp. 3449–62, 2014.

Index

1,4-butanediol diglycidyl ether, 26
316L SS, 187, 191–192
3D environment, 92, 99, 115–124
3D scaffolds, 200, 220–222
510 (k), 325

Aarbodiimine, 207–209, 211
Ablation, 173–176, 192
Absorptivity, 169
Acid
   acrylic, 207, 209, 213–217, 220
   carboxylic, 206, 209, 211
   hyaluronic, 207, 209
   polylactic, 215, 220
Additive manufacturing, 235, 244
Adipose, 429, 436, 441, 444
Adipose tissue, 390
Adult stem cells, 428, 429
Alcohol, 207, 209, 213
Aldehyde, 207, 210, 217
Allograft, 234, 243, 244
Alloplastic, 234, 245
Amines, 206–214, 218–220
Angiogenesis, 282–283, 287–288, 291
Anhydride, 210, 211
Anterior cruciate ligament (ACL), 390, 393
Arginine–glycine–aspartic acid (RGD), 9, 14
Arginylglycylaspartic acid (RGD), 90, 103, 106–110, 112, 114, 122–123

Atherosclerosis, 34
Atomic force microscopy, 155–159
Autograft, 233, 234, 244

Barrier membranes, 290
Beam homogenization, 180–182, 186
Beam quality, 169, 172, 178
Biochemical stimuli,
   biochemical gradients, 89, 101, 105, 109, 111–114, 122
   microppaterns, 102–106, 109, 118
   nanopatterns, 102, 106–110, 114, 121
Biocompatibility, 167, 192
Biocompatible, 182, 186
Biodegradable, 25–27
Bioengineering, 283, 296–297
Bioink, 236, 259, 262
Bioinspired materials, 438, 439, 444
Biomaterials, 425, 426, 433, 434, 436, 439, 440, 442, 443
Biomolecules coatings, 78–79
Biopolymer surface modification, 402
   extracellular Matrix protein-based, 406
   for materials used in bone reconstruction, 408
   nitric oxide producing, 405
   role in cardiac prostheses, 407
Bioprinting, 259, 260, 262, 264

Bioreactor, 50–53, 56, 291,
    293–309, 311–312
  compressive bioreactor, 52
Bioresorbable, 235, 236
Bone Ingrowth, 312
Bone marrow (BM), 2–4, 10–13
Bone marrow msc, 393–394
Bone morphogenetic protein
    (BMP), 387, 389, 391–392
Bone morphogenetic proteins
    (BMPs), 8, 49, 52, 55,
    282, 289–290
Bone regeneration, 280, 283–284,
    286, 288, 290–293, 296,
    301–302, 304–305,
    307, 439–441
Bone TE, 71, 72, 74, 77–80
Bone tissue, 389, 391
Breast implant, 395

Cadherin, 32
Cardiac regeneration, 439,
    440, 443, 444
Cardiac tissue, 390
Cardiomyocytes, 344, 356, 358, 361,
    362, 366, 393, 443, 444
Cardiovascular device
    polymers, 402, 403
  history, 403
Cardiovascular diseases, 402
Cartilage, 27, 35, 390–392, 442
Cartilage regeneration, 439,
    440, 442, 443
Cartilage TE, 75, 78
Cell
  adhesion, 167–168, 191, 200–201,
    204–209, 211–213, 215–216,
    218, 220
  differentiation, 200, 204–205,
    212–214, 216–221
  proliferation, 167–168, 190–193
  spreading, 200, 209, 212–213,
    216, 218
Cell therapy products, 325

Cell–surface interaction,
    167–168, 188–189
Cellulose, 440, 442
Center for biologics evaluation and
    research (CBER), 329
Chitosan, 26
Chondocytes, 392
Chondrocytes, 344, 429, 434,
    439, 440, 442, 443
Chondrogenesis, 392
Co-Cr Alloy, 297, 310–311
Collagen, 4–9, 11–14, 25, 26,
    31, 32, 35, 348, 350–352,
    357, 359–364, 431, 433,
    434, 439, 441, 442
Common lymphoid progenitors
    (CLP), 3–4
Common myeloid progenitors
    (CMP), 3
Computational modeling, 312
Contact angle, 206, 209, 211, 213, 214
Cord blood, 2
Craniomaxillofacial, 235, 259
Critical sized bone defect,
    233, 234, 236, 264
CSD, 233, 234, 236, 237, 239,
    243, 248, 250, 251, 264

Decellularized matrices, 343, 344, 352,
    353, 364, 365
Depth of focus (DOF), 175, 178–179
Differentiation, 167–168, 190–193
Direct laser interference patterning
    (DLIP), 182–183, 193

EBM, 239, 245–247
Ectoderm, 2–3, 8
Elastin, 25
Electron beam melting, 239
Electron microscopy, 151–155
Electrospinning, 73
Embryonic, 24, 33, 35
Embryonic stem cells (ESCs), 2–3, 8,
    14, 428, 429, 432, 433, 436, 437

Endoderm, 2-3, 8
Endothelial, 31
Endothelial cell, 395
Epidermal growth factor
    (EGF) receptor, 10
Epoxide, 210
Ester, 210
European medicines agency
    (EMA), 330
European union (EU), 330
European free trade association
    (EFTA), 330
European regulatory system, 330
Extracellular matrix (ECM), 2-14, 24,
    27, 32-35, 46, 89-92, 94-95,
    98, 105-106, 115, 117, 121,
    123, 425, 426, 434, 437-441
Extracellular signal-regulated
    kinase, 49
Extrusion based technologies,
    233, 237, 240, 242, 243,
    245, 247, 249-251, 253,
    254, 259, 260, 264

FA kinase (FAK), 8-10
Fabrication techniques
    electrospinning, 100, 115-116
    inkjet printing, 103, 109, 112, 121
    micellar lithography, 106-108
    microcontact printing (μCP),
        103, 105
    nanoimprint lythography, 95, 100
    photolythography, 65, 103, 106
    plasma polymerization, 111-112
FDA, 331
FDM, 235, 240, 243, 247,
    249-251, 256
Ferroscaffolds, 54
Fibers, 386, 388, 390, 391, 393
Fibrinogen, 9, 12, 14
Fibroblast, 393, 395
Fibroblast growth factor
    (FGF), 8-9, 15
Fibroblasts, 2, 4, 9, 31, 429-430

Fibroin, 382-384, 386, 388-389
Fibronectin, 4-9, 11-13, 31,
    348, 350-352, 359,
    362-364, 431, 433, 443
Films, 383, 386, 388, 390, 394
Fluorescence microscopy, 150, 151
Focal adhesion, 431, 432, 434
Functionalization, 258

Gaussian, 172-174, 178, 180
Gene therapy products, 331
Glass, 434, 435
Glycosaminoglycans, 4, 9,
    14-15, 434, 439, 442
Good laboratory practice (GLP), 331
Good manufacturing practices
    (GMPs), 331
Gradient, 200, 217-219
Growth factors (GFs), 200, 204-205,
    208, 211-212, 220-221,
    280, 282-283, 285, 286, 289,
    291-292, 296, 301-305,
    312, 430-432, 439, 441
Guidance documents, 332
Guided bone regeneration (GBR),
    280, 284, 286, 290,
    292-293, 296

HA, 238, 245, 248, 249, 251, 253-257,
    260, 261
Hard segment (HS), 27
Hematopoietic stem cells
    (HSCs), 2-4, 11, 13
Hepatocyte growth factor
    (HGF) receptor, 10
Hepatocytes, 344, 361, 365
Hexamethylene diisocyanate
    (HDI), 27
Human aortic endothelial cell, 394
Human bone marrow (hBM), 192
Human bone marrow stromal
    cell, 388
Human bone mesencyhmal
    stem cells, 388

Human coronary artery smooth muscle cells, 394
Human embryonic stem cells (hESCs), 324
Human mesenchymal stem cells (hMSCs), 29
Humanitarian use device (HUD), 332
Hyaluronan/Hyaluronic acid (HA), 4, 14–15, 348, 352, 439, 442
Hydrogels, 69, 70, 77, 79, 92, 95, 99, 101, 106, 108–109, 114–124, 220, 383, 386–387, 390–392, 434, 439–441, 443
Hydrophilic, 24–28, 203, 206, 209, 211–215
Hydrophilicity, 436
Hydrophobic, 25, 203, 206, 216
Hydrophobicity, 436, 438
Hydroxyapatite, 26, 435, 440, 441
Hydroxyl, 201, 206–210, 213, 217

Implants, 167–169, 171, 176, 189, 190, 192, 233, 236, 239, 243, 244, 246, 247, 248, 264, 283, 291, 296–298, 307–312
Induced pluripotent stem cells, 428, 430, 440, 443
Inkjet, 236, 240–243, 251, 252, 255, 259, 260, 262
Inkjet 3D printing, 241, 242, 243, 251, 252, 255, 259, 260, 262
Insulin growth factor (IGF-I), 389
Insulin receptor, 10
Integrins, 3–4, 6–14, 47, 54–56, 431, 432, 434, 437, 441, 443

Keratinocytes, 344, 364, 395

Laminin, 4–14, 348, 350–353, 359–363, 368, 431, 433, 439
Laser, 167–194
Laser based technologies, 233, 237, 242, 245, 246, 249, 256, 257, 259

Laser surface alloying (LSA), 186–187
Lithography, 68
Liver, 390

Magnetic particles, 53–54
Magnetic resonance, 55
Material, 167–169, 171–176, 180, 183–192
Matrigel, 345, 347–352, 354–356, 358–361, 363, 365
Mechanoregulation,
 micro-nanotopography, 100–101, 111
 nanofibers, 92, 95, 106, 115–117, 121
 physical gradients, 98–101
 roughness, 92–94, 97, 99–101
 stiffness, 89, 92, 96, 98–101, 108, 115, 119
Mechanosome, 48
 theory, 48
Mechanotransduction, 47–56
Megakaryocyte (MK), 3–4, 12–13
Membranes, 383, 386, 391
Mesenchymal stem cells (MSCs), 3–4, 190–193, 346, 347, 357, 361, 429, 434–444
Mesenchymal stromal cells, 393
Mesoderm, 2–4, 8
MHDS, 242, 243, 250
Micro and nanopatterning, 72
Microenvironment, 23–25, 27–28, 30–34, 36, 285–286, 295, 297, 301
Microfluidics, 73
Microlens array, 179–182
Microstructure, 25, 26
Micro-structure/microscale, 167–168, 187, 189–192
Mimicked substrates, 438, 439, 441, 444
Modulus, 26, 27, 31, 32
Multiple-head deposition systems, 242
Myoblast, 25, 32, 34

INDEX 457

Myocardial, 394
Myocardium, 394
Myogenic, 35

Nanofibers, 75–77
Nanoparticles, 75, 79
Nanopillars, 441, 442
Nanostructure, 25, 28, 214–215, 218–219, 221
Nanostructure/nanoscale, 167–168, 187, 189–192
Nanostructured surface, 146
Nanotechnology based surface modification, 411
　polysaccharide-based, 415
　protein nanoparticles, 412
Nerve tissue, 390
Net-mat-fiber, 386
Neural stem, 29, 33
Neural stem cells, 434, 439
Neural TE, 71, 75, 76, 77
Neurite growth, 395
Neuron, 33, 34
Neuron/Neuronal, 344, 355, 357–360, 363, 365, 367, 368
Neuronal TE, 73
Neutrophil, 31
Niche, 2–5, 11–13, 23, 24, 28, 32, 33
Nitinol (NiTi), 187, 191–192
Nitrogen rich surfaces, 210–212

Office of Cellular, Tissue and Gene Therapies (OCTGT), 333
Osteoarthritis, 34
Osteoblasts, 34, 344, 364, 389–391, 429, 432, 434, 437, 438, 441
Osteoclast, 389
Osteoconduction, 244, 250
Osteocytes, 389
Osteogenesis, 234, 251
Osteoporosis, 34
Osteoprogenitor, 389
Oxazoline, 208, 212
Oxygen rich surfaces, 206–210

PCL, 236, 238, 240, 241, 249–251, 253, 254–257, 260, 263
Peptide grafting, 79
Perichondrium, 34
Periosteum, 34
Peripheral nerve injury, 79
PHBV, 238, 257
Physicochemical cues, 205
Plasma
　physics, 201, 202
　polymer deposition, 203
　surface modification, 202–204
Plasma treatment, 147–149
Platelet-derived growth factor-β (PDGF-β) receptor, 10
Platelet-rich plasma (PRP), 286, 288, 302–304
PLGA, 250, 253, 254, 257, 263
PLLA, 238, 249, 254, 257
Pluripotent Stem Cells (PSC), 2–3, 343–368
Polarization, 169–171, 190, 192
Poly(ethylene glycol) (PEG), 27
Poly(ethylene oxide) (PEO), 25
Poly(ethylene terephthalate) (PET), 191–192
Poly(lactic-co-glycolic acid) (PLGA), 25–27, 29, 30
Poly(ε-caprolactone) (PCL), 25, 27
Polydimethylsiloxane (PDMS), 95–97, 105
Polyglycolide (PGA), 25
Polyurethane, 27
Porous sponges, 383, 387, 390
Postprocessing, 238, 241, 249, 251, 252, 258
Premarket notification (PMN), 338
Preprocessing, 252
Progenitor, 3, 10, 12–13
Progenitor cells, 29
Protein adsorption, 205–207, 209–213, 215, 217

Proteoglycans (HSPG), 4, 9, 15
Pulse
  ablation, 175
  duration (length/width), 167–168, 171–174, 176, 184–185, 190
  energy, 169, 173–174, 182
Pulsed electromagnetic field, 52, 55
Pulsed laser deposition (PLD), 183–184, 186, 191, 193

Rayleigh length, 175
Reconstruction, 237, 242, 243, 248
Reflectivity, 169–170, 185
Refractive index, 169, 182
Regeneration, 233, 234, 236–239, 248, 249, 258, 259, 262, 264
Regenerative medicine, 284–286, 296, 425–429, 433, 438, 441–443
Regulations, 338
Regulatory systems, 338
Rho A, 6, 10
Roughness, 24, 167–168, 189–190, 192–193

Scaffold, 144, 145, 147, 233, 234, 236, 238, 240, 247–252, 258, 259, 264, 426, 427, 433, 434, 438, 440–443
Scaffolds, 69–80, 233–236, 238, 240, 247–254, 256–260, 264, 283, 286, 292, 294–295, 297, 302–305, 307, 383
Scanning electron micrographs (SEM), 29
Self-renewal, 23, 24
Sericin, 382–383
Signaling factors, 427, 430–432
Silk, 381
Silk fibroin, 381–382, 384
Sintering, 258
Skin, 390, 395
SLA, 235–238, 256
SLM, 235, 238, 239, 245, 246
SLS, 235, 238, 239, 242, 249, 256

Soft segment (SS), 27
Soft tissue, 390, 395
Spinal cord injury, 77, 79
Statues, 338
Stem cell, 167–168, 170, 172, 189–191, 193–194, 233, 236, 250, 258, 262, 264
Stem cell fate
  adipogenic, 93, 95, 97, 100–101, 103, 105–110, 112
  chondrogenic, 100, 107–108, 116–118, 123
  myogenic, 92, 100, 110, 112, 116
  neurogenic, 89, 92, 96–97, 100–101, 103, 105, 109, 112–113, 116, 118–119
  osteogenic, 92–93, 95, 97–101, 103, 105–113, 116–117, 119
Stem cell niche, 88–91, 94, 115, 117–119, 121, 429–432
Stem cells, 2, 5, 7, 10, 14, 68–71, 74–80, 145, 148, 284–285, 288, 294, 307, 427
Stereolithography, 236, 237
Stiffness, 24, 31–34, 200, 204, 216–218, 220, 432, 437, 439–443
STL, 237, 239, 241, 242, 249, 256, 259
Stoichiometry, 183
Substrates Chemical Modifications, 77
Surface
  chemistry, 167–168, 192–193
  topography, 168, 171, 190, 193
Surface charge, 206, 211, 217
Surface chemistry, 205–214, 432, 433, 438–440, 443
Surface modification, 167–169, 176–177, 183, 191, 193
Surface topography, 437–444
Surface treatment, 297–299, 301–302
Surfaces, 68–74, 77–80
Surgical guides, 233, 236, 240–244, 264
Syndecan, 4

Tenascin, 9, 14
Tendon, 390, 393
Tendon TE, 72
Tenocyte, 393
Tensegrity theory, 47
Tensile strain, 50, 52
Thermally induced phase separation (TIPS), 26
Three-dimensional printing, 233, 236, 237, 239, 252, 262, 264
Ti_6Al_4V, 239, 244, 247
Ti-6Al-4V, 187, 190–191
Tissue engineering (TE), 68, 74–78, 80, 81, 144, 284–285, 292–296, 302, 304–305, 426, 428, 429, 433, 438, 441–443
Tissue-engineering, 26–28, 33, 35
Titanium, 169, 187–192
Titanium alloy, 239, 244–248
Titanium dioxide (TiO_2), 187, 192
Topography, 24, 28–30, 46, 47, 56, 147, 200, 205, 214–215, 217–219
Transcription factors, 429, 430
Transforming growth factor (TGF), 8–10, 15
  receptor, 10
Transforming growth factor beta (TGF-β), 280–281, 288–290
Transplantation, 234
Type 1 insulin-like growth factor (IGF1) receptor, 10
Tyrosine kinase, 9

Ultrasound, 55
Umbilical cord, 2, 4
Umbilical cord blood (UCB), 2, 13
US regulatory system, 338

Vascular endothelial growth factor (VEGF), 9–10, 15
  receptor, 10
Vascular epidermal growth factor (VEGF), 387, 389
Vascular TE, 72, 75
Vascular tissue, 391–392
Vibrational force, 55
Vitronectin, 7, 9, 14, 348, 351, 352, 359–361

Wavelength, 167–170, 172–173, 175–176, 178, 181–182, 184–185, 188–190, 193
Wax like sericin, 381–383
Wettability, 24, 25, 200–201, 205–206, 213–215, 217
Wetting, 167, 168
Wharton's Jelly, 2, 4
Wolff's law, 49
Wound healing, 395

β-sheets, 382–384
β-TCP, 238, 240, 245, 249–256, 258, 260, 263
β-tubulin, 33

Also of Interest

Check out these published volumes in the Advanced Materials Series

Advanced Magnetic and Optical Materials
Edited by Ashutosh Tiwari, Parameswar K. Iyer, Vijay Kumar and Hendrik Swart
Forthcoming 2016. ISBN 978-1-119-24191-1

Advanced Surfaces for Stem Cell Research
Edited by Ashutosh Tiwari, Bora Garipcan and Lokman Uzun
Forthcoming 2016. ISBN 978-1-119-24250-5

Advanced Electrode Materials
Edited by Ashutosh Tiwari, Filiz Kuralay and Lokman Uzun
Published 2016. ISBN 978-1-119-24252-9

Advanced Molecularly Imprinting Materials
Edited by Ashutosh Tiwari and Lokman Uzun
Published 2016. ISBN 978-1-119-33629-7

Intelligent Nanomaterials (2nd edition)
Edited by Tiwari, Yogendra Kumar Mishra, Hisatoshi Kobayashi and Anthony P. F. Turner
Published 2016. ISBN 978-1-119-24253-6

Advanced Composite Materials
Edited by Ashutosh Tiwari, Mohammad Rabia Alenezi and Seong Chan Jun
Published 2016. ISBN 978-1-119-24253-6

Advanced Surface Engineering Materials
Edited by Ashutosh Tiwari, Rui Wang, and Bingqing Wei
Published 2016. ISBN 978-1-119-24244-4

Advanced Ceramic Materials
Edited by Ashutosh Tiwari, Rosario A. Gerhardt and Magdalena Szutkowska
Published 2016. ISBN 978-1-119-24244-4

Advanced Engineering Materials and Modeling
Edited by Ashutosh Tiwari, N. Arul Murugan and Rajeev Ahuja
Published 2016. ISBN 978-1-119-24246-8

Advanced 2D Materials
Ashutosh Tiwari and Mikael Syväjärvi
Published 2016. ISBN 978-1-119-24249-9

Advanced Materials Interfaces
Edited by Ashutosh Tiwari, Hirak K. Patra and Xumei Wang
Published 2016. ISBN 978-1-119-24245-1

Advanced Bioelectronics Materials
Edited by Ashutosh Tiwari, Hirak K. Patra and Anthony P.F. Turner
Published 2015. ISBN 978-1-118-99830-4

Graphene
An Introduction to the Fundamentals and Industrial Applications
By Madhuri Sharon and Maheswar Sharon
Published 2015. ISBN 978-1-118-84256-0

Advanced Theranostic Materials
Edited by Ashutosh Tiwari, Hirak K. Patra and Jeong-Woo Choi
Published 2015. ISBN: 978-1-118-99829-8

Advanced Functional Materials
Edited by Ashutosh Tiwari and Lokman Uzun
Published 2015. ISBN 978-1-118-99827-4

Advanced Catalytic Materials
Edited by Ashutosh Tiwari and Salam Titinchi
Published 2015. ISBN 978-1-118-99828-1

Graphene Materials
Fundamentals and Emerging Applications
Edited by Ashutosh Tiwari and Mikael Syväjärvi
Published 2015. ISBN 978-1-118-99837-3

DNA Engineered Noble Metal Nanoparticles
Fundamentals and State-of-the-art-of Nanobiotechnology
By Ignác Capek
Published 2015. ISBN 978-1-118-07214-1

Advanced Electrical and Electronics Materials
Process and Applications
By K.M. Gupta and Nishu Gupta
Published 2015. ISBN: 978-1-118-99835-9

Advanced Materials for Agriculture, Food and Environmental Safety
Edited by Ashutosh Tiwari and Mikael Syväjärvi
Published 2014. ISBN: 978-1-118-77343-7

Advanced Biomaterials and Biodevices
Edited by Ashutosh Tiwari and Anis N. Nordin
Published 2014. ISBN 978-1-118-77363-5

Biosensors Nanotechnology
Edited by Ashutosh Tiwari and Anthony P. F. Turner
Published 2014. ISBN 978-1-118-77351-2

Advanced Sensor and Detection Materials
Edited by Ashutosh Tiwari and Mustafa M. Demir
Published 2014. ISBN 978-1-118-77348-2

Advanced Healthcare Materials
Edited by Ashutosh Tiwari
Published 2014. ISBN 978-1-118-77359-8

Advanced Energy Materials
Edited by Ashutosh Tiwari and Sergiy Valyukh
Published 2014. ISBN 978-1-118-68629-4

Advanced Carbon Materials and Technology
Edited by Ashutosh Tiwari and S.K. Shukla
Published 2014. ISBN 978-1-118-68623-2

Responsive Materials and Methods
State-of-the-Art Stimuli-Responsive Materials and Their Applications
Edited by Ashutosh Tiwari and Hisatoshi Kobayashi
Published 2013. ISBN 978-1-118-68622-5

Other Scrivener books edited by Ashutosh Tiwari

Nanomaterials in Drug Delivery, Imaging, and Tissue Engineering
Edited by Ashutosh Tiwari and Atul Tiwari
Published 2013. ISBN 978-1-118-29032-3

Biomedical Materials and Diagnostic Devices Devices
Edited by Ashutosh Tiwari, Murugan Ramalingam, Hisatoshi Kobayashi and Anthony P.F. Turner
Published 2012. ISBN 978-1-118-03014-1

Intelligent Nanomaterials (first edition)
Processes, Properties, and Applications
Edited by Ashutosh Tiwari Ajay K. Mishra, Hisatoshi Kobayashi and Anthony P.F. Turner
Published 2012. ISBN 978-0-470-93879-9

Integrated Biomaterials for Biomedical Technology
Edited by Murugan Ramalingam, Ashutosh Tiwari, Seeram Ramakrishna and Hisatoshi Kobayashi
Published 2012. ISBN 978-1-118-42385-1